Study Guide and Solutions Ma

Organic Chemistry
A Short Course

THIRTEENTH EDITION

David J. Hart
The Ohio State University

Christopher M. Hadad
The Ohio State University

Leslie E. Craine
Central Connecticut State University

Harold Hart
Michigan State University

Prepared by

David J. Hart
The Ohio State University

Christopher M. Hadad
The Ohio State University

Leslie E. Craine
Central Connecticut State University

Harold Hart
Michigan State University

BROOKS/COLE
CENGAGE Learning™

Australia • Brazil • Japan • Korea • Mexico • Singapore • Spain • United Kingdom • United States

ISBN-13: 978-1-111-42585-2
ISBN-10: 1-111-42585-X

Brooks/Cole
20 Davis Drive
Belmont, CA 94002-3098
USA

Cengage Learning is a leading provider of customized learning solutions with office locations around the globe, including Singapore, the United Kingdom, Australia, Mexico, Brazil, and Japan. Locate your local office at: **www.cengage.com/global**

Cengage Learning products are represented in Canada by Nelson Education, Ltd.

To learn more about Brooks/Cole, visit **www.cengage.com/brookscole**

Purchase any of our products at your local college store or at our preferred online store **www.cengagebrain.com**

Printed in the United States of America
2 3 4 5 6 7 15 14 13 12

Contents

Introduction to the Student

This study guide and solutions book was written to help you learn organic chemistry. The principles and facts of this subject are not easily learned by simply reading them, even repeatedly. Formulas, equations, and molecular structures are best mastered by *written* practice. To help you become thoroughly familiar with the material, we have included many problems within and at the end of each chapter in the text.

It is our experience that such questions are not put to their best use unless correct answers are also available. Indeed, answers alone are not enough. If you know how to work a problem and find that your answer agrees with the correct one, fine. But what if you work conscientiously, yet cannot solve the problem? You then give in to temptation, look up the answer, and encounter yet another dilemma–how in the world did the author get that answer? This solutions book has been written with this difficulty in mind. For many of the problems, all of the reasoning involved in getting the correct answer is spelled out in detail. Many of the answers also include cross-references to the text. If you cannot solve a particular problem, these references will guide you to parts of the text that you should review.

Each chapter of the text is briefly summarized. Whenever pertinent, the chapter summary is followed by a list of all the new reactions and mechanisms encountered in that chapter. These lists should be especially helpful to you as you review for examinations.

When you study a new subject, it is always useful to know what is expected. To help you, we have included in this study guide a list of learning objectives for each chapter—that is, a list of what you should be able to do after you have read and studied that chapter. Your instructor may want to delete items from these lists of objectives or add to them. However, we believe that if you have mastered these objectives—and the problems should help you to do this—you should have no difficulty with examinations. Furthermore, you should be very well prepared for further courses that require this course as a prerequisite.

Near the end of this study guide you will find additional sections that may help you to study for the final examination in the course. The SUMMARY OF SYNTHETIC METHODS lists the important ways to synthesize each class of compounds discussed in the text. It is followed by the SUMMARY OF REACTION MECHANISMS. Both of these sections have references to appropriate portions of the text, in case you feel that further review is necessary. Finally, you will find two lists of sample test questions. The first deals with synthesis, and the second is a list of multiple-choice questions. Both of these sets should help you prepare for examinations.

In addition, we offer you a brief word of advice about how to learn the many reactions you will study during this course. First, learn the nomenclature systems thoroughly for each new class of compounds that is introduced. Then, rather than memorizing the particular examples of reactions given in the text, study reactions as being typical of a class of compounds. For example, if you are asked how compound A will react with compound B, proceed in the following way. First ask yourself: to what class of compounds does A belong? How does this class of compounds react with B (or with compounds of the general class to which B belongs)? Then proceed from the general reaction to the specific case at hand. This approach will probably help you to eliminate some of the memory work often associated with organic chemistry courses. We urge you to study regularly, and hope that this study guide and solutions book will make it easier for you to do so.

Great effort has been expended to ensure the accuracy of the answers in this book and we wish to acknowledge the helpful comments provided by David Ball (Cleveland State University) in this regard. It is easy for errors to creep in, however, and we will be particularly grateful to anyone who will call them to our attention. Suggestions for improving the book will also be welcome. Send them to:

Christopher M. Hadad
Department of Chemistry
The Ohio State University
Columbus, Ohio 43210

David J. Hart
Department of Chemistry
The Ohio State University
Columbus, Ohio 43210

1

Bonding and Isomerism

Chapter Summary*

An **atom** consists of a nucleus surrounded by **electrons** arranged in **orbitals**. The electrons in the outer shell, or the **valence electrons**, are involved in bonding. **Ionic bonds** are formed by electron transfer from an **electropositive** atom to an **electronegative** atom. Atoms with similar electronegativities form **covalent bonds** by sharing electrons. A **single bond** is the sharing of one electron pair between two atoms. A covalent bond has specific **bond length** and **bond energy**.

Carbon, with four valence electrons, mainly forms covalent bonds. It usually forms four such bonds, and these may be with itself or with other atoms such as hydrogen, oxygen, nitrogen, chlorine, and sulfur. In pure covalent bonds, electrons are shared equally, but in **polar covalent bonds**, the electrons are displaced toward the more electronegative element. **Multiple bonds** consist of two or three electron pairs shared between atoms.

Structural (or **constitutional**) **isomers** are compounds with the same **molecular formulas** but different **structural formulas** (that is, different arrangements of the atoms in the molecule). **Isomerism** is especially important in organic chemistry because of the capacity of carbon atoms to be arranged in so many different ways: continuous chains, branched chains, and rings. Structural formulas can be written so that every bond is shown, or in various abbreviated forms. For example, the formula for *n*-pentane (*n* stands for normal) can be written as:

```
   H  H  H  H  H
   |  |  |  |  |
H—C—C—C—C—C—H    or   CH3CH2CH2CH2CH3   or   (zigzag line structure)
   |  |  |  |  |
   H  H  H  H  H
```

$CH_3CH_2CH_2CH_2CH_3$

Some atoms, even in covalent compounds, carry a **formal charge**, defined as the number of valence electrons in the neutral atom minus the sum of the number of unshared electrons and half the number of shared electrons. **Resonance** occurs when we can write two or more structures for a molecule or ion with the same arrangement of atoms but different arrangements of the electrons. The correct structure of the molecule or ion is a **resonance hybrid** of the **contributing structures**, which are drawn with a double-headed arrow ($\leftrightarrow$) between them. Organic chemists use a curved arrow (↷) to show the movement of an electron pair.

A **sigma** (σ) **bond** is formed between atoms by the overlap of two atomic orbitals along the line that connects the atoms. Carbon uses ***sp*^{3}-hybridized orbitals** to form four such bonds. These bonds are directed from the carbon nucleus toward the corners of a tetrahedron. In **methane**, for example, the carbon is at the center and the four hydrogens are at the corners of a regular tetrahedron with H–C–H bond angles of 109.5°.

* In the chapter summaries, terms whose meanings you should know appear in boldface type.

Carbon compounds can be classified according to their molecular framework as **acyclic** (*not* cyclic); **carbocyclic** (containing rings of carbon atoms), or **heterocyclic** (containing at least one ring atom that is not carbon). They may also be classified according to **functional group** (Table 1.6).

Learning Objectives*

1. Know the meaning of: nucleus, electrons, protons, neutrons, atomic number, atomic weight, shells, orbitals, valence electrons, valence, kernel.
2. Know the meaning of: electropositive, electronegative, ionic and covalent bonds, radical, catenation, polar covalent bond, single and multiple bonds, nonbonding or unshared electron pair, bond length, bond energy.
3. Know the meaning of: molecular formula, structural formula, structural (or constitutional) isomers, continuous and branched chain, formal charge, resonance, contributing structures, sigma (σ) bond, sp^3-hybrid orbitals, tetrahedral carbon.
4. Know the meaning of: acyclic, carbocyclic, heterocyclic, functional group.
5. Given a periodic table, determine the number of valence electrons of an element and write its electron-dot formula.
6. Know the meaning of the following symbols:

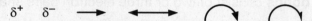

7. Given two elements and a periodic table, tell which element is more electropositive or electronegative.
8. Given the formula of a compound and a periodic table, classify the compound as ionic or covalent.
9. Given an abbreviated structural formula of a compound, write its electron-dot formula.
10. Given a covalent bond, tell whether it is polar. If it is, predict the direction of bond polarity from the electronegativities of the atoms.
11. Given a molecular formula, draw the structural formulas for all possible structural isomers.
12. Given a structural formula abbreviated on one line of type, write the complete structure and clearly show the arrangement of atoms in the molecule.
13. Given a line formula, such as (pentane), write the complete structure and clearly show the arrangement of atoms in the molecule. Tell how many hydrogens are attached to each carbon, what the molecular formula is, and what the functional groups are.
14. Given a simple molecular formula, draw the electron-dot formula and determine whether each atom in the structure carries a formal charge.

* Although the objectives are often worded in the form of imperatives (i.e., determine …,write …, draw …), these verbs are all to be preceded by the phrase "be able to …". This phrase has been omitted to avoid repetition.

15. Draw the electron-dot formulas that show all important contributors to a resonance hybrid and show their electronic relationship using curved arrows.

16. Predict the geometry of bonds around an atom, knowing the electron distribution in the orbitals.

17. Draw in three dimensions, with solid, wedged, and dashed bonds, the tetrahedral bonding around sp^3-hybridized carbon atoms.

18. Distinguish between acyclic, carbocyclic, and heterocyclic structures.

19. Given a series of structural formulas, recognize compounds that belong to the same class (same functional group).

20. Begin to recognize the important functional groups: alkene, alkyne, alcohol, ether, aldehyde, ketone, carboxylic acid, ester, amine, nitrile, amide, thiol, and thioether.

ANSWERS TO PROBLEMS

Problems Within the Chapter

1.1 The sodium atom donates its valence electron to the chlorine atom to form the ionic compound, sodium chloride.

$$Na\cdot + \cdot\ddot{\underset{..}{Cl}}: \longrightarrow Na^+ + :\ddot{\underset{..}{Cl}}:^-$$

1.2 Elements with fewer than four valence electrons tend to give them up and form positive ions: Al^{3+}, Li^+. Elements with more than four valence electrons tend to gain electrons to complete the valence shell, becoming negative ions: S^{2-}, O^{2-}.

1.3 Within any horizontal row in the periodic table, the most electropositive element appears farthest to the left. Na is more electropositive than Al, and C is more electropositive than N. In a given column in the periodic table, the lower the element, the more electropositive it is. Si is more electropositive than C.

1.4 In a given column of the periodic table, the higher the element, the more electronegative it is. F is more electronegative than Cl, and N is more electronegative than P. Within any horizontal row in the periodic table, the most electronegative element appears farthest to the right. F is more electronegative than O.

1.5 As will be explained in Sec. 1.3, carbon is in Group IV and has a half-filled (or half-empty) valence shell. It is neither strongly electropositive nor strongly electronegative.

1.6 The unpaired electrons in the fluorine atoms are shared in the fluorine molecule.

$$:\ddot{\underset{..}{F}}\cdot + \cdot\ddot{\underset{..}{F}}: \longrightarrow :\ddot{\underset{..}{F}}:\ddot{\underset{..}{F}}: + \text{heat}$$

fluorine atoms — fluorine molecule

1.7 dichloromethane (methylene chloride) — trichloromethane (chloroform)

dichloromethane: H:C:Cl: (electron-dot, with H above and :Cl: below C) or $H-\underset{Cl}{\overset{H}{C}}-Cl$

trichloromethane: H:C:Cl: (electron-dot, with :Cl: above and :Cl: below C) or $H-\underset{Cl}{\overset{Cl}{C}}-Cl$

1.8 If the C–C bond length is 1.54 Å and the Cl–Cl bond length is 1.98 Å, we expect the C–Cl bond length to be about 1.76 Å: (1.54 + 1.98)/2. In fact, the C–Cl bond (1.75 Å) is longer than the C–C bond.

1.9 Propane

```
   H  H  H
   |  |  |
H–C––C––C–H
   |  |  |
   H  H  H
```

1.10 $N^{\delta+}$–$Cl^{\delta-}$; $S^{\delta+}$–$O^{\delta-}$ The key to predicting bond polarities is to determine the relative electronegativities of the elements in the bond. In Table 1.4, Cl is more electronegative than N. The polarity of the S–O bond is easy to predict because both elements are in the same column of the periodic table, and the more electronegative atom appears nearer the top.

1.11 Both Cl and F are more electronegative than C.

```
          δ−
          F
          |δ+
δ− Cl–––C–––Cl δ−
          |
          F
          δ−
```

1.12 Both the C–O and H–O bonds are polar, and the oxygen is more electronegative than either carbon or hydrogen.

```
   H
   |
H–C–O–H
   |
   H
```

1.13 H:C:::N: H:C:::N: H:C:::N:

1.14

a. The carbon shown has 12 electrons around it, 4 more than are allowed.

b. There are 20 valence electrons shown, whereas there should only be 16 (6 from each oxygen and 4 from the carbon).

c. There is nothing wrong with this formula, but it does place a formal charge of –1 on the "left" oxygen and +1 on the "right" oxygen (see Sec. 1.11). This formula is one possible contributor to the resonance hybrid structure for carbon dioxide (see Sec. 1.12); it is less important than the structure with two carbon–oxygen double bonds, because it takes energy to separate the + and – charges.

1.15 Methanal (formaldehyde), H_2CO. There are 12 valence electrons altogether (C = 4, H = 1, and O = 6). A double bond between C and O is necessary to put 8 electrons around each of these atoms.

```
  H                H
H:C::O:     or      C=O
                   H
```

1.16 There are 10 valence electrons, 4 from C and 6 from O. An arrangement that puts 8 electrons around each atom is shown below. This structure puts a formal charge of –1 on C and +1 on O (see Sec. 1.11).

:C: : :O: or :C≡O:

1.17 If the carbon chain is linear, there are two possibilities:

```
             H  H                       H   H   H   H
H            |  |                       |   |   |   |
 \C=C-C-C-H     and     H-C-C=C-C-H
H/    |  |  |                           |               |
      H  H  H                           H               H
```

But the carbon chain can be branched, giving a third possibility:

```
          H   H
           \ /
H          C-H
 \        /
  C=C
 /        \
H          C-H
           / \
          H   H
```

1.18 a.

```
     H
     |  ..
H-C-F:
     |  ..
     H
```

b.

```
     H  H
     |  |
H-C-N:
     |  |
     H  H
```

Notice that the fluorine has three non-bonded electron pairs (part a) and nitrogen has one non-bonded electron pair (part b).

1.19 No, it does not. We cannot draw any structure for C_2H_5 that has four bonds to each carbon and one bond to each hydrogen.

1.20 First write the alcohols (compounds with an O–H group).

```
   H  H  H                                H   H   H
   |  |  |    ..                          |   |   |
H-C-C-C-O-H      and      H-C-C-C-H
   |  |  |    ..                          |   |   |
   H  H  H                                H :O: H
                                              |
                                              H
```

Then write the structures with a C–O–C bond (ethers).

```
   H       H  H
   |   ..  |  |
H-C-O-C-C-H
   |   ..  |  |
   H       H  H
```

There are no other possibilities. For example,

```
 ..  H  H  H                              H   H   H
     |  |  |                              |   |   |
H-O-C-C-C-H      and      H-C-C-C-H
 ..  |  |  |                              |   |   |
     H  H  H                              H   H :O:
                                                  |
                                                  H
```

are the same as

```
   H  H  H
   |  |  |  ..
H-C-C-C-O-H
   |  |  |  ..
   H  H  H
```

They all have the same bond connectivities and represent a single structure. Similarly,

```
  H     H  H                           H  H     H
  |     |  |                           |  |     |
H-C-Ö-C-C-H    is the same as    H-C-C-Ö-C-H
  |     |  |                           |  |     |
  H     H  H                           H  H     H
```

1.21 From left to right: *n*-pentane, isopentane, and isopentane.

1.22 a.

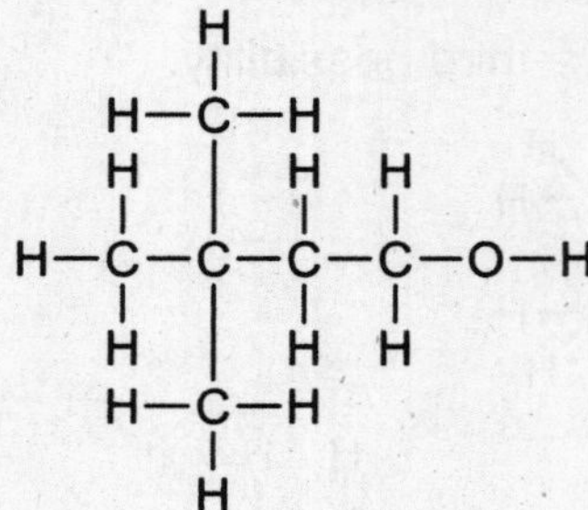

b.

```
Cl     Cl
  C=C
Cl     Cl
```

Notice that the non-bonded electron pairs on oxygen and chlorine are not shown. Non-bonded electron pairs are frequently omitted from organic structures, but it is important to know that they are there.

1.23 First draw the carbon skeleton showing all bonds between carbons.

```
     C-C
C-C
     C-C
```

Then add hydrogens to satisfy the valency of four at each carbon.

```
         H  H
         |  |
       H-C-C-H
  H     /   |                      CH2-CH3
  |    /    H                     /
H-C-C-H          or        H3C-CH
  |    \    H                     \
  H     \   |                      CH2-CH3
       H-C-C-H
         |  |
         H  H
```

1.24 [skeletal structure] stands for the carbon skeleton

```
             C
              \
C             C-C
 \           /
  C=C
 /
C
```

Addition of the appropriate number of hydrogens on each carbon completes the valence of 4.

1.25 ammonia

```
   H
   |
H-N:
   |
   H
```

formal charge on nitrogen = 5 – (2 + 3) = 0

ammonium ion

```
 ┌    H    ┐+
 |    |    |
 |  H-N-H  |
 |    |    |
 └    H    ┘
```

formal charge on nitrogen = 5 – (0 + 4) = +1

amide ion $\left[H-\ddot{N}-H \right]^-$ formal charge on nitrogen = 5 – (4 + 2) = –1

The formal charge on hydrogen in all three cases is zero [1 – (0 + 1) = 0].

1.26 For the singly bonded oxygens, formal charge = 6 – (6 + 1) = –1.
For the doubly bonded oxygen, formal charge = 6 – (4 + 2) = 0.
For the carbon, formal charge = 4 – (0 + 4) = 0.

$$\left[:\ddot{O}-\overset{\overset{:O:}{\|}}{C}-\ddot{O}: \right]^{2-}$$

1.27 There are 24 valence electrons to use in bonding (6 from each oxygen, 5 from the nitrogen, and one more because of the negative charge). To arrange the atoms with 8 valence electrons around each atom, we must have one nitrogen–oxygen double bond:

$$\left[:\ddot{O}-\overset{\overset{:O:}{\|}}{N}-\ddot{O}: \longleftrightarrow :\ddot{O}=\overset{\overset{:\ddot{O}:}{|}}{N}-\ddot{O}: \longleftrightarrow :\ddot{O}-\overset{\overset{:\ddot{O}:}{|}}{N}=\ddot{O}: \right]^-$$

The formal charge on nitrogen is 5 – (0 + 4) = +1.
The formal charge on singly bonded oxygen is 6 – (6 + 1) = –1.
The formal charge on doubly bonded oxygen is 6 – (4 + 2) = 0.
The net charge of the ion is –1 because each resonance structure has one positively charged nitrogen atom and two negatively charged oxygen atoms. In the resonance hybrid, the formal charge on the nitrogen is +1; on the oxygens, the charge is –2/3 at each oxygen, because each oxygen has a –1 charge in two of the three structures and a zero charge in the third structure.

1.28 There are 16 valence electrons (five from each N plus one for the negative charge). The formal charges on each nitrogen are shown below the structures.

$$\left[\underset{-1}{:\ddot{N}}=\underset{+1}{N}=\underset{-1}{\ddot{N}:} \longleftrightarrow \underset{0}{:N}\equiv\underset{+1}{N}-\underset{-2}{\ddot{\underset{\cdot\cdot}{N}}:} \right]$$

1.29 In tetrahedral methane, the H–C–H bond angle is 109.5°. In “planar” methane, this angle would be 90° and bonding electrons would be closer together. Thus, repulsion between electrons in different bonds would be greater in “planar” methane than in tetrahedral methane. Consequently, “planar” methane would be less stable than tetrahedral methane.

1.30 a. C=O, ketone; C=C, alkene; O–H, alcohol
b. arene; C(=O)NH, amide; C–S–C, thioether, C(=O)O–H, carboxylic acid
c. C=O, ketone
d. C=C, alkene

ADDITIONAL PROBLEMS

1.31 The number of valence electrons is the same as the number of the group to which the element belongs in the periodic table (Table 1.3).

a. :C̤̈l: b. :Ṅ· c. ·S̤̈· d. ·Ċ·

e. ·Ö̤· f. K·

1.32

a.	covalent	b.	covalent	c.	covalent	d.	covalent
e.	ionic	f.	ionic	g.	ionic	h.	covalent

1.33 The bonds in sodium chloride are ionic; Cl is present as chloride ion (Cl^-); Cl^- reacts with Ag^+ to give AgCl, a white precipitate. The C–Cl bonds in CCl_4 are covalent, no Cl^- is present to react with Ag^+.

1.34

		Valence Electrons	Common Valence
a.	N	5	3
b.	C	4	4
c.	F	7	1
d.	O	6	2
e.	P	5	3
f.	S	6	2

Note that the *sum* of the number of valence electrons and the common valence is 8 in each case. (The only exception is H, where it is 2, the number of electrons in the completed first shell.)

1.35

```
a.     H              b.    H  H             c.    H  H  H
       |                    |  |  ..               |  |  |
    H—C—Ö—H               H—C—C—Cl:             H—C—C—C—H
       |   ..               |  |  ..               |  |  |
       H                    H  H                   H  H  H

d.     H  H  H        e.    H  H             f.  H
       |  |  |              |  |  ..               \
    H—C—C—N:              H—C—C—F:                 C=O
       |  |  |              |  |  ..               /   ..
       H  H  H              H  H                  H
```

1.36 a.

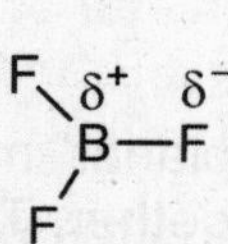

Fluorine is more electronegative than boron.

b.

```
       H
       | δ+  δ−
    H—C—F
       |
       H
```

Fluorine is more electronegative than carbon.

c.

```
    δ−   δ+  δ−
    :O=C=O:
```

The C=O bond is polar, and the oxygen is more electronegative than carbon.

d. :C̤̈l—C̤̈l:

Since the bond is between identical atoms, it is pure covalent (nonpolar).

e. 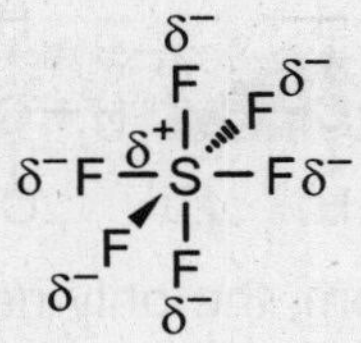 $\delta^-F-S^{\delta+}(F^{\delta-})_5$ (structure: S bonded to six F, each F δ^-, S δ^+)

Fluorine is more electronegative than sulfur. Indeed, it is the most electronegative element. Note that the S has 12 electrons around it. Elements below the first full row sometimes have more than 8 valence electrons around them.

f. 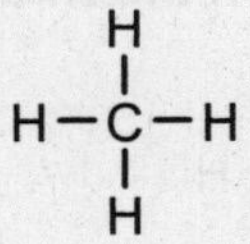 CH_4 (H–C–H with H above and below)

Carbon and hydrogen have nearly identical electronegativities, and the bonds are essentially nonpolar.

g. 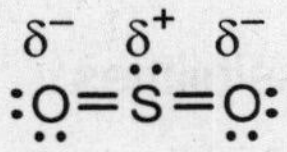 $\overset{\delta-}{:\ddot{O}}=\overset{\delta+}{\ddot{S}}=\overset{\delta-}{\ddot{O}:}$

Oxygen is more electronegative than sulfur.

h. $H_3\overset{\delta+}{C}-\overset{\delta-}{O}-\overset{\delta+}{H}$

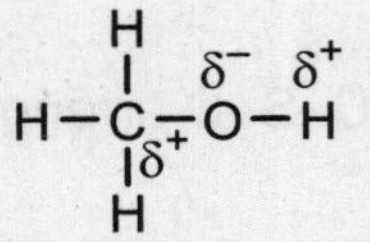

Oxygen is more electronegative than carbon, or hydrogen, so the C–O and O–H bonds are polar covalent.

1.37 The O–H bond is polar (there is a big difference in electronegativity between oxygen and hydrogen), with the hydrogen δ^+. The C–H bonds in acetic acid are not polar (there is little electronegativity difference between carbon and hydrogen). The negatively charged oxygen of the carbonate deprotonates the acetic acid.

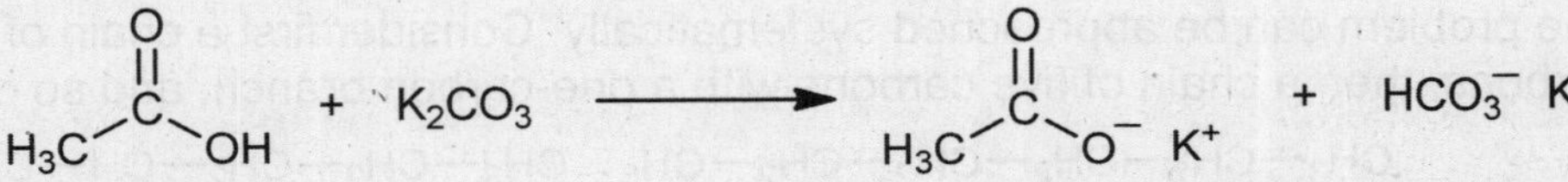

$$H_3C-C(=O)-OH + K_2CO_3 \longrightarrow H_3C-C(=O)-O^-\ K^+ + HCO_3^-\ K^+$$

1.38 a. C_3H_8 — The only possible structure is $CH_3CH_2CH_3$.

b. C_3H_7F — The fluorine can replace a hydrogen on an end carbon or on the middle carbon in the answer to a: $CH_3CH_2CH_2F$ or $CH_3CH(F)CH_3$

c. $C_2H_2Br_2$ — The sum of the hydrogens and bromines is 4, not 6. Therefore there must be a carbon–carbon double bond: $Br_2C{=}CH_2$ or $BrCH{=}CHBr$. No carbocyclic structure is possible because there are only two carbon atoms. (In Chapter 3, Sec. 3.5, we will see that the relative orientations of the Br in the latter structure can lead to additional isomers.)

d. C_3H_6 — There must be a double bond or, with three carbons, a ring:

$CH_3CH{=}CH_2$ or cyclopropane (CH_2 bonded to H_2C—CH_2)

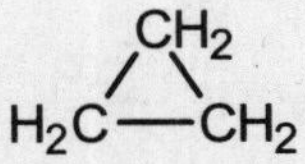

e. C_4H_9Cl — The carbon chain may be either linear or branched. In each case there are two possible positions for the chlorine.

$CH_3CH_2CH_2CH_2Cl$ $CH_3CH(Cl)CH_2CH_3$

$(CH_3)_2CHCH_2Cl$ $(CH_3)_3CCl$

f. $C_3H_6Cl_2$ — Be systematic. With one chlorine on an end carbon, there are three possibilities for the second chlorine.

```
    Cl H  H          H  Cl H          H  H  Cl
    |  |  |          |  |  |          |  |  |
  H-C--C--C-H      H-C--C--C-H      H-C--C--C-H
    |  |  |          |  |  |          |  |  |
    Cl H  H          Cl H  H          Cl H  H
```

If one chlorine is on the middle atom, the only new structure arises with the second chlorine also on the middle carbon:

```
    H  Cl H
    |  |  |
  H-C--C--C-H
    |  |  |
    H  Cl H
```

g. C_3H_8S — With an S–H bond, there are two possibilities:

$CH_3CH_2CH_2\text{–}SH$ $\quad (CH_3)_2CH\text{–}SH$

There is also one possibility in a C–S–C arrangement:

$CH_3\text{–}S\text{–}CH_2CH_3$

h. $C_2H_4F_2$ — The fluorines can either be attached to the same carbon or to different carbons:

CH_3CHF_2 or FCH_2CH_2F

1.39 The problem can be approached systematically. Consider first a chain of six carbons, then a chain of five carbons with a one-carbon branch, and so on.

```
CH3—CH2—CH2—CH2—CH2—CH3     CH3—CH2—CH2—CH—CH3
                                         |
                                         CH3

CH3—CH2—CH—CH2—CH3
           |
           CH3                          CH3
                                         |
CH3—CH—CH—CH3                  CH3—CH2—C—CH3
     |   |                               |
     CH3 CH3                             CH3
```

1.40 a.

```
    H  H  H  H  H  H
    |  |  |  |  |  |
  H-C--C--C--C--C--C-H
    |  |  |  |  |  |
    H  H  H  H  H  H
```

b.

```
          H
          |
        H-C-H
    H     |  H  H  H
    |     |  |  |  |
  H-C-----C--C--C--C-H
    |     |  |  |  |
    H     |  H  H  H
        H-C-H
          |
          H
```

c.

```
    H  H     H  H
    |  |     |  |
  H-C--C--N--C--C-H
    |  |     |  |
    H  H     H  H
```

d.

```
    H  H     H  H
    |  |     |  |
  H-C--C--S--C--C-H
    |  |     |  |
    H  H     H  H
```

e.

```
    H   H
    |   |
H — C — C — H
    |   |
    Cl  O
        |
        H
```

f.

```
            H
            |
    H   H   O   H   H
    |   |   |   |   |
H — C — C — C — C — C — H
    |   |   |   |   |
    H   H   H   H   H
```

1.41 a. $CH_3CH_2C(CH_3){=}CH_2$

b.

```
        H
  H     |
   C == C
  |      \   H
  |       C <
  |      /   H
   C == C
  H     |
        H
```

c. $(CH_3)_3CCH(CH_3)_2$

d. $CH_3CH_2CH_2CHCH_2CHCH_2CH_2CH_3$ (with CH_3 on the fourth carbon and CH_2CH_3 on the sixth carbon)

```
CH3CH2CH2CHCH2CHCH2CH2CH3
         |    |
        CH3  CH2CH3
```

e. $CH_3CH_2OCH_2CH_3$

f. $(CH_3)_2CHCH_2CH_2CH_3$

g.

```
      H     H
  H    \   /    H
H — C — C — C — H  ... 
```

(ring of six carbons: C_6H_{10} with one C=C double bond; each CH_2 carbon bears two H, each double-bond carbon bears one H)

h.

```
H2C —— CH2
 |       |
H2C     CH2
   \   /
     O
```

1.42 a. In line formulas, each line segment represents a C–C bond, and C–H bonds are not shown (see Sec. 1.10). There are a number of acceptable line structures for $CH_3(CH_2)_4CH_3$, three of which are shown here. The orientation of the line segments on the page is not important, but the number of line segments connected to each point is important.

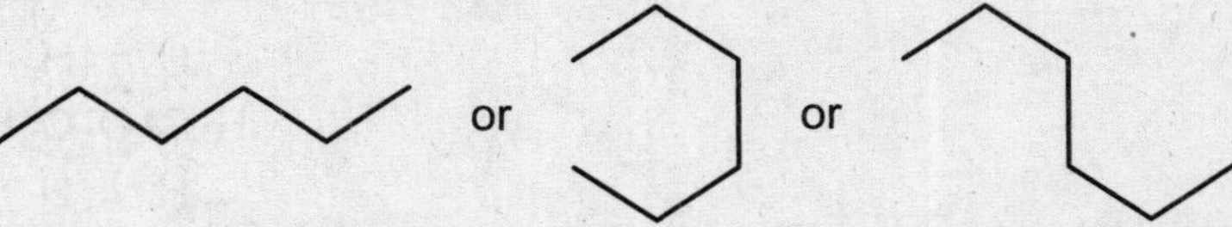

b. Bonds between carbon atoms and non-carbon atoms are indicated by line segments that terminate with the symbol for the non-carbon atom. In this case, the non-carbon atom is an oxygen.

O or O

c. Bonds between hydrogens and non-carbon atoms are shown.

OH or OH

d. The same rules apply for cyclic structures.

or

1.43 a. 7 b. $C_7H_{14}O$

c.

```
    H  O  H  H  H  H  H
    |  ‖  |  |  |  |  |
H — C — C — C — C — C — C — C — H
    |     |  |  |  |  |
    H     H  H  H  H  H
```

1.44 First count the carbons, then the hydrogens, and finally the remaining atoms.

a. $C_{10}H_{14}N_2$ b. $C_5H_5N_5$ c. $C_{10}H_{16}$
d. $C_9H_6O_2$ e. C_6H_6

1.45 a. CN^- There are 10 valence electrons (C = 4, N = 5, plus 1 more because of the negative charge).

$^-$:C:::N:

The carbon has a –1 formal charge [4 – (2 + 3) = –1]. Some might say that it would be better to write cyanide as NC^-.

b. $HONO_2$ There are 24 valence electrons. The nitrogen has a +1 formal charge [5 – (0 + 4) = +1], and the singly bonded oxygen has a –1 formal charge [6 – (6 + 1)] = –1. The whole molecule is neutral.

```
      :Ö: ← (–1)
   H:Ö:N::Ö:
        ↑
      (+1)
```

c. H_3COCH_3 There are no formal charges.

```
   H    H
 H:C:Ö:C:H
   H    H
```

d. NH_4^+ The nitrogen has a +1 formal charge; see the answer to Problem 1.25.

```
   H ↙(+1)
 H:N:H
   H
```

e. HONO First determine the total number of valence electrons; H = 1, O = 2 x 6 = 12, N = 5, for a total of 18. These must be arranged in pairs so that the hydrogen has 2 and the other atoms have 8 electrons around them.

H:Ö:N::Ö: or H—Ö—N=Ö:

Using the formula for formal charge given in Sec. 1.11, it can be determined that none of the atoms has a formal charge.

f. CO There are 10 valence electrons.

(–1) → :C:::O: ← (+1)

The carbon has a –1 formal charge, and the oxygen has a +1 formal charge. Carbon monoxide is isoelectronic with cyanide ion but has no net charge (–1 + 1 = 0).

g. BF_3 There are 24 valence electrons (B = 3, F = 7). The structure is usually written with only 6 electrons around the boron.

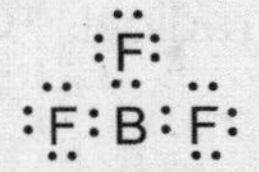

In this case, there are no formal charges. This structure shows that BF_3 is a Lewis acid, which readily accepts an electron pair to complete the octet around the boron.

h. H_2O_2 There are 14 valence electrons and an O–O bond. There are no formal charges.

H:Ö:Ö:H

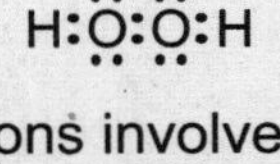

i. HCO_3^- There are 24 valence electrons involved. The hydrogen is attached to an oxygen, not to carbon.

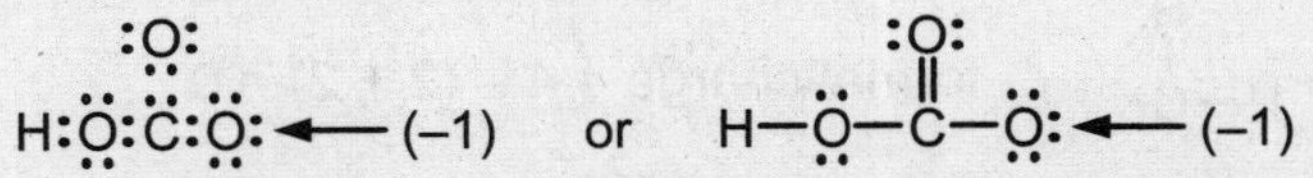

← (–1) or ← (–1)

The indicated oxygen has a formal charge: 6 – (6 + 1) = –1. All other atoms are formally neutral.

1.46 There are 18 valence electrons (6 from each of the oxygens, 5 from the nitrogen, and 1 from the negative charge).

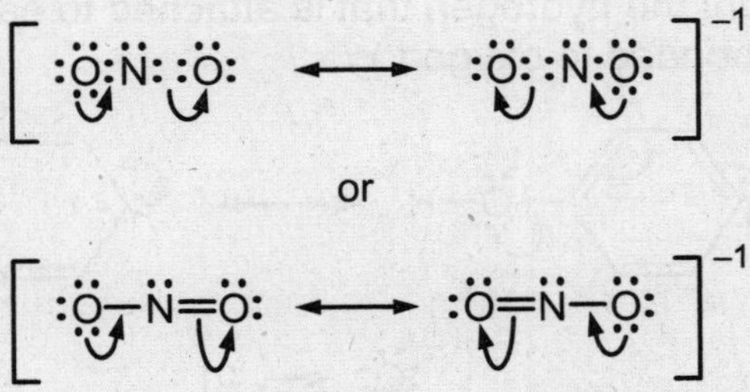

–1

or

–1

The negative charge in each contributor is on the singly bonded oxygen [6 – (6 + 1) = –1. The other oxygen and the nitrogen have no formal charge. In the resonance hybrid, the negative charge is spread equally over the two oxygens; the charge on each oxygen is –1/2.

1.47 Each atom in both structures has a complete valence shell of electrons. There are no formal charges in the first structure, but in the second structure, the oxygen is formally positive and the carbon is formally negative.

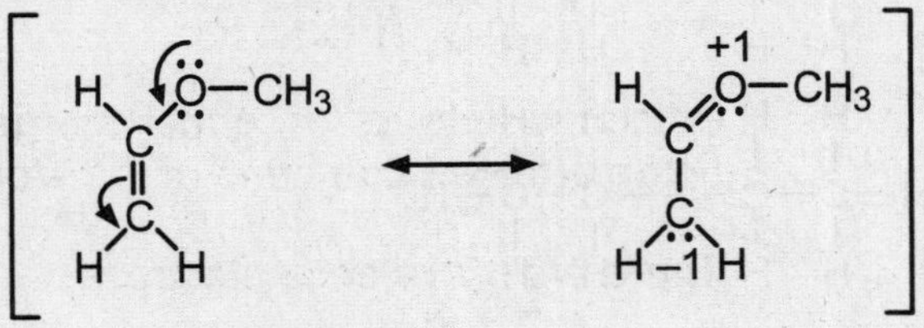

1.48 This is a methyl carbocation, CH_3^+.

$$\begin{array}{c} \text{H} \\ | \\ \text{H}-\text{C} \\ | \\ \text{H} \end{array} \qquad \text{formal charge} = 4 - (0 + 3) = +1$$

This is a methyl free radical, $\cdot CH_3$.

$$\begin{array}{c} \text{H} \\ | \\ \text{H}-\text{C}\cdot \\ | \\ \text{H} \end{array} \qquad \text{formal charge} = 4 - (1 + 3) = 0$$

This is a methyl carbanion, $:CH_3^-$.

$$\begin{array}{c} \text{H} \\ | \\ \text{H}-\text{C}: \\ | \\ \text{H} \end{array} \qquad \text{formal charge} = 4 - (2 + 3) = -1$$

This is a methylene or carbene, $:CH_2$.

$$\begin{array}{c} \text{H} \\ | \\ \text{H}-\dot{\text{C}}\cdot \end{array} \qquad \text{formal charge} = 4 - (2 + 2) = 0$$

All of these fragments are extremely reactive. They may act as intermediates in organic reactions.

1.49 In the first structure, there are no formal charges. In the second structure, the oxygen is formally +1, and the ring carbon bearing the unshared electron pair is formally –1. (Don't forget to count the hydrogen that is attached to each ring carbon except the one that is doubly bonded to oxygen.)

[$C_6H_5-\ddot{O}-H$ ⟷ $^{-1}:C_6H_5{=}\overset{+1}{\dot{O}}-H$]

1.50

$CH_3C(=O)OCH_3 + CH_3-\ddot{O}-H \longrightarrow$ $CH_3-C(\overset{..}{\underset{..}{O}}:)(OCH_3)-\overset{+}{O}(H)-CH_3$ (O⁻: –1; O–H: +1)

1.51 a.

$$\text{H}-\underset{\text{H}}{\overset{\text{H}}{\text{C}}}-\underset{\text{H}}{\overset{\text{H}}{\text{C}}}-\underset{..}{\overset{\text{H}}{\text{N}}}-\underset{\text{H}}{\overset{\text{H}}{\text{C}}}-\underset{\text{H}}{\overset{\text{H}}{\text{C}}}-\text{H}$$

b.

$$CH_3-\overset{\ddot{O}:}{\overset{\|}{C}}-\ddot{\underset{..}{O}}-OH$$

c.

$$\text{H}-\underset{\text{H}}{\overset{\text{H}}{\text{C}}}-\underset{\text{H}}{\overset{\text{H}}{\text{C}}}-\ddot{\underset{..}{\text{S}}}-\underset{\text{H}}{\overset{\text{H}}{\text{C}}}-\underset{\text{H}}{\overset{\text{H}}{\text{C}}}-\text{H}$$

d.

$$CH_3-\ddot{\underset{..}{O}}-CH_2CH_2-\ddot{\underset{..}{O}}-$$

1.52 If the *s* and *p* orbitals were hybridized to sp^3, two electrons would go into one of these orbitals and one electron would go into each of the remaining three orbitals.

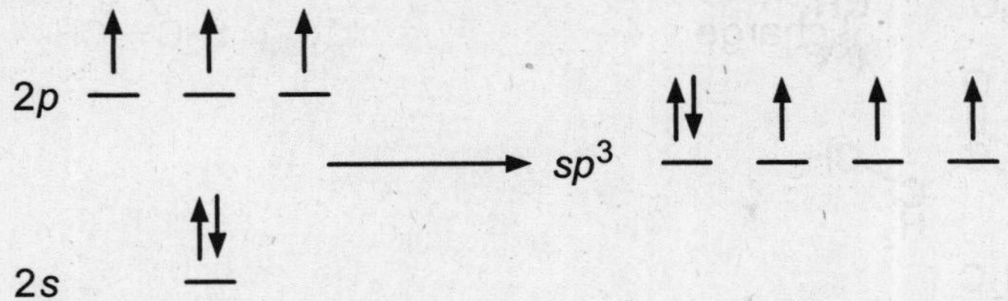

The predicted geometry of ammonia would then be tetrahedral, with one hydrogen at each of three corners, and the unshared pair at the fourth corner. In fact, ammonia has a pyramidal shape, a somewhat flattened tetrahedron. The H–N–H bond angle is 107º.

1.53 The ammonium ion is, in fact, isoelectronic (the same arrangement of electrons) with methane, and consequently has the same geometry. Four sp^3 orbitals of nitrogen each contain one electron. These orbitals then overlap with the 1*s* hydrogen orbitals, as in Figure 1.9.

1.54

The geometry is tetrahedral at carbon. It does not matter whether we draw the wedge bonds to the right or to the left of the carbon, or indeed "up" or "down".

1.55 The bonding in SiF_4 is exactly as in carbon tetrafluoride (CF_4). The geometry of silicon tetrafluoride is tetrahedral.

1.56 a.

A B C

b. Structures **A** and **B** are identical. Structures **A** and **C** are isomers. They both have the molecular formula C_3H_8O, but **A** is an alcohol and **C** is an ether.

1.57 Many correct answers are possible; a few are given here for part (a).

a. $H_3C-C(=O)-CH_3$

b. H_2C-CH_2 ring with $CH(OH)$ (cyclopropanol-type structure: OH on CH, bonded to H_2C-CH_2)

$H-C(=O)-CH_2-CH_3$

$H_2C=CH-O-CH_3$

c. CH_3-CH in ring with H_2C-O (epoxide); H_2C-O / H_2C-CH_2 (oxetane)

1.58 The structure of diacetyl (2,3-butanedione) is:

$H_3C-C(=O)-C(=O)-CH_3$

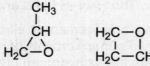

1.59 Compounds c, e, h, and j all have hydroxyl groups (—OH) and belong to a class of compounds called alcohols. Compounds b and k are ethers. They have a C—O—C unit. Compounds f, i and l contain amino groups ($—NH_2$) and belong to a family of compounds called amines. Compounds a, d, and g are hydrocarbons.

1.60 The more common functional groups are listed in Table 1.6. Often more than one answer may be possible.

a. $CH_3CH_2CH_2OH$ b. $CH_3CH_2OCH_2CH_3$

c. $H-C(=O)-CH_2-CH_3$ d. $H_3C-C(=O)-CH_3$

e. $CH_3CH_2-C(=O)-OH$

f. There are a number of possible answers. Five are shown below. Can you think of more?

$CH_3CH_2CH_2CH_2-C(=O)-H$ $(CH_3)_2CHCH_2O-C(=O)-H$ $(CH_3)_3CO-C(=O)-H$

$CH_3CH_2CH_2O-C(=O)-CH_3$ $CH_3CH_2O-C(=O)-CH_2CH_3$

1.61 a. carbonyl group (carboxylic acid), amino group (amine), aromatic group (arene)

b. $HO-C(=O)-CH(NH_2)CH_2-C_6H_5$

c. $C_9H_{11}NO_2$

d. The isomer of phenylalanine shown below is both an alcohol and an amide.

$H_2N-C(=O)-CH(OH)CH_2-C_6H_5$

2

Alkanes and Cycloalkanes; Conformational and Geometric Isomerism

Chapter Summary

Hydrocarbons contain only carbon and hydrogen atoms. **Alkanes** are acyclic **saturated hydrocarbons** (contain only single bonds); **cycloalkanes** are similar but have carbon rings.

Alkanes have the general molecular formula $\mathbf{C_nH_{2n+2}}$. The first four members of this **homologous series** are **methane**, **ethane**, **propane**, and **butane**; each member differs from the next by a $-CH_2-$, or **methylene group**. The **IUPAC** (International Union of Pure and Applied Chemistry) nomenclature system is used worldwide to name organic compounds. The IUPAC rules for naming alkanes are described in Secs. 2.3–2.5. **Alkyl groups**, alkanes minus one hydrogen atom, are named similarly except that the *-ane* ending is changed to *-yl*. The letter **R** stands for any alkyl group.

The two main natural sources of alkanes are **natural gas** and **petroleum**. Alkanes are insoluble in and less dense than water. Their boiling points increase with molecular weight and, for isomers, decrease with chain branching.

Conformations are different structures that are interconvertible by rotation about single bonds. For ethane (and in general), the **staggered** conformation is more stable than the eclipsed conformation (Figure 2.5).

The prefix *cyclo-* is used to name cycloalkanes. **Cyclopropane** is planar, but larger carbon rings are puckered. **Cyclohexane** exists mainly in a **chair conformation** with all bonds on adjacent carbons staggered. One bond on each carbon is **axial** (perpendicular to the mean carbon plane); the other is **equatorial** (roughly in that plane). The conformations can be interconverted by "flipping" the ring, which requires only bond rotation and occurs rapidly at room temperature for cyclohexane. Ring substituents usually prefer the less crowded, equatorial position.

Stereoisomers have the same order of atom attachments but different arrangements of the atoms in space. *Cis–trans* isomerism is one kind of stereoisomerism. For example, two substituents on a cycloalkane can be on either the same (*cis*) or opposite (*trans*) sides of the mean ring plane. Stereoisomers can be divided into two groups, **conformational isomers** (interconvertible by bond rotation) and **configurational isomers** (not interconvertible by bond rotation). ***Cis–trans* isomers** belong to the latter class. Alkanes are fuels; they burn in air if ignited. Complete combustion gives carbon dioxide and water; less complete combustion gives carbon monoxide or other less oxidized forms of carbon. Alkanes react with **halogens** (chlorine or bromine) in a reaction initiated by heat or light. One or more hydrogens can be replaced by halogens. This **substitution reaction** occurs by a **free-radical chain mechanism**.

Reaction Summary

Combustion

$$C_nH_{2n+2} + \left(\frac{3n+1}{2}\right)O_2 \longrightarrow nCO_2 + (n+1)H_2O$$

Halogenation (Substitution)

$$R{-}H + X_2 \xrightarrow{\text{heat or light}} R{-}X + H{-}X \quad (X = Cl, Br)$$

Mechanism Summary

Halogenation

Initiation $:\ddot{X}{-}\ddot{X}: \xrightarrow{\text{heat or light}} 2\,:\ddot{X}\cdot$

Propagation $R{-}H + \cdot\ddot{X}: \longrightarrow R\cdot + H{-}\ddot{X}:$

$R\cdot + :\ddot{X}{-}\ddot{X}: \longrightarrow R{-}\ddot{X}: + :\ddot{X}\cdot$

Termination $2\,:\ddot{X}\cdot \longrightarrow :\ddot{X}{-}\ddot{X}:$

$2R\cdot \longrightarrow R{-}R$

$R\cdot + \cdot\ddot{X}: \longrightarrow R{-}\ddot{X}:$

Learning Objectives

1. Know the meaning of: saturated hydrocarbon, alkane, cycloalkane, homologous series, methylene group.
2. Know the meaning of: conformation, staggered, eclipsed, "dash-wedge" projection, Newman projection, "sawhorse" projection, rotational isomers, rotamers.
3. Know the meaning of: chair conformation of cyclohexane, equatorial, axial, geometric or *cis–trans* isomerism, conformational and configurational isomerism.
4. Know the meaning of: substitution reaction, halogenation, chlorination, bromination, free-radical chain reaction, chain initiation, propagation, termination, combustion.
5. Given the IUPAC name of an alkane or cycloalkane, or a halogen-substituted alkane or cycloalkane, draw its structural formula.
6. Given the structural formula of an alkane or cycloalkane or a halogenated derivative, write the correct IUPAC name.
7. Know the common names of the alkyl groups, cycloalkyl groups, methylene halides, and haloforms.
8. Tell whether two hydrogens in a particular structure are identical or different from one another by determining whether they give the same or different products by monosubstitution with some group X.

9. Know the relationship between boiling points of alkanes and (a) their molecular weights and (b) the extent of chain branching.

10. Write all steps in the free-radical chain reaction between a halogen and an alkane, and identify the initiation, propagation, and termination steps.

11. Write a balanced equation for the complete combustion of an alkane or cycloalkane.

12. Draw, using dash-wedge, sawhorse, or Newman projection formulas, the important conformations of ethane, propane, butane, and various halogenated derivatives of these alkanes.

13. Recognize, draw, and name *cis–trans* isomers of substituted cycloalkanes.

14. Draw the chair conformation of cyclohexane and show clearly the distinction between axial and equatorial bonds.

15. Identify the more stable conformation of a monosubstituted cyclohexane; also, identify substituents as axial or equatorial when the structure is "flipped" from one chair conformation to another.

16. Classify a pair of isomers as structural (constitutional) isomers or stereoisomers, and if the latter, as conformational or configurational (see Figure 2.8).

ANSWERS TO PROBLEMS

Problems Within the Chapter

2.1 $C_{12}H_{26}$; use the formula C_nH_{2n+2}, where $n = 12$.

2.2 The formulas in parts a and d fit the general formula C_nH_{2n+2} and are alkanes. C_7H_{12} (part b) and C_8H_{16} (part c) have two fewer hydrogens than called for by the alkane formula and must be either an alkene or a cycloalkane.

2.3
a. 2-methylbutane (number the longest chain from left to right)
b. 2-methylbutane (number the longest chain from right to left)

The structures in parts a and b are identical, as the names show.

c. 2,2-dimethylpropane

2.4 Chlorofluoromethane; the substituents are alphabetized when more than one is present.

2.5
a. $(CH_3)_2CH_2CHCl$ b. $(CH_3)_2CHBr$ c. $CH_3CH(Cl)CH_2CH_3$
d. $(CH_3)_3C–I$ e. $CH_3CH_2CH_2F$
f. RBr; the letter R stands for any alkyl group

2.6
a. 2-fluorobutane
b. 3-bromo-2,2,4-trimethylpentane (*not* 3-bromo-2,4,4-trimethylpentane, which has higher numbers for the substituents)

2.7 H_3C CH_3 H_3C CH_3

2.8 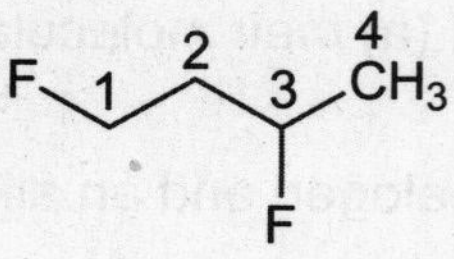 1,3-difluorobutane

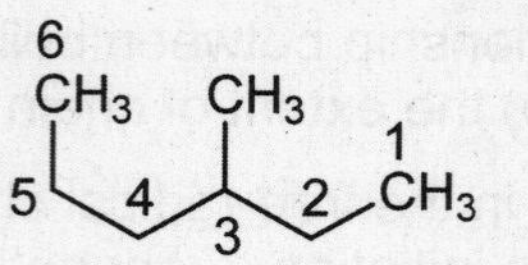

Correct name: 3-methylhexane; the longest chain is 6 carbon atoms.

2.9 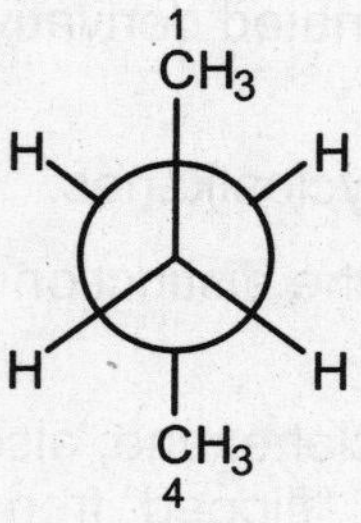 (anti) 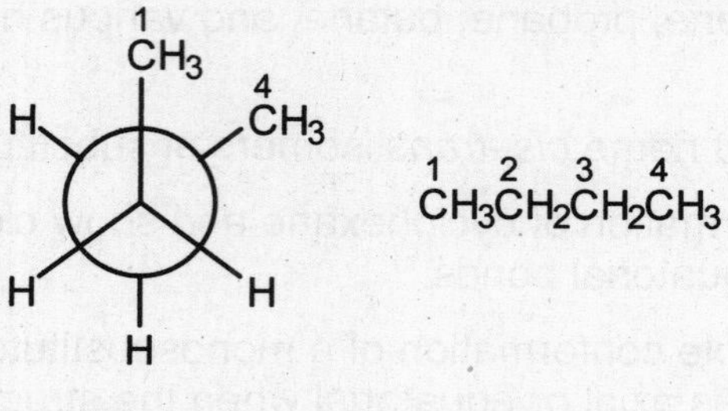 (gauche)

$\overset{1}{C}H_3\overset{2}{C}H_2\overset{3}{C}H_2\overset{4}{C}H_3$

Carbon-2 and carbon-3 in the chain each have two hydrogens and a methyl group attached. The staggered conformation at the left is the more stable. In the staggered conformation at the right, the two methyl substituents (large groups) come close to one another, a less stable arrangement than in the structure to the left. These conformations are referred to as *anti* and *gauche*, respectively.

2.10 A cycloalkane has two fewer hydrogens than the corresponding alkane. Thus, the general formula for a cycloalkane is C_nH_{2n}.

2.11 a. 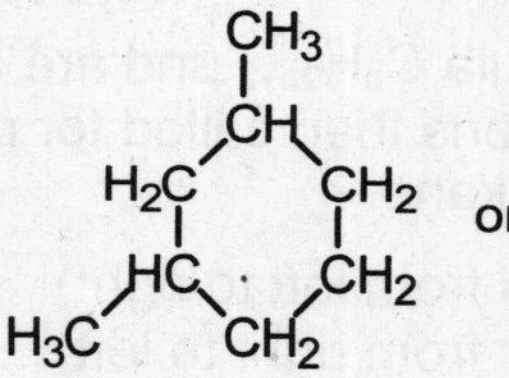 or 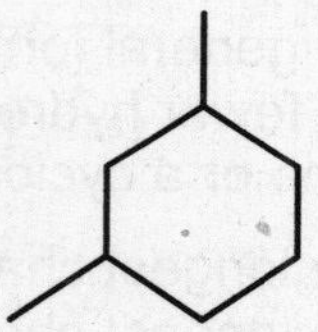 b. 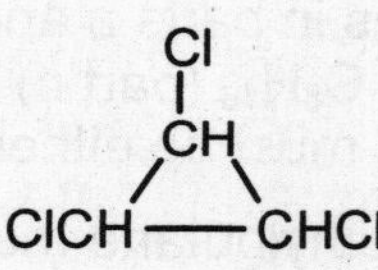 or 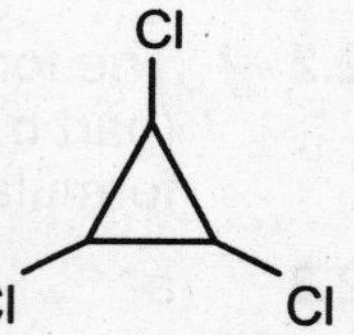 c. 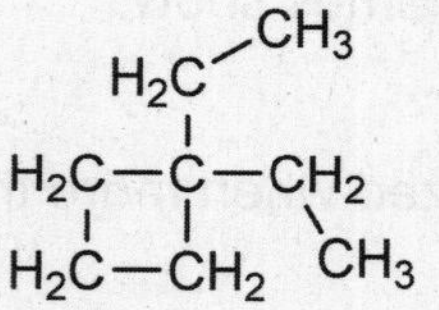 or 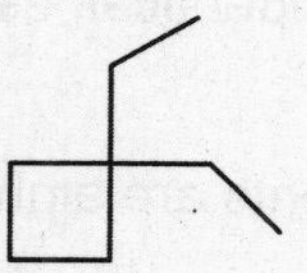 d. 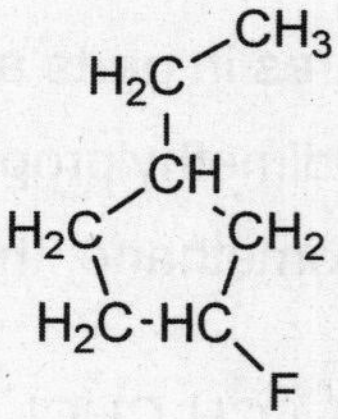 or

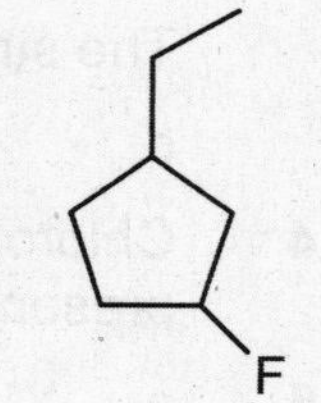

2.12
a. ethylcyclopentane
b. 1,1-dichlorocyclopropane
c. 1-bromo-3-methylcyclobutane

2.13 Since each ring carbon is tetrahedral, the H–C–H plane and the C–C–C plane at any ring carbon are mutually perpendicular.

2.14 If you sight down the bond joining carbon-2 and carbon-3, you will see that the substituents on these carbons are eclipsed. The same is true for the bond between carbon-5 and carbon-6. Also, two of the hydrogens on carbon-1 and carbon-4, the "inside" hydrogens, come quite close to one another. All these factors destabilize boat cyclohexane compared to chair cyclohexane.

2.15 The *tert*-butyl group is much larger than a methyl group. Therefore, the conformational preference is essentially 100% for an equatorial *tert*-butyl group:

2.16 a.

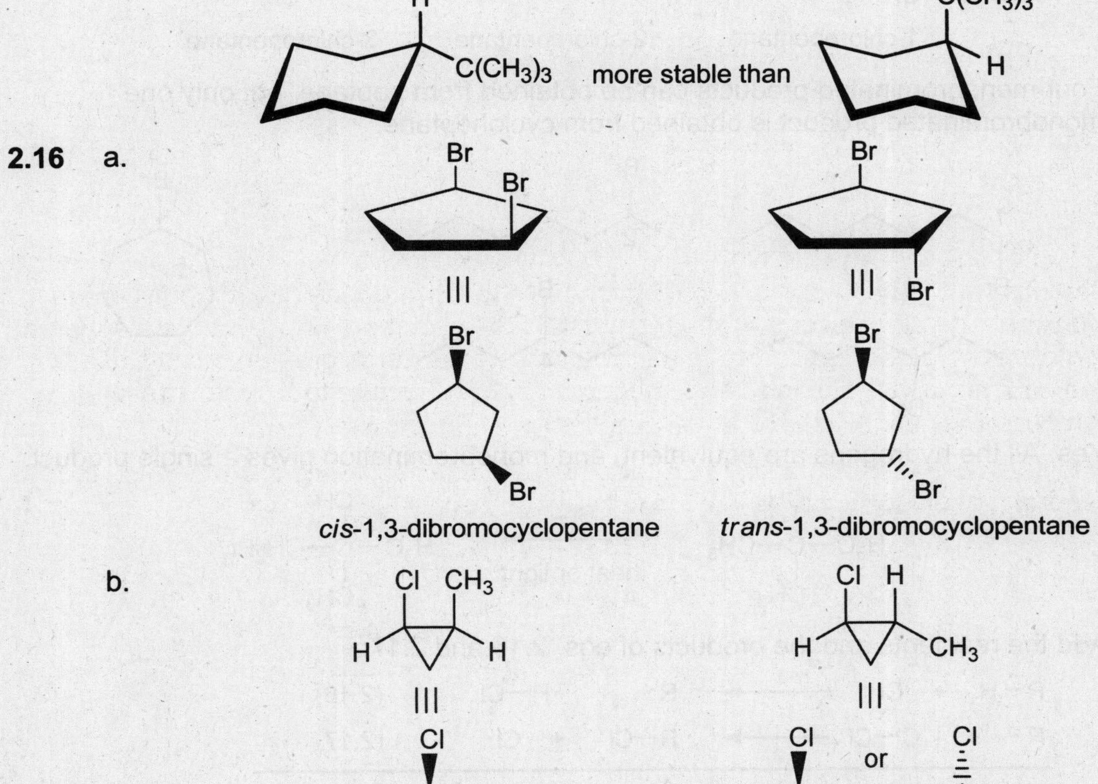

cis-1,3-dibromocyclopentane *trans*-1,3-dibromocyclopentane

b.

cis-1-chloro-2-methylcyclopentane *trans*-1-chloro-2-methylcyclopentane

2.17 a. configurational isomers (same bond pattern, but not interconvertible by σ-bond rotations)
b. conformational isomers
c. structural (constitutional) isomers

2.18 a. Formaldehyde (two C–O bonds) is more highly oxidized than methanol (one C–O bond).
b. The carbons in methanol (CH_3OH) and dimethyl ether (CH_3OCH_3) are equally oxidized. Each carbon has one C–O bond and three C–H bonds.
c. Methyl formate (three C–O bonds) is more highly oxidized than formaldehyde (two C–O bonds).

2.19 Follow eq. 2.12, but replace chloro with bromo:

CH_3Br	bromomethane (methyl bromide)
CH_2Br_2	dibromomethane (methylene bromide)
$CHBr_3$	tribromomethane (bromoform)
CBr_4	tetrabromomethane (carbon tetrabromide)

2.20

Cl Cl Cl

1-chloropentane 2-chloropentane 3-chloropentane

2.21 Four monobrominated products can be obtained from heptane, but only one monobrominated product is obtained from cycloheptane.

Br 1 Br 2 Br Br 3 Br 4

2.22 Yes. All the hydrogens are equivalent, and monobromination gives a single product.

$$H_3C{-}C(CH_3)_2{-}CH_3 \xrightarrow[\text{heat or light}]{Br_2} H_3C{-}C(CH_3)_2{-}CH_2Br$$

2.23 Add the reactants and the products of eqs. 2.16 and 2.17:

$$R{-}H \;+\; Cl\cdot \longrightarrow R\cdot \;+\; H{-}Cl \qquad (2.16)$$

$$R\cdot \;+\; Cl{-}Cl \longrightarrow R{-}Cl \;+\; Cl\cdot \qquad (2.17)$$

$$R{-}H \;+\; Cl\cdot \;+\; R\cdot \;+\; Cl{-}Cl \longrightarrow R{-}Cl \;+\; Cl\cdot \;+\; R\cdot \;+\; H{-}Cl$$

The intermediate radicals ($R\cdot$ and $Cl\cdot$) cancel and what is left is eq. 2.9:

$$R{-}H \;+\; Cl{-}Cl \longrightarrow R{-}Cl \;+\; H{-}Cl \qquad (2.9)$$

2.24 *initiation*

$$:\ddot{Cl}{-}\ddot{Cl}: \xrightarrow{\text{heat or light}} 2\;:\ddot{Cl}\cdot$$

propagation

$$CH_4 \;+\; :\ddot{Cl}\cdot \longrightarrow \cdot CH_3 \;+\; H{-}\ddot{Cl}:$$

$$\cdot CH_3 \;+\; :\ddot{Cl}{-}\ddot{Cl}: \longrightarrow CH_3{-}\ddot{Cl}: \;+\; :\ddot{Cl}\cdot$$

termination

$$2\,:\ddot{Cl}\cdot \longrightarrow :\ddot{Cl}{-}\ddot{Cl}:$$

$$2\,\cdot CH_3 \longrightarrow CH_3{-}CH_3$$

$$\cdot CH_3 \;+\; :\ddot{Cl}\cdot \longrightarrow CH_3{-}Cl$$

2.25 The second termination step in the answer to Problem 2.24 accounts for the formation of ethane in the chlorination of methane. Ethane can also be chlorinated, which explains the formation of small amounts of chloroethane in this reaction.

ADDITIONAL PROBLEMS

2.26 a. 2-methylhexane: First, note the root of the name (in this case, hex) and write down and number the carbon chain.

```
1   2   3   4   5   6
C — C — C — C — C — C
```

Next, locate the substituents (2-methyl).

```
1   2   3   4   5   6
C — C — C — C — C — C
    |
    C
```

Finally, fill in the remaining hydrogens.

$$H_3C{-}\underset{\displaystyle CH_3}{\underset{|}{CH}}{-}CH_2CH_2CH_2CH_3$$

b. $$CH_3{-}\underset{\displaystyle CH_3}{\underset{|}{CH}}{-}\underset{\displaystyle CH_3}{\underset{|}{CH}}{-}CH_3$$

c. (skeletal structure)

d. $$H_3C{-}\underset{\displaystyle Br}{\underset{|}{CH}}{-}CH_2\overset{\displaystyle CH_3}{\overset{|}{C}}H{-}CH_2CH_2CH_2CH_3$$

e. I, I (skeletal structure)

f. Cl, CH_3, H_3C (skeletal structure)

g. H_3C, H_3C, CH_3, CH_3 (skeletal structure)

h. F, F, CH_3, F (skeletal structure)

i. Cl, Cl, Cl, Cl (skeletal structure)

2.27 a. $$CH_3{-}\underset{\displaystyle CH_3}{\underset{|}{\overset{\displaystyle CH_3}{\overset{|}{C}}}}{-}CH_2{-}CH_2{-}CH_3$$

2,2-dimethylpentane

b. $CH_3{-}CH_2{-}CH_2{-}CH_3$

butane

c. $$CH_3{-}\underset{\displaystyle CH_3}{\underset{|}{CH}}{-}CH_2{-}CH_2{-}CH_3$$

2-methylpentane

d.

$$H_3C-\underset{Cl}{\overset{Cl}{C}}-\underset{F}{\overset{F}{C}}-F$$

2,2-dichloro-1,1,1-trifluoropropane (The placement of hyphens and commas important. Commas are used to separate numbers from numbers. Hyphens are used to separate numbers from letters. Also notice that "dichloro" comes before "trifluoro" because prefixes are ignored when alphabetizing substituents.)

e.

$$\begin{array}{c} CH_2-CH_2 \\ | \quad\quad | \\ CH_2-CH_2 \end{array}$$

cyclobutane

f.

$$CH_3-CH_2-\underset{F}{CH}-CH_3$$

2-fluorobutane

g.

$$H_3C-\underset{H}{\overset{H}{C}}-Br$$

bromoethane

h.

$$Cl-CH_2-CH_2-Cl$$

1,2-dichloroethane

i.

$$CH_3-\underset{Cl}{CH}-CH_3$$

2-chloropropane

2.28

	Common	IUPAC
a.	methyl bromide	bromomethane
b.	ethyl chloride	chloroethane
c.	methylene chloride (CH_2 = methylene)	dichloromethane
d.	isopropyl bromide	2-bromopropane
e.	iodoform	triiodomethane
f.	*t*-butyl chloride	2-chloro-2-methylpropane
g.	*n*-butyl fluoride	1-fluorobutane

2.29 a. The numbering started at the wrong end. The name should be 1,2-difluoropropane.

$$\overset{F}{C}H_2\overset{F}{C}H\text{-}CH_3$$

b. $CH_3CH_2CH_2CH_2CH_3$

The longest consecutive chain has five carbon atoms; the correct name is pentane.

c. $\overset{1}{CH_3}-\overset{2}{CH}(CH_2CH_3)-\overset{3}{CH_2}-\overset{4}{CH_3}$

The longest chain was not selected when the compound was numbered. The correct numbering is

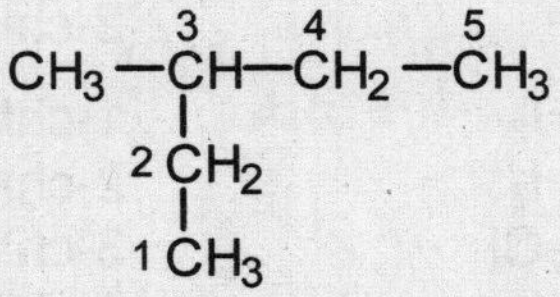

$$\overset{3}{CH_3}\text{—}\overset{4}{CH}\text{—}\overset{5}{CH_2}\text{—}CH_3$$

(with ${}^{2}CH_2$ and ${}^{1}CH_3$ on the branch)

and the correct name is 3-methylpentane.

d. The ethyl substituent should get the lower number. The correct name is 1-ethyl-2-methylcyclopropane.

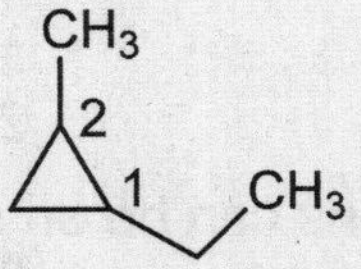

e. The longest chain is actually 7 carbons, hence a heptane. The correct name is 2,4-dimethylheptane.

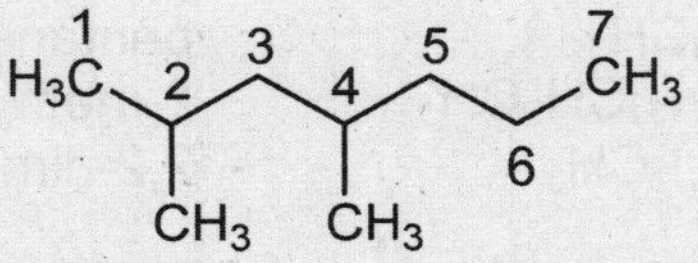

f. The ring was numbered the wrong way to give the lowest substituent. The correct name is 1-bromo-3-methylbutane.

g. The ring was numbered the wrong way to give the lowest substituent numbers. The correct name is 1,2-dimethylcyclopropane.

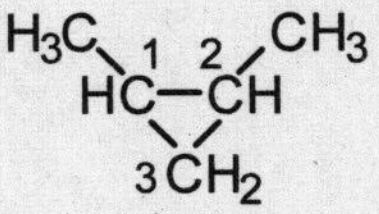

2.30 The root of the name, heptadec, indicates a 17-carbon chain. The correct formula is

$$CH_3CH(CH_3)CH_2CH_2CH_2CH_2CH_2CH_2CH_2CH_2CH_2CH_2CH_2CH_2CH_2CH_2CH_3$$

or $(CH_3)_2CH(CH_2)_{14}CH_3$

2.31 Approach each problem systematically. Start with the longest possible carbon chain and shorten it one carbon at a time until no further isomers are possible. To conserve space, the formulas below are written in condensed form, but you should write them out as expanded formulas.

a. $CH_3(CH_2)_2CH_3$ butane
$(CH_3)_3CH$ 2-methylpropane

b. $CH_3CH_2CH_2CH_2Cl$ 1-chlorobutane
$CH_3CH(Cl)CH_2CH_3$ 2-chlorobutane
$(CH_3)_2CHCH_2Cl$ 1-chloro-2-methylpropane
$(CH_3)_3Cl$ 2-chloro-2-methylpropane

c. $CHFClCH_2CH_3$ 1-chloro-1-fluoropropane
$CH_2FCHClCH_3$ 2-chloro-1-fluoropropane
$CH_2FCH_2CH_2Cl$ 3-chloro-1-fluoropropane
$CH_2ClCHFCH_3$ 1-chloro-2-fluoropropane
$CH_3CFClCH_3$ 2-chloro-2-fluoropropane

d. $CH_3CH_2CHBr_2$ 1,1-dibromopropane
$CH_3CHBrCH_2Br$ 1,2-dibromopropane
$CH_2BrCH_2CH_2Br$ 1,3-dibromopropane
$CH_3CBr_2CH_3$ 2,2-dibromopropane

e. The two hydrogens can be on either the same or different carbon atoms:
CH_2ClCBr_3 2-chloro-1,1,1-tribromoethane
$CH_2BrCClBr_2$ 1-chloro-1,1,2-tribromoethane
$CHBrClCHBr_2$ 2-chloro-1,1,2-tribromoethane

f. $CH_3(CH_2)_3CH_3$ pentane
$CH_3CH(CH_3)CH_2CH_3$ 2-methylbutane
$CH_3C(CH_3)_2CH_3$ 2,2-dimethylpropane

2.32 a. Start with a five-membered ring, then proceed to the four- and three-membered rings.

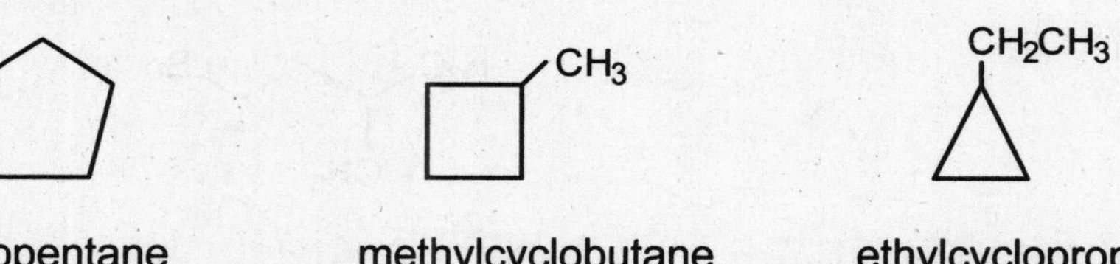

cyclopentane methylcyclobutane ethylcyclopropane

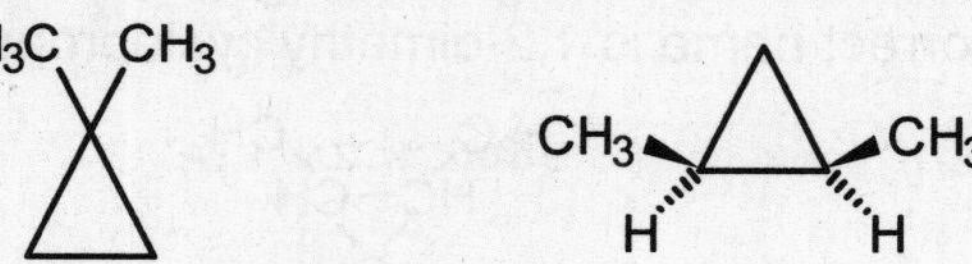

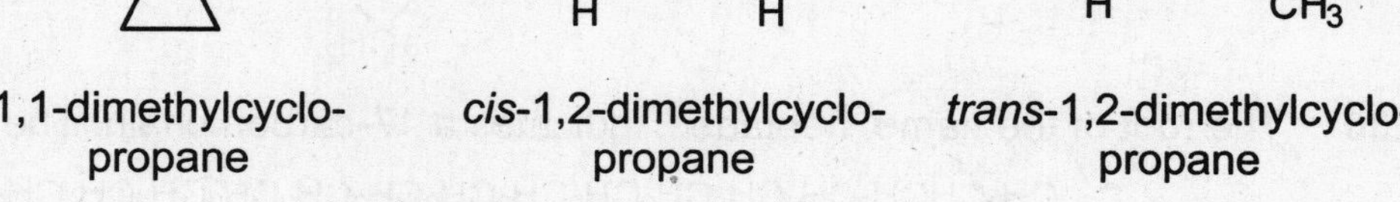

1,1-dimethylcyclopropane *cis*-1,2-dimethylcyclopropane *trans*-1,2-dimethylcyclopropane

b.

CH_2CH_3 CH_3

cyclohexane methylcyclopentane ethylcyclobutane

CH_3 CH_3 CH_3 CH_3 CH_3 CH_3

1,1-dimethylcyclobutane *trans*-1,2-dimethylcyclobutane *cis*-1,2-dimethylcyclobutane

CH_3 CH_3 CH_3 CH_3 $CH_2CH_2CH_3$

trans-1,3-dimethylcyclobutane *cis*-1,3-dimethylcyclobutane 1-propylcyclopropane

CH_3 CH_3 CH_3 CH_2CH_3 CH_3 CH_3CH_2

2-propylcyclopropane 1-ethyl-1-methylcyclopropane *cis*-1-ethyl-2-methylcyclopropane

CH_3 CH_3CH_2 CH_3 CH_3 CH_3 CH_3 CH_3 CH_3

trans-1-ethyl-2-methylcyclopropane 1,1,2-trimethylcyclopropane *cis,cis*-1,2,3-trimethylcyclopropane

CH_3 CH_3 CH_3

cis,trans-1,2,3-trimethylcyclopropane

2.33 **A** = methylcyclohexane and **B** = 2,2-dimethylbutane

2.34 The correct name would be 3,3-diethylpentane.

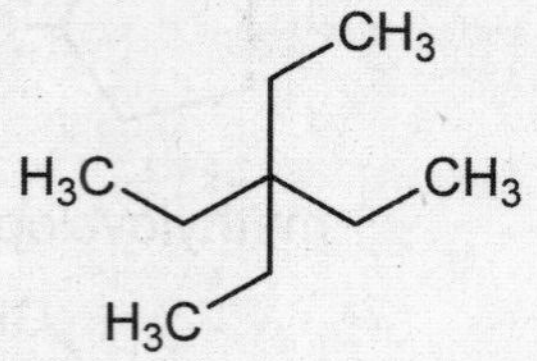

2.35 See Sec. 2.7 for a discussion of underlying principles. When comparing a series of alkanes, in general, the lower the molecular weight, the lower the intermolecular contact, the lower the van der Waals attractions, the less energy required to separate molecules from one another and the lower the boiling point. Thus, the hexane isomers (e and d) are expected to have lower boiling points than the heptane isomers (a, b, and c). Within a series of isomeric compounds, the greater the branching, the lower the intermolecular contact and the lower the boiling point. On these grounds, the expected order from the lowest to highest boiling point should be e, d, c, a, b. The actual boiling points are as follows: e, 2-methylpentane (60°C); c, 3,3-dimethylpentane (86°C); d, *n*-hexane (69°C); a, 2-methylhexane (90°C); b, *n*-heptane (98.4°C).

2.36 See Sec. 2.7 for a discussion of the underlying principles. The expected order of solubility in hexane is b (water) less than c (methyl alcohol) less than a (octane). Water does not dissolve in hexane because, to do so, it would have to break up hydrogen-bonding interactions bsetween water molecules, an energetically unfavorable process. Methyl alcohol is only sparingly soluble in hexane for the same reason. It is slightly soluble because it has a small hydrocarbon-like portion that can enter into weak van der Waals attractions with hexane. Octane is very soluble with hexane because van der Waals attractions between octane and hexane and between octane and itself are nearly equal.

2.37 The four conformations are as follows:

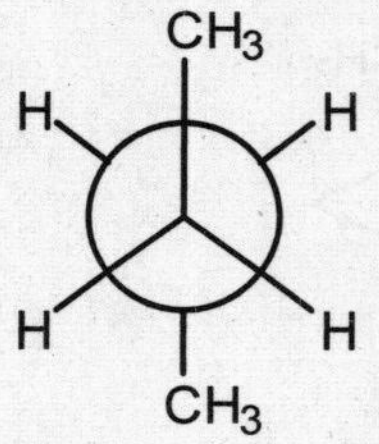

A: most stable staggered conformation

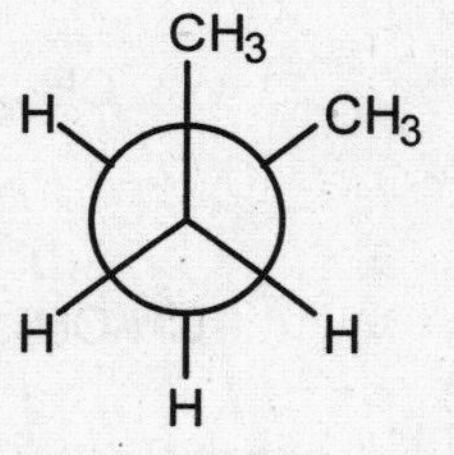

B: less stable staggered conformation (large groups are closer together)

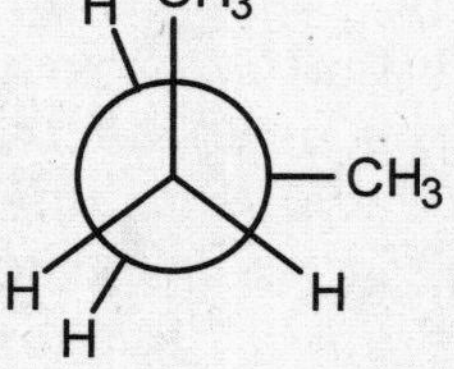

C: less stable than the staggered but more stable than the eclipsed conformation with two methyls eclipsed

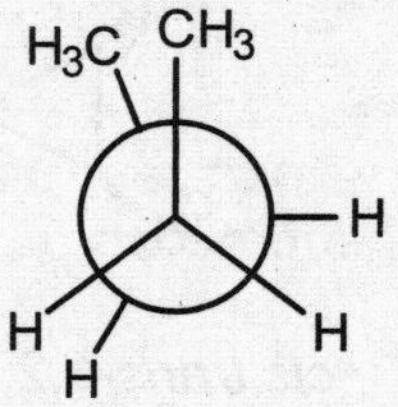

D: least stable of all four conformations

Staggered conformations are more stable than eclipsed conformations. Therefore A and B are more stable than C or D. Within each pair, CH_3-CH_3 interactions (for methyls on adjacent carbons) are avoided because of the large size of these groups.

2.38 By sighting down the C2–C3 bond, we have:

2.39

The stability decreases from left to right in each series of structures.

2.40 a.

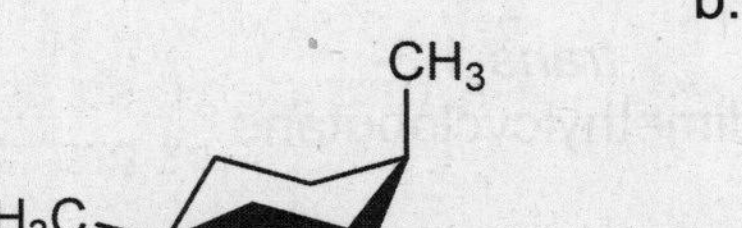

For *trans*-1,4-substituents, one substituent is axial and the other is equatorial.

b.

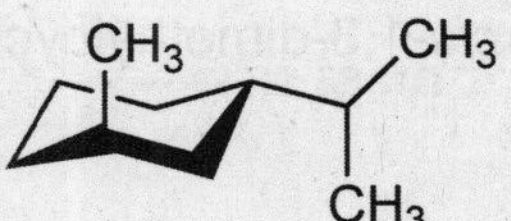

For 1,3-substitution and *trans*, one substituent is axial and the other is equatorial. The larger isopropyl group would be equatorial.

c.

d.

One ethyl group would be equatorial.

2.41 In each case, the formula at the left is named; the right-hand structure shows the other isomer.

a. *cis*-1,3-dichlorocyclohexane

b. *trans*-1-bromo-2-methylcyclohexane

2.42 *cis*-1,3-Dimethylcyclohexane can exist in a conformation in which both methyl substituents are equatorial:

CH_3 CH_3 H H

In the *trans* isomer, one methyl group must be axial. On the other hand, in 1,2- or 1,4-dimethylcyclohexane, only the *trans* isomer can have both methyls equatorial.

H CH_3 CH_3 H H CH_3 CH_3 H

2.43 We can have 1,1- or 1,2- or 1,3-dimethylcyclobutanes. Of these, the last two can exist as *cis–trans* isomers.

CH_3 CH_3 1,1-dimethylcyclobutane

CH_3 CH_3 *cis*-1,2-dimethylcyclobutane

CH_3 CH_3 *trans*-1,2-dimethylcyclobutane

CH_3 H_3C *cis*-1,3-dimethylcyclobutane

CH_3 H_3C *trans*-1,3-dimethylcyclobutane

2.44 The *trans* isomer is very much more stable than the *cis* isomer because both *t*-butyl groups, which are *huge*, can be equatorial.

H $C(CH_3)_3$ $(CH_3)_3C$ H *trans (e,e)*

axial $C(CH_3)_3$ H $(CH_3)_3C$ H *cis (e,a)*

2.45 a. All pairs have the same bond patterns and are stereoisomers. Since they are *not* interconvertible by σ-bond rotations, they are configurational isomers.

b. conformational isomers

c. conformational isomers

d. The bond patterns are *not* the same; the first Newman projection is for 1,1-dichloropropane and the second is for 1,2-dichloropropane. Thus, these are structural (or constitutional) isomers.

e. The structures are *identical* (both represent 2-methylpentane) and are not isomers at all.

2.46 We can have 1,1- or 1,2- or 1,3 or 1,4-difluorocyclohexanes. Of these, the last three can exist as *cis–trans* isomers.

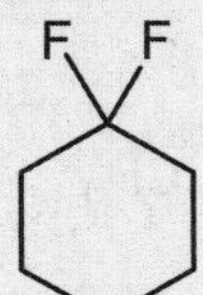

1,1-difluorocyclohexane

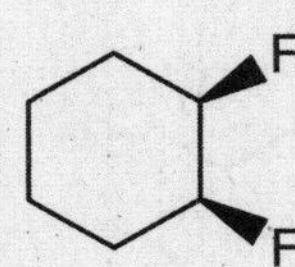

cis-1,2-difluorocyclohexane

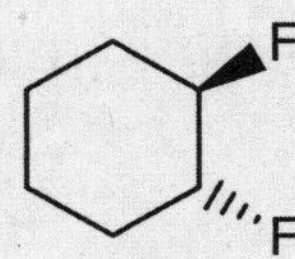

trans-1,2-difluorocyclohexane

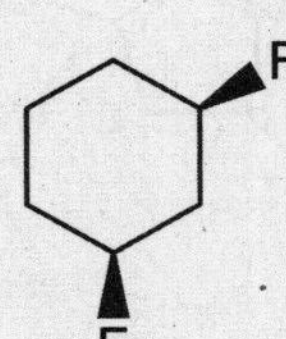

cis-1,3-difluorocyclohexane

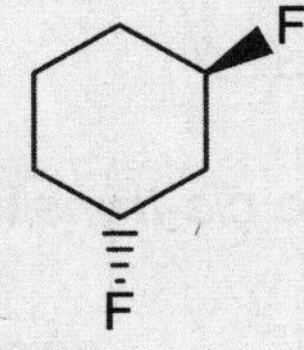

trans-1,3-difluorocyclohexane

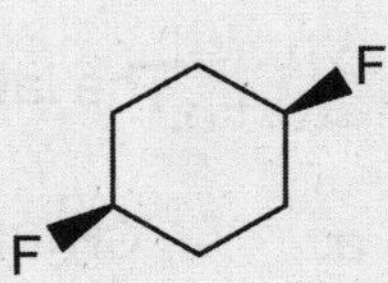

cis-1,4-difluorocyclohexane

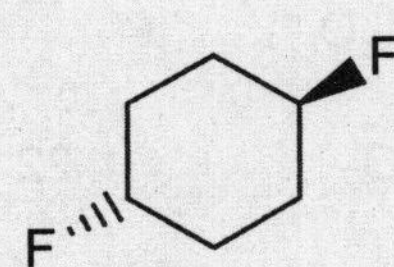

trans-1,4-difluorocyclohexane

2.47 a. Three monobrominated structural isomers are possible.

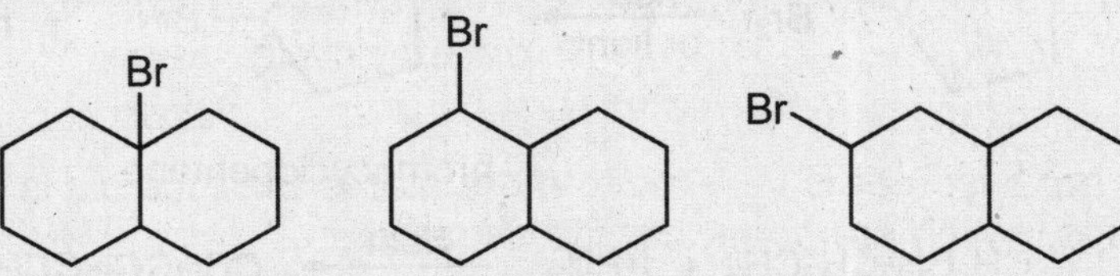

Several stereoisomers (not shown here) are possible for each structural isomer.

b. Three structural isomers

Br Br Br

c. Two structural isomers

Br Br

d. Two structural isomers

$Br\text{–}CH_2CH_2CH_2\text{–}CH_3$ $H_3C\text{–}CH(Br)\text{–}CH_2\text{–}CH_3$

e. Four structural isomers

Br Br CH_3 CH_3 CH_3 Br Br

The latter two can have *cis-trans* isomers.

2.48 a. $CH_3CH_2CH_3 + 5O_2 \longrightarrow 3CO_2 + 4H_2O$

b. $CH_3(CH_2)_3CH_3 + 8O_2 \longrightarrow 5CO_2 + 6H_2O$

c. $2CH_3CH_2CH_2CH_3 + 13O_2 \longrightarrow 8CO_2 + 10H_2O$

2.49 $2\,C_{10}H_{22} + 31\,O_2 \longrightarrow 20\,CO_2 + 22\,H_2O$

2.50 a. $CH_3CH_2CH_3 + Cl_2 \xrightarrow[\text{or light}]{\text{heat}} CH_3CH_2CH_2Cl + CH_3CH(Cl)CH_3 + HCl$

1-chloropropane 2-chloropropane

b. cyclopentane $+ Br_2 \xrightarrow[\text{or light}]{\text{heat}}$ cyclopentyl–Br $+ HBr$

bromocyclopentane

c. $CH_3CH_2CH_2CH_3 + 10\,Cl_2 \xrightarrow[\text{or light}]{\text{heat}} CCl_3CCl_2CCl_2CCl_3 + 10\,HCl$

decachlorobutane

d. (methylcyclohexane, CH_3) $\xrightarrow[\text{heat or light}]{Br_2}$

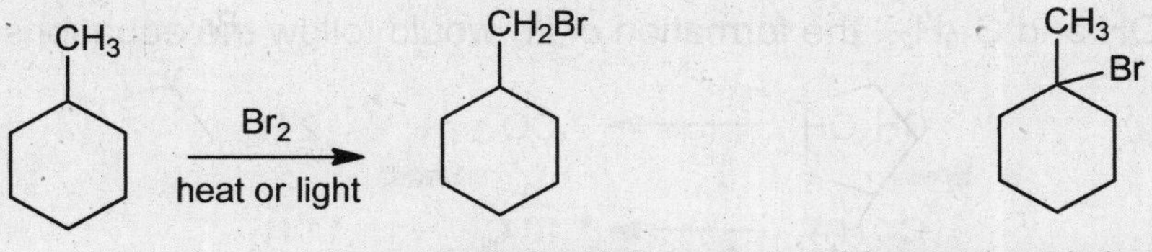

1-(bromomethyl)cyclohexane 1-bromo-1-methylcyclohexane

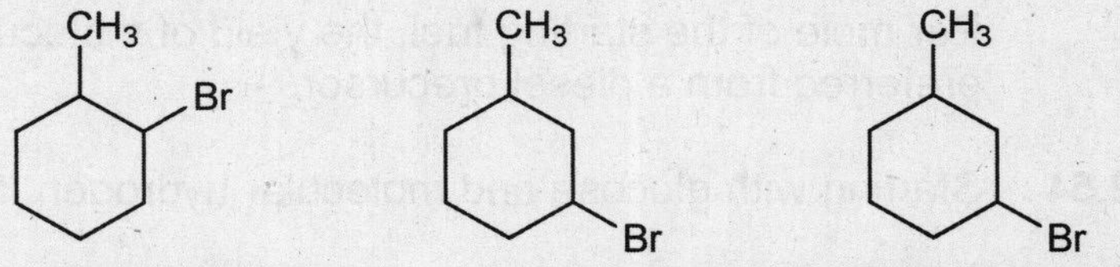

1-bromo-2-methylcyclohexane 1-bromo-4-methylcyclohexane

1-bromo-3-methylcyclohexane

2.51 The four possible structures are

$CH_3CH_2CHCl_2$ (1,1-dichloropropane)
$CH_3CH_2ClCH_2Cl$ (1,2-dichloropropane)
$CH_2ClCH_2CH_2Cl$ (1,3-dichloropropane)
$CH_3CCl_2CH_3$ (2,2-dichloropropane)

Only the last structure has all hydrogens equivalent and can give only *one* trichloro compound. This structure must, therefore, be C:

$$CH_3CCl_2CH_3 \xrightarrow{Cl_2} CH_3CCl_2CH_2Cl$$

1,3-Dichloropropane has only two different "kinds" of hydrogens. It must be D:

$$CH_2ClCH_2CH_2Cl \xrightarrow{Cl_2} CH_2ClCHClCH_2Cl \text{ and } CH_2ClCH_2CHCl_2$$

Next, A must be capable of giving 1,2,2-trichloropropane (the product from C). This is not possible for the 1,1-isomer since it already has two chlorines on carbon-1. Therefore, A must be 1,2-dichloropropane; it can give the 1,2,2-trichloro product (as well as 1,1,2- and 1,2,3-). By elimination, B is $CH_3CH_2CHCl_2$.

2.52 The equations follow the same pattern as in eqs. 2.16 through 2.21.

initiation $:\ddot{\underset{..}{Cl}}-\ddot{\underset{..}{Cl}}: \xrightarrow[\text{or light}]{\text{heat}} 2\ :\ddot{\underset{..}{Cl}}\cdot$

propagation $CH_3CH_3 + :\ddot{\underset{..}{Cl}}\cdot \longrightarrow CH_3CH_2\cdot + H{-}Cl$

$CH_3CH_2\cdot + :\ddot{\underset{..}{Cl}}-\ddot{\underset{..}{Cl}}: \longrightarrow CH_3CH_2Cl + :\ddot{\underset{..}{Cl}}\cdot$

termination $2\ :\ddot{\underset{..}{Cl}}\cdot \longrightarrow :\ddot{\underset{..}{Cl}}-\ddot{\underset{..}{Cl}}:$

$2\ CH_3CH_2\cdot \longrightarrow CH_3CH_2CH_2CH_3$

$CH_3CH_2\cdot + :\ddot{\underset{..}{Cl}}\cdot \longrightarrow CH_3CH_2Cl$

By-products resulting from the chain-termination steps would be butane and any chlorination products derived from it.

2.53 For CH_3OH and $C_{10}H_{22}$, the formation of H_2 would follow the equations:

$$CH_3OH \longrightarrow CO + 2\ H_2$$

$$C_{10}H_{22} \longrightarrow 10\ C + 11\ H_2$$

Per mole of the starting fuel, the yield of molecular hydrogen (H_2) is significantly preferred from a diesel precursor.

2.54 Starting with glucose and molecular hydrogen, the balanced equation is:

$$C_6H_{12}O_6 + 12\ H_2 \longrightarrow 6\ CH_4 + 6\ H_2O$$

3

Alkenes and Alkynes

Chapter Summary

Alkenes have a carbon-carbon double bond and **alkynes** have a carbon–carbon triple bond. Nomenclature rules are given in Sec. 3.2. Each carbon of a double bond is **trigonal**, and connected to only *three* other atoms, all of which lie in a plane with bond angles of 120°. Ordinarily, rotation around double bonds is restricted. All six atoms of **ethylene** lie in a single plane. The C=C bond length is 1.34 Å, shorter than a C–C bond (1.54 Å). These facts can be explained by an orbital model with three sp^2 hybrid orbitals (one electron in each) and one *p* orbital perpendicular to these (containing the fourth electron). The double bond is formed by end-on overlap of sp^2 orbitals to form a σ bond and lateral overlap of aligned *p* orbitals to form a π **bond** (Figures 3.4 and 3.5). Since rotation around the double bond is restricted, ***cis–trans* isomerism** is possible if each carbon atom of the double bond has two different groups attached to it.

Alkenes react mainly by **addition**. Typical reagents that add to the double bond are halogens, hydrogen (metal catalyst required), water (acid catalyst required), and various acids. If either the alkene or the reagent is **symmetrical** (Table 3.2), only one product is possible. If *both* the alkene and reagent are unsymmetrical, however, two products are possible, in principle. In this case, **Markovnikov's rule** (Secs. 3.8–3.10) allows us to predict the product obtained.

Electrophilic additions occur by a two-step mechanism. In the first step, the **electrophile** adds in such a way as to form the most stable **carbocation** (the stability order is tertiary > secondary > primary). Then the carbocation combines with a **nucleophile** to give the product.

The energetics of electrophilic additions, and all other reactions, can be described using **reaction energy diagrams** (Figures 3.10–3.12). Such diagrams show each step in the reaction mechanism, and indicate the relative energies of **reactants**, **products**, **intermediates**, and **transition states**. They indicate whether the **enthalpy** of a step in a reaction is **exothermic** or **endothermic**, or whether the step has a high or low **energy of activation**. In general, reactions that are exothermic and have low energies of activation proceed at relatively fast rates (Secs. 3.11 and 3.12).

Conjugated dienes have alternating single and double bonds. They may undergo **1,2-** or **1,4-addition**. **Allylic carbocations**, which are stabilized by resonance, are intermediates in both the 1,2- and 1,4-additions (Sec. 3.15a). Conjugated dienes also undergo cycloaddition reactions with alkenes (**Diels–Alder reaction**), a useful synthesis of six-membered rings (Sec. 3.15b).

Addition to double bonds may also occur by a **free-radical mechanism**. Polyethylene can be made in this way from the **monomer** ethylene.

Alkenes undergo a number of other reactions, such as **hydroboration**, permanganate oxidation, and **ozonolysis**.

Triple bonds are **linear** and the carbons are ***sp*-hybridized** (Figure 3.16). Alkynes, like alkenes, undergo addition reactions. A hydrogen connected to a triply bonded carbon is weakly acidic and can be removed by a very strong base such as **sodium amide**, $NaNH_2$, to give acetylides.

Reaction Summary

Additions to the Double Bond

$$\mathrm{>C{=}C<} + X_2 \xrightarrow{(X = Cl \text{ or } Br)} \mathrm{-\underset{X}{\overset{|}{C}}-\underset{X}{\overset{|}{C}}-}$$

$$\mathrm{>C{=}C<} + H_2 \xrightarrow{\text{Pd, Pt, or Ni}} \mathrm{-\underset{H}{\overset{|}{C}}-\underset{H}{\overset{|}{C}}-}$$

$$\mathrm{>C{=}C<} + H{-}OH \xrightarrow{H^+} \mathrm{-\underset{H}{\overset{|}{C}}-\underset{OH}{\overset{|}{C}}-}$$

(an alcohol)

$$\mathrm{>C{=}C<} + H{-}X \xrightarrow[\substack{(X = -F,\ -Cl,\\ -Br,\ -I,\ -OSO_3H,\\ -OCR \\ \ \ \ \ \| \\ \ \ \ \ O)}]{} \mathrm{-\underset{H}{\overset{|}{C}}-\underset{X}{\overset{|}{C}}-}$$

Conjugated Dienes

$$C{=}C{-}C{=}C \xrightarrow{X_2} \underset{\text{1,2-addition}}{\mathrm{\underset{X}{C}-\underset{X}{C}-C{=}C}} + \underset{\text{1,4-addition}}{\mathrm{\underset{X}{C}-C{=}C-\underset{X}{C}}}$$

$$C{=}C{-}C{=}C \xrightarrow[X = Cl \text{ or } Br]{H-X} \underset{\text{1,2-addition}}{\mathrm{\underset{H}{C}-\underset{X}{C}-C{=}C}} + \underset{\text{1,4-addition}}{\mathrm{\underset{H}{C}-C{=}C-\underset{X}{C}}}$$

Cycloaddition (Diels–Alder)

$$\text{C=C–C=C (diene)} + \text{C=C} \xrightarrow{\text{heat}} \text{cyclohexene (C=C in six-membered ring of C–C bonds)}$$

Polymerization of Ethylene

$$n\ H_2C{=}CH_2 \xrightarrow{\text{catalyst}} (\!-CH_2-CH_2-\!)_n$$

Hydroboration–Oxidation

$$3\ R{-}CH{=}CH_2 \xrightarrow{BH_3} (RCH_2CH_2)_3B \xrightarrow[HO^-]{H_2O_2} 3\ RCH_2CH_2OH$$

Permanganate Oxidation

$$RCH{=}CHR \xrightarrow{KMnO_4} RCH(OH){-}CH(OH)R + MnO_2$$

Ozonolysis

$$>C{=}C< \xrightarrow{O_3} >C{=}O + O{=}C<$$

Additions to the Triple Bond

$$R{-}C{\equiv}C{-}R' + H_2 \xrightarrow[\text{catalyst}]{\text{Lindlar's}} (R)(H)C{=}C(R')(H)$$

$$R{-}C{\equiv}C{-}H + X_2 \longrightarrow (R)(X)C{=}C(X)(H) \xrightarrow{X_2} RCX_2CHX_2$$

$$R{-}C{\equiv}C{-}H + H{-}X \longrightarrow (R)(X)C{=}C(H)(H) \xrightarrow{H{-}X} RCX_2CH_3$$

$$R{-}C{\equiv}C{-}H \xrightarrow{H_2O,\ Hg^{2+},\ H^+} R{-}C({=}O){-}CH_3$$

Alkyne Acidity

$$RC{\equiv}CH + NaNH_2 \xrightarrow{NH_3} RC{\equiv}C^-\ Na^+ + NH_3$$

Mechanism Summary

Electrophilic Addition (E^+ = electrophile and $Nu:^-$ = nucleophile)

$$>C{=}C< \xrightarrow{E^+} {-}\overset{|}{\underset{E}{C}}{-}\overset{+}{C}{-} \xrightarrow{Nu:^-} {-}\overset{|}{\underset{E}{C}}{-}\overset{|}{\underset{Nu}{C}}{-}$$

carbocation

1,4-Addition

$$C{=}C{-}C{=}C \xrightarrow{E^+} \left[\underset{E}{C}{-}\overset{+}{C}{-}C{=}C \longleftrightarrow \underset{E}{C}{-}C{=}C{-}\overset{+}{C} \right] \xrightarrow{Nu:^-} E{-}C{-}C{=}C{-}C{-}Nu$$

Free-Radical Polymerization of Ethylene

$$R\cdot + C{=}C \longrightarrow R{-}C{-}C\cdot$$

$$R{-}C{-}C\cdot + C{=}C \longrightarrow R{-}C{-}C{-}C{-}C\cdot \text{ (and so on)}$$

Learning Objectives

1. Know the meaning of: saturated and unsaturated; alkene, alkyne, and diene; conjugated, cumulated, and isolated double bonds; vinyl and allyl groups.
2. Know the meaning of: trigonal carbon, sp^2 hybridization, restricted rotation, σ and π bonds, *cis* and *trans* double-bond isomers.
3. Know the meaning of: addition reaction, Markovnikov's rule, electrophile, nucleophile, symmetrical and unsymmetrical double bonds and reagents, carbocation.
4. Know the meaning of: equilibrium constants, reaction enthalpy, energy of activation, transition state.
5. Given information about the energetics of a reaction, be able to construct an appropriate reaction energy diagram.
6. Know the meaning of: 1,2-addition, 1,4-addition, cycloaddition, diene, dienophile, polymer, monomer, polymerization.
7. Know the meaning of: hydroboration, glycol, ozone, ozonolysis, combustion.
8. Given the structure of an acyclic or cyclic alkene, alkyne, diene, and so on, state the IUPAC name.
9. Given the IUPAC name of an alkene, alkyne, diene, and so on, write the structural formula.
10. Given the molecular formula of a hydrocarbon and the number of double bonds, triple bonds, or rings, draw the possible structures.
11. Given the name or abbreviated structure of an unsaturated compound, tell whether it can exist in *cis* and *trans* isomeric forms and, if so, how many. Draw them.
12. Given an alkene, alkyne, or diene, and one of the following reagents, draw the structure of the product reagents: acids such as HCl, HBr, HI, and H_2SO_4; water in the presence of an acid catalyst; halogens such as Br_2 and Cl_2; hydrogen and Pd, Pt or Ni.
13. Given the structure or name of a compound that can be prepared by an addition reaction, deduce what unsaturated compound and what reagent react to form it.
14. Write the steps in the mechanism of an electrophilic addition reaction.
15. Given an unsymmetrical alkene and an unsymmetrical electrophilic reagent, give the structure of the predominant product (that is, apply Markovnikov's rule).
16. Given a conjugated diene and a reagent that adds to it, write the structures of the 1,2-and 1,4-addition products.

17. Given a diene and dienophile, write the structure of the resulting cycloaddition (Diels–Alder) adduct.

18. Given the structure of a cyclic compound that can be synthesized by the Diels–Alder reaction, deduce the structures of the required diene and dienophile.

19. Given an alkyne, write the structures of products obtained by adding one or two moles of a particular reagent to it.

20. Write the steps in the mechanism of ethylene polymerization catalyzed by a free radical.

21. Write the structure of the alcohol produced from the hydroboration–oxidation sequence when applied to a particular alkene.

22. Given an alkene or cycloalkene (or diene, and so on), write the structures of the expected ozonolysis products.

23. Given the structures of ozonolysis products, deduce the structure of the unsaturated hydrocarbon that produced them.

24. Draw orbital pictures for a double bond and a triple bond.

25. Draw conventional structures for the contributors to the resonance hybrid of an allyl cation.

26. Describe simple chemical tests that can distinguish an alkane from an alkene or alkyne.

27. Know the meaning of: cracking, alkylation, isomerization, platforming, and octane number as applied to petroleum refining.

ANSWERS TO PROBLEMS

Problems Within the Chapter

3.1 The formula C_4H_6 corresponds to C_nH_{2n-2}. The possibilities are one triple bond, two double bonds, one double bond and one ring, or two rings.

acyclic

$HC{\equiv}C{-}CH_2CH_3$ $H_3C{-}C{\equiv}C{-}CH_3$

$H_2C{=}CH{-}CH{=}CH_2$ $H_2C{=}C{=}CHCH_3$

cyclic

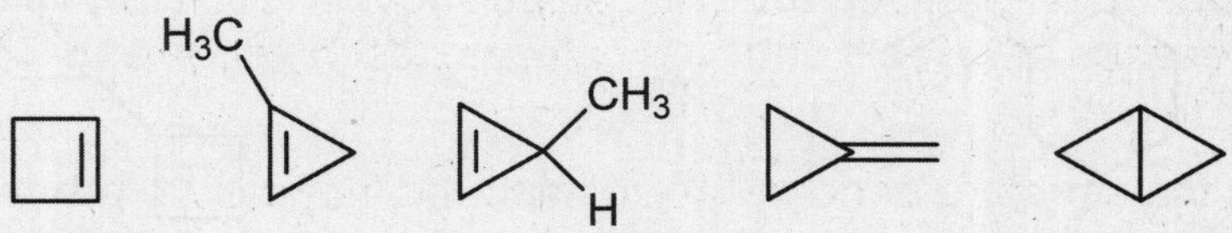

H3C

CH3

H

3.2 Compounds a and c have alternating single and double bonds, and are conjugated. In b and d, the double bonds are isolated.

3.3 a.

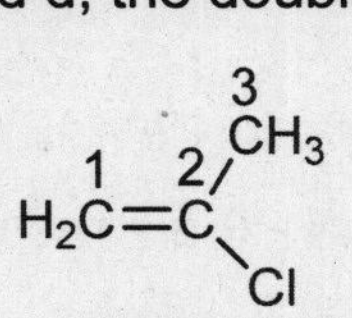

3 CH_3

1 2

$H_2C{=}C$

Cl

2-chloropropene

b.

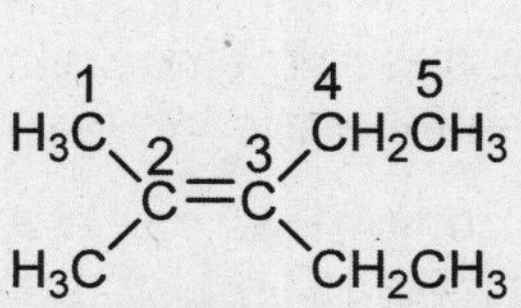

1 4 5

H_3C 2 3 CH_2CH_3

$C{=}C$

H_3C CH_2CH_3

2-methyl-3-ethyl-2-butene

c.
1 2 3 4
$FCH{=}CHCH_2CH_3$

1-fluorobutene

d.
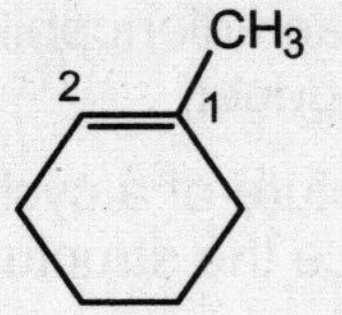

1-methylcyclohexene

e.
1 2 3 4
$H_2C{=}\underset{|\atop Br}{C}{-}CH{=}CH_2$

2-bromo-1,3-butadiene

f.
7 6 5 4 3 2 1
$CH_3CH_2CH_2CH_2{-}C{\equiv}C{-}CH_3$

1-heptyne

3.4 a. First write out the five-carbon chain, with a double bond between carbon-2 and carbon-3:

1 2 3 4 5
$C{-}C{=}C{-}C{-}C$

Add the substituents:

1 2 3 4 5
$\underset{|\atop Cl}{C}{-}C{=}C{-}\underset{|\atop Cl}{C}{-}C$

Fill in the hydrogens:

1 2 3 4 5
$H_2\underset{|\atop Cl}{C}{-}\underset{H}{C}{=}\underset{H}{C}{-}\underset{|\atop Cl}{CH}{-}CH_3$ or $ClCH_2CH{=}CH{-}\underset{|\atop Cl}{CH}{-}CH_3$

b. $CH_3CH_2{-}C{\equiv}C{-}CH_2CH_3$

c.
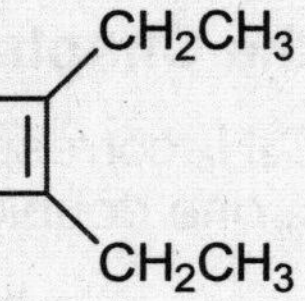

d. $H_2C{=}\underset{|\atop Br}{C}{-}CH{=}CHCH_3$

3.5 a.
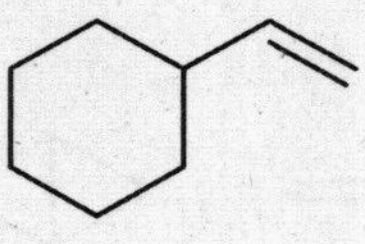

b.
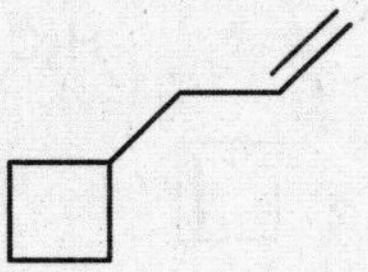

3.6 Compounds a and c have only one possible structure because in each case one of the carbons of the double bond has two identical substituents:

a. $(H)_2C{=}C(H)(CH_2CH_2CH_3)$

b. $H(CH_3CH_2)C{=}C(CH_2CH_2CH_3)H$ and $H(CH_3CH_2)C{=}C(H)CH_2CH_2CH_3$

trans-3-heptene *cis*-3-heptene

c. $(H_3C)_2C{=}C(CH_2CH_3)H$

d. $H(CH_3)C{=}C(CH_2CH_2CH_3)H$ and $H(CH_3)C{=}C(H)CH_2CH_2CH_3$

trans-2-hexene *cis*-2-hexene

3.7 The electron pair in a σ bond lies directly between the nuclei it joins. In a π bond, the electron pair is further from the two nuclei that it joins. Therefore more energy is required to break a σ bond than a π bond.

3.8 a. One bromine atom adds to each doubly bonded carbon, and the double bond in the starting material becomes a single bond in the product:

$$H_2C{=}CH{-}CH_3 + Br_2 \longrightarrow H_2C(Br){-}CH(Br){-}CH_3$$

b. 4-methylcyclohexene (H₃C–cyclohexene) + Br_2 ⟶ 1,2-dibromo-4-methylcyclohexane (H₃C–cyclohexane with two Br)

3.9 a.

$$CH_3CH_2CH{=}CHCH_2CH_3 + H{-}OH \xrightarrow{H^+} CH_3CH_2CH_2{-}CH(OH)CH_2CH_3$$

b.

$$CH_3CH{=}CHCH_3 + H{-}OH \xrightarrow{H^+} CH_3CH_2{-}CH(OH)CH_3$$

c. H₃C–cyclopentene + $H{-}OH \xrightarrow{H^+}$ H₃C–cyclopentane–OH

3.10 a.

$$CH_3CH{=}CHCH_3 + H{-}Cl \longrightarrow CH_3CH_2{-}CH(Cl)CH_3$$

b.

$$CH_3CH_2CH{=}CHCH_2CH_3 + H{-}I \longrightarrow CH_3CH_2CH_2{-}CH(I)CH_2CH_3$$

c. H₃C–cyclopentene + $H{-}Br \longrightarrow$ H₃C–cyclopentane–Br

3.11 a. $H_2C{=}CHCH_2CH_2CH_3$

↑

This doubly bonded carbon has the most hydrogens.

H—Br

↑

This is the more electropositive part of the reagent.

$$H_2C{=}CHCH_2CH_2CH_3 + \overset{\delta+}{H}{-}\overset{\delta-}{Br} \longrightarrow CH_3{-}\underset{}{\overset{Br}{CH}}CH_2CH_2CH_3$$

b.

$$(H_3C)_2C{=}CH(CH_2CH_2CH_3) + \overset{\delta+}{H}{-}\overset{\delta-}{OH} \xrightarrow{H^+} H_3C{-}\underset{CH_3}{\overset{OH}{C}}{-}CH_2CH_2CH_2CH_3$$

3.12 Since both carbons of the double bond have the same number of attached hydrogens (one), application of Markovnikov's rule is ambiguous. Both products are formed, and the reaction is not regiospecific.

$$CH_3CH{=}CHCH_2CH_2CH_2CH_2CH_3 + H{-}Cl \longrightarrow CH_3CH_2{-}\overset{Cl}{CH}CH_2CH_2CH_2CH_2CH_3 + CH_3\overset{Cl}{CH}{-}CH_2CH_2CH_2CH_2CH_2CH_3$$

3.13 a. Secondary; there are two carbons bonded to the positively charged carbon.
b. Primary; one carbon is bonded to the positively charged carbon.
c. Tertiary; three carbons are bonded to the positively charged carbon.

3.14 The order of stability is c > a > b.

3.15 eq. 3.15:

$$CH_3{-}\underset{CH_3}{C}{=}CH_2 + H^+ \longrightarrow CH_3{-}\underset{CH_3}{\overset{+}{C}}{-}CH_3$$

$$CH_3{-}\underset{CH_3}{\overset{+}{C}}{-}CH_3 + H{-}\ddot{O}{-}H \longrightarrow CH_3{-}\underset{CH_3}{\overset{H\overset{+}{O}H}{C}}{-}CH_3$$

$$CH_3{-}\underset{CH_3}{\overset{H\overset{+}{O}H}{C}}{-}CH_3 \longrightarrow CH_3{-}\underset{CH_3}{\overset{HO:}{C}}{-}CH_3 + H^+$$

The intermediate carbocation is tertiary. If, in the first step, the proton had added to the other carbon of the double bond, the carbocation produced would have been primary:

$$CH_3-C(CH_3)=CH_2 + H^+ \not\longrightarrow CH_3-CH(CH_3)-\overset{+}{C}H_2$$

eq. 3.16:

1-methylcyclopentene $+ H^+ \longrightarrow$ 1-methylcyclopentyl cation (CH_3, +)

1-methylcyclopentyl cation $+ I^- \longrightarrow$ 1-iodo-1-methylcyclopentane

The intermediate carbocation is tertiary. Had the proton added to the methyl-bearing carbon, the intermediate carbocation would have been secondary:

3.16

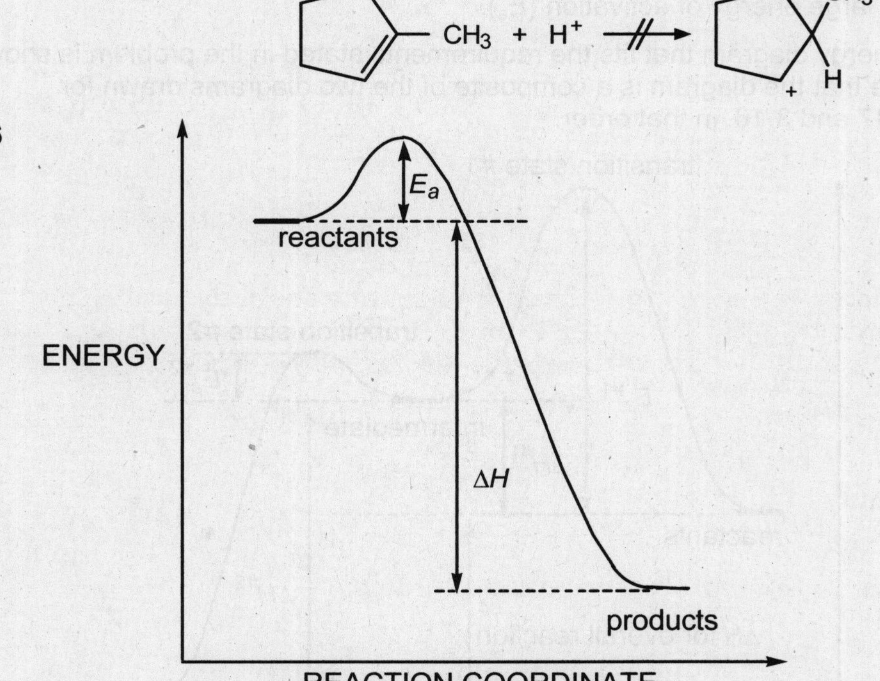

Because we are told that the reaction is very exothermic, the diagram is drawn with the energy of the reactants much higher than the energy of the products (large and negative ΔH). Because we are told that the reaction is very fast, the diagram is drawn with a small energy of activation (E_a).

3.17

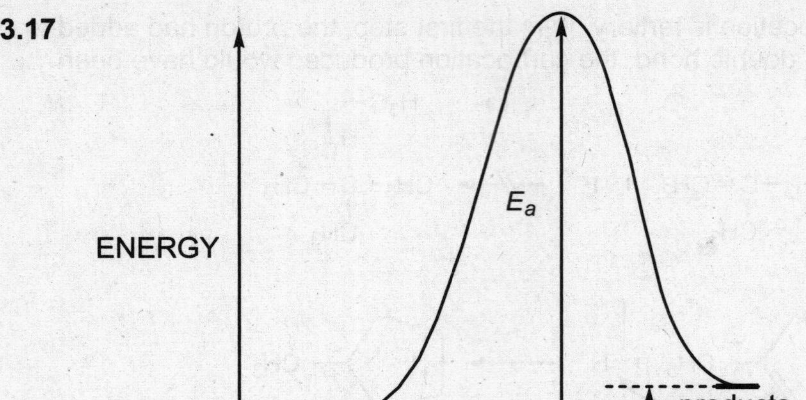

Because we are told that the reaction is slightly endothermic, the diagram is drawn with the energy of the reactants slightly lower than the energy of the products (small and positive ΔH). Because we are told that the reaction is very slow, the diagram is drawn with a large energy of activation (E_a).

3.18 A reaction energy diagram that fits the requirements stated in the problem is shown below. Notice that the diagram is a composite of the two diagrams drawn for Problems 3.17 and 3.16, in that order.

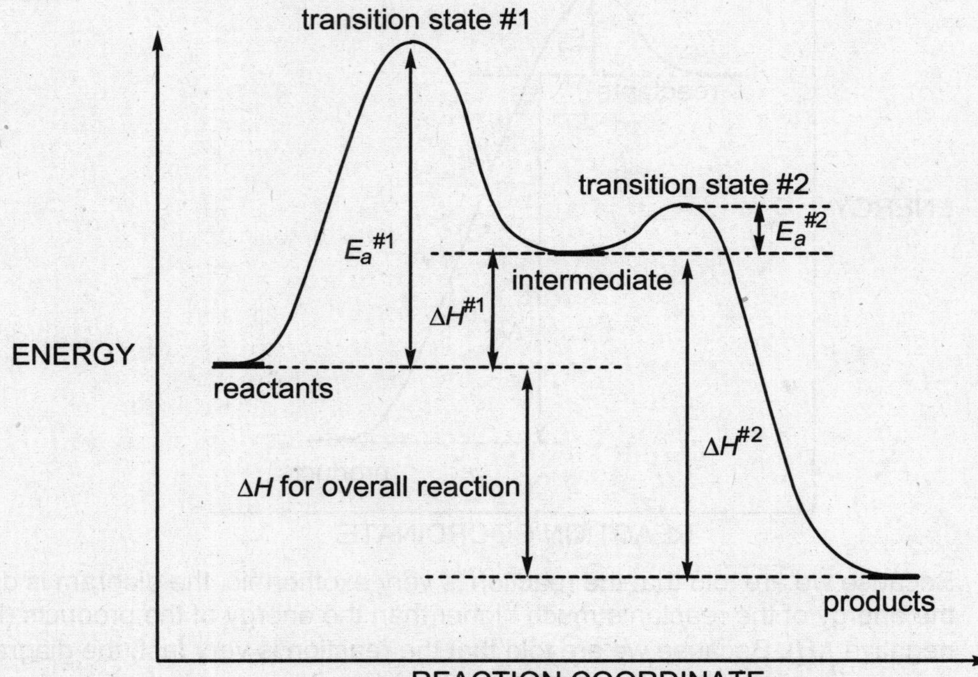

3.19 The boron adds to the *less* substituted carbon of the double bond, and, in the oxidation, the boron is replaced by an OH group. Note that the acid-catalyzed hydration of the same alkene would give the alcohol $(CH_3)_2C(OH)CH_2CH_2CH_2CH_3$ instead, according to Markovnikov's rule.

$$(CH_3)_2C{=}CHCH_2CH_2CH_3 \xrightarrow[2.\ H_2O_2,\ HO^-]{1.\ \overset{\delta^-}{H}{-}\overset{\delta^+}{BH_2}} H_3C{-}\underset{CH_3}{\overset{H}{C}}{-}\overset{OH}{C}HCH_2CH_2CH_3$$

3.20

$$C_6H_{11}{-}CH{=}CH_2 \xrightarrow[2.\ H_2O_2,\ NaOH]{1.\ BH_3} C_6H_{11}{-}CH_2CH_2OH$$

Acid-catalyzed hydration of vinylcyclohexane occurs in the Markovnikov sense:

$$\underset{\text{vinylcyclohexane}}{C_6H_{11}{-}CH{=}CH_2} \xrightarrow{H_2O,\ H^+} C_6H_{11}{-}\overset{OH}{C}H{-}CH_3$$

3.21 a.

$$(H_3C)_2C{=}C(H)CH_2CH_3 + H_2 \xrightarrow{\text{catalyst}} CH_3{-}\overset{CH_3}{C}HCH_2CH_2CH_3$$

b.

$$H_3C{-}\text{cyclopentene} + H_2 \xrightarrow{\text{catalyst}} H_3C{-}\text{cyclopentane}$$

c.

$$\text{4-methylcyclohexene } (CH_3) + H_2 \xrightarrow{\text{catalyst}} \text{methylcyclohexane } (CH_3)$$

d.

$$\text{vinylcyclobutane} + H_2 \xrightarrow{\text{catalyst}} \text{ethylcyclobutane } (CH_3)$$

3.22

$$\overset{1}{C}H_2{=}CH{-}CH{=}CH_2 + H^+ \xrightarrow[\text{C-1}]{\text{adds to}} \overset{1}{C}H_3{-}\overset{+}{C}H{-}CH{=}CH_2$$

allylic carbocation, stabilized by resonance

$$CH_2{=}\overset{2}{C}H{-}CH{=}CH_2 + H^+ \xrightarrow[\text{C-2}]{\text{adds to}}\!\!\!\!\!\!/\!/\ \ \overset{+}{C}H_2{-}\overset{2}{C}H_2{-}CH{=}CH_2$$

primary carbocation, much less stable than the allylic carbocation

3.23

$$\text{cyclopentenyl cation (+)} \longleftrightarrow \text{cyclopentenyl cation (+)}$$

3.24

$$CH_2{=}CH{-}CH{=}CH_2 \xrightarrow{Br_2} \underset{\text{1,2-addition}}{\underset{Br}{C}H_2{-}\underset{Br}{C}H{-}CH{=}CH_2} + \underset{\text{1,4-addition}}{\underset{Br}{C}H_2{-}CH{=}CH{-}\underset{Br}{C}H_2}$$

3.25 To find the structures of the diene and dienophile, break the cyclohexene ring just beyond the ring carbons that are connected to the double bond (the allylic carbons):

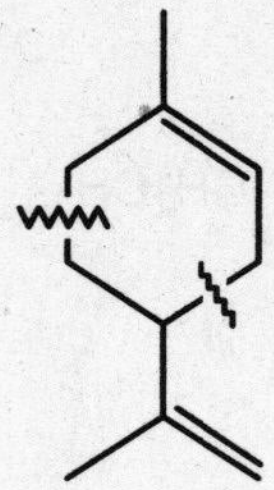

The equation is: heat

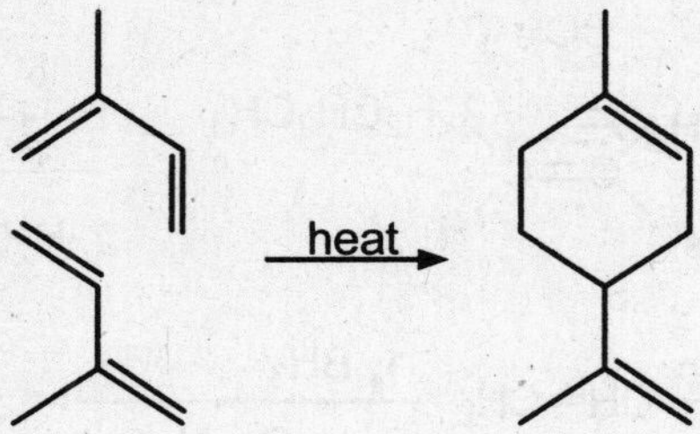

3.26 a. O CN b. CN CN

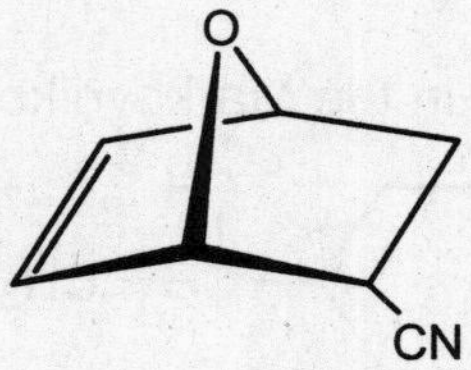

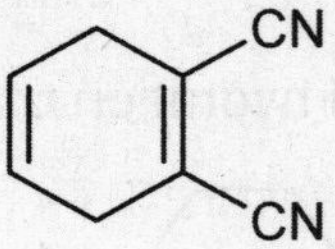

c. NO_2

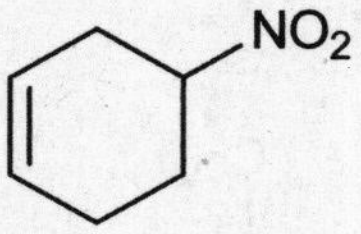

3.27

$$CH_3CH{=}CHCH_2CH_2CH_3 + KMnO_4 \longrightarrow CH_3\overset{\displaystyle HO}{\overset{|}{C}}H{-}\overset{\displaystyle OH}{\overset{|}{C}}HCH_2CH_2CH_3 + MnO_2$$

brown precipitate

3.28 $(CH_3)_2C{=}C(CH_3)_2$

3.29 a. $CH_3CH_2C{\equiv}CH + Br_2 \longrightarrow CH_3CH_2CBr{=}CHBr$

b. $CH_3CH_2C{\equiv}CCH_3 + 2\ Cl_2 \longrightarrow CH_3CH_2C(Cl)_2{-}C(Cl)_2CH_3$

c. $CH_3CH_2CH_2CH_2C{\equiv}CH + HBr \longrightarrow CH_3CH_2CH_2CH_2(Br)C{=}CH_2$

$CH_3CH_2CH_2CH_2C{\equiv}CH + 2\ HBr \longrightarrow CH_3CH_2CH_2CH_2(Br)_2C{-}CH_3$

d.

$$CH_3CH_2C{\equiv}CH + H{-}OH \xrightarrow[H^+]{Hg^{2+}} CH_3CH_2{-}\overset{\displaystyle O}{\overset{\|}{C}}{-}CH_3$$

3.30 Follow eq. 3.53 as a guideline:

$$CH_3(CH_2)_3C{\equiv}C{-}H + NaNH_2 \longrightarrow CH_3(CH_2)_3C{\equiv}C^-\ Na^+ + NH_3$$

3.31 $HC{\equiv}C^-\ Na^+ + H_2O \longrightarrow HC{\equiv}CH + NaOH$

3.32 2-Hexyne has no hydrogens on the triple bond:

$$CH_3C{\equiv}CCH_2CH_2CH_3$$

Therefore, it does not react with sodium amide.

ADDITIONAL PROBLEMS

3.33 In each case, start with the longest possible chain and determine all possible positions for the double bond. Then shorten the chain by one carbon and repeat, and so on.

a. $HC{\equiv}CCH_2CH_3$ 1-butyne

$CH_3C{\equiv}CCH_3$ 2-butyne

b. $H_2C{=}CHCH_2CH_2CH_3$ 1-pentene

$H_2C{=}CHCH(CH_3)_2$ 3-methyl-1-butene

$CH_3CH{=}CHCH_2CH_3$ 2-pentene (*cis* and *trans*)

c. $CH_2{=}C{=}CHCH_2CH_3$ 1,2-pentadiene

$CH_2{=}CH{-}CH{=}CHCH_3$ 1,3-pentadiene

$CH_2{=}CHCH_2CH{=}CH_2$ 1,4-pentadiene

$CH_3CH{=}C{=}CHCH_3$ 2,3-pentadiene

$CH_2{=}C{=}C(CH_3)_2$ 3-methyl-1,2-butadiene

$CH_2{=}C(CH_3)CH{=}CH_2$ 2-methyl-1,3-butadiene

3.34 a. 3-propyl-2-hexene (number the chain such that the double bond carbons get the lowest numbers possible)
b. 4-methyl-2-pentene
c. 3,5-dimethylcyclohexene (number the ring starting with the double bond such that the substituent methyl group get the lowest numbers possible)
d. 4-chloro-2-hexyne
e. 2-hepten-5-yne (the double bond receives the lowest number)
f. 2-bromo-1,3-butadiene (number from the end nearest the bromine substituent)
g. 1-penten-3-yne (the double bond receives the lowest number)
h. *cis*-2-pentene
i. *trans*-2-pentene
j. 3-methylpentyne
k. cyclobutene
l. 2-methylpropene

3.35 a. $BrCH_2CH_2CH{=}CHCH_2CH_2CH_3$

b. [structure: H_3C– on cyclopentene, ring numbered 1, 2, 4]

c. $H_2C{=}CHBr$

d. [structure: vinylcyclohexane]

e. [structure: CH_3, F, F substituted alkene]

f.
$$CH_3C{\equiv}C{-}\underset{}{CH}(CH_3){-}CH(CH_3){-}CH_2CH_3$$

g. [structure: CH_2CH_3 on cyclohexadiene]

h. [structure: cyclopentene with H_3C and CH_3]

i. [structure: vinylcyclopentane]

j. $H_2C{=}CHCH_2Cl$

k. (1,2-dichlorocyclopentadiene structure: Cl, Cl)

3.36

a. $\overset{8}{C}H_3\overset{7}{C}H_2\overset{6}{C}H_2\overset{5}{C}H_2-\overset{4}{C}\equiv\overset{3}{C}-\overset{2}{C}H_2\overset{1}{C}H_3$ — 3-octyne; number the chain from the other end.

b. $\overset{1}{C}H_3\overset{2}{C}H=\overset{3}{C}H\overset{4}{C}H_2\overset{5}{C}H_3$ — 2-pentene; use the lower of the two numbers for the double bond.

c. $H_2\overset{1}{C}=\overset{2}{C}H-\overset{3}{C}\equiv\overset{4}{C}H$ — 1-buten-3-yne; use the lower of the two numbers for the double bond.

d. $CH_3CH_2CH=CHCH_2CH_3$ — 3-hexene; the "1-methyl" substituent lengthens the chain.

e. (cyclopentene ring, C1 bearing CH_3, C2) — 1-methylcyclopentene

f. $H_2C=C(CH_3)CH_2CH_2CH_3$ — 2-methyl-1-pentene; number the longest chain.

g. $H_2\overset{1}{C}=\overset{2}{C}H-\overset{3}{C}\equiv\overset{4}{C}\overset{5}{C}H_3$ — 1-penten-3-yne; name as an enyne, *not* an ynene

h. $\overset{4}{H_2C}=\overset{3}{C}H-\overset{2}{C}(CH_3)=\overset{1}{C}H_2$ — 2-methyl-1,3-butadiene; number to give the substituent the lowest possible number.

3.37

a. The average values are 1.54 Å, 1.34 Å, and 1.21 Å, respectively.

b. These single bonds are shorter than the usual 1.54 Å because they are between sp^2–sp^2 (1.47 Å), and sp^2–sp (1.43 Å), and sp–sp (1.37 Å) hybridized carbons. The more *s*-character the orbitals have, the more closely the electrons are pulled in toward the nuclei and the shorter the bonds.

3.38 Review Sec. 3.5 if you have difficulty with this question.

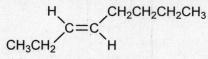

a. $CH_3CH_2(H)C=C(H)CH_2CH_2CH_2CH_3$ (H and $CH_2CH_2CH_2CH_3$ on opposite sides) — *trans*-3-octene

$CH_3CH_2(H)C=C(H)CH_2CH_2CH_2CH_3$ (H atoms on the same side) — *cis*-3-octene

b. Only one structure is possible since one of the doubly bonded carbons has two identical groups (hydrogens):

$H_2C=CHCH_2Cl$

c. Only one structure: $H_2C=CHCH_2CH_2CH_2CH_3$

d.

$$\underset{cis\text{-1-bromopropene}}{\text{Br(H)C}=\text{CH}(\text{CH}_3)} \qquad \underset{trans\text{-1-bromopropene}}{\text{H(Br)C}=\text{CH}(\text{CH}_3)}$$

cis-1-bromopropene *trans*-1-bromopropene

e. H_3C–(cyclohexene)

Only one structure is possible.

f. (cyclooctatetraene)

Only one structure is possible.

g. F(H_3C)C=C(CH_3)F *trans* H_3C(F)C=C(CH_3)F *cis*

3.39

$$\overset{13}{\text{HC}}\equiv\underset{12}{\text{C}}-\overset{11}{\text{C}}\equiv\underset{10}{\text{C}}-\overset{9}{\text{CH}}=\underset{8}{\text{C}}=\overset{7}{\text{CH}}-\underset{6}{\text{CH}}=\overset{5}{\text{CH}}-\underset{4}{\text{CH}}=\overset{3}{\text{CH}}-\overset{2}{\text{CH}_2}-\underset{1}{\overset{\text{O}}{\overset{\|}{\text{C}}}}-\text{OH}$$

a. The 3-4, 5-6, 7-8 double bonds are conjugated. Also, the 8-9, 10-11, and 12-13 multiple bonds are conjugated (alternate single and multiple bonds).

b. The 7-8 and 8-9 double bonds are cumulated.

c. Only the C=O bond is isolated (separated from the nearest multiple bond by two single bonds).

3.40

CH_3 H_3C OH

only this double bond can exist as *cis-trans* isomers

H_3C CH_3 CH_3 CH_3 CH_3

only this double bond can exist as *cis-trans* isomers

3.41 a.

$$CH_3CH_2CH{=}CHCH_2CH_3 + Br_2 \longrightarrow CH_3CH_2\overset{Br}{\overset{|}{C}H}-\overset{Br}{\overset{|}{C}H}CH_2CH_3$$

3,4-dibromohexane

b.

$$H_2C{=}CHCH_2Cl + Br_2 \longrightarrow H_2\overset{Br}{\overset{|}{C}}-\overset{Br}{\overset{|}{C}H}CH_2Cl$$

1,2-dibromo-3-chloropropane

c. [cyclohexa-1,4-diene] + Br_2 → [cyclohexene ring with Br on C-4 and C-5]

4,5-dibromocyclohexene

The double bonds are not conjugated, so only 1,2-addition is possible.

d. [cyclohexa-1,3-diene] + Br_2 → [cyclohexene ring with Br on C-3 and C-6] + [cyclohexene ring with Br on C-3 and C-4]

3,6-dibromocyclohexene (1,4-addition) 3,4-dibromocyclohexene (1,2-addition)

Compare with eqs. 3.31–3.33. The 1,4-addition product predominates.

e. $(H_3C)_2C{=}C(CH_3)_2$ + Br_2 → $CH_3{-}C(Br)(H_3C){-}C(Br)(CH_3){-}CH_3$

2,3-dibromo-2,3-dimethylbutane

f. $CH_3CH{=}C(CH_3)(CH_2CH_3)$ + Br_2 → $CH_3CH(Br){-}C(CH_3)(Br)(CH_2CH_3)$

2,3-dibromo-3-methylpentane

3.42 To work this kind of problem, try to locate (on adjacent carbons) the atoms or groups that must have come from the small molecule or reagent. Then remove them from the structure and insert the multiple bond appropriately.

a. $CH_3CH{=}CHCH_3$ + Br_2

b. $CH_2{=}CHCH_3$ + $H{-}OSO_3H$

c. $CH_2{=}C(CH_3)_2$ + HOH + H^+

d. [cyclohexene] + H−Cl

e. $CH_2{=}CH{-}CH{=}CH_2$ + H−Cl (1,4-addition)

f. $CH_3CH_2C{\equiv}CCH_3$ + $2Cl_2$

g. [cyclopentyl]$-CH{=}CH_2$ + H−Br

h. [1-methylcyclohexene] + H_2O + H^+

3.43 a. Electrophile; hydronium ion is a proton donor.

b. Electrophile; HCl can donate a proton to a nucleophile.

c. Nucleophile; Br^- can donate an electron pair to an electrophile.

d. Electrophile; the iron can accept an electron pair.

e. Electrophile; sulfuric acid is a proton donor.

f. Electrophile; the aluminum has only a sextet of electrons. It is electron-deficient and can accept an electron pair.

g. Nucleophile; the negatively charged oxygen reacts with electrophiles.

3.44

1-methylcyclohexene (CH_3) + H^+ → tertiary carbocation (+, CH_3) + $H{-}\ddot{O}{-}H$ → oxonium ion (CH_3, $:\overset{+}{O}{-}H$, H) → $-H^+$ → 1-methylcyclohexanol (CH_3, OH)

The oxygen carries a formal +1 charge

In the first step, the proton adds in such a way as to give the tertiary carbocation intermediate.

3.45 a.

$$CH_3{-}\underset{H}{\overset{CH_3}{C}}{-}\overset{+}{C}H_2 \xleftarrow[\text{protonation of C-2}]{H^+} CH_3{-}\overset{CH_3}{\underset{2}{C}}{=}\underset{1}{C}H_2 \xrightarrow[\text{protonation of C-1}]{H^+} CH_3{-}\overset{CH_3}{\underset{+}{C}}{-}CH_3$$

1° carbocation 3° carbocation

The 3° carbocation is more stable than the 1° carbocation.

b. Both protonations are endothermic, but the protonation which gives the tertiary carbocation is less endothermic, and thus energetically more favorable than the reaction leading to the primary carbocation.

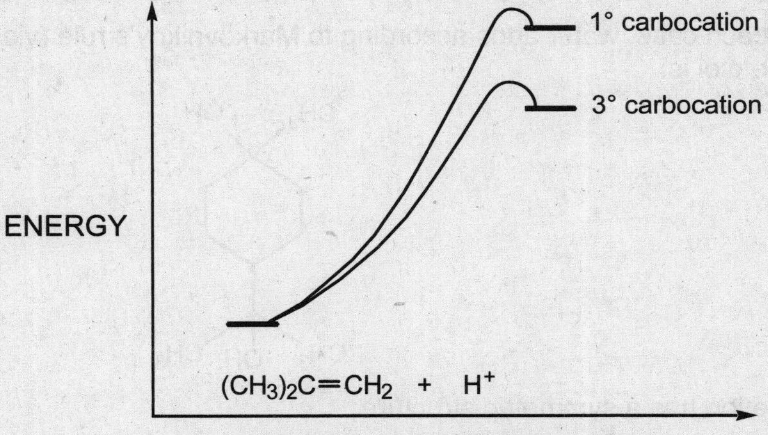

The tertiary carbocation then reacts with water to give $(CH_3)_3C{-}OH$ (2-methyl-2-propanol).

3.46 If the saturated hydrocarbon contained no rings, it would have the molecular formula $C_{15}H_{32}$. Since there are fewer hydrogens in $C_{15}H_{28}$, it must have two rings (two hydrogens are deleted per ring). Since caryophyllene absorbed 2 moles of H_2 ($C_{15}H_{24} + 2\ H_2 \rightarrow C_{15}H_{28}$), it must also have two double bonds or one triple bond. The structure of caryophyllene follows:

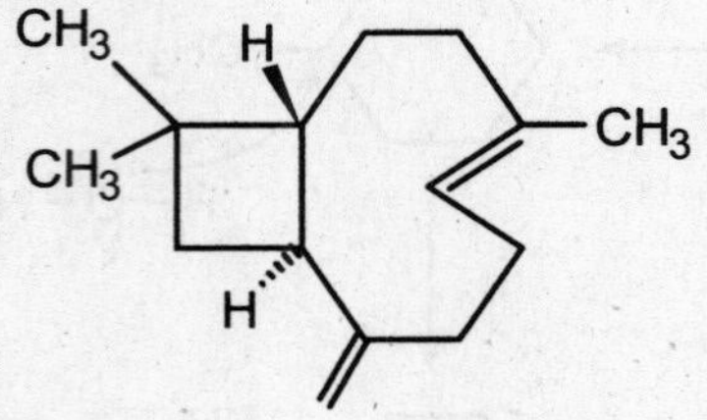

CH3 H CH3 CH3 H

3.47 Water can donate a proton to a nucleophile:

$$H{-}\ddot{O}{-}H \;+\; Nu{:}^- \longrightarrow Nu{:}H \;+\; H{-}\ddot{O}{:}^-$$

or an oxygen lone pair to an electrophile:

$$H{-}\ddot{O}{-}H \;+\; E^+ \longrightarrow H{-}\overset{+}{\dot{O}}(E){-}H$$

3.48 CH3 CH2 CH3 $\xrightarrow{H_2O,\ H^+}$ CH3 OH CH3 CH2 + CH3 CH3 OH CH2

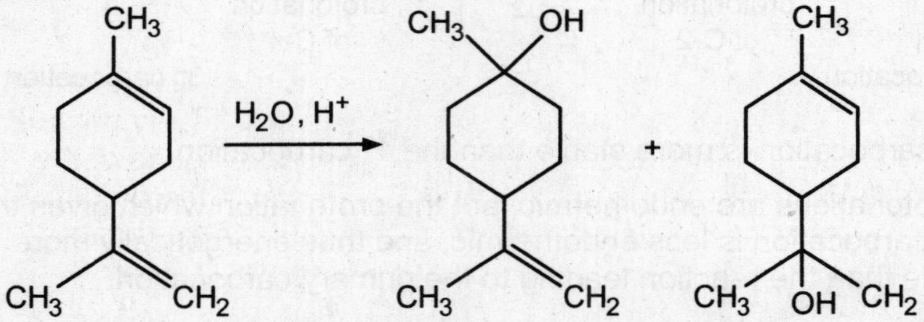

In each case, water adds according to Markovnikov's rule (via a tertiary carbocation). The diol is:

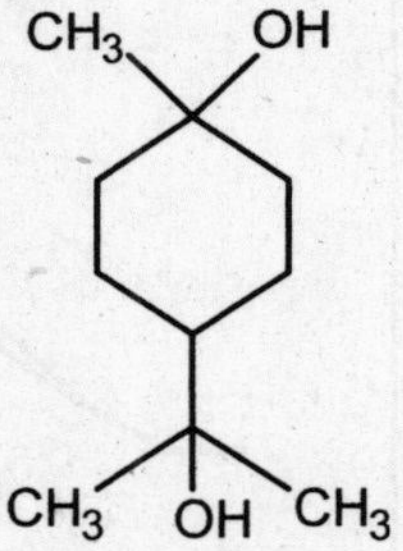

CH3 OH CH3 OH CH3

3.49 The ion has a symmetric structure.

$$(CH_3)_2CH\overset{+}{C}H{-}CH{=}CHCH(CH_3)_2 \longleftrightarrow (CH_3)_2CHCH{=}CH{-}\overset{+}{C}HCH(CH_3)_2$$

3.50

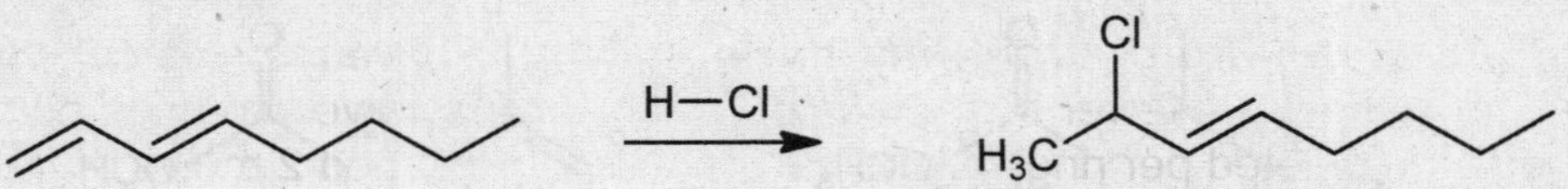

$\xrightarrow{H-Cl}$

1,2-addition (2-chloro-3-octene, *cis* and *trans*)

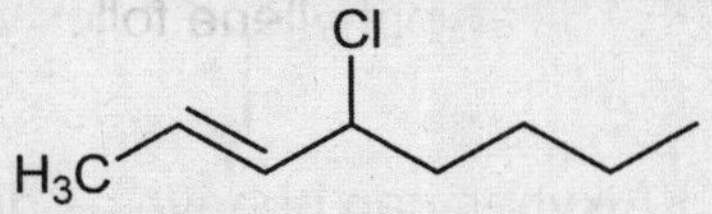

1,4-addition (4-chloro-2-octene, *cis* and *trans*)

Mechanism:

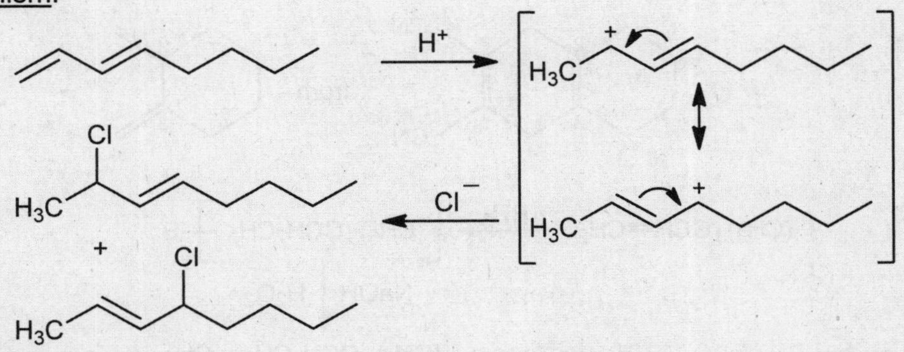

In the final step, the nucleophile Cl^- can react with the allylic carbocation at either of the two positive carbons.

3.51 a. The product of a Diels–Alder reaction is a cyclohexene. The terminal carbons of the diene (C-1 and C-4) form single bonds with the doubly bonded carbons of the dienophile (marked with an asterisk). The double bond of the dienophile becomes a single bond. The double bond in the product forms between the carbons that originated as C-2 and C-3 of the diene.

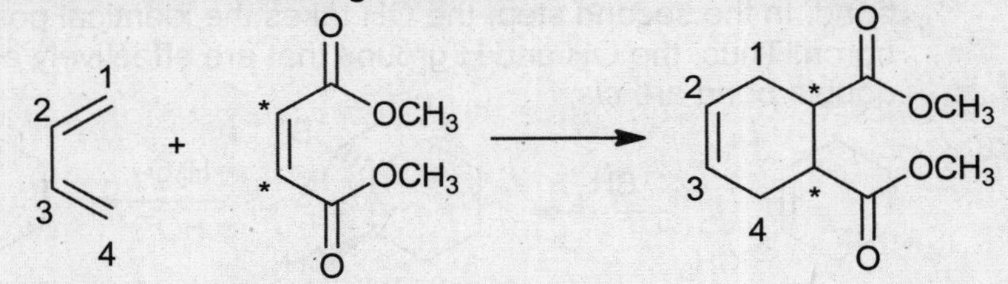

diene dienophile

b. The methyl groups at the ends of the 1,3-diene simply come along for the ride.

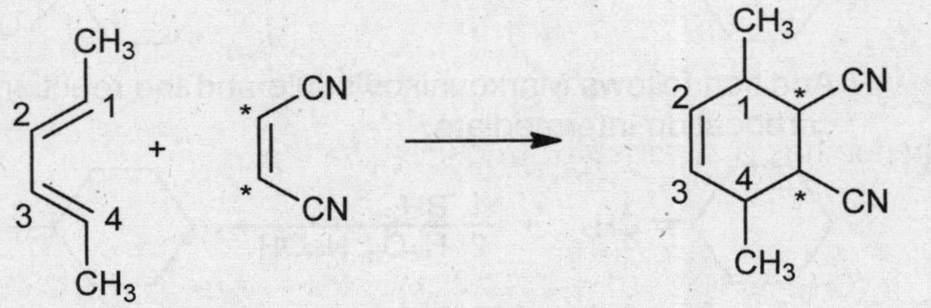

3.52 a. Use the reverse of the strategy used in Problem 3.51:

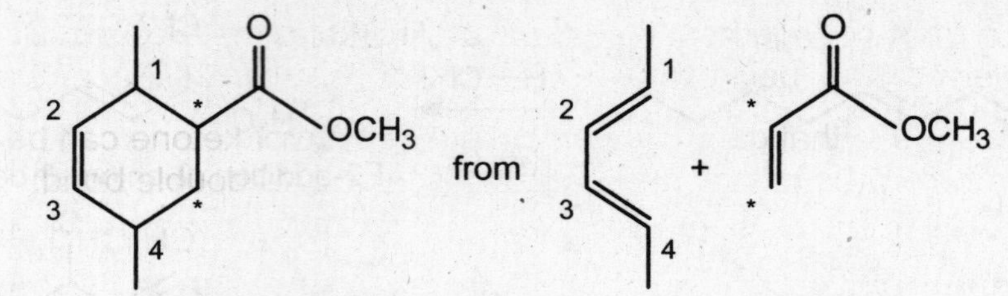

from +

b. Alkynes can behave as dienophiles in Diels–Alder reactions. One of the π bonds reacts with the diene and the other π bond remains in the product.

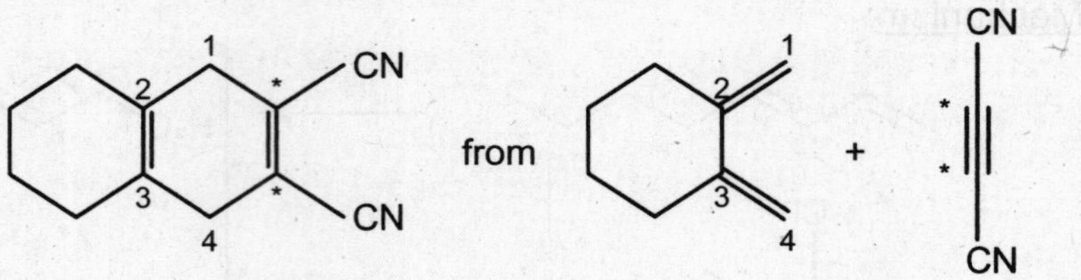

from +

3.53 a. $(CH_3)_3CCH{=}CH_2 \xrightarrow{BH_3} [(CH_3)_3CCH_2CH_2]_3B \xrightarrow{NaOH,\ H_2O_2} (CH_3)_3CCH_2CH_2{-}OH$

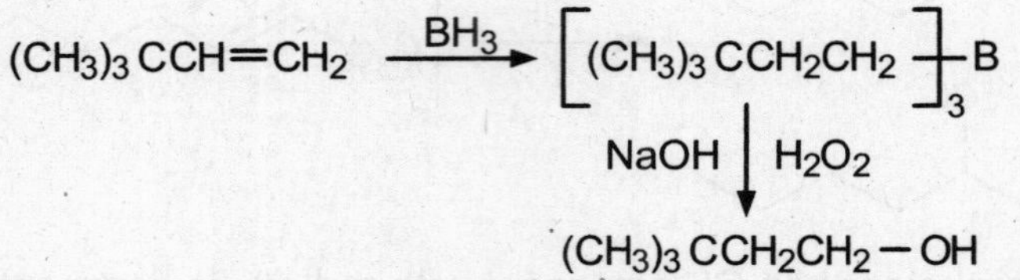

b. $\xrightarrow{BH_3}$ $\xrightarrow[HO^-]{H_2O_2}$

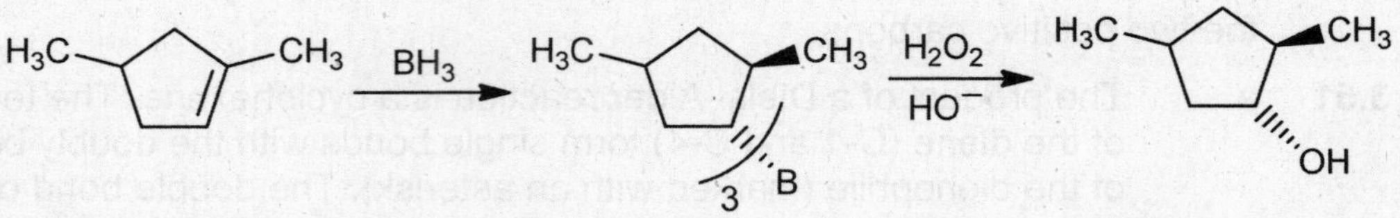

In the first step, the boron and hydrogen add to the same face of the double bond. In the second step, the OH takes the identical position occupied by the boron. Thus, the OH and H groups that are effectively added to the original double bond are *cis*.

c. $\xrightarrow{BH_3}$ $\xrightarrow[HO^-]{H_2O_2}$

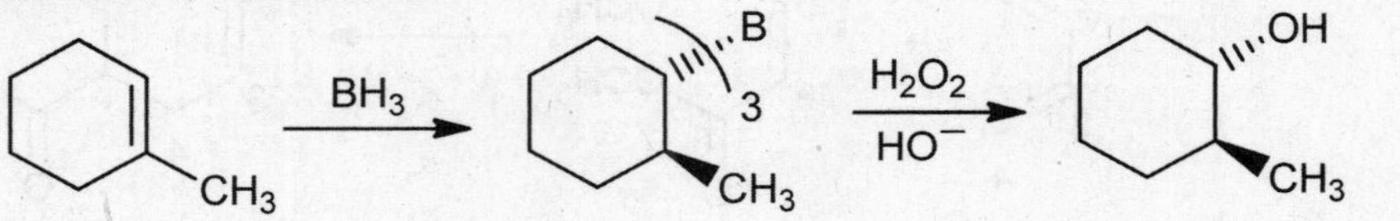

3.54 a. $=CH_2 + H{-}OH \xrightarrow{H^+}$

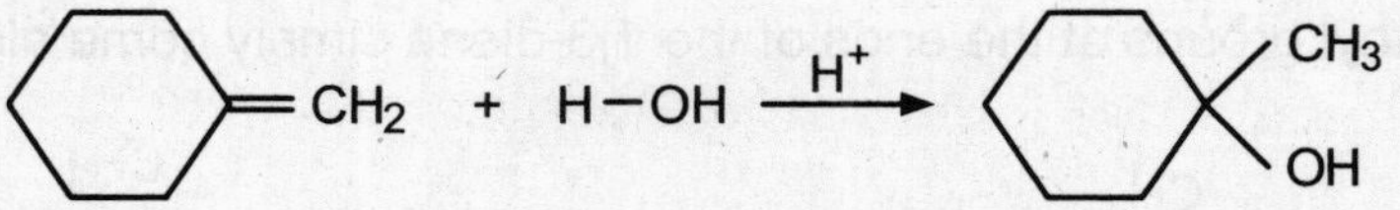

Addition follows Markovnikov's rule and the reaction proceeds *via* a tertiary carbocation intermediate.

b. $=CH_2 + \xrightarrow[2.\ H_2O_2,\ NaOH]{1.\ BH_3} -CH_2OH$

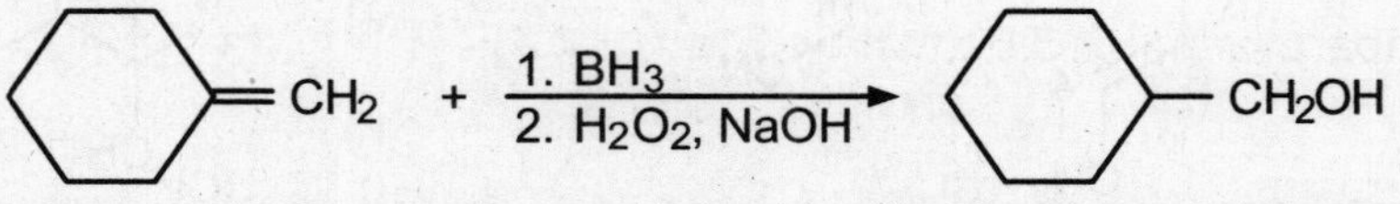

The boron adds to the less substituted, or CH_2, carbon. In the second step, the boron is replaced by the hydroxyl group.

3.55 Cyclohexene will rapidly decolorize a dilute solution of bromine in carbon tetrachloride (Sec. 3.7a) and will be oxidized by potassium permanganate, resulting in a color change from the purple of $KMnO_4$ to the brown solid MnO_2 (Sec. 3.17a). Cyclohexane, being saturated, does not react with either of these reagents.

3.56 The alkene that gave the particular aldehyde or ketone can be deduced by joining the two carbons attached to oxygens by a C=C double bond:

a. $(CH_3)_2C{=}CH_2$ b. $(CH_3CH_2)_2C{=}C(CH_2CH_3)_2$
c. $CH_3CH{=}CHCH_2CH_3$ d. cyclopentene

In the case of compound a, where *cis* and *trans* are possible, either isomer gives the same ozonolysis products.

3.57 Each radical adds to a molecule of propene to give the more stable radical (secondary rather than primary):

$$RO\cdot + H_2C{=}CH(CH_3) \longrightarrow RO{-}CH_2{-}\dot{C}H(CH_3)$$

$$RO{-}CH_2{-}\dot{C}H(CH_3) + H_2C{=}CH(CH_3) \longrightarrow ROCH_2CH(CH_3){-}CH_2{-}\dot{C}H(CH_3)$$

and so on

$$\left(CH_2{-}\underset{CH_3}{CH}\right)_n \quad \text{polypropylene}$$

3.58 a.

$$CH_3C{\equiv}CCH_2CH_2CH_2CH_2CH_3 + H_2 \xrightarrow[\text{catalyst}]{\text{Lindlar's}} \underset{H_3C}{\overset{H}{}}C{=}C\underset{CH_2CH_2CH_2CH_2CH_3}{\overset{H}{}}$$

This catalyst limits the addition to 1 mole of H_2, which adds to the same face of the double bond.

b.

$$CH_3CH_2C{\equiv}CCH_2CH_3 + 2\,Br_2 \longrightarrow CH_3CH_2{-}CBr_2{-}CBr_2{-}CH_2CH_3$$

c.

$$HC{\equiv}CCH_2CH_2CH_2CH_3 \xrightarrow{NaNH_2} Na^{+}\ {}^{-}C{\equiv}CCH_2CH_2CH_2CH_3 + NH_3$$

Compare with eq. 3.53, where R = $CH_3CH_2CH_2CH_2$.

d.

$$CH_3CH_2C{\equiv}CH + H_2O \xrightarrow[Hg^{2+}]{H^+} \left[CH_3CH_2\overset{OH}{C}{=}CH_2\right] \longrightarrow CH_3CH_2\overset{O}{\overset{\|}{C}}CH_3$$

Compare with eq. 3.52, where R = CH_3CH_2.

3.59 a.

$$CH_3CH_2C{\equiv}CH \xrightarrow{2\,HCl} CH_3CH_2{-}CCl_2{-}CH_3 \xleftarrow{2\,HCl} CH_3C{\equiv}CCH_3$$

Either 1- or 2-butyne will add HCl to give the same product.

b.

$$CH_3C{\equiv}CCH_3 + 2\ Cl_2 \longrightarrow CH_3CCl_2{-}CCl_2CH_3$$

The triple bond must be between C-2 and C-3 if the chlorines are to be attached to those carbons in the product.

3.60 a.

$$ClCH_2CH_2{-}P(=O)(OH){-}OH + H_2O \longrightarrow HCl + H_2C{=}CH_2 + HO{-}P(=O)(OH){-}OH$$

b. The phosphonic acid ($RP(=O)(OH)_2$) group is very acidic and will provide solubility for 2-chloroethylphosphonic acid in water.

3.61 a. Follow Markovnikov's rule:

$$H_2C{=}CHCH_2CH_3 + H{-}Cl \longrightarrow CH_3{-}CH(Cl)CH_2CH_3$$

b.

$$H_2C{=}CHCH_2CH_3 + Br_2 \longrightarrow CH_2(Br){-}CH(Br)CH_2CH_3$$

c.

$$H_2C{=}CHCH_2CH_3 + H_2 \xrightarrow{Pt} CH_3{-}CH_2CH_2CH_3$$

d. Follow Markovnikov's rule:

$$H_2C{=}CHCH_2CH_3 + H{-}OH \xrightarrow{H^+} CH_3{-}CH(OH)CH_2CH_3$$

The product is the regioisomer of that obtained by the hydroboration–oxidation sequence (part g).

e.

$$H_2C{=}CHCH_2CH_3 \xrightarrow[\text{2. Zn, }H^+]{\text{1. }O_3} O{=}CHCH_2CH_3 + H_2C{=}O$$

f.

$$H_2C{=}CHCH_2CH_3 \xrightarrow[HO^-]{KMnO_4} CH_2(OH){-}CH(OH)CH_2CH_3$$

g.

$$H_2C{=}CHCH_2CH_3 \xrightarrow[\text{2. }H_2O_2,\ HO^-]{\text{1. }B_2H_6} HOCH_2{-}CH_2CH_2CH_3$$

h.

$$H_2C{=}CHCH_2CH_3 + 6\ O_2 \longrightarrow 4\ CO_2 + 4\ H_2O$$

3.62 This is an electrophilic addition. The triple bond is protonated in the first step to give the vinylic carbocation, which reacts with chloride ion in the second step to provide the alkene as an intermediate product. Then the alkene is protonated to give the most stable carbocation and then chloride ion attacks the positively charged center to form the final product.

$$HC\equiv CCH_2CH_3 \xrightarrow{H-Cl} H_2C=\overset{+}{C}CH_2CH_3 \xrightarrow{Cl^-} H_2C=C(Cl)CH_2CH_3$$

$$H_2C=C(Cl)CH_2CH_3 \xrightarrow{H^+} H_3C-\overset{+}{C}(Cl)CH_2CH_3 \xrightarrow{Cl^-} H_3C-C(Cl)_2CH_2CH_3$$

3.63 For nerolidol, the products will be:

a.

b. Bromination will provide a mixture as the two trisubstituted double bonds will react faster than the monosubstituted double bond:

c.

These 3 double bonds will be cut to form the carbonyl products.

For 4,8-dimethyl-1,3,7-nonatriene, the products will be:

a.

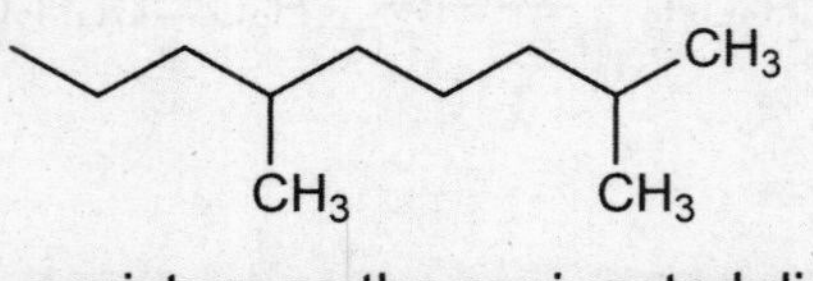

CH3 CH3 CH3

b. Bromination will provide a mixture as the conjugated diene will react faster than the isolated double bond, leading to 1,2- and 1,4-addition of Br_2 across the diene.

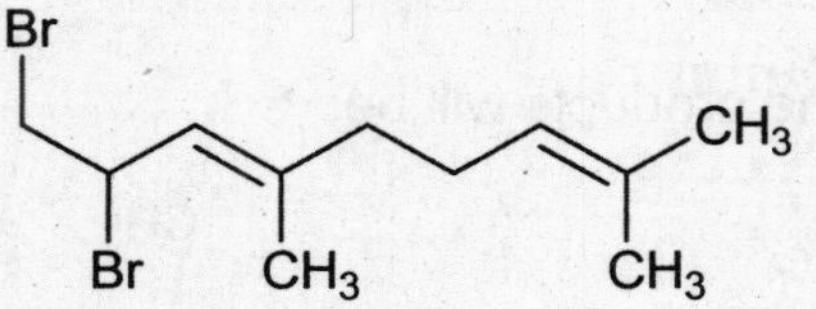

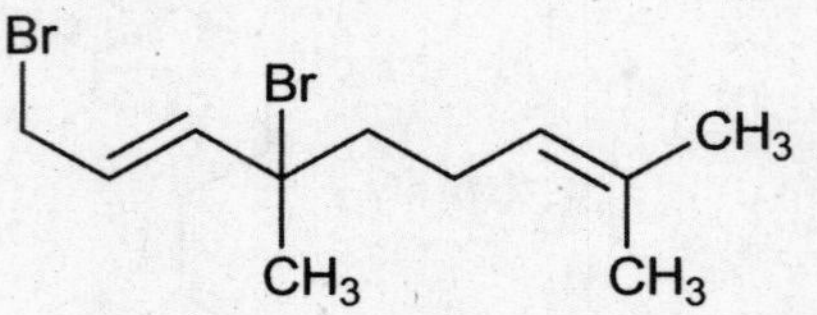

Br Br CH3 CH3 CH3

Br Br CH3 CH3 CH3

c.

CH3 CH3 CH3

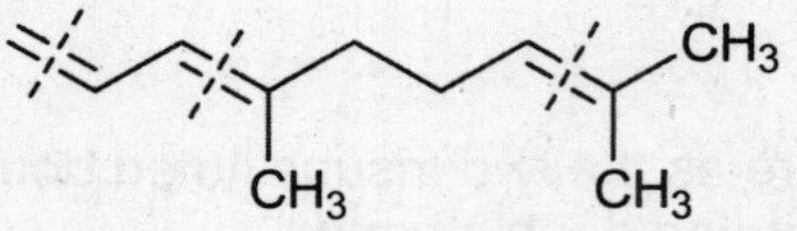

These 3 double bonds will be cut to form the carbonyl products.

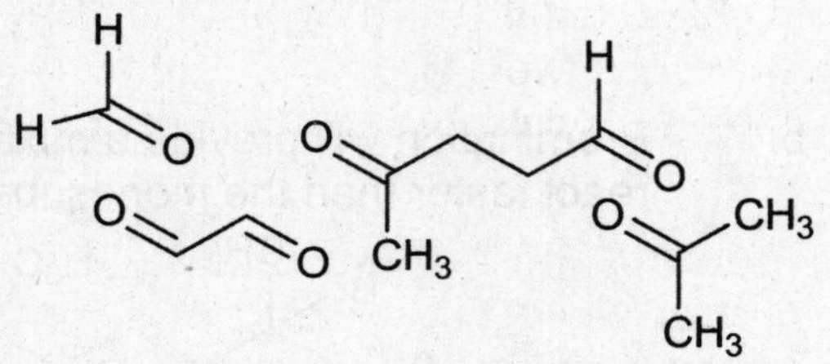

H H O O O H O O CH3 CH3 CH3

3.64 a. This reaction is similar to a hydroboration–oxidation reaction. The elements of water are added to 1-butyne (C_4H_6) to give butanol (C_4H_8O).

b. The hydroboration–oxidation reaction initially provides the following enol, which tautomerizes to give the product.

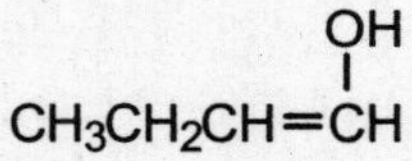

$$CH_3CH_2CH{=}CH(OH)$$

3.65 The Lewis acid ($AlCl_3$) and Bronsted acid (HCl) catalyzed isomerization of pentane to a more highly branched product would form 2,3-dimethylpropane:

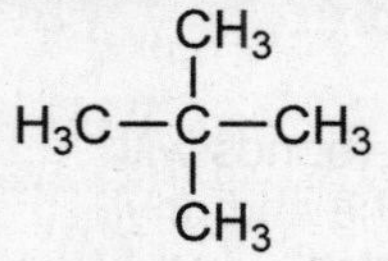

$$H_3C-C(CH_3)_2-CH_3$$

4

Aromatic Compounds

Chapter Summary

Benzene is the parent of the family of **aromatic hydrocarbons**. Its six carbons lie in a plane at the corners of a regular hexagon and each carbon has one hydrogen attached. Benzene is a resonance hybrid of two contributing Kekulé structures:

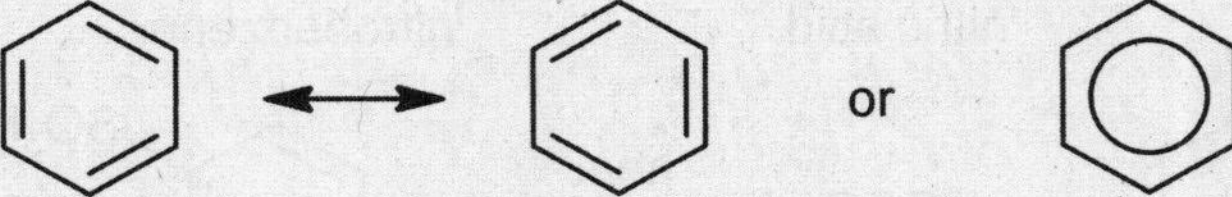

In orbital terms, each carbon is sp^2-hybridized. These orbitals form σ bonds to the hydrogen and the two neighboring carbons are all in the ring plane. A *p* orbital at each carbon is perpendicular to this plane, and the six electrons, one from each carbon, form an electron cloud of π bonds which lie above and below the ring plane.

The bond angles in benzene are 120°. All C–C bond distances are equal (1.39 Å). The compound is more stable than either of the contributing Kekulé structures and has a **resonance** or **stabilization energy** of about 36 kcal/mol.

The nomenclature of benzene derivatives is described in Sec. 4.6. Common names and structures to be memorized include those of **toluene**, **styrene**, **phenol**, **aniline**, and **xylene**. Monosubstituted benzenes are named as benzene derivatives (bromobenzene, nitrobenzene, and so on). Disubstituted benzenes are named as *ortho*- (1,2-), *meta*- (1,3-), or *para*- (1,4-), depending on the relative positions of the substituents on the ring. Two important groups are **phenyl** (C_6H_5–) and **benzyl** ($C_6H_5CH_2$–).

Aromatic compounds react mainly by **electrophilic aromatic substitution**, in which one or more ring hydrogens are replaced by various electrophiles. Typical reactions are **chlorination**, **bromination**, **nitration**, **sulfonation**, **alkylation**, and **acylation** (the last two are Friedel–Crafts reactions). The mechanism involves two steps: addition of the electrophile to a ring carbon, to produce an intermediate **benzenonium ion**, followed by proton loss to again achieve the (now substituted) aromatic system.

Substituents already present on the ring affect the rate of further substitution and the position taken by the next substituent. Most groups are either *meta-directing* and *ring-deactivating* or *ortho,para-directing* and *ring-activating* (Table 4.1). Exceptions are the halogens, which are *ortho,para-directing* but *ring-deactivating*. These effects must be taken into account in devising syntheses of aromatic compounds.

Benzene is a major commercial chemical–a source of styrene, phenol, other aromatics, acetone, and cyclohexane.

Polycyclic aromatic hydrocarbons, which are built of *fused* benzene rings, include **naphthalene**, **anthracene**, and **phenanthrene**. Some, such as benzo[*a*]pyrene, are carcinogens. Graphite, a common allotrope of carbon, consists of layers of planar, hexagonal rings separated by 3.4 Å. The **fullerenes** are a recently discovered novel polycyclic form of carbon.

Reaction Summary

Electrophilic Aromatic Substitution

Halogenation

$$C_6H_6 + X_2 \xrightarrow{FeX_3} C_6H_5X + HX$$

(X = Cl, Br)

Nitration

$$C_6H_6 + HNO_3 \xrightarrow{H_2SO_4} C_6H_5NO_2 + H_2O$$

nitric acid — nitrobenzene

Sulfonation

$$C_6H_6 + HOSO_3H \xrightarrow{heat} C_6H_5SO_3H + H_2O$$

sulfuric acid — benzenesulfonic acid

Alkylation

$$C_6H_6 + R{-}Cl \xrightarrow{AlCl_3} C_6H_5R + HCl$$

Friedel-Crafts reaction; R = alkyl group

Alkylation

$$C_6H_6 + H_2C{=}CH_2 \xrightarrow{H^+} C_6H_5CH_2CH_3$$

Acylation

$$C_6H_6 + RC(=O)Cl \xrightarrow{AlCl_3} C_6H_5C(=O)R + HCl$$

Catalytic Hydrogenation

$$C_6H_6 + 3\,H_2 \xrightarrow[\text{catalyst}]{\text{metal}} C_6H_{12}$$

cyclohexane

Mechanism Summary

Electrophilic Aromatic Substitution

$$C_6H_6 + E^+ \rightleftharpoons \text{benzenonium ion} \xrightarrow{-H^+} C_6H_5E$$

benzenonium ion

Learning Objectives

1. Know the meaning of: Kekulé structure, benzene resonance hybrid, resonance or stabilization energy.
2. Know the meaning of: *ortho*, *meta*, *para*, phenyl group (C_6H_5– or Ph–), benzyl group ($C_6H_5CH_2$–), aryl group (Ar–), benzene, toluene, styrene, phenol, aniline, xylene, arene, fullerene.
3. Know the meaning of: electrophilic aromatic substitution, halogenation, nitration, sulfonation, alkylation, acylation, Friedel–Crafts reaction.
4. Know the meaning of: benzenonium ion, *ortho*, *para*-directing group, *meta*-directing group, ring-activating substituent, ring-deactivating substituent.
5. Know the meaning of: polycyclic aromatic hydrocarbon, naphthalene, anthracene, phenanthrene, carcinogenic, graphite, fullerene.
6. Name and write the structures for aromatic compounds, especially monosubstituted and disubstituted benzenes and toluenes.
7. Given the reactants, write the structures of the main organic products of the common electrophilic aromatic substitution reactions (halogenation, nitration, sulfonation, alkylation, and acylation).
8. Write the steps in the mechanism for an electrophilic aromatic substitution reaction.
9. Draw the structures of the main contributors to the benzenonium ion resonance hybrid.
10. Draw the structures of the main contributors to substituted benzenonium ions, and tell whether the substituent stabilizes or destabilizes the ion.
11. Know which groups are *ortho*, *para*-directing, which are *meta*-directing, and explain why each group directs the way that it does.
12. Know which groups are ring-activating, ring-deactivating, and explain why each group affects the rate of electrophilic aromatic substitution as it does.
13. Given two successive electrophilic aromatic substitution reactions, write the structure of the product, with substituents in the correct locations on the ring.

14. Given a disubstituted or trisubstituted benzene, deduce the correct sequence in which to carry out electrophilic substitutions to give the product with the desired orientation.

ANSWERS TO PROBLEMS

Problems Within the Chapter

4.1 The formula C_6H_6 (or C_nH_{2n-6}) corresponds to any of the following possibilities: two triple bonds; one triple bond and two double bonds; one triple bond, one double bond, and one ring; one triple bond and two rings; four double bonds; three double bonds and one ring; two double bonds and two rings; one double bond and three rings; four rings. Obviously there are very many possibilities. One example from each of the above categories is shown below.

$HC{\equiv}C-C{\equiv}C-CH_2CH_3$ $\quad HC{\equiv}C-CH{=}CH-CH{=}CH_2$ $\quad HC{\equiv}C-C{=}CH$ (ring closed by H_2C-CH_2)

$HC{\equiv}C-C$ (bridged by two CH_2 groups to) CH $\quad H_2C{=}C{=}CH-CH{=}C{=}CH_2$ $\quad H_2C{=}C{=}C{=}C$ (ring with two CH_2)

4.2 eq. 4.2:

$$C_6H_6 + Br_2 \xrightarrow{FeBr_3} C_6H_5Br + HBr$$

only one possible structure

eq. 4.4:

Br + Br_2 $\xrightarrow{FeBr_3}$ (1,2-dibromobenzene) ⇌ (1,2-dibromobenzene) + HBr

or

(1,3-dibromobenzene) ⇌ (1,3-dibromobenzene) + HBr

or

(1,4-dibromobenzene) ⇌ (1,4-dibromobenzene) + HBr

Yes, Kekulé's explanation accounts for one monobromobenzene and three dibromobenzenes, if the equilibrium between the two Kekulé structures for each dibromo isomer is taken into account. As you will see in Sec. 4.3, however, this picture is not quite correct.

4.3 Kekulé would have said that the structures are in such rapid equilibrium with one another that they cannot be separated. We know now, however, that there is only one such structure, which is not accurately represented by either Kekulé formula.

4.4 Although *two* Kekulé structures can be drawn for each of the *three* dibromobenzenes (see the 2 x 3 = 6 structures shown in the answer to Problem 4.2), each pair of structures represents only *one* dibrombenzene. The structure of each dibromobenzene is actually a resonance hybrid of the two contributing resonance structures.

4.5

$C_6H_5-CH_2OH$ benzyl alcohol

$C_6H_5-CH_3$ toluene

$C_6H_5-CO_2H$ benzoic acid

4.6

CH_3 (1), CH_3 (2) or H_3C (1), CH_3 (2)

ortho -xylene

It does not matter whether we write the benzene ring standing on one corner or lying on a side.

H_3C 1 3 CH_3 or CH_3 1 3 CH_3 or H_3C 1 3 CH_3

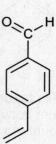

meta-xylene

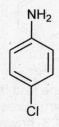

The important feature of the structure is that the two methyl substituents are in a 1,3-relationship.

4.7 a. NO_2, NO_2 b. CH_3, NO_2 c. OH, Br

d. NH_2, Cl e. O, C, H f. Cl, Cl

4.8 a. OH, H_3C, CH_3, CH_3 b. CH_3, F, F, CH_2CH_3

4.9 a. CH_2CH_2 b.

c. CH_2I d.

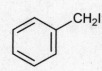

4.10 a. phenylcyclohexane or cyclohexylbenzene
b. *o*-benzylphenol

4.11 The first step has a high activation energy (E_a1) and is endothermic because the aromatic ring is disrupted. The second step has a low activation energy (E_a2) and is exothermic because the aromatic ring is regenerated.

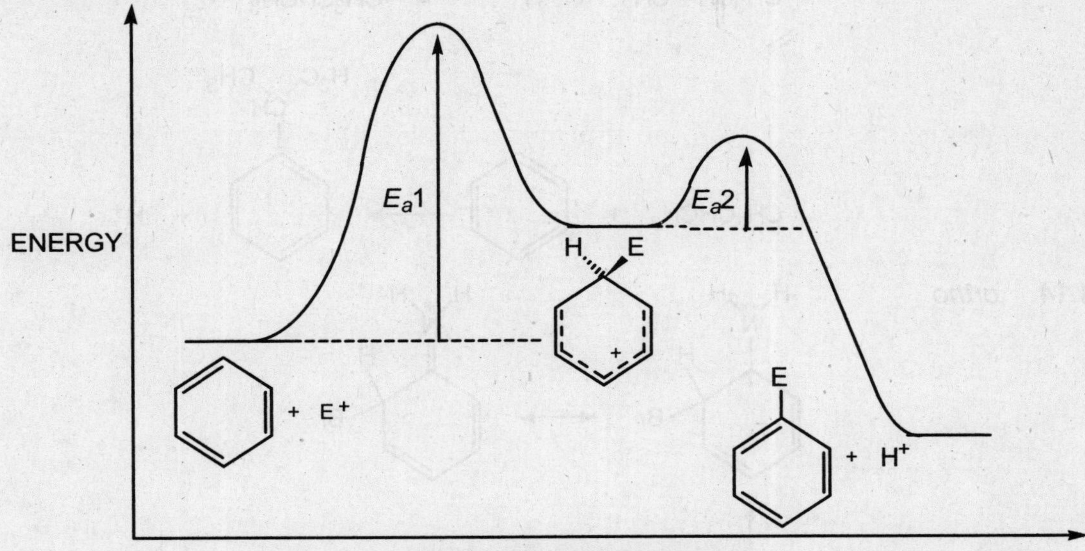

REACTION COORDINATE

4.12 The electrophile is formed according to the following equilibria, beginning with the protonation of one sulfuric acid molecule by another:

$$\text{H}-\ddot{\underset{\cdot\cdot}{\text{O}}}-\text{SO}_3\text{H} + \text{H}-\ddot{\underset{\cdot\cdot}{\text{O}}}-\text{SO}_3\text{H} \rightleftharpoons {}^-\text{OSO}_3\text{H} + \text{H}-\overset{\text{H}}{\overset{|}{\text{O}^+}}-\text{SO}_3\text{H}$$

$$\text{H}-\overset{\text{H}}{\overset{|}{\text{O}^+}}-\text{SO}_3\text{H} \xrightarrow{-\text{H}_2\text{O}} {}^+\text{SO}_3\text{H}$$

$$\text{H}^+ + \text{SO}_3 \rightleftharpoons {}^+\text{SO}_3\text{H}$$

Using ${}^+SO_3H$ as the electrophile, we can write the sulfonation mechanism as follows:

$$\text{C}_6\text{H}_6 + {}^+\text{SO}_3\text{H} \longrightarrow \left[\text{C}_6\text{H}_6\text{SO}_3\text{H}^+ \longleftrightarrow \text{C}_6\text{H}_6\text{SO}_3\text{H}^+ \longleftrightarrow \text{C}_6\text{H}_6\text{SO}_3\text{H}^+\right] \xrightarrow{-\text{H}^+} \text{C}_6\text{H}_5\text{SO}_3\text{H}$$

4.13 The product would be isopropylbenzene because the proton of the acid catalyst would add to propene according to Markovnikov's rule to give the isopropyl cation:

$$CH_3CH{=}CH_2 + H^+ \longrightarrow CH_3\overset{+}{C}HCH_3$$

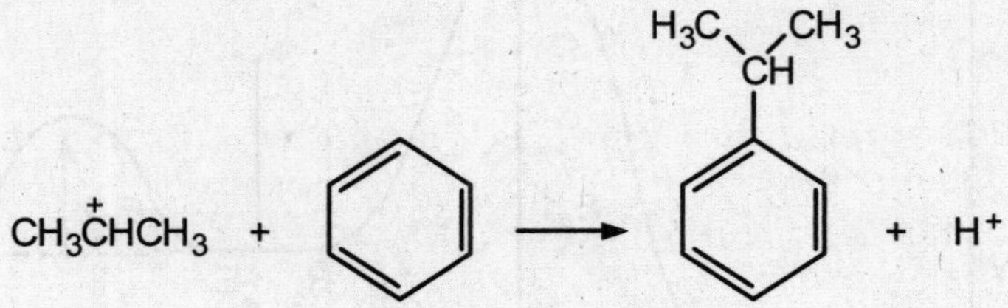

$$CH_3\overset{+}{C}HCH_3 + C_6H_6 \longrightarrow C_6H_5CH(CH_3)_2 + H^+$$

4.14 *ortho*

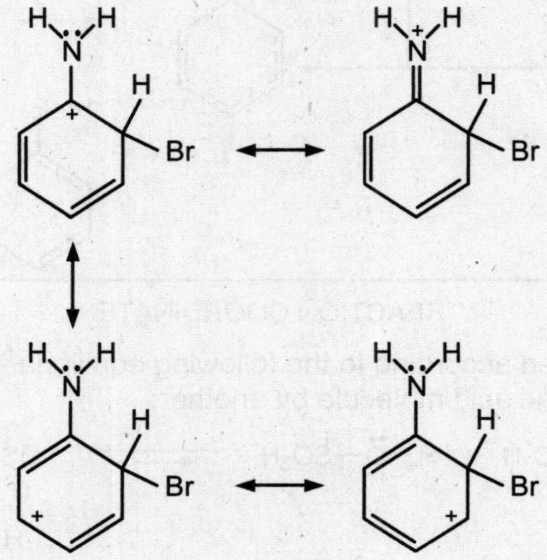

para

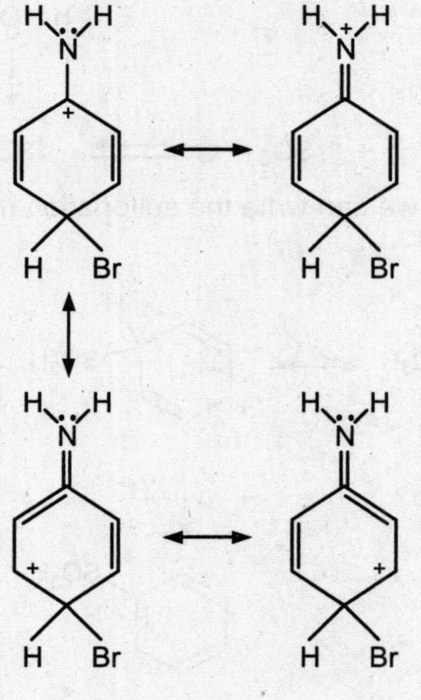

meta

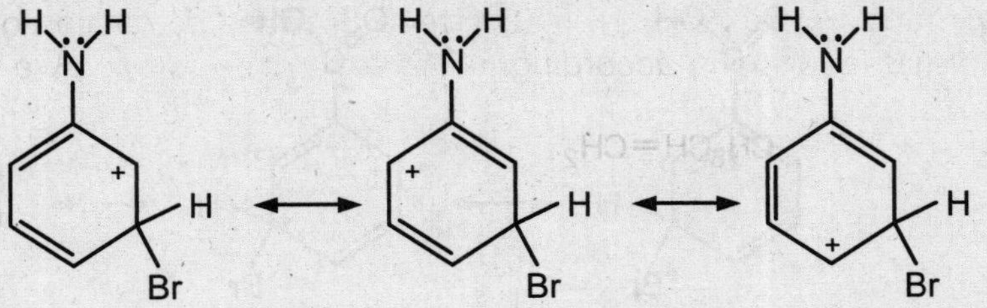

Note that in *ortho* or *para* substitution the carbocation can be stabilized by delocalization of the positive charge to the nitrogen atom. This is not possible for *meta* substitution. Therefore, *ortho* and *para* substitution are preferred.

4.15 *ortho*

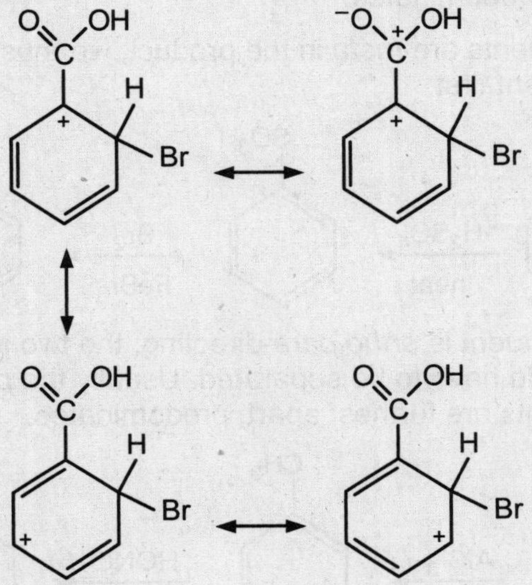

para

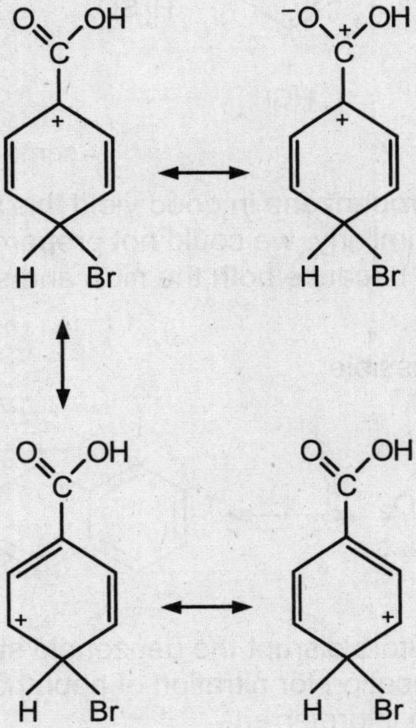

meta

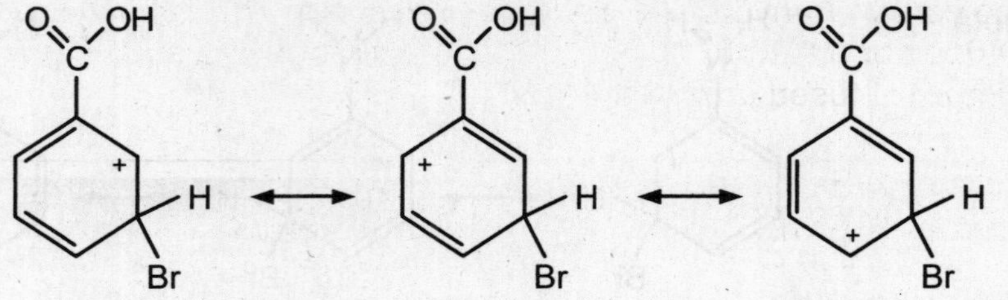

In *ortho* or *para* substitution, the positive charge of the benzenonium ion is adjacent to the partially positive carboxyl carbon. It is energetically unfavorable to have two adjacent like charges. In *meta* substitution, this unfavorable possibility does not exist. Thus, *meta* substitution predominates.

4.16 a. Since the substituents are *meta* in the product, we must introduce the *meta*-directing substituent first:

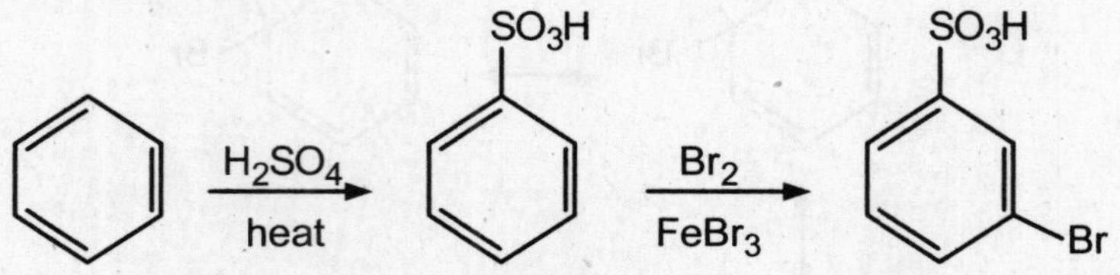

SO_3H SO_3H

H_2SO_4, heat; Br_2, $FeBr_3$; Br

b. The methyl substituent is *ortho,para*-directing; the two isomers obtained in the final step would have to be separated. Usually the *para* isomer, in which the two substituents are furthest apart, predominates.

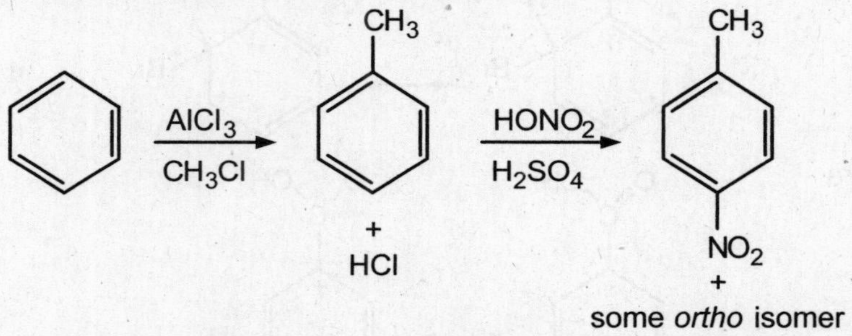

CH_3 CH_3

$AlCl_3$, CH_3Cl; + HCl; $HONO_2$, H_2SO_4; NO_2 + some *ortho* isomer

4.17 We could not make *m*-bromochlorobenzene in good yield this way because both Br and Cl are *ortho,para*-directing. Similarly, we could not prepare *p*-nitrobenzenesulfonic acid directly because both the nitro and sulfonic acid substituents are *meta*-directing.

4.18 Only two such contributors are possible:

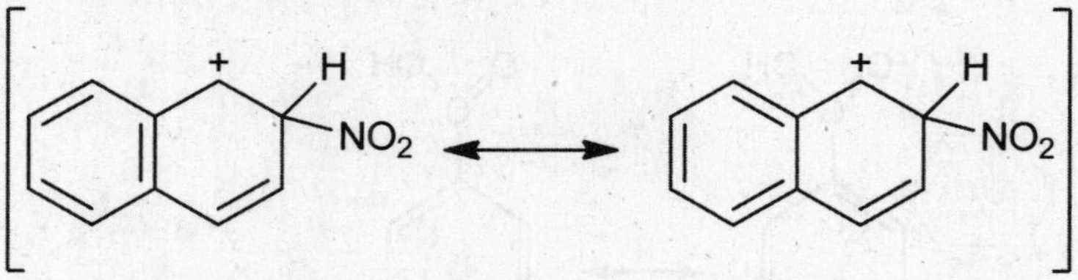

+ H NO_2 ⟷ + H NO_2

Any additional resonance contributors disrupt the benzenoid structure in the “left” ring. Since the intermediate carbocation for nitration of naphthalene at C-1 is more stable, substitution at that position is preferred.

4.19 The carbon-to-hydrogen ratios are: benzene = 1; naphthalene = 1.25; anthracene = 1.40; pyrene = 1.60. The percentage of carbon in a structure increases with the number of fused aromatic rings.

ADDITIONAL PROBLEMS

4.20 a. Cl

b. NH_2 F F F

c. OH Cl

d. $CH_2CH{=}CH_2$

e. OH H_3C CH_3

f. CH_3 CH_3

g.

h. OCH_3 I NO_2 CH_3

i. SO_3H Cl

j. OH NO_2

k. CH_2Br

l. CH_3 CH_3CH_2 CH_2CH_3 CH_2CH_3

m.

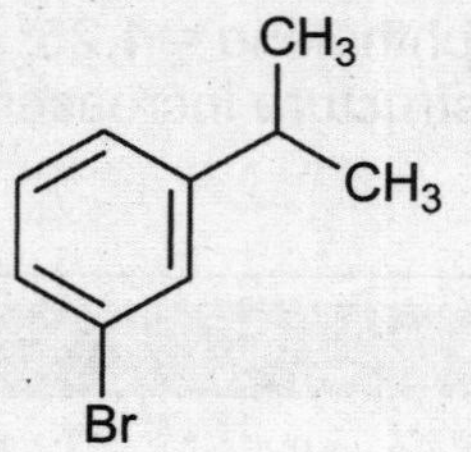

CH3
CH3
Br

n.

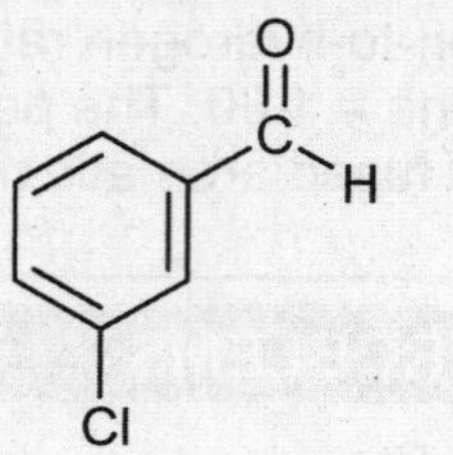

O
C
H
Cl

o.

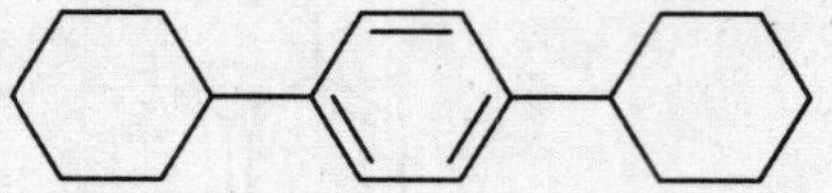

p.

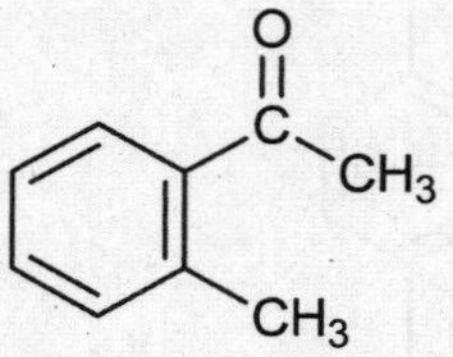

O
C
CH3
CH3

4.21 a. isopropylbenzene (or 2-phenylpropane)
b. *m*-bromobenzaldehyde
c. 1,8-dichloronaphthalene
d. 2,5-dichlorotoluene (start numbering with the carbon bonded to the methyl group and go around the ring such that substituents get the lowest possible numbers)
e. *p*-(*t*-butyl)phenol or 4-(1,1-dimethylethyl)phenol
f. *o*-nitrotoluene
g. hexafluorobenzene
h. 3,5-dibromostyrene
i. 1-ethyl-1-phenylcyclopropane (substituents in alphabetical order)

4.22 a.

CH3
CH3
CH3

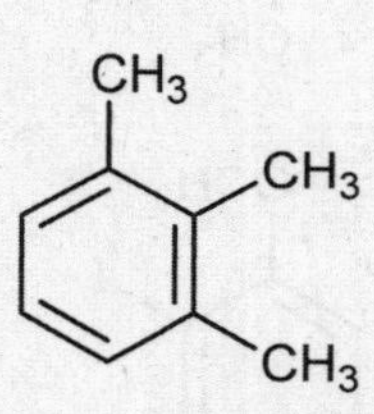

1,2,3-trimethylbenzene

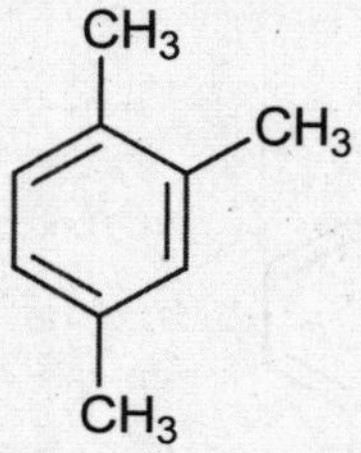

CH3
CH3
CH3

1,2,4-trimethylbenzene

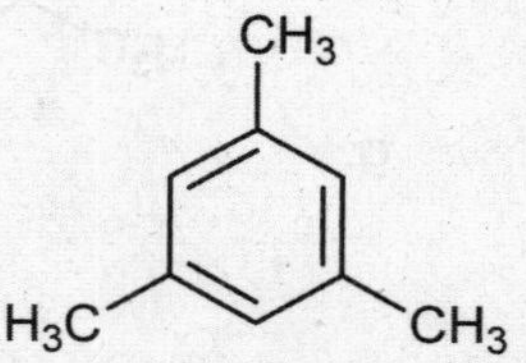

CH3
H3C
CH3

1,3,5-trimethylbenzene

b.

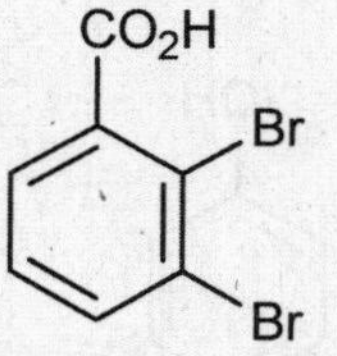

CO2H
Br
Br

2,3-dibromobenzoic acid

CO2H
Br
Br

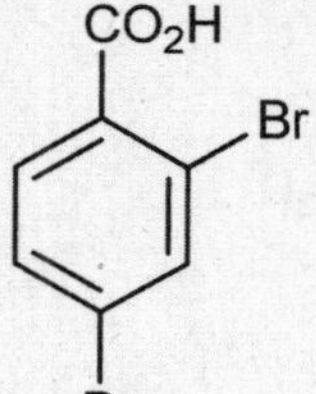

2,4-dibromobenzoic acid

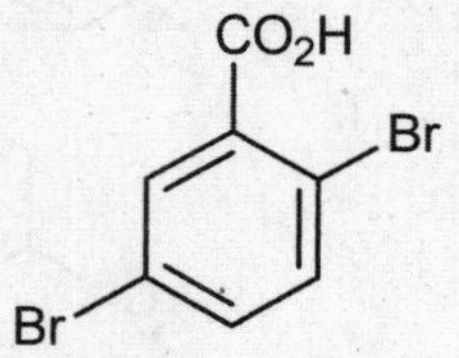

CO2H
Br
Br

2,5-dibromobenzoic acid

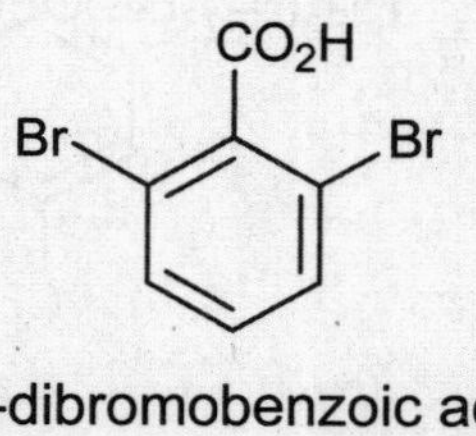

CO2H
Br
Br

2,6-dibromobenzoic acid

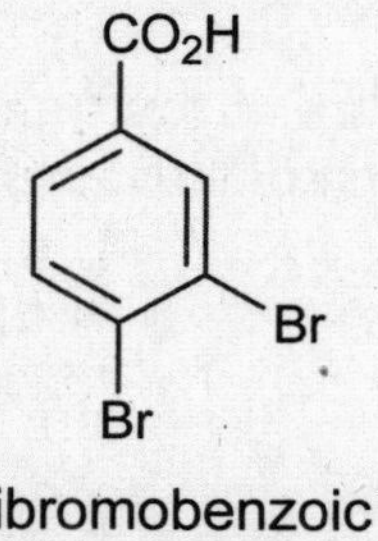

CO2H
Br
Br

3,4-dibromobenzoic acid

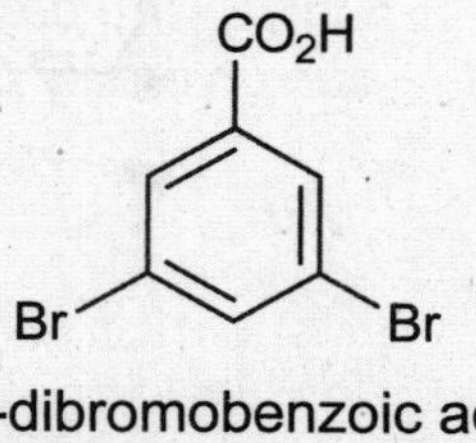

CO2H
Br
Br

3,5-dibromobenzoic acid

c.

2,3-dinitroanisole 2,4-dinitroanisole 2,5-dinitroanisole

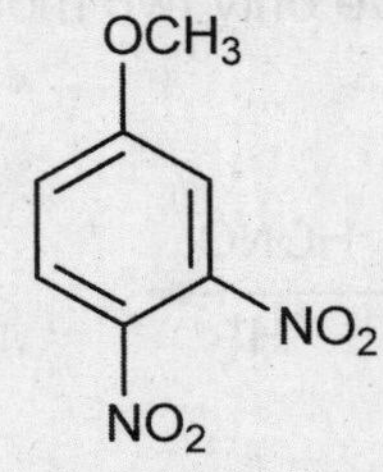

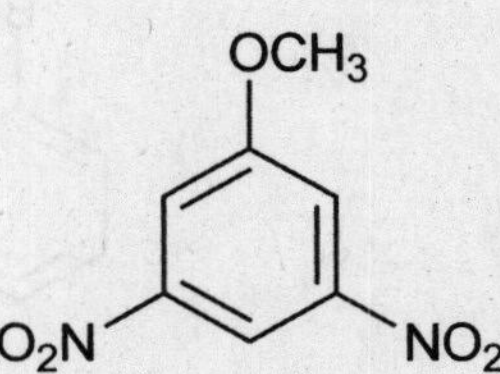

2,6-dinitroanisole 3,4-dinitroanisole 3,5-dinitroanisole

d.

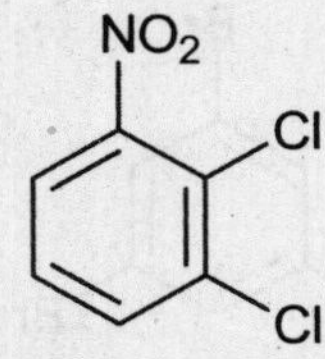

2,3-dichloronitrobenzene 2,4-dichloronitrobenzene 2,5-dichloronitrobenzene

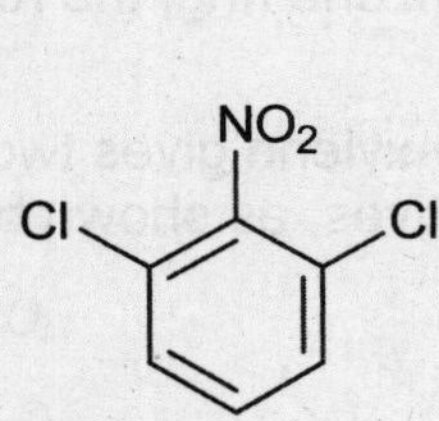

2,6-dichloronitrobenzene 3,4-dichloronitrobenzene 3,5-dichloronitrobenzene

4.23 The three possible structures are:

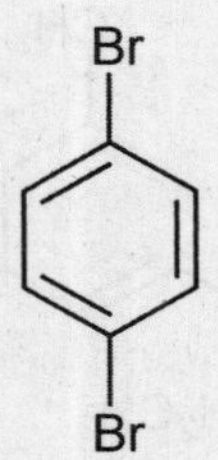

Only in the *para* isomer are all four remaining hydrogens equivalent. Therefore, it can give only *one* mononitro derivative and must be **A**:

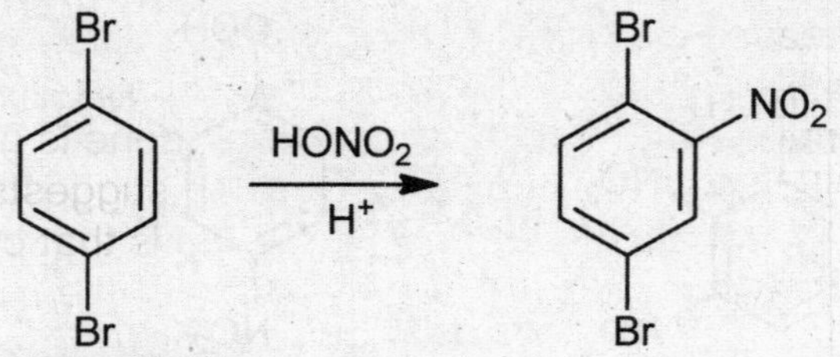

A (mp 87°C)

The *ortho* isomer can give only *two* mononitration products and must be **B**:

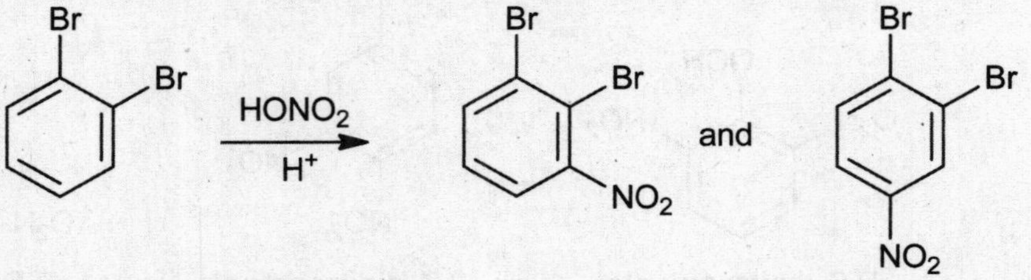

The *meta* isomer is, therefore, **C**:

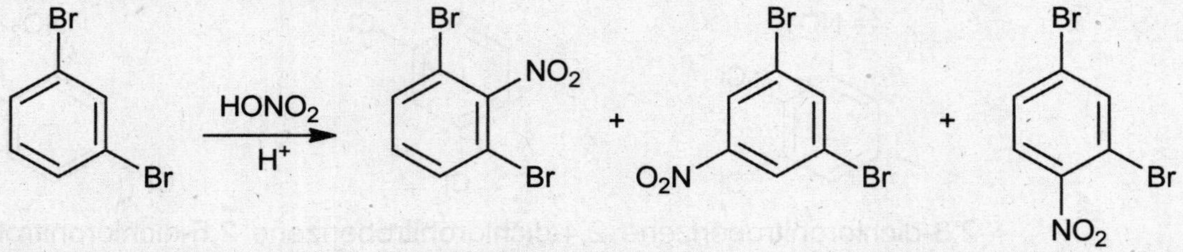

4.24 In each case, six carbons are required for the benzene ring; the remaining carbons must be present as alkyl substituents:

a. Only *o*-xylene gives two monobromo derivatives, as shown by the arrows.

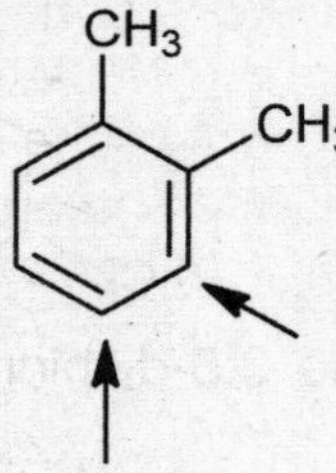

b. The structure is symmetrical, and all three positions for aromatic substitution are equivalent.

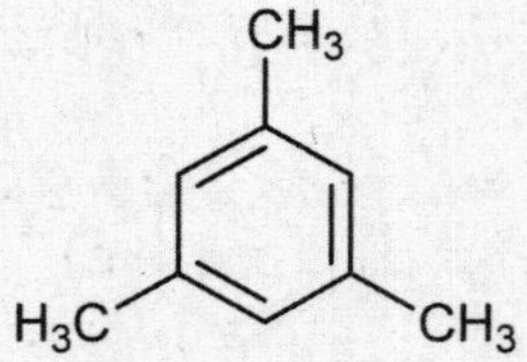

c. Substitution at each unoccupied ring position gives a different product.

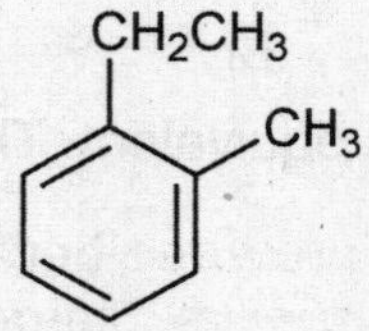

The compound in part b is 1,3,5-trimethylbenzene and that in part c is *o*-ethyltoluene or 2-ethyltoluene.

4.25 The energy released on hydrogenating a carbon–carbon double bond is 26–30 kcal/mol (eq. 4.5). With four double bonds, we can calculate that 104–120 kcal/mol should be liberated when 1,3,5,7-cyclooctatetraene is hydrogenated. The observed value (110 kcal/mol) falls within this range and suggests that cyclooctatetraene has no appreciable resonance energy. One reason is that cyclooctatetraene is not planar, and its tub-like shape prevents overlap of the *p* orbitals around the ring.

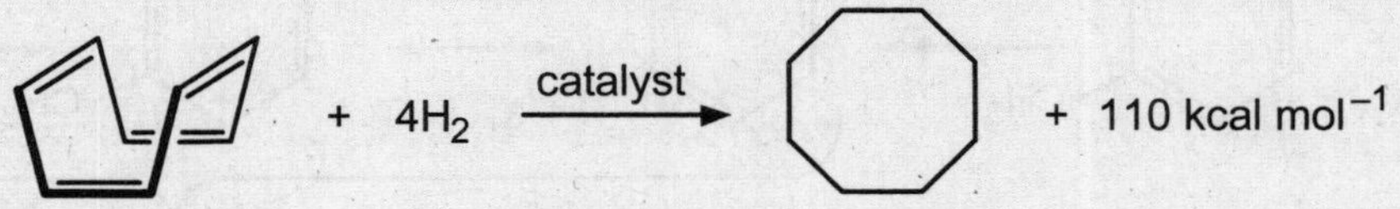

$$\text{cyclooctatetraene} + 4H_2 \xrightarrow{\text{catalyst}} \text{cyclooctane} + 110 \text{ kcal mol}^{-1}$$

4.26 The nitro group has two main contributing resonance structures:

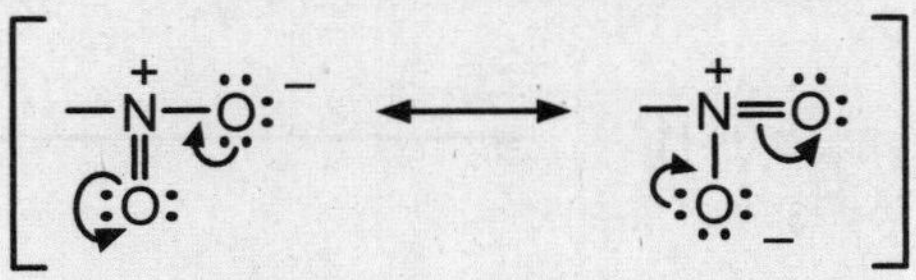

$$\left[-\overset{+}{N}(=O)-\ddot{O}:^{-} \longleftrightarrow -\overset{+}{N}(-\ddot{O}:^{-})=\ddot{O}: \right]$$

Since they are identical and contribute equally, there is only one type of nitrogen–oxygen bond, intermediate between double and single in length.

4.27 For naphthalene ($C_{10}H_8$), there are a number of resonance forms:

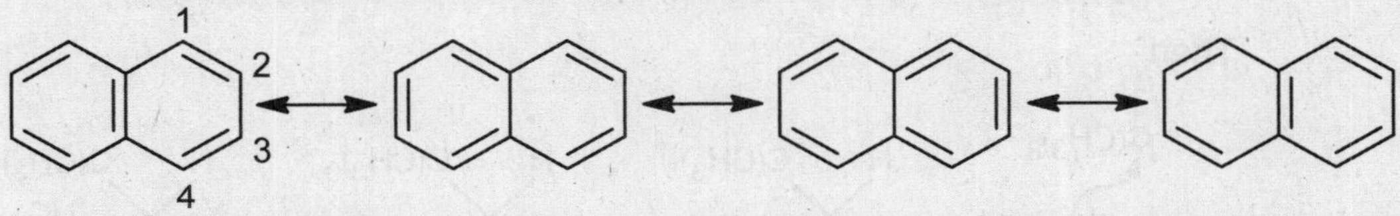

(resonance structures of naphthalene, positions 1, 2, 3, 4 labeled)

The C2–C3 bond is more often a single bond in these different resonance forms, so one would expect that the C2–C3 bond would be longer than the C1–C2 bond.

4.28 The energy released on hydrogenating a carbon–carbon double bond in cyclohexene is 28.6 kcal/mol. With five double bonds, we can calculate that 143 kcal/mol should be liberated when naphthalene is hydrogenated. The observed value (80 kcal/mol) is less than this value and suggests that naphthalene has 63 kcal/mol of resonance energy. For reference, benzene has 36 kcal/mol of resonance energy so napthalene's resonance energy is less than twice that of benzene. As we can see from the resonance forms of naphthalene in problem 4.27, both rings do not have perfect delocalization and in general, only one ring can be perfectly aromatic at a time; hence, the reduced resonance energy of naphthalene relative to twice that of benzene.

4.29 NO_2^+ There are 16 valence electrons available (N = 5, and 2 × O = 2 × 6 = 12 for a total of 17, but we must subtract 1 electron since the ion is positive).

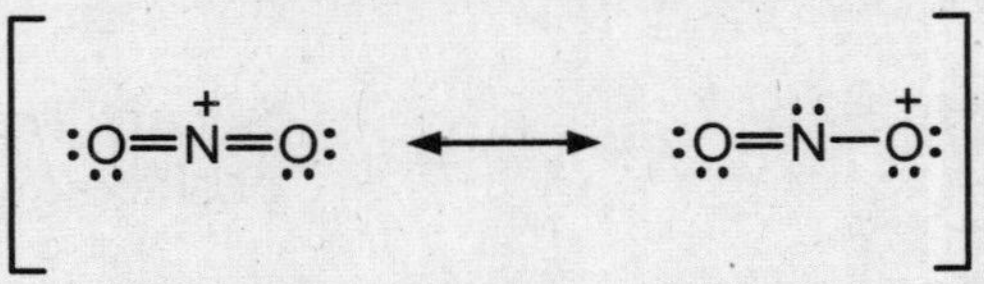

$$\left[:\ddot{O}=\overset{+}{N}=\ddot{O}: \longleftrightarrow :\ddot{O}=\ddot{N}-\overset{+}{\ddot{O}}: \right]$$

The structure with the positive charge on the nitrogen is preferred because each atom has an octet of electrons. In the structure with the positive charge on the oxygen, the oxygen atom has only six electrons around it. Note that in aromatic nitrations, it is the nitrogen atom of NO_2^+ that is attacked and becomes attached to the aromatic ring.

4.30 a. The electrophile, NO_2^+, is formed as in eq. 4.18. Then the mechanism follows the same steps as the solution to Example 4.2.

NO_2^+ H NO2 H NO2 H NO2
H3C H3C H3C H3C
+ + +
CH3 CH3 CH3 CH3

NO2
H3C
+ H^+
CH3

b. The electrophile is formed as in eq. 4.20:

$$(CH_3)_3CCl + AlCl_3 \longrightarrow (CH_3)_3C^+ + AlCl_4^-$$

Then

$\overset{+}{C}(CH_3)_3$ H C(CH3)3 H C(CH3)3 H C(CH3)3
+ + +
CH3 CH3 CH3 CH3

C(CH3)3
+ H^+
CH3

4.31 *ortho*

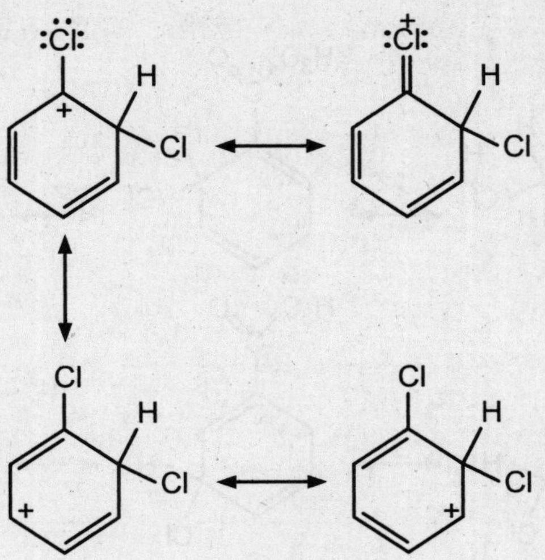

meta

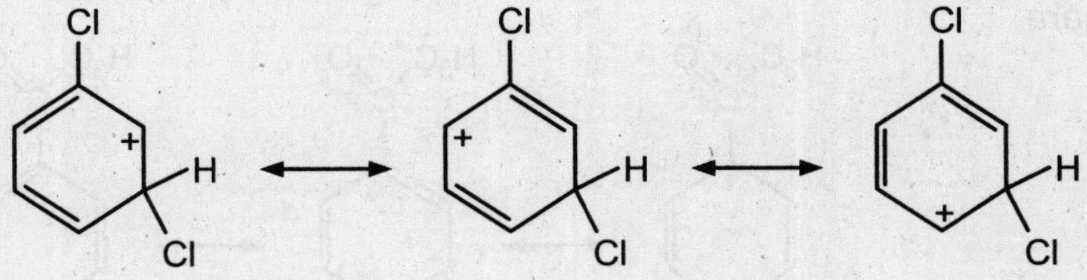

para

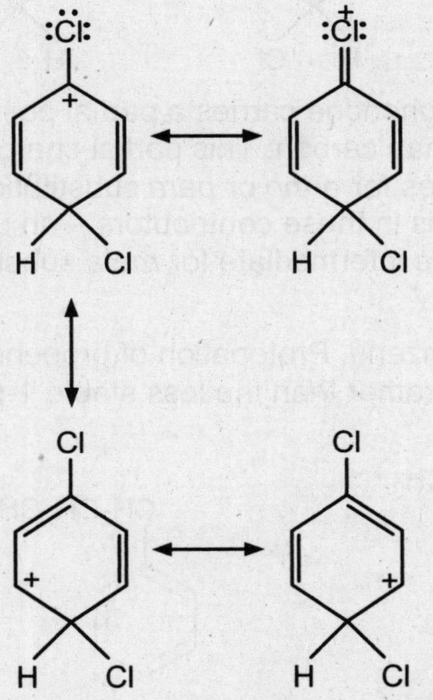

In the intermediate for *ortho* or *para* substitution, the positive charge can be delocalized to the initial chloro substituent. This delocalization is not possible with *meta* substitution. Therefore, *ortho,para*-substitution predominates.

4.32 *ortho*

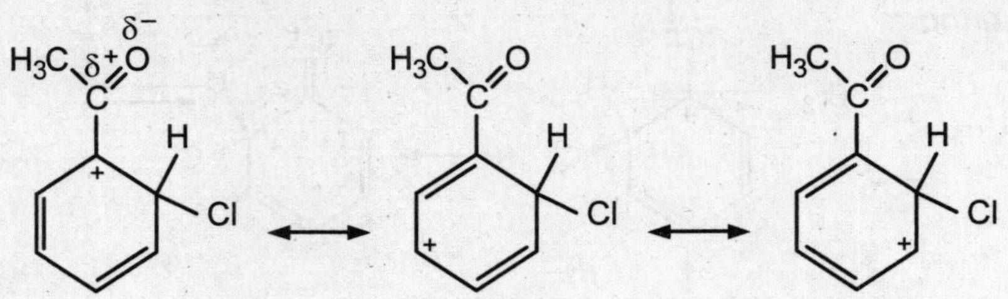

meta

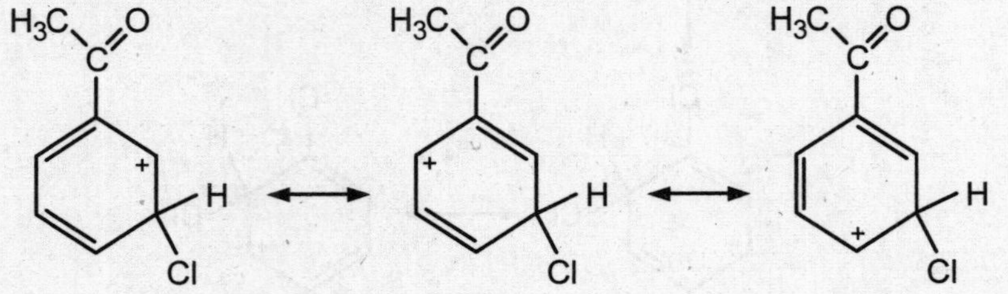

para

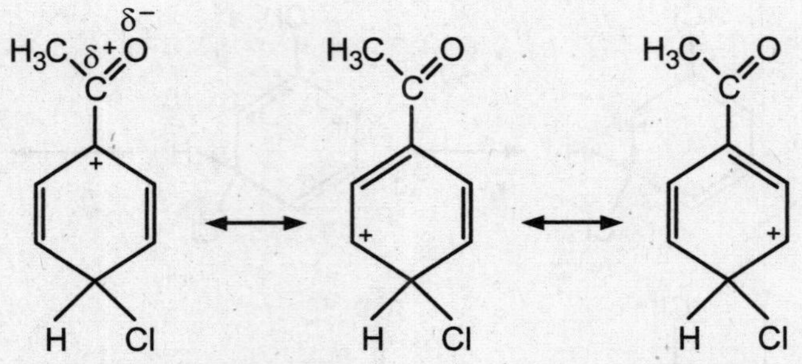

The carbonyl carbon in the acetophenone carries a partial positive charge because oxygen is more electronegative than carbon. This partial charge is illustrated in the first contributor to the intermediates for *ortho* or *para* substitution. Note that positive charges appear on adjacent atoms in these contributors – an unfavorable situation. No such contributor appears in the intermediate for *meta* substitution. Thus *meta* substitution predominates.

4.33 The major product is isopropylbenzene. Protonation of propene gives the 2-propyl cation, a secondary carbocation, rather than the less stable 1-propyl cation, a primary carbocation.

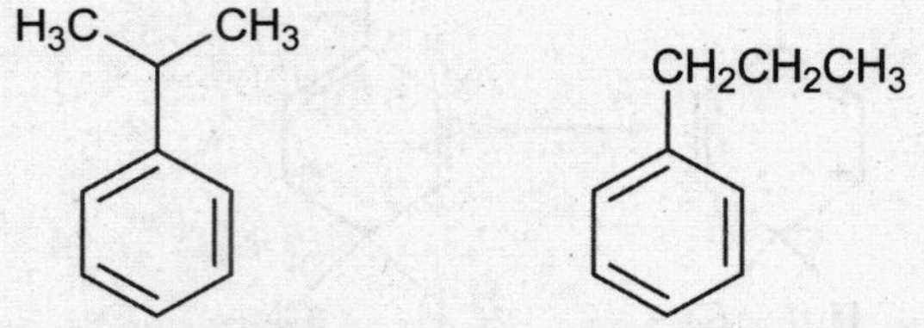

isopropylbenzene propylbenzene

4.34 If the halides were not identical, the following kind of exchange could occur (see eq. 4.13):

$$Cl_2 + FeBr_3 \rightleftharpoons \delta^+ Cl\text{---}Cl\text{—}Fe^{\delta-}Br_3 \rightleftharpoons \delta^+ Cl\text{---}Br\text{—}Fe^{\delta-}ClBr_2$$

$$\rightleftharpoons FeBr_2Cl + Cl\text{—}Br \rightleftharpoons \delta^+ Br\text{---}Cl\text{—}Fe^{\delta-}Br_2Cl$$

In this way, electrophilic bromine (Br^+) could be formed. Consequently, a mixture of chlorinated and brominated aromatic products would be obtained.

4.35 D_2SO_4 is a strong acid and a source of the electrophile D^+ (analogous to H^+ from H_2SO_4).

$$C_6H_6 + D^+ \rightleftharpoons [C_6H_6D]^+ \rightleftharpoons C_6H_5D + H^+$$

Loss of H^+ from the intermediate benzenonium results in replacement of H by D. With a large excess of D_2SO_4, these equilibria are shifted to the right, eventually resulting in fully deuterated benzene, C_6D_6.

4.36 The carbon bonded to the electrophile (E) is sp^3-hybridized. The remaining carbons are sp^2-hybridized.

4.37 a. *Meta*-directing and ring-deactivating because of the positive charge on the nitrogen (electron-withdrawing).

$$\text{—}\overset{+}{N}H(CH_3)_2$$

b. *Meta*-directing and ring-deactivating because of the partial positive charge on the carbonyl carbon, due to contributors such as

$$\left[\ \text{—}\overset{\delta-}{\overset{:O:}{\overset{\|}{C}}}\text{—}OCH_3\ (\delta^+) \longleftrightarrow \text{—}\overset{:\ddot{O}:^-}{\overset{|}{\underset{+}{C}}}\text{—}OCH_3\ \right]$$

c. *Ortho,para*-directing and ring-activating because of the unshared electron pairs on the sulfur.

$$\text{—}\ddot{\underset{..}{S}}CH_3$$

d. *Ortho,para*-directing and ring-activating because of the unshared electron pair on the nitrogen.

$$—\ddot{N}{=}O$$

e. *Meta*-directing and ring-deactivating because of the partial positive charge on the carbon.

$$—\overset{\delta^+}{C}{\equiv}\overset{\delta^-}{N}$$

f. *Ortho,para*-directing because of the lone pairs on bromine, but ring-deactivating because of the electronegativity of Br.

$$—\ddot{B}r:$$

g. *Ortho,para*-directing and ring-activating because of the unshared electron pairs on the oxygen.

$$—\ddot{O}CH(CH_3)_2$$

4.38 See Sec. 4.11 for a discussion of the orienting influence of substituents.

a. Cl— —OCH_3
(and *ortho*)

b. Br
NO_2

c. Br— —CH_3
(and *ortho*)

d. 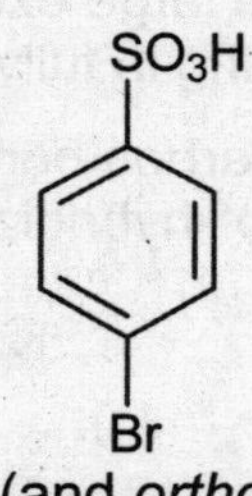

(and *ortho*)

e. Cl— —I
(and *ortho*)

f. SO_3H
NO_2

g. O, CH_3, C
CH_3
(and *ortho*)

h. O_2N— —CH_2CH_3
(and *ortho*)

4.39 The 2-methyl-2-butene can be protonated by sulfuric acid to make a tertiary carbocation (an electrophile). The electrophile can be attacked from the *ortho* and *para* positions of toluene to provide the two products. For steric congestion reasons, the *para* isomer will be the major product.

CH3 H+ CH3 H3C CH3 H3C + CH3 CH3 CH3 *ortho* attack *para* attack CH3 CH3 + H H + CH3 H3C −H+ −H+ CH3 CH3 H3C CH3

4.40 a. Anisole; the $-OCH_3$ group is ring-activating, whereas the $-CO_2H$ group is ring-deactivating.

b. Ethylbenzene; although both substituents ($-CH_2CH_3$ and $-Cl$) are *ortho,para*-directing, the ethyl group is ring-activating, whereas the chlorine is ring-deactivating.

4.41 Nitro groups are ring-deactivating. Thus, as we substitute nitro groups for hydrogens, we make the ring less and less reactive toward further electrophilic substitution and, therefore, we must increase the severity of the reaction conditions.

4.42 The α-tocotrienol has a number of isolated double bonds on the alkyl chain and also has the double bonds of the aromatic ring. The isolated double bonds are easier to reduce by catalytic hydrogenation as the aromatic ring would prefer to retain its resonance energy from aromaticity. Thus, low-pressure hydrogenation and the use of simple catalysts like Pt will selectively reduce the isolated double bonds of α-tocotrienol to α-tocopherol.

4.43 a. Since the two substituents must end up in a *para* relationship, the first one introduced into the benzene ring must be *para*-directing. Therefore, if we start with toluene, we can then brominate the ring and benefit from the *ortho,para*-directing of the methyl group.

CH_3 (toluene) $\xrightarrow[FeBr_3]{Br_2}$ p-bromotoluene (CH_3, Br)

b. First make the ethylbenzene and then nitrate it:

benzene + $H_2C{=}CH_2$ $\xrightarrow{H^+}$ $C_6H_5CH_2CH_3$

or

benzene + CH_3CH_2Cl $\xrightarrow{AlCl_3}$ $C_6H_5CH_2CH_3$

$C_6H_5CH_2CH_3$ $\xrightarrow[H^+]{HONO_2}$ O_2N–C_6H_4–CH_2CH_3 + some *ortho* isomer

c. Since the two substituents must end up in a *para* relationship, the first one introduced into the benzene ring must be *para*-directing. Therefore, brominates first and then nitrate.

benzene $\xrightarrow[FeBr_3]{Br_2}$ bromobenzene (Br) $\xrightarrow[H^+]{HONO_2}$ p-bromonitrobenzene (Br, NO_2)

d. Alkyl groups are *ortho,para*-directing but the $-SO_3H$ group is *meta*-directing. So, first we need to install the isopropyl group in a similar manner as part b, perhaps using isopropyl chloride, $(CH_3)_2CHCl$, with $AlCl_3$ as a catalyst, and then sulfonate the resulting isopropylbenzene (cumene).

benzene $\xrightarrow[AlCl_3]{(CH_3)_2CHCl}$ cumene (H_3C, CH_3) $\xrightarrow{HOSO_3H}$ p-isopropylbenzenesulfonic acid (H_3C, CH_3, SO_3H)

e. Arenes can be converted to substituted cyclohexanes by catalytic hydrogenation.

benzene $\xrightarrow[AlCl_3]{(CH_3)_3CCl}$ $C_6H_5C(CH_3)_3$ $\xrightarrow[\text{metal catalyst}]{H_2}$ $C_6H_{11}C(CH_3)_3$

4.44 a. The nitro substituent must be introduced first, to block the *para* position from the chlorination.

toluene (CH_3) $\xrightarrow[H^+]{HONO_2}$ 4-nitrotoluene (CH_3, NO_2) $\xrightarrow[FeCl_3]{2\ Cl_2}$ 2,6-dichloro-4-nitrotoluene (CH_3, Cl, Cl, NO_2)

+ some *ortho* isomer

b. The nitration must be performed first. The nitro group is a *m*-director while a chloro group is an *o,p*-director.

benzene $\xrightarrow[H^+]{HONO_2}$ nitrobenzene (NO_2) $\xrightarrow[FeCl_3]{Cl_2}$ 3-chloronitrobenzene (NO_2, Cl)

c. If the chlorination were performed first, a considerable portion of product would have the chlorine *para* to the methyl group. Also, note that in the second step, both substituents direct the chlorine to the desired position (–CH_3 is *o,p*-directing, –NO_2 is *m*-directing).

toluene (CH_3) $\xrightarrow[H^+]{HONO_2}$ 4-nitrotoluene (CH_3, NO_2) $\xrightarrow[FeCl_3]{Cl_2}$ 2-chloro-4-nitrotoluene (CH_3, Cl, NO_2)

d. If the bromination were performed first, the Friedel–Crafts acylation would occur at the 2 and 4 positions.

$C_6H_5CH_3$ $\xrightarrow[AlCl_3]{H_3C-C(=O)-Cl}$ $C_6H_5C(=O)CH_3$ $\xrightarrow[FeBr_3]{Br_2}$ 3-bromo-$C_6H_4C(=O)CH_3$ (Br)

4.45 3-Nitrobenzoic acid is better because both substituents are *meta*-directing:

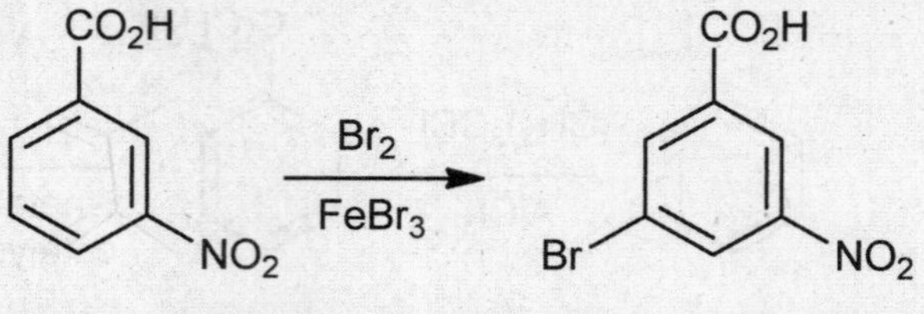

CO_2H $\xrightarrow[FeBr_3]{Br_2}$ CO_2H

NO_2 Br NO_2

On the other hand, 3-bromobenzoic acid has an *ortho,para*-directing and a *meta*-directing substituent and, on nitration, would give a mixture of isomers:

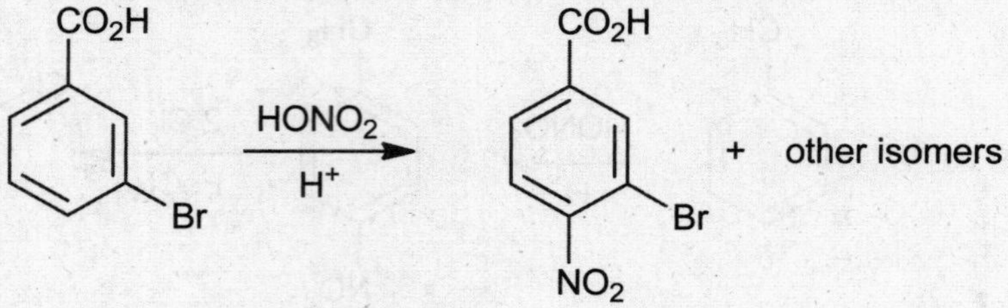

CO_2H $\xrightarrow[H^+]{HONO_2}$ CO_2H + other isomers

Br Br

NO_2

4.46 The *meta*-directing effect on a nitro group would facilitate installation of the chlorine substituent:

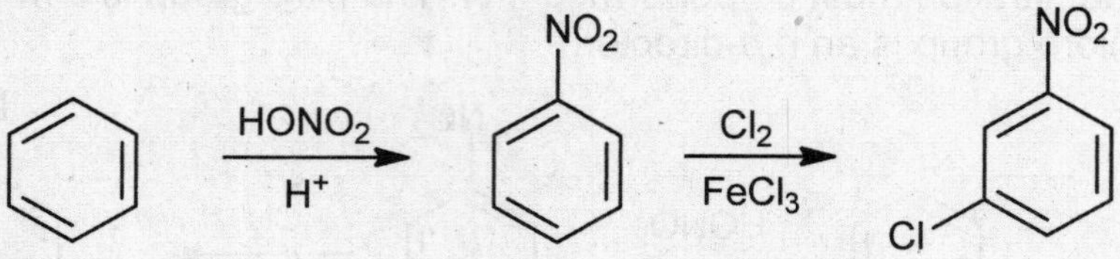

$\xrightarrow[H^+]{HONO_2}$ NO_2 $\xrightarrow[FeCl_3]{Cl_2}$ NO_2

Cl

4.47 The *meta*-directing effect on both nitro groups reinforces substitution in the position shown.

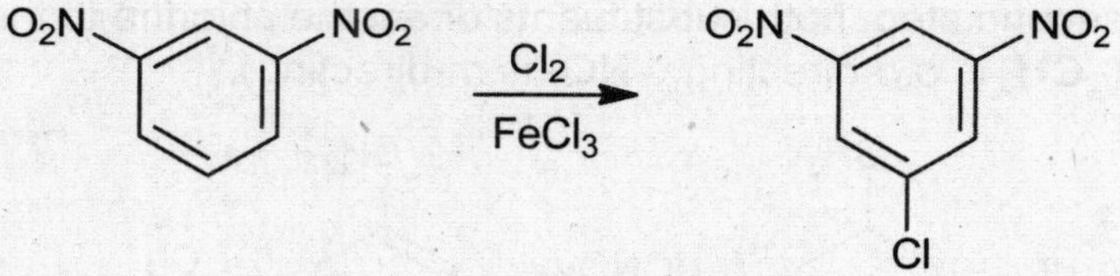

O_2N NO_2 $\xrightarrow[FeCl_3]{Cl_2}$ O_2N NO_2

Cl

4.48 If corannulene would undergo dibromination, there would be 5 unique dibrominated isomers:

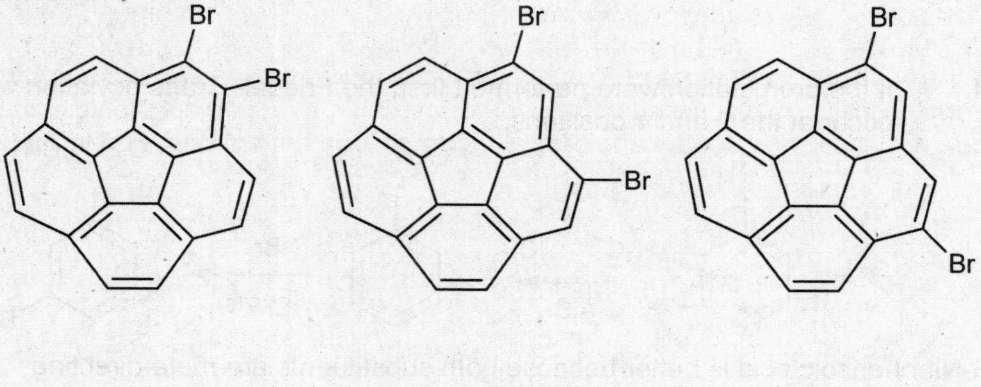

Br Br Br Br Br Br

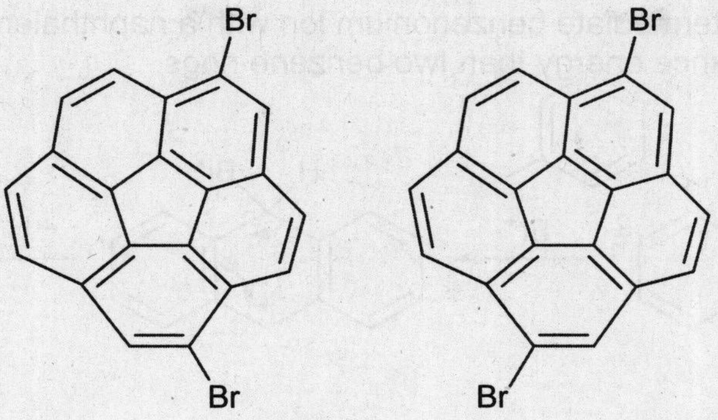

Br Br

Br Br

4.49 a. Three different monosubstitution products are possible: at C-1 (equivalent to C-4, C-5, and C-8); at C-2 (equivalent to C-3, C-6, and C-7); and at C-9 (equivalent to C-10).

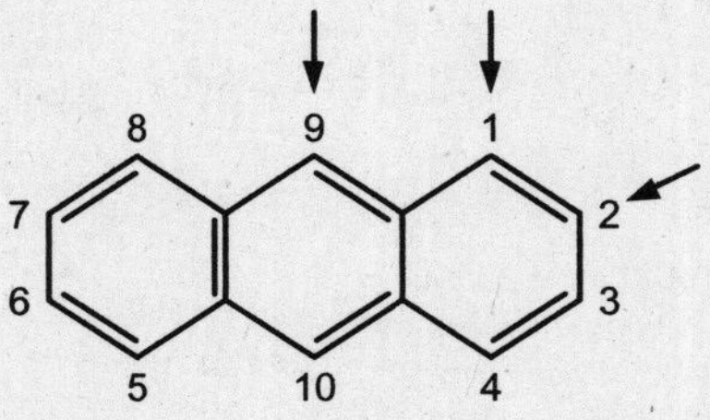

b. Five different monosubstitution products are possible: at C-1 (equivalent to C-8); at C-2 (equivalent to C-7); at C-3 (equivalent to C-6); at C-4 (equivalent to C-5); and at C-9 (equivalent to C-10).

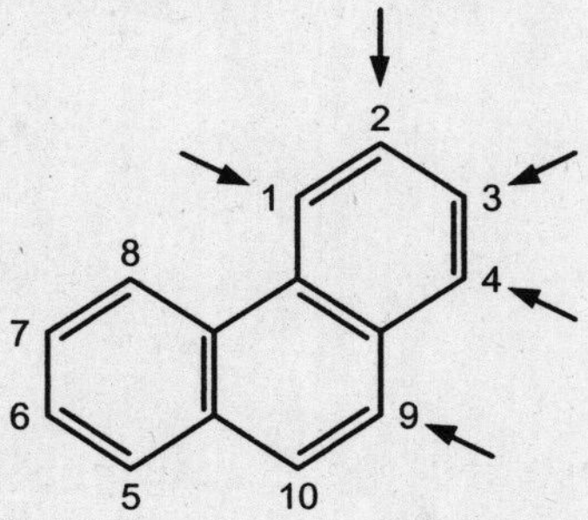

4.50 Starting with benzaldehyde, we have a ring-deactivating substituent on the benzene ring, so that group will be *meta*-directing. However, we need to put on the hydroxyl unit of a phenol, and there are limited ways to install that functional group on an aromatic ring. As discussed in Sec. 4.9.c, we can first put on a sulfonic acid and then with the use of NaOH and heat, we can convert the sulfonic acid to a phenol.

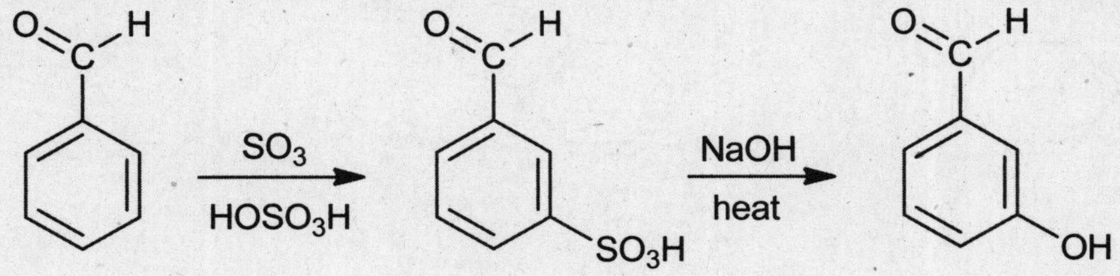

$\xrightarrow[HOSO_3H]{SO_3}$ SO_3H $\xrightarrow[heat]{NaOH}$ OH

4.51 Bromination at C-9 is preferred over the other two possibilities (C-1 or C-2) because the intermediate benzenonium ion retains two benzenoid rings. Substitution at C-1 or

C-2 gives an intermediate benzenonium ion with a naphthalene substructure, which has less resonance energy than two benzene rings.

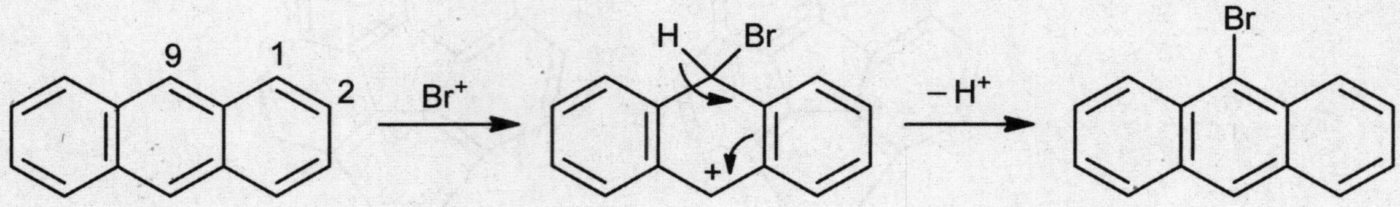

5

Stereoisomerism

Chapter Summary

Stereoisomers have the same atom connectivities but different arrangements of the atoms in space. They may be **chiral** or **achiral**. A stereoisomer is chiral if its mirror image is not identical or superimposable on the original molecule. It is achiral if the molecule and its mirror image are identical or superimposable. **Enantiomers** are a pair of molecules related as nonsuperimposable mirror images.

A carbon atom with four different groups attached is called a **stereogenic center** *because it gives rise to stereoisomers*. Any molecule with a **plane of symmetry** is achiral. **Configuration** refers to the arrangement of groups attached to a stereogenic center. Enantiomers have opposite configurations. Configuration can be designated by the ***R–S* convention**. Groups attached to the stereogenic center are ranked in a priority order according to decreasing atomic number. When the stereogenic center is viewed from the side *opposite* the lowest-priority group, the center is said to be *R* if the other three groups, in decreasing priority order, form a *clockwise* array. If the three-group array is counterclockwise, the configuration is *S*. A similar convention (*E–Z*) has been applied to alkene *cis–trans* isomers.

Chiral molecules are **optically active**. They rotate a beam of plane-polarized light. They are **dextrorotatory** (+) or **levorotatory** (-), depending on whether they rotate the beam to the right or left, respectively. The rotations are measured with a **polarimeter** and are expressed as **specific rotations**, defined as

$$[\alpha]_{\lambda}^{t} = \frac{\alpha}{l \times c} \quad \text{(solvent)}$$

where α = observed rotation, l = length of sample tube in decimeters, c = concentration in g/mL, and the measurement conditions of temperature (t), wavelength of polarized light (λ), and solvent are given. Achiral molecules are **optically inactive**.

Pasteur showed that optical activity was related to molecular right- or left-handedness (chirality). Later, van't Hoff and LeBel proposed that the four valences of carbon are directed toward the corners of a tetrahedron. If the four attached groups are different, two arrangements are possible and are related as an object and its nonsuperimposable mirror image. Enantiomers differ *only* in chiral (or handed) properties, such as the *direction* of rotation of plane-polarized light. They have identical **achiral properties**, such as melting and boiling points.

Fischer projection formulas show three-dimensional structures in two dimensions. In such formulas, horizontal groups project toward the viewer, and vertical groups project away from the viewer.

Diastereomers are stereoisomers that are not mirror images of one another. They may differ in all types of properties, and may be chiral or achiral.

Compounds with n different stereogenic centers may exist in a maximum of 2^n forms. Of these, there will be $2^n/2$ pairs of enantiomers. Compounds from different enantiomeric pairs are diastereomers. If two (or more) of the stereogenic centers are identical, certain isomers will be achiral. A **meso form** is an optically inactive, achiral form of a compound with stereogenic centers. **Tartaric acid**, which has two identical stereogenic centers, exists in three forms: the *R,R* and *S,S* forms (a pair of enantiomers) and the achiral *meso* form.

Stereoisomers may be classified as conformational or configurational, chiral or achiral, and enantiomers or diastereoisomers.

The stereochemistry of organic reactions depends on the nature of the reactants. Achiral molecules can react to give a chiral product. In such reactions, both enantiomers of the product will always be formed in equal amounts. When chiral molecules react with achiral reagents to create a new stereogenic center, diastereoisomers are formed in unequal amounts.

A **racemic form** is 50:50 mixture of enantiomers. It is optically inactive. A racemic mixture of configurational isomers cannot be separated (resolved) by ordinary chemical means (distillation, crystallization, chromatography) unless the reagent is chiral. One way to separate a pair of enantiomers is to first convert them to diastereomers by reaction with a chiral reagent, then separate the diastereomers and regenerate the (now separate) enantiomers.

Learning Objectives

1. Know the meaning of: chiral, achiral, enantiomers, plane of symmetry, superimposable and nonsuperimposable mirror images, racemic mixture.
2. Know the meaning of: stereogenic carbon atom, stereogenic center, *R*–*S* convention, priority order, *E*–*Z* convention, Fischer projection.
3. Know the meaning of: diastereomer, *meso* compound, lactic acid, tartaric acid, resolution.
4. Know the meaning of: plane-polarized light, polarimeter, optically active or optically inactive, observed rotation, specific rotation, dextrorotatory, levorotatory.
5. Given the concentration of an optically active compound, length of the polarimeter tube, and observed rotation, calculate the specific rotation. Given any three of the four quantities mentioned, calculate the fourth.
6. Given a structural formula, draw it in three dimensions and locate any plane of symmetry.
7. Given the structure of a compound, determine if any stereogenic centers are present.
8. Given the structure or name of a compound, tell whether it is capable of optical activity.
9. Know the rules for establishing priority orders of groups in the *R*–*S* convention.
10. Given a compound with a stereogenic center, assign the priority order of groups attached to it.

11. Given a stereogenic center in a molecule, assign the *R* or S configuration to it.

12. Draw the three-dimensional formula of a molecule with a particular configuration, *R* or *S*.

13. Given a pair of *cis–trans* isomers, assign the *E* or *Z* configuration.

14. Draw the formula of an alkene with a particular configuration, *E* or *Z*.

15. Given a structure with one or more stereogenic centers, draw a Fischer projection.

16. Given Fischer projections of two isomers, tell their relationship (for example, same structure, enantiomers, diastereomers).

17. Assign *R* or *S* configuration to each stereogenic center in a Fischer projection.

18. Given a structure with more than one stereogenic center, tell how many stereoisomers are possible and draw the structure of each. Tell what relationship the stereoisomers have to one another (for example, enantiomeric, diastereomeric).

19. Tell whether a particular structure can exist as a *meso* form.

20. Given a structure with two or more identical stereogenic centers, draw the structure of the *meso* form.

21. Given a pair of stereoisomers, classify them as configurational or conformational, chiral or achiral, and enantiomers or diastereomers.

22. Given a chemical reaction that gives a chiral product, tell the stereochemistry of the products.

ANSWERS TO PROBLEMS

Problems Within the Chapter

5.1

a.	chiral	b.	achiral	c.	achiral	d.	chiral
e.	achiral	f.	chiral	g.	chiral	h.	achiral

The achiral objects (teacup, football, tennis racket, and pencil) can be used with equal ease by right- or left-handed persons. Their mirror images are superimposable on the objects themselves. On the other hand, a golf club must be either left- or right-handed and is chiral; a shoe will fit a left or a right foot; a corkscrew may have a right- or left-handed spiral. These objects, as well as a portrait, have mirror images that are *not* identical with the objects themselves, and thus they are chiral.

5.2 The stereogenic centers are marked with an asterisk. Note that each stereogenic center has four different groups attached.

a. $CH_3CH_2-\overset{*}{C}H(Br)-CH_2CH_2CH_2CH_3$ (H above, Br below the starred C)

b. Cyclohexene ring with the starred carbon bearing H and CH_3

c. $Cl-\overset{*}{C}H(F)-CH_3$ (H above, F below the starred C)

d. $CH_3-\overset{*}{C}H(Br)-\overset{*}{C}H(Br)-CH_3$ (H above, Br below each starred C)

5.3 a.

H
|
Br—C*—CH$_3$

chiral

b.

H H
Br—C—C—
H H

no stereogenic centers; achiral

5.4 Note that if the right structure is rotated 180° about the carbon-phenyl bond, the methyl and phenyl groups can be superimposed on those of the left structure, but the positions of the hydrogen and bromine will be interchanged.

H$_3$C C Br H   Br C CH$_3$ H

5.5 The planes of symmetry are (a) the three planes that pass through any pair of eclipsed hydrogens and (b) the perpendicular bisector of the C–C bond. Ethane in this conformation is achiral.

5.6 There are three planes of symmetry that pass through any pair of anti-hydrogens. Ethane in this conformation is achiral.

5.7 *cis*-1,2-Dichloroethene has a plane of symmetry that bisects the double bond. The molecular plane is also a symmetry plane. *trans*-1,2-Dichloroethene is planar. That plane is a symmetry plane. Both *cis*- and *trans*-1,2-dichloroethene are achiral.

5.8 In each case, proceed from high to low priority.

a. $-NH_2 > -CH(CH_3)_2 > -CH_3 > -H$

b. $-Br > -OH > CH_2OH > -CH_3$

c. $-OCH_3 > -OH > -N(CH_3)_2 > -CH_2NH_2$

The oxygen in the methoxy group ($-OCH_3$) is bonded to carbon, whereas the hydroxyl oxygen (–OH) is bonded only to hydrogen.

d. $-C(CH_3)_3 > -CH(CH_3)_2 > -CH_2CH_2CH_3 > -CH_2CH_3$

5.9 a. $-C{\equiv}CH > -CH{=}CH_2$

The acetylenic carbon (–C≡) is treated as though it is bonded to three carbons, while the olefinic carbon (–CH=) is treated as though it is bonded to two carbons and a hydrogen (see Sec. 5.3).

b. 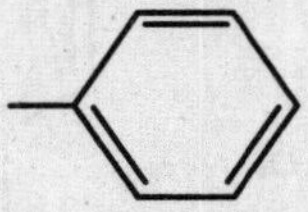 $-C_6H_5 > -CH{=}CH_2$

The phenyl carbon is treated as though it is bonded to three carbons.

c. $-CH{=}O > -CH_2OH > -CH{=}CH_2 > -CH_2CH_3$

The aldehyde carbon is treated as though it is bonded to two oxygens.

5.10 a. Priority order: $OH > CH{=}O > CH_3 > H$. Configuration is *R*.

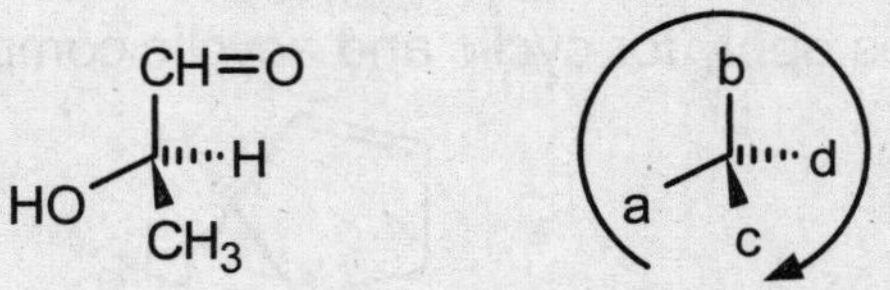

CH=O, H, HO, CH_3 b, d, a, c   clockwise or *R*

b. Priority order: $NH_2 > -C_6H_5 > CH_3 > H$. The configuration is *S*.

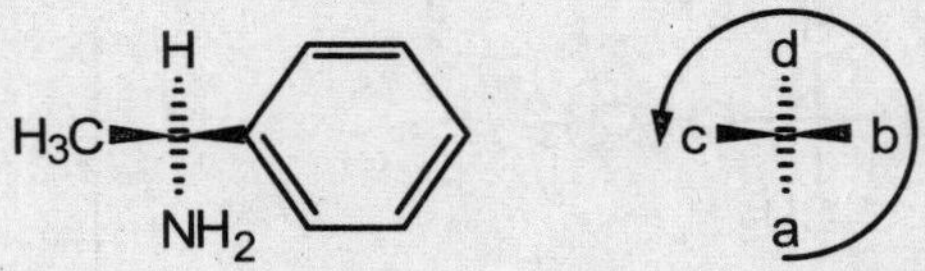

H, H_3C, NH_2 d, c, b, a   counterclockwie or *S*

5.11 a. The priority of groups around the stereogenic center is as follows:

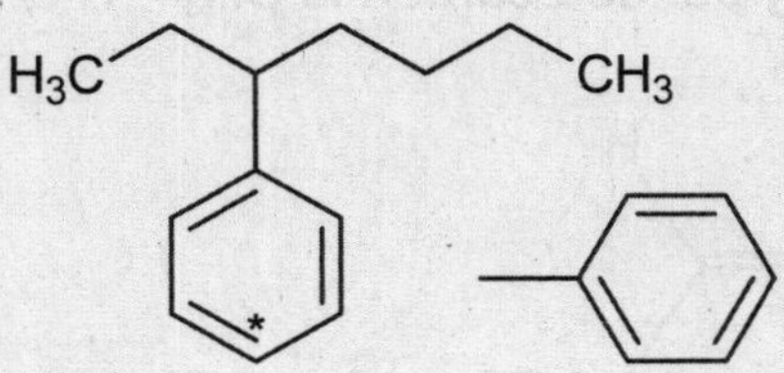

H_3C, CH_3, *

$-C_6H_5 > -CH_2CH_2CH_2CH_3 > -CH_3 > -H$

First, draw the lowest-priority group pointing away from you:

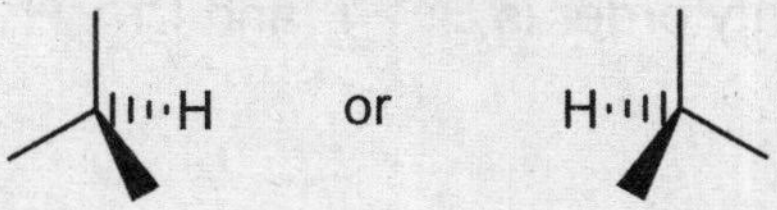

H or H

Then fill in the groups in priority order, counterclockwise (*S*),

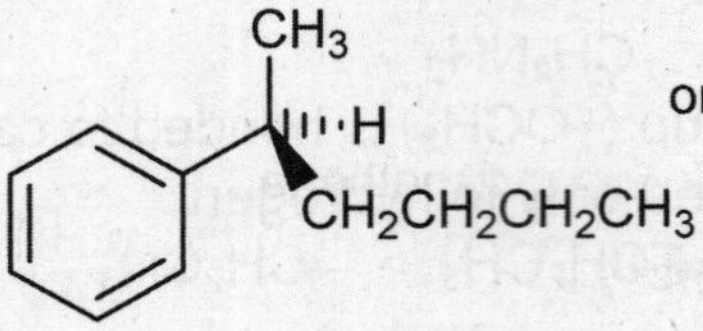

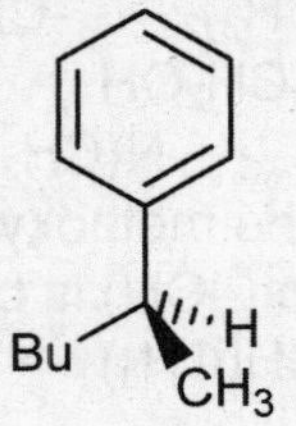

CH3, H, CH2CH2CH2CH3 or Bu, H, CH3 or CH3, Bu, H

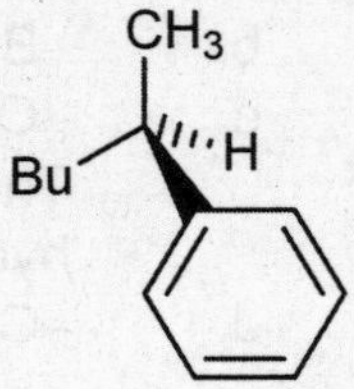

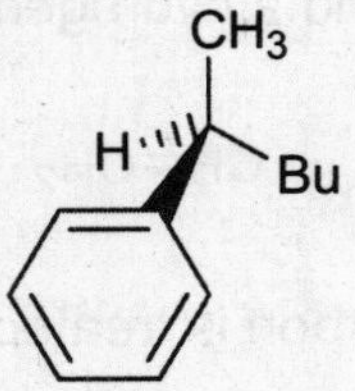

$Bu = CH_2CH_2CH_2CH_3$

Similarly,

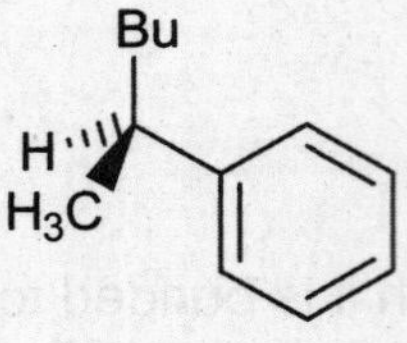

Bu, H, H3C or CH3, H, Bu or H, CH3, Bu

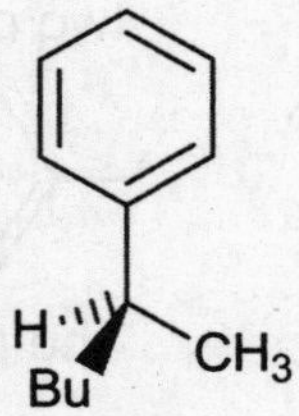

As you can see, there are many ways to write the correct answer. In subsequent problems, only one correct way will be shown. Work with models if you have difficulty.

b. The same rules apply for cyclic and acyclic compounds.

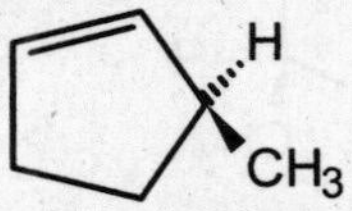

H, CH3

c. The priority around the stereogenic center is as follows:

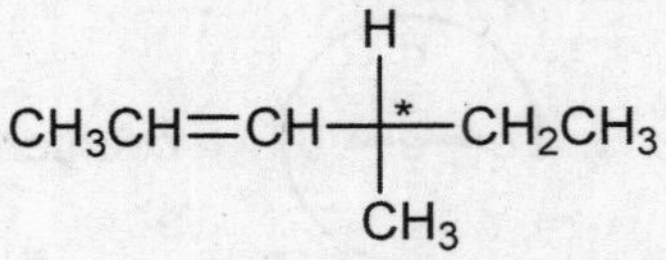

$CH_3CH{=}CH-\overset{H}{\underset{CH_3}{C^*}}-CH_2CH_3$

$-CH{=}CHCH_3 > -CH_2CH_3 > -CH_3 > -H$

CH3, *R*, H, CH=CHCH3, CH3CH2

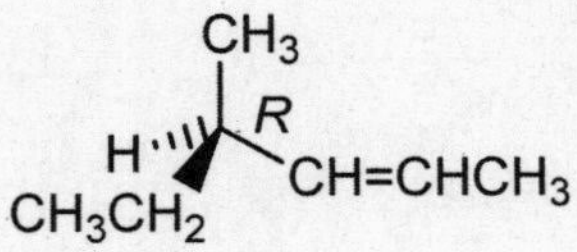

5.12 a. The priority order at each doubly bonded carbon is $CH_3 > H$ and $CH_3CH_2 > H$. The configuration is *E*.

CH3, H, H, CH2CH3

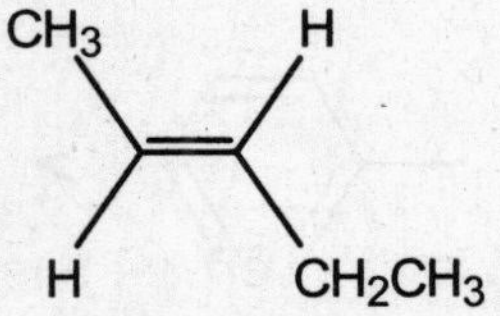

(*E*)-2-pentene

b. The priority order is Br > F and Cl > H. The configuration is *E*.

F Cl Br H

(E)-1-bromo-2-chloro-1-fluoroethene

5.13 a. The priorities are CH_2=C– > H and CH_3CH_2– > H. The two highest-priority groups, CH_2=CH– and CH_3CH_2–, are *entgegen* (opposite).

H H CH_3

(E)-1,3-hexadiene

b. The two highest-priority groups, CH_3 and CH_3, are *zusammen*, or together.

H_3C CH_3 H H

(Z)-2-butene s

5.14 $$[\alpha]_D^{20} = \frac{+0.66}{0.5 \times \frac{1.5}{50}} = 44^o \quad \text{(ethanol)}$$

5.15 CO_2H HO–H CH_3 ≡ CO_2H HO–H CH_3 ≡ CO_2H H_3C H OH

(S)-lactic acid

5.16 There are several ways to approach this problem. Here's one. Assign priorities to the four groups.

4 H; 1 CH_3; 2 CH_2CH_3; OH 1

If the lowest priority group is in a vertical position, determine whether the sequence of the remaining three groups (1→2→3) is clockwise (R) or counterclockwise (S). In this case, it is clockwise and the Fischer projection represents (S)-2-butanol.

5.17

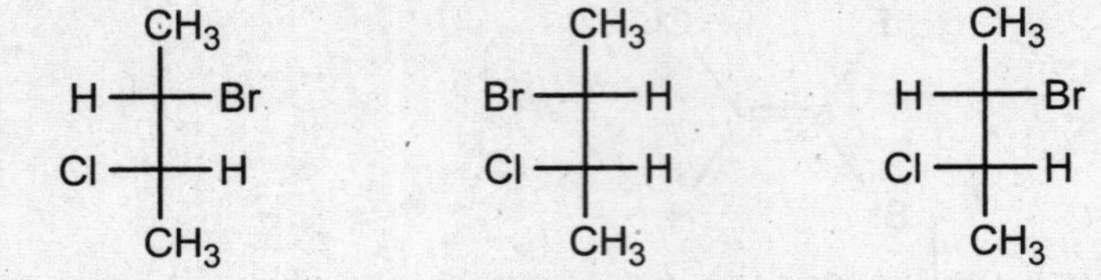

(2*S*,3*S*)-2-bromo-3-chlorobutane

(2*R*,3*S*)-2-bromo-3-chlorobutane

(2*S*,3*R*)-2-bromo-3-chlorobutane

These compounds are enantiomers.

This is the enantiomer of (2*R*,3*R*)-2-bromo-3chlorobutane (Fig. 5.12)

5.18 The (2*R*,3*R*) and (2*S*,3*S*) isomers are a pair of enantiomers. Their specific rotations will be equal in magnitude and opposite in sign.

The (2*R*,3*R*) and (2*S*,3*R*) isomers are a pair of diastereomers. Their specific rotations will be unequal in magnitude and may or may not differ in sign.

5.19 There are *four* different stereogenic centers, marked below with asterisks. There are, therefore, $2^4 = 16$ possible stereoisomers.

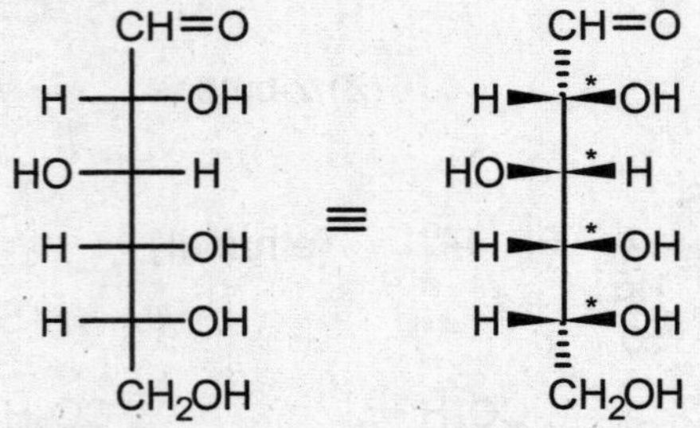

5.20 When the three CH_2 groups are superimposed, the methyl groups of one mirror image are superimposed on the hydrogens of the other mirror image. The mirror images are nonsuperimposable and are therefore enantiomers.

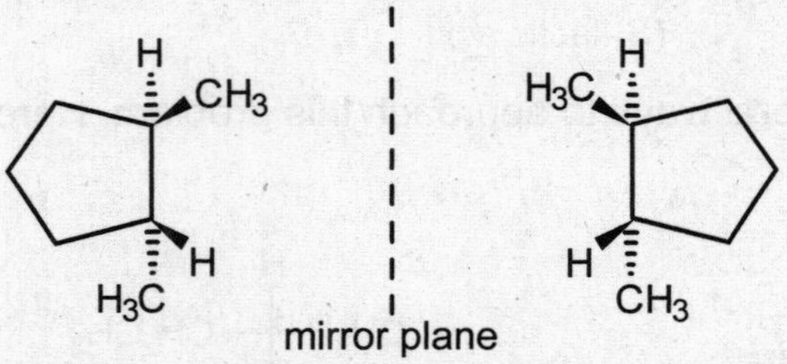

mirror plane

5.21 *cis*-1,2-Dimethylcyclopentane is achiral. It has a mirror plane of symmetry and is a *meso* compound.

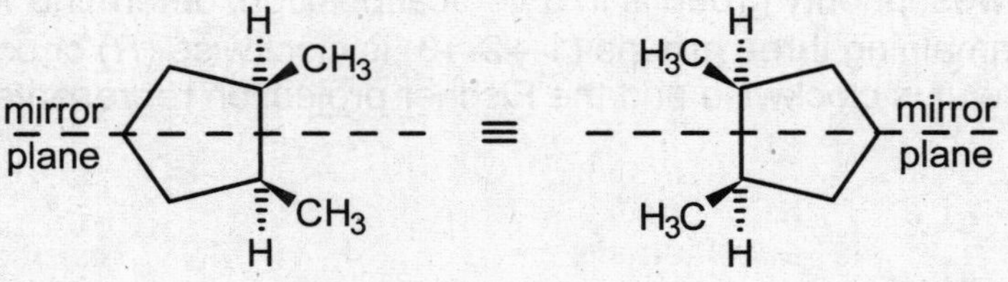

mirror plane

5.22

cis-1,3-dimethylcyclobutane *trans*-1,3-dimethylcyclobutane

There are *no* stereogenic centers. Both molecules have planes of symmetry. The *cis* isomer has two such planes, through opposite corners of the ring. The *trans* isomer has one such plane, through the opposite methyl-bearing corners. Both compounds are optically inactive and achiral. They are not *meso* compounds because there are no chiral centers. To summarize, the two isomers are configurational, achiral and diastereomers.

5.23

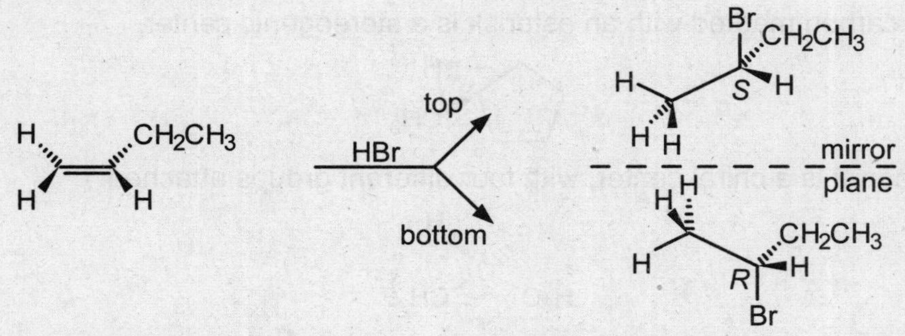

One-step addition of H–Br to the top face of the double bond gives (*S*)-2-bromobutane. Addition of H–Br to the bottom face gives (*R*)-2-bromobutane. Since 1-butene is achiral, the probability of addition to either face of the double bond is equal, and the product will be racemic (an equal mixture of enantiomers).

5.24 If we draw the structure in three dimensions, we see that either hydrogen at C-2 can be replaced with equal probability:

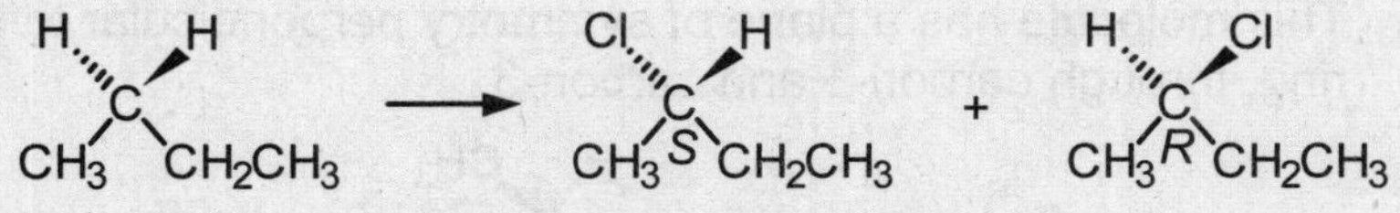

Thus a 50:50 mixture of the two enantiomers is obtained.

5.25 (*S*)-3-chloro-1-butene is the enantiomer of (*R*)-3-chloro-1-butene, and it will react with HBr to give the enantiomers of the products shown in eq. 5.7 [(2*S*, 3*S*)- and (2*R*, 3*S*)-2-bromo-3-chlorobutane], in the same 60:40 ratio. Therefore a racemic mixture of 3-chloro-1-butene will react with HBr to give the following mixture of 2-bromo-3-chlorobutanes:

enantiomers		enantiomers
(2*R*,3*R*) 30%	← diastereomers →	(2*S*,3*R*) 20%
(2*S*,3*S*) 30%		(2*R*,3*S*) 20%

In other words, a 60:40 mixture of the two diastereomeric products will be obtained, each as a racemic mixture.

ADDITIONAL PROBLEMS

5.26 Each of these definitions can be found explicitly or implicitly in the following sections of the text:

a.	5.8	b.	5.2	c.	5.2	d.	5.9
e.	5.6	f.	5.5	g.	5.5	h.	5.1
i.	5.1	j.	5.12				

5.27 a. The molecule has a plane of symmetry and has no stereogenic centers.

Br, Br, C, H_3C, CH_2CH_3

b. The carbon marked with an asterisk is a stereogenic center.

*, H, CH_3

c. Carbon-2 is a chiral center, with four different groups attached.

F, H, C, H_3C * CH_2F

d. There are two carbons (marked with an asterisk) which are stereogenic centers.

H CH_3, H_3C * * CH_3, H_3C H

e. This molecule has a plane of symmetry perpendicular to the four-membered ring, through carbon-1 and carbon-3.

H, CH_3, 1, 2, 3

f. The carbon marked with an asterisk is a stereogenic center, with four different groups attached. Even isotopes of the same element are sufficiently different to lead to optical activity.

$CH_3CH_2-C^*(H)(D)-OH$

5.28 a. $HOCH_2-C^*H(OH)-C^*H(OH)-C(=O)-H$

b. $HO-C^*(CO_2H)(H)-H_2C-C_6H_5$

c. $Br{-}\overset{*}{C}(CH_3)(CF_3){-}H$

d. (cyclopropyl)$-\overset{*}{C}H(OH)-CH_3$

e. (cyclopent-2-enyl*)$-\overset{*}{C}H(OH)-CH_3$

f. H_3C-(cyclohexane)$-CH_3$

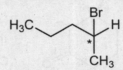

no stereogenic centers

5.29 In each case the *observed* rotation would be doubled, but the specific rotation would remain constant. For example, if *c* is doubled, α will also double, but the fraction α/c in the formula for specific rotation will remain constant.

5.30

$$\left[\alpha\right]_D^{25} = \frac{1.33}{0.5 \times \frac{1}{100}} = 66.5^{\circ} \text{ (water)}$$

5.31 a. enantiomers b. identical

5.32 The following are examples. There may be other possibilities.

a. $CH_3CH_2\overset{*}{C}H(CH_3)CH_2CH_2CH_3$

b. H_3C–CH_2–CH_2–$\overset{*}{C}(Br)(H)CH_3$

c. H_3C–CH_2–$\overset{*}{C}(OH)(H)CH_3$

d. $H_3C-\overset{*}{C}(CH_2CH_2Cl)(Cl)-H$

In each case the stereogenic carbon atom is marked with an asterisk.

5.33 All structures must contain only one double bond and no rings because if the monovalent bromine were replaced by a hydrogen, we would have C_5H_{10}, corresponding to the general formula C_nH_{2n} (a molecule with one double bond or one ring). Since the bromide is described as unsaturated, there can be no ring present.

a. $CH_2{=}CHCH_2CH_2CH_2Br$ or $(CH_3)_2{=}CHCH_2Br$

b. $CH_3CH{=}C(CH_3)CH_2Br$ or $CH_3CH_2CH{=}CHCH_2Br$ or $CH_3CH_2CH_2CH{=}CHBr$

c. $H_2C{=}CH\overset{*}{C}H(Br)CH_2CH_3$ or $H_2C{=}CHCH_2\overset{*}{C}H(Br)CH_3$

d. $CH_3CH{=}CH\overset{*}{C}H(Br)CH_3$

5.34 The rules for priority order are given in Sec. 5.3.

a. $C_6H_5– > CH_3CH_2– > CH_3– > H–$

b. $–CH{=}O > CH_2{=}CH– > CH_3CH_2CH_2– > CH_3CH_2–$

c. $Br– > HS– > CH_3CH_2– > H–$

d. $HS– > BrCH_2– > HOCH_2– > CH_3–$

5.35 a.

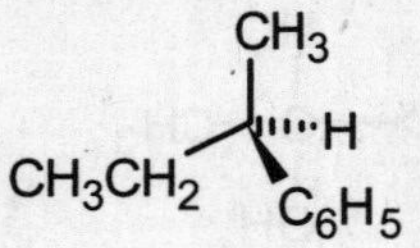

CH_3CH_2, CH_3, H, C_6H_5 and HS, CH_3, H, I

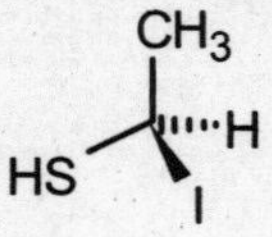

b.

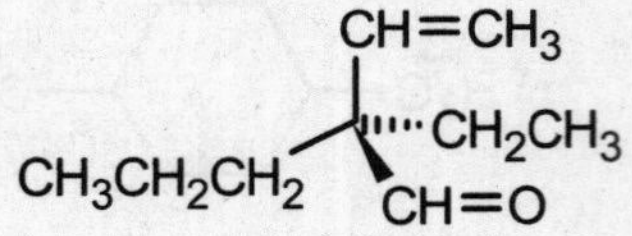

$CH_3CH_2CH_2$, CH=CH_3, CH_2CH_3, CH=O and $HOCH_2$, CH_2Br, CH_3, OH

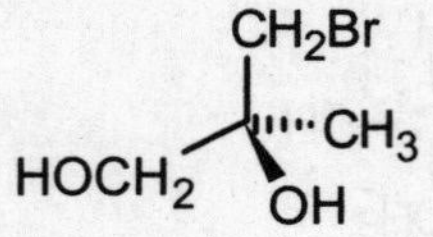

There are many ways to write these structures. They have been drawn here by putting the lowest-priority group receding away from the viewer and the remaining three groups in a clockwise (a) or counterclockwise (b) array, with the highest-priority group at the lower right extending toward the viewer.

5.36 In each case, write down the groups in the proper priority order. Then view the stereogenic center from the face opposite the lowest-priority group and determine whether the remaining array is clockwise (*R*) or counterclockwise (*S*). If you have difficulty, construct and examine molecular models.

a.

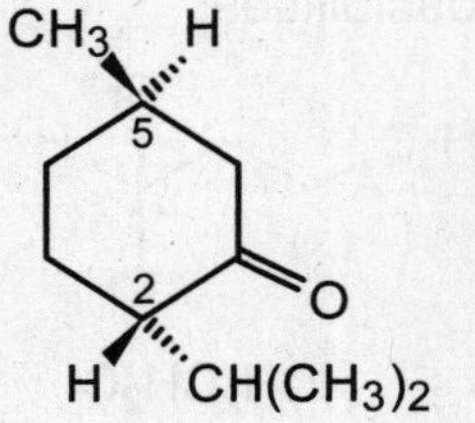

Carbon-5

$-CH_2C(=O) > -CH_2CH_2 > -CH_3 > -H$

therefore *R*

Carbon-2

$-C(=O) > -CH(CH_3)_2 > -CH_2 > -H$

therefore *S*

b.

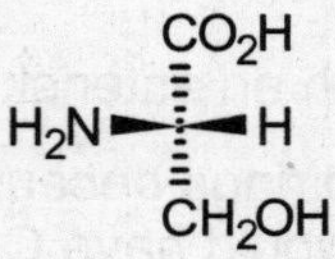

$-NH_2 > -CO_2H > -CH_2OH > -H$

therefore *S*

c.

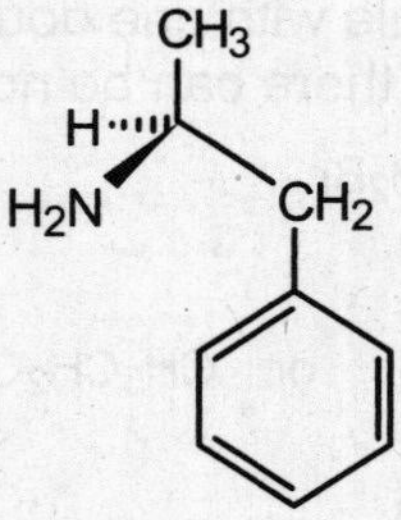

$-NH_2 > -CH_2C_6H_5 > -CH_3 > -H$

therefore *S*

5.37

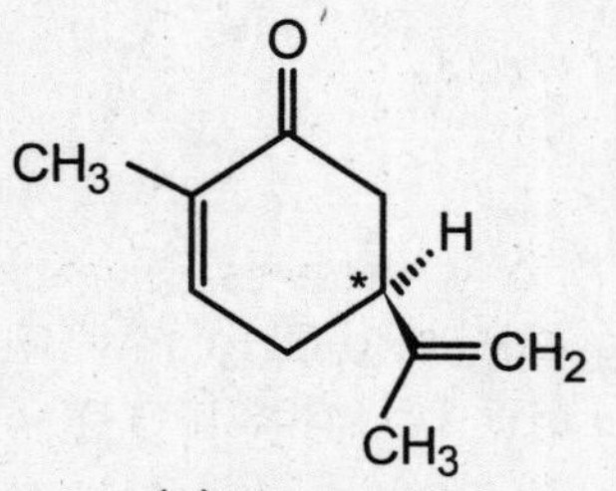

(+)-carvone

The stereogenic center is marked with an asterisk. The priority order of groups at this center is

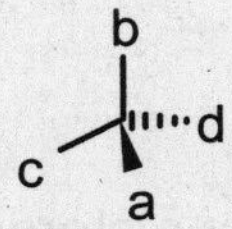

b
c ─C···d
a

and the configuration is *S*. A word about the priority order may be helpful. The

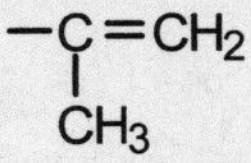

$$-\underset{\displaystyle CH_3}{\underset{|}{C}}=CH_2$$

group has three bonds from the attached carbon atom to the next atoms "out" and is therefore of the highest priority. The remaining groups both begin with $-CH_2$, so we must proceed further. One group is

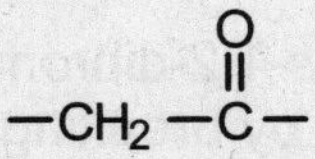

$$-CH_2-\overset{\displaystyle O}{\overset{\|}{C}}-$$

and the other is

$$-CH_2-\underset{\displaystyle H}{\underset{|}{C}}=C$$

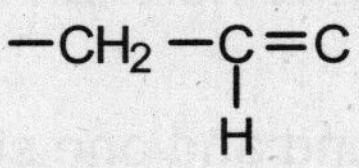

Of these, the group with C=O has the higher priority because oxygen has a higher atomic number than carbon.

5.38 The priority of the four groups is Br > Cl > F > H. The structure of the *R* enantiomer is:

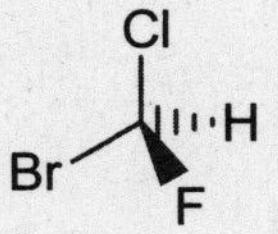

Cl
Br─C···H
F

5.39 a.

3
2 4
1 5
6

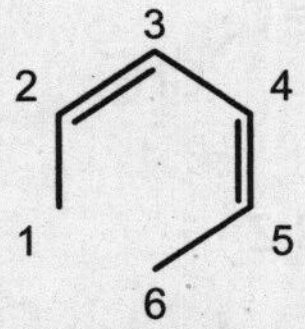

(*Z*,*Z*)-2,4-hexadiene or more precisely, (2*Z*,4*Z*)-2,4-hexadiene

If you have difficulty, draw the full structure:

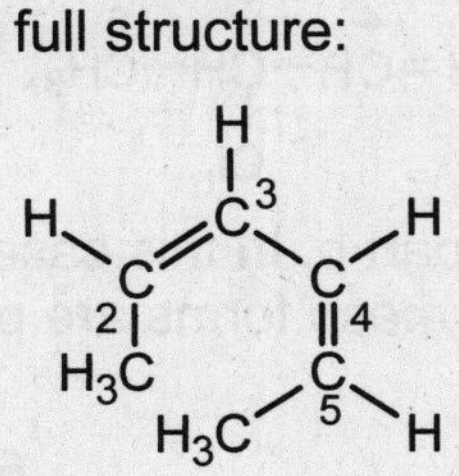

H
H C3 H
C2 C
H3C C4
H3C C5 H

At the double bond between C-2 and C-3, the priority order is CH_3 > H and $CH_3CH=CH-$ > H. The two high-priority groups, CH_3 and $CH_3CH=CH-$, are *Z* or *zusammen*. The same is true at the double bond between C-4 and C-5.

b.

(*E*,*E*)-2,4-hexadiene

c.

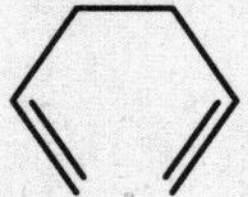

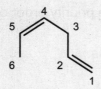

(Z)-1,4-hexadiene; there is no stereochemistry at the double bond joining C-1 and C-2 because both substituents at C-1 are identical ($=CH_2$).

d.

There is no stereochemistry at either double bond. The IUPAC name is 1,5-hexadiene.

5.40 *Trans*-1,2-dibromoethene and *cis*-1,2-dibromoethene are configurational isomers (they can only be interconverted by breaking and remaking a bond, in this case the pi–bond of the alkene), they are both achiral (their mirror images are identical to themselves), and they are diastereomers (although they are stereoisomers, they are not enantiomers).

5.41 The structure has one double bond and one stereogenic center (C-4). Four isomers are possible (R or S at C-4 and E or Z at the double bond).

(*R,E*) (*S,E*)

(*R,Z*) (*S,Z*)

The upper and lower sets form two pairs of enantiomers.

5.42 a.

$$\overset{1}{CH_3}-\underset{Cl}{\overset{2\,*}{CH}}-\overset{3}{CH}=\overset{4}{CH}-\underset{Cl}{\overset{5\,*}{CH}}-\overset{6}{CH_3}$$

Compare with part b. In this case, both stereogenic centers are identical. Therefore, two *meso* forms are possible, and the total number of isomers is reduced to six:

R, *Z*, *R*
R, *Z*, *S* (meso)
R, *E*, *R*
R, *E*, *S* (*meso*)
S, *Z*, *S*
S, *E*, *S*

There are two sets of enantiomers:

R, *Z*, *R* and *S*, *Z*, *S*
R, *E*, *R* and *S*, *E*, *S*

And there are two optically inactive, *meso* forms: *R*, *Z*, *S* and *R*, *E*, *S*. The *R*, *Z*, *R* isomer is shown below.

The other five structures can be derived from this one by interchanging groups. For example, the *R*, *Z*, *S* *meso* form is:

b. $CH_3-\overset{*}{C}H(Cl)-CH=CH-\overset{*}{C}H(F)-CH_3$ (carbons numbered 1–6)

There are two stereogenic centers, marked with asterisks. Each can be either *R* or *S*. Also, the double bond joining C-3 and C-4 can be *E* or *Z*. Thus eight isomers are possible:

R, *Z*, *R*	*S*, *Z*, *R*
R, *Z*, *S*	*S*, *Z*, *S*
R, *E*, *R*	*S*, *E*, *R*
R, *E*, *S*	*S*, *E*, *S*

The first of these is shown below:

The other seven isomers can be drawn by interchanging one or more groups, using this structure as a guide. For example, the *S*, *Z*, *R* isomer is

and so on.

c. $CH_2=CH-CH(CH_3)-CH=CH_2$ (carbons numbered 1–5)

There are no stereogenic centers and no *cis–trans* possibilities at either double bond. Only one structure is possible.

d. $H_2C=CH-CH(CH_3)-CH=CH-CH_2CH_3$ (carbons numbered 1–7)

Carbon-3 is a stereogenic center, and *cis–trans* isomers are possible at the double bond between C-4 and C-5. Therefore, four structures are possible (*R* or *S* at C-3, and *E* or *Z* at the double bond).

CH_3 H $H_2C{=}CH$ CH_2CH_3 C=C H H
(*R,Z*)-3-methyl-1,4-heptadiene

CH_3 $H_2C{=}CH$ H CH_2CH_3 C=C H H
(*S,Z*)-3-methyl-1,4-heptadiene

CH_3 H $H_2C{=}CH$ H C=C H CH_2CH_3
(*R,E*)-3-methyl-1,4-heptadiene

CH_3 $H_2C{=}CH$ H H C=C H CH_2CH_3
(*S,E*)-3-methyl-1,4-heptadiene

5.43 a. The projections are enantiomers. There are several ways to work this problem. One involves assigning and comparing absolute configurations (*R* or *S*) at each stereogenic carbon. First, convert Fischer projection **A** into a wedge-dash drawing that depicts the arrangement of groups at the stereogenic carbon (use Figure 5.11 as a guide). Next, assign a priority order to the four groups ($OH > C_2H_5 > CH_3 > H$) and determine the arrangement of the three top-priority groups (clockwise or counterclockwise). In this case, they are clockwise, so the configuration is *R*.

A

C_2H_5 HO—+—CH_3 H ≡ C_2H_5 HO—CH_3 H ≡ CH_3 *R* C_2H_5 HO H

Manipulate the Fischer projection in part **a** in the same manner. In this case, the configuration is *S*.

part **a**

CH_3 H—+—OH C_2H_5 ≡ CH_3 H—OH C_2H_5 ≡ OH *S* CH_3 H C_2H_5

b. enantiomers
c. enantiomers

5.44

CO_2H H OH HO H CO_2H

This is the *meso* form. As drawn, this conformation has a center of symmetry, the midpoint of the central C–C bond. The plane of symmetry is readily seen if we rotate the "rear" carbon 180°:

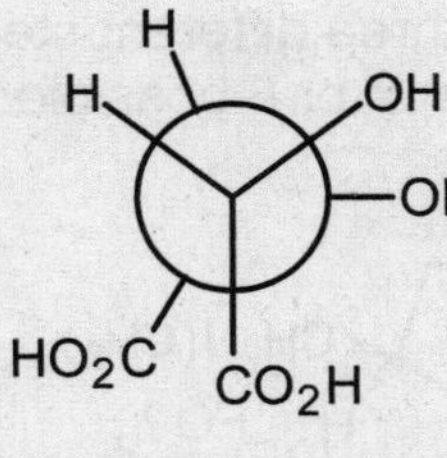

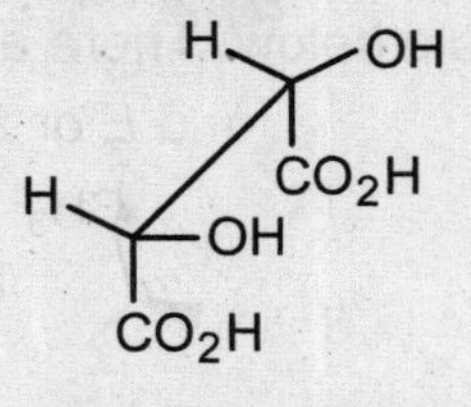

The remaining two structures correspond to the *S*,*S* and *R*,*R* isomers, respectively:

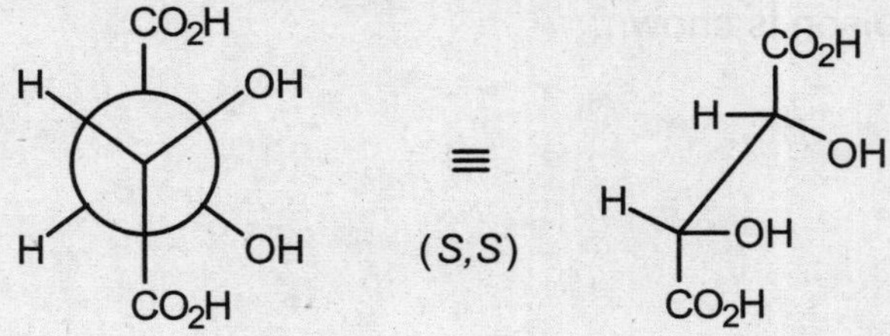

(*S*,*S*)

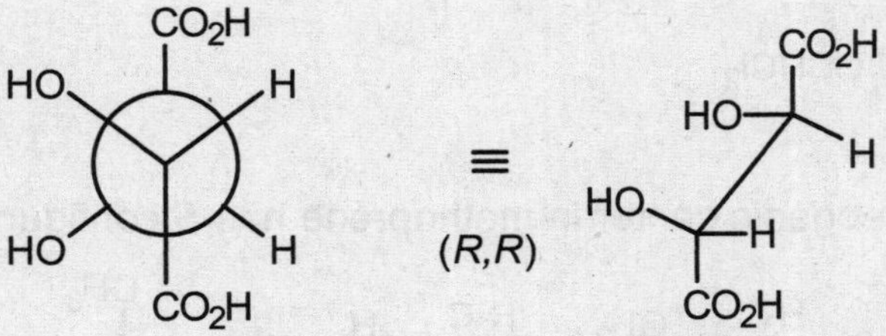

(*R*,*R*)

The priority order at each stereogenic center is OH > CO_2H > $CH(OH)CO_2H$ > H.

5.45 Rotate 180°

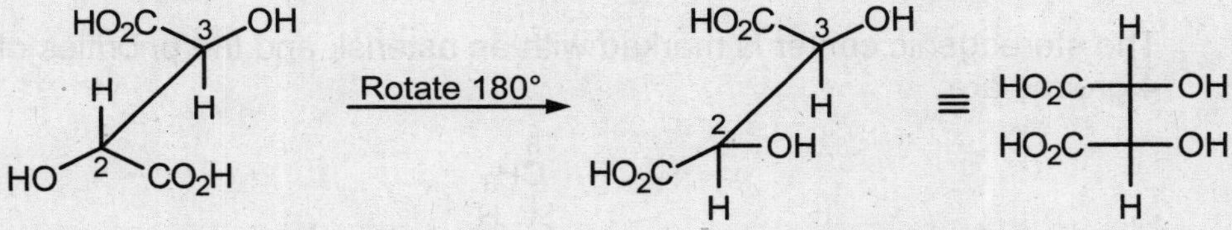

Rotate around the C2-C3 bond to give an eclipsed sawhorse projection. This can be converted into a Fischer projection. Remember that vertical lines project behind the plane of the page and horizontal lines project out of the page. There are several correct Fischer projections in this case. The plane of symmetry, which is easily seen in the projection shown above, shows that this is *meso*-tartaric acid.

5.46 The structures are conformational isomers. Both are achiral. They are diastereomers (stereoisomers but not mirror images).

5.47 For three different stereogenic carbons, $2^3 = 8$. The possibilities are as follows:

R–R–R	*S–R–R*
R–R–S	*S–R–S*
R–S–R	*S–S–R*
R–S–S	*S–S–S*

For four different stereogenic carbons, $2^4 = 16$. The possibilities are as follows:

R–R–R–R	*R–S–R–R*	*S–R–R–R*	*S–S–R–R*
R–R–R–S	*R–S–R–S*	*S–R–R–S*	*S–S–R–S*
R–R–S–R	*R–S–S–R*	*S–R–S–R*	*S–S–S–R*
R–R–S–S	*R–S–S–S*	*S–R–S–S*	*S–S–S–S*

5.48 Muscarine is chiral. The compound has three different stereogenic centers, indicated in the formula below. There are therefore 2^3 or 8 possible stereoisomers.

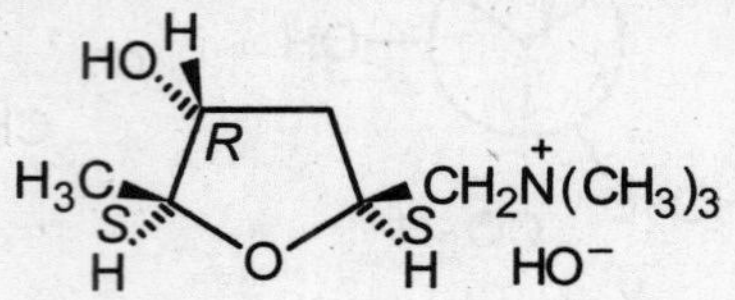

The configurational designation (*R* or *S*) at each stereogenic center in the naturally occurring poison is shown.

5.49

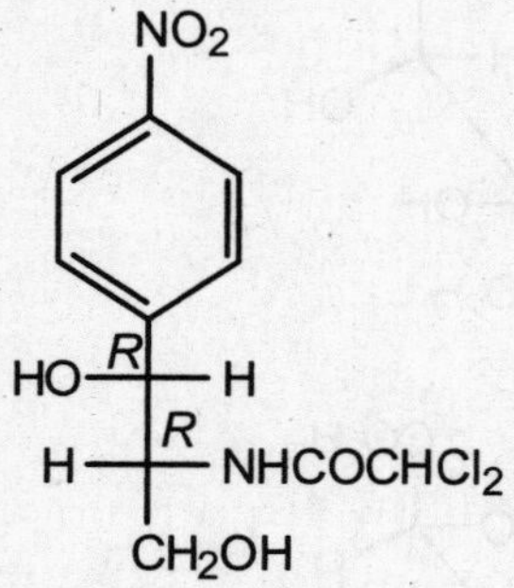

5.50 The one stereogenic center in methoprene has *S* configuration:

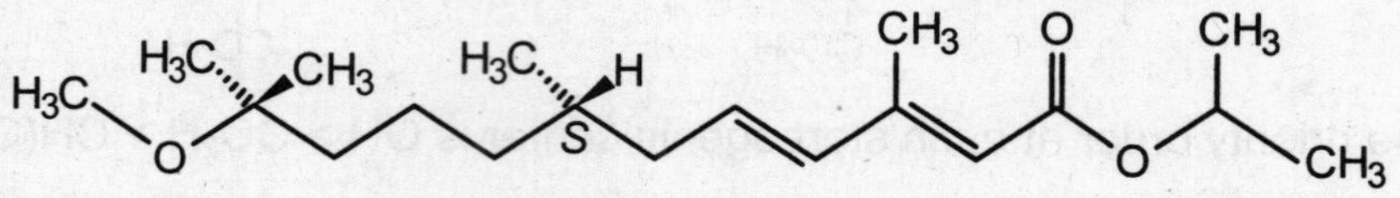

5.51 a. The stereogenic center is marked with an asterisk and the priorities of the 4 groups are:

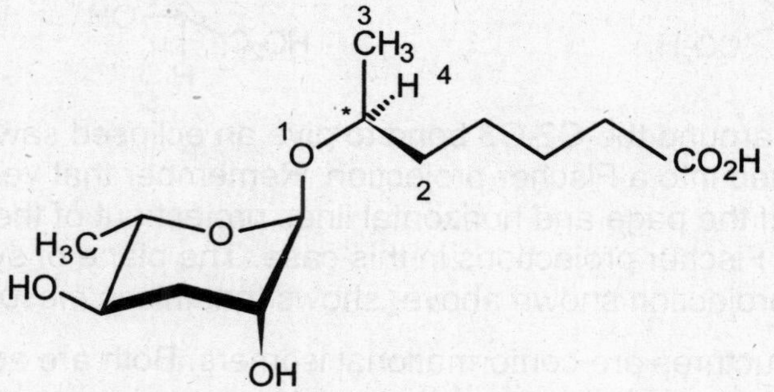

So, the absolute configuration is *S*.

b. The additional stereogenic carbons, and their configuration, in daumone are:

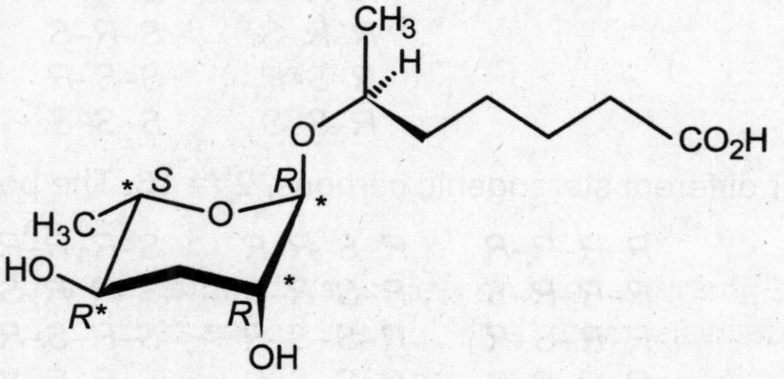

c. The easiest way is to switch two of the groups attached to each of the 5 stereogenic centers, and then confirm by assigning the configuration for each center.

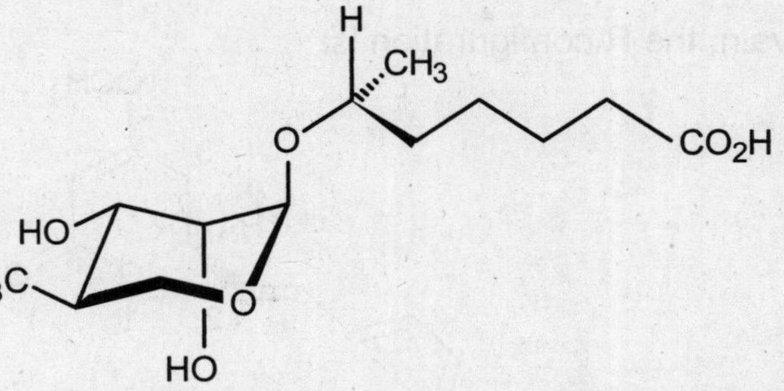

d. The enantiomers will probably not induce hibernation. Enzymes are chiral, and the active site of the enzyme will interact with one enantiomers selectively.

5.52 a. There are two stereogenic centers, one at C-3 and one at C-7.

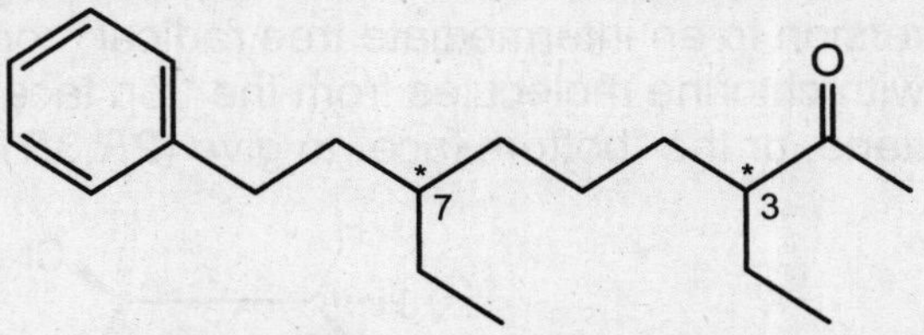

b. Since there are two stereogenic centers, there are $2^2 = 4$ possible stereoisomers (two pairs of enantiomers). These would be the (3*R*,7*R*), (3*S*,7*R*), (3*R*,7*S*) and (3*S*,7*R*) stereoisomers.

5.53 The absolute configuration for the stereogenic center of nerolidol is *S*. (–OH > –CH=CH_2 > –CH_2CH_2CH= > –CH_3)

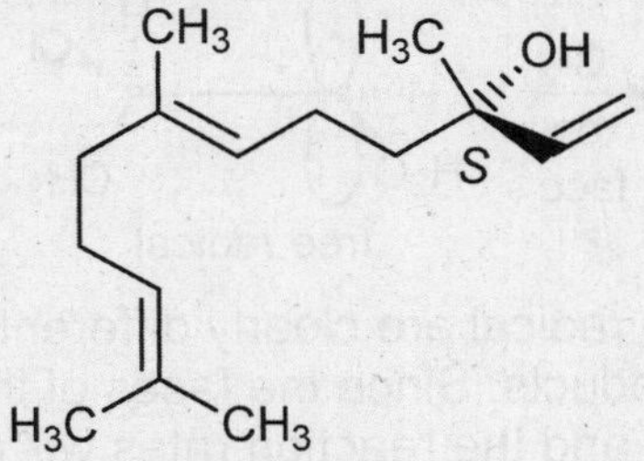

5.54 a. The absolute configuration at C-1 is *R* (OH > CH($NHCH_3$)CH_3 > C_6H_5 > H). The absolute configuration at C-2 is *S* ($NHCH_3$ > CH(OH)(C_6H_5) > CH_3 > H).

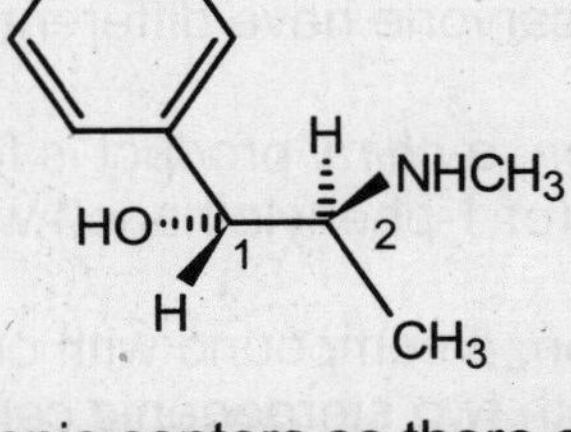

b. Ephedrine has two stereogenic centers so there are $2^2 = 4$ possible stereoisomers (including ephedrine itself).

c. The aromatic units in (–)-ephedrine and (–)-epinephrine are slightly different and (–)-epinephrine lacks the C-2 stereogenic center present in (–)-ephedrine. They both have the same configuration at C-1.

5.55 For kavain, the R configuration is:

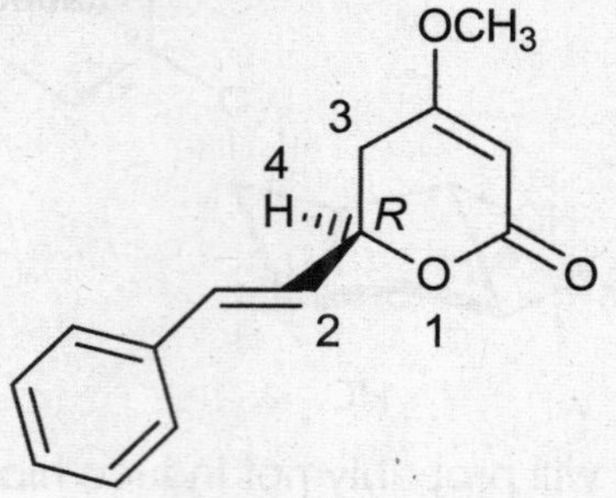

and the group priorities are noted as 1, 2, 3, and 4.

5.56 The reaction of (*R*)-2-chlorobutane with chlorine to give 2,3-dichlorobutane involves initial conversion to an intermediate free radical (consult Sec. 2.13). This free radical can react with chlorine molecules from the "top face" to give *meso*-2,3-dichlorobutane, or the "bottom face" to give (2*R*,3*R*)-dichlorobutane.

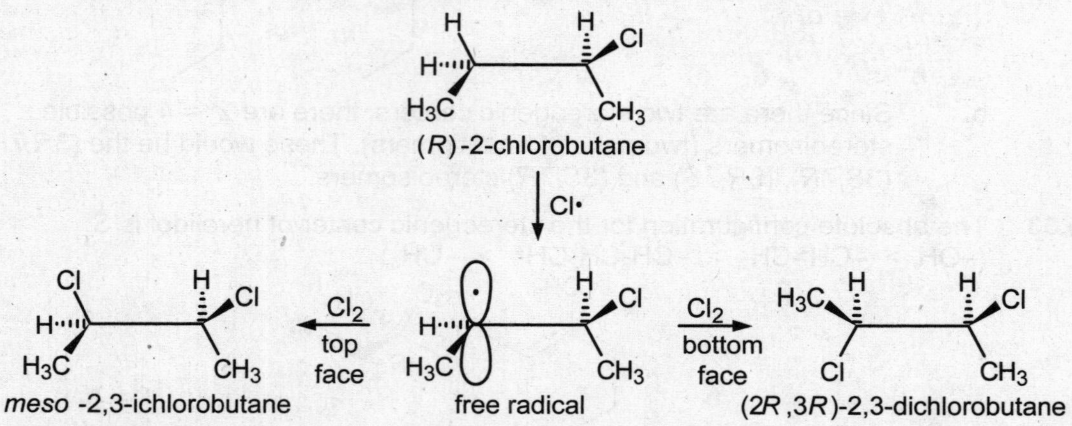

(*R*)-2-chlorobutane

meso -2,3-ichlorobutane free radical (2*R* ,3*R*)-2,3-dichlorobutane

The two faces of the free radical are clearly different because their reactions with chlorine give different products. Since the faces of the radical differ, the steric environments must differ and the reaction rates will differ. This will lead to formation of an unequal ratio of the two diastereomeric products.

5.57 The receptor site for (+)-carvone and (–)-carvone must be chiral such that the complex of the (+) and (–) isomers with the receptor are diastereomers. Therefore (+)-carvone and (–)-carvone have different odors.

5.58 a. In this reaction, a *chiral* product is formed from *achiral* reactants, so both enantiomers (of 1-phenylethanol) will be formed in equal amounts (see Sec. 5.11).

b. In this reaction, a compound with one stereogenic center is converted to a compound with two stereogenic centers. The new stereogenic center can be either *R* or *S* and the latter is a *meso* compound.

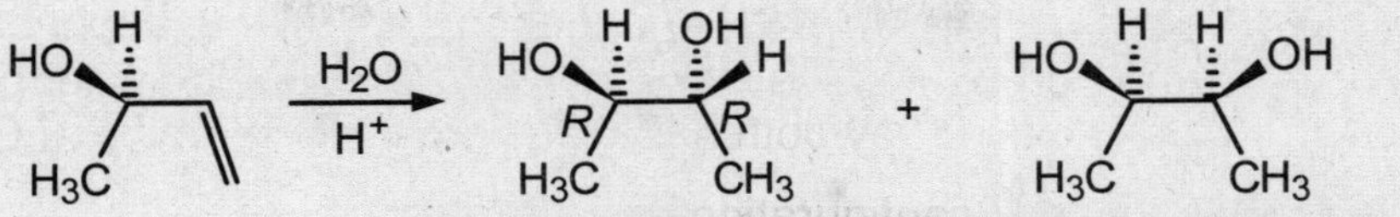

HO H H3C → (H_2O, H^+) HO H OH H R R H3C CH3 (2*R*,3*R*)-2,3-butanediol + HO H H OH H3C CH3 *meso* -2,3-butanediol

5.59 The absolute configuration of the stereogenic center in L-DOPA is *S*.

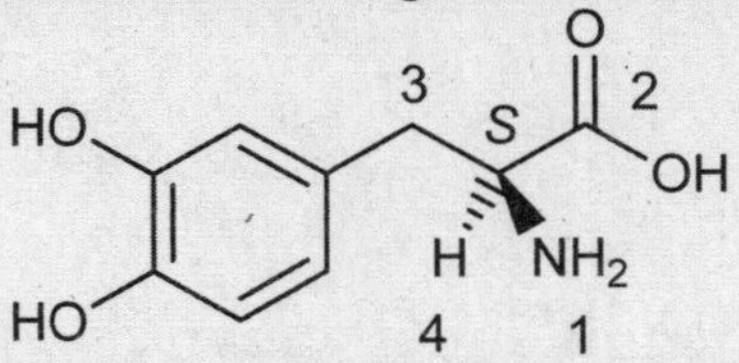

HO HO 3 S 2 O OH H NH2 4 1

As noted above, the NH_2 group has the highest priority, followed by the carboxylic acid, then the alkyl group and finally H.

If (*R*,*R*)-DiPAMP would produce the *S* enantiomer of L-DOPA, then the (*S*,*S*)-DiPAMP ligand would produce the opposite DOPA enantiomer, thus (*R*).

5.60 The (*R*,*R*) form of tartaric acid has a specific rotation of +12° while the (*S*,*S*) isomer has a value of –12°. For a 75% (*R*,*R*) and 25% (*S*,*S*) mixture, the specific rotation of the final solution would be (0.75)(+12) + (0.25)(–12) = 6°.

6

Organic Halogen Compounds; Substitution and Elimination Reactions

Chapter Summary

Alkyl halides react with **nucleophiles**, reagents that can supply an electron pair to form a covalent bond, to give a product in which the nucleophile takes the place of the halogen. Table 6.1 gives fifteen examples of such **nucleophilic substitution reactions**, which can be used to convert alkyl halides to alcohols, ethers, esters, amines, thiols, alkyl cyanides, or acetylenes.

Nucleophilic substitution may occur by two mechanisms. The **S_N2** mechanism is a one-step process. Its rate depends on the concentrations of substrate and nucleophile. If the halogen-bearing carbon is stereogenic, substitution occurs with inversion of configuration. The reaction is fastest for primary halides and slowest for tertiary halides.

The **S_N1 mechanism** is a two-step process. In the first step, the alkyl halide ionizes to a carbocation and a halide ion. In the second, fast step, the carbocation combines with the nucleophile. The overall rate is independent of nucleophile concentration. If the halogen-bearing carbon is stereogenic, substitution occurs with racemization. The reaction is fastest for tertiary halides and slowest for primary halides. The two mechanisms are compared in Table 6.2.

Elimination reactions often compete with substitution. They involve elimination of the halogen and a hydrogen from adjacent carbons to form an alkene. Like substitution, they occur by two main mechanisms. The **E2 mechanism** is a one-step process. The nucleophile acts as a base to remove the adjacent proton. The preferred form of the transition state is planar, with the hydrogen and the **leaving group** in an *anti* conformation. The **E1 mechanism** has the same first step as the S_N1 mechanism. The resulting carbocation then loses a proton from a carbon atom adjacent to the positive carbon to form the alkene.

Several polyhalogen compounds have useful properties. Among them, carbon tetrachloride, chloroform, and methylene chloride are useful solvents. Other important polyhalogen compounds include **Halons** ($CBrClF_2$ and $CBrF_3$) used as fire extinguishers, and **hydrochlorofluorocarbons** (HCFCs), used as refrigerants, blowing agents, aerosol propellants, and solvents. **Teflon** is a polymer of tetrafluoroethene. It is used in nonstick coatings, Gore-Tex fabrics, insulators, and many other things. Certain **perfluorochemicals** dissolve high percentages of oxygen and can be used as artificial blood. Many halogen-containing compounds are important pesticides.

Reaction Summary

Nucleophilic Substitution

$$Nu\!: + R{-}X \longrightarrow R{-}Nu^{+} + X^{-}$$

$$Nu\!:^{-} + R{-}X \longrightarrow R{-}Nu + X^{-}$$

(See Table 6.1 for examples)

Elimination

$$H{-}\overset{|}{\underset{|}{C}}{-}\overset{|}{\underset{|}{C}}{-}X \xrightarrow[\text{base}]{B:} \;>C{=}C<\; + BH^{+} + X^{-}$$

Preparation of Teflon

$$F_2C{=}CF_2 \xrightarrow{\text{peroxide}} {-}\!(CF_2CF_2)_n\!{-}$$

Mechanism Summary

S_N2 (Bimolecular Nucleophilic Substitution)

$$\underset{\text{(nucleophile)}}{Nu\!:} + \underset{\text{(subtrate)}}{C{-}L} \longrightarrow \left[\overset{\delta+}{Nu}\cdots C \cdots \overset{\delta-}{L} \right] \longrightarrow \overset{+}{Nu}{-}C + \underset{\text{(leaving group)}}{^{-}\!:\!L}$$

S_N1 (Unimolecular Nucleophilic Substitution)

$$C{-}L \underset{}{\overset{\text{slow}}{\rightleftharpoons}} \underset{\text{(carbocation)}}{C^{+}} + {^{-}\!:\!L} \xrightarrow[Nu:^{-}]{\text{fast}} C{-}Nu \text{ or } Nu{-}C$$

E2 (Bimolecular Elimination)

$$B\!: + H{-}C{-}C{-}L \longrightarrow BH^{+} + C{=}C + {^{-}\!:\!L}$$

E1 (Unimolecular Elimination)

$$-\overset{H}{\underset{|}{\overset{|}{C}}}-\underset{|}{\overset{|}{C}}-L \longrightarrow -\overset{H}{\underset{|}{\overset{|}{C}}}-\underset{|}{\overset{|}{C}}^{+} + L:^{-} \longrightarrow \;\rangle C=C\langle\; + H^{+}$$

Learning Objectives

1. Know the meaning of: nucleophilic substitution reaction, nucleophile, substrate, leaving group.
2. Be familiar with the examples of nucleophilic substitution reactions listed in Table 6.1.
3. Know the meaning of: S_N2 mechanism, inversion of configuration, S_N1 mechanism, racemization, rate-determining step, E2 and E1 mechanisms.
4. Know the formulas of carbon tetrachloride, chloroform, methylene chloride, Freons, Halons, Teflon.
5. Given the name of an alkyl halide or a polyhalogen compound, write its structural formula.
6. Given the structural formula of an alkyl halide, write a correct name for it.
7. Write the equation for the reaction of an alkyl halide with any of the nucleophiles listed in Table 6.1. Recognize the class of organic compound to which the product belongs.
8. Given the structure of an alkyl halide, predict whether it is most likely to react with nucleophiles by an S_N1 or an S_N2 mechanism.
9. Given the structure of an alkyl halide and a nucleophile, write the equations that illustrate the formation of both the substitution and elimination products and be able to predict which path is likely to be favored.
10. Know the stereochemical outcome of S_N1 and S_N2 substitutions and E1 and E2 eliminations.
11. Given an alkyl halide with a particular stereochemistry, a nucleophile, and reaction conditions, predict the stereochemistry of the product of nucleophilic substitution.
12. Combine nucleophilic substitutions with previously studied reactions to devise a multi-step synthesis of a given product.

ANSWERS TO PROBLEMS

Problems Within the Chapter

6.1 a. $NaOH + CH_3CH_2CH_2Br \longrightarrow CH_3CH_2CH_2OH + Na^+Br^-$
(item 1, Table 6.1)

b. $(CH_3CH_2)_3N + CH_3CH_2Br \longrightarrow (CH_3CH_2)_4N^+ \ Br^-$
(item 9, Table 6.1)

c. $NaSH + C_6H_5\text{-}CH_2Br \longrightarrow C_6H_5\text{-}CH_2SH + Na^+ Br^-$

(item 10, Table 6.1)

6.2 a. $(CH_3CH_2)_2NH + CH_3CH_2Br \longrightarrow (CH_3CH_2)_3\overset{+}{N}H + Br^-$
nucleophile substrate leaving group
(item 8, Table 6.1) This reaction is followed by the acid-base equilibrium:

$$(CH_3CH_2)_2NH + (CH_3CH_2)_3\overset{+}{N}H \rightleftharpoons (CH_3CH_2)_2\overset{+}{N}H_2 + (CH_3CH_2)_3N$$

b. $HO^- + CH_3CH_2CH_2CH_2Br \longrightarrow CH_3CH_2CH_2CH_2OH + Br^-$
nucleophile substrate leaving group

(item 1, Table 6.1)

c. $^-C{\equiv}N + (CH_3)_2CHCH_2Br \longrightarrow (CH_3)_2CHCH_2CN + Br^-$
nucleophile substrate leaving group
(item 14, Table 6.1)

d.
$CH_3CH_2CH_2CH_2O^- + CH_3Br \longrightarrow CH_3CH_2CH_2CH_2OCH_3 + Br^-$
nucleophile substrate leaving group

or

$CH_3O^- + CH_3CH_2CH_2CH_2Br \longrightarrow CH_3OCH_2CH_2CH_2CH_3 + Br^-$
nucleophile substrate leaving group
(item 2, Table 6.1) Iodides or chlorides could also be used as substrates.

e. $(CH_3CH_2)_2S\colon + CH_3CH_2Br \longrightarrow (CH_3CH_2)_3\overset{+}{S}\colon + Br^-$
nucleophile substrate leaving group
(item 12, Table 6.1)

f. $I^- + CH_2{=}CHCH_2Br \longrightarrow CH_2{=}CHCH_2I + Br^-$
nucleophile substrate leaving group
(item 13, Table 6.1)

6.3 Use Figure 6.1 as a guide:

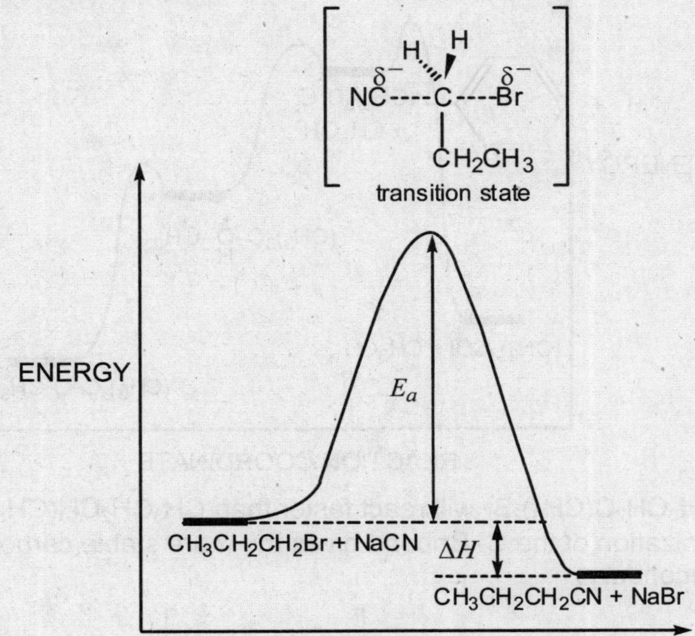

6.4 a. CH_3-cyclohexyl-Br $\xrightarrow[\text{Note: } trans \rightarrow cis]{^{-}CN}$ H_3C-cyclohexyl-CN + Br^-

b. CH_3CH_2(H)(CH_3)C–Br $\xrightarrow[\text{Note: inversion}]{^{-}CN}$ NC–C(CH_3)(H)CH_2CH_3 + Br^-

c. H(CH_3)(CH_3CH_2)C–Br $\xrightarrow[\text{Note: inversion}]{Na^{+}\ ^{-}SH}$ HS–C(CH_3)(H)CH_2CH_3 + Br^-

6.5 $CH_3CH_2CH_2CH_2Br$ > $(CH_3)_2CHCH_2Br$ > $CH_3CH_2CH(CH_3)Br$
The more crowded the carbon where displacement occurs, the slower the reaction rate.

6.6 The reaction takes place via an S_N1 mechanism because the substrate is a tertiary halide. The expected product is *tert*-butyl methyl ether. The reaction energy diagram resembles that for the S_N1 reaction shown in Figure 6.2:

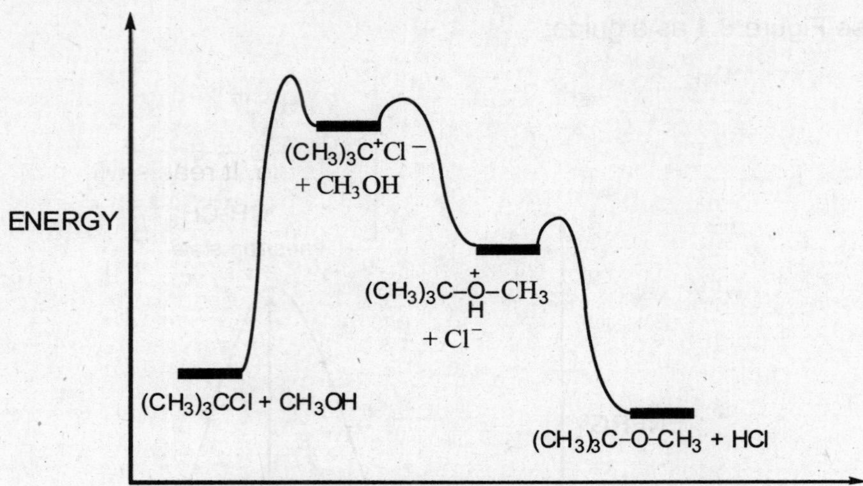

6.7 a. $CH_3CH_2C(CH_3)_2Br$ will react faster than $CH_3CH_2CH(CH_3)Br$ because ionization of the C–Br bond gives the more stable carbocation (tertiary versus secondary).

$$CH_3CH_2-C(CH_3)_2-Br \longrightarrow \left[CH_3CH_2-\overset{+}{C}(CH_3)_2 \quad Br^- \right] \xrightarrow{CH_3OH} CH_3CH_2-C(CH_3)_2-OCH_3 + HBr$$

b. Allyl bromide, $CH_2{=}CHCH_2Br$, will react faster than $CH_3CH_2CH_2Br$ because ionization of the C–Br bond gives the more stable carbocation (allylic versus primary).

$$H_2C{=}CH-CH_2-Br \longrightarrow \left[H_2C{=}CH-\overset{+}{C}H_2 \longleftrightarrow \overset{+}{C}H_2CH{=}CH_2 \quad Br^- \right] \xrightarrow{CH_3OH} H_2C{=}CH-CH_2-OCH_3 + HBr$$

allylic carbocation

6.8 a. S_N2. The substrate is a secondary halide and may react by either S_N2 or S_N1. The nucleophile HS^- is a strong nucleophile, favoring S_N2.

b. S_N1. The substrate is a secondary halide and may react by either mechanism. The nucleophile (CH_3OH) is relatively weak and also polar, favoring the ionization mechanism.

6.9 Two products are possible. Removal of the hydrogen from carbon-1 gives 2-methyl-1-pentene. Removal of the hydrogen from carbon-3 gives 2-methyl-2-pentene.

$$Cl-C(\overset{1}{C}H_3)(CH_3)-\overset{3}{C}H_2CH_2CH_3 \longrightarrow H_2C{=}C(CH_3)-CH_2CH_2CH_3 + (H_3C)_2C{=}C(H)CH_2CH_3$$

6.10

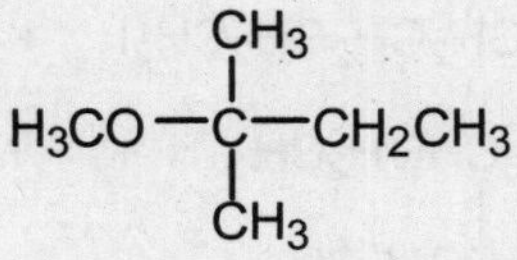

$$H_3CO-\overset{\displaystyle CH_3}{\underset{\displaystyle CH_3}{\overset{|}{\underset{|}{C}}}}-CH_2CH_3$$

2-Bromo-2-methylbutane is a tertiary alkyl halide. It reacts with methanol, a weak nucleophile and polar solvent, to give an ether by an S_N1 mechanism.

ADDITIONAL PROBLEMS

6.11 a. Structures of alkyl halides can be generated by taking a hydrocarbon and replacing one of the hydrogens with a halogen. Thus, taking butane (C_4H_{10}) and replacing one of the hydrogens with a chlorine will generate an alkyl halide with the molecular formula C_4H_9Cl. Replacement of a primary hydrogen gives a primary halide.

$CH_3CH_2CH_2CH_2Cl$ 1-chlorobutane

b. Start with a hydrocarbon with the molecular formula C_5H_{12} that also has a tertiary hydrogen.

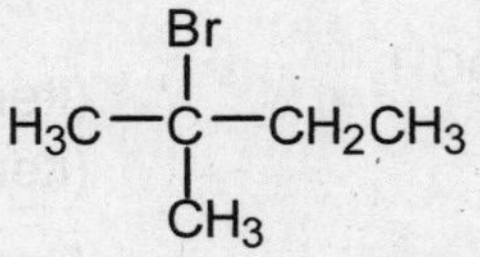

$$H_3C-\overset{\displaystyle Br}{\underset{\displaystyle CH_3}{\overset{|}{\underset{|}{C}}}}-CH_2CH_3$$ 2-bromo-2-methylbutane

c. In this case, we must start with a hydrocarbon with the molecular formula C_6H_{12}. This formula indicates that the structure must also contain a double bond or a ring. Three of the many possibilities are shown below.

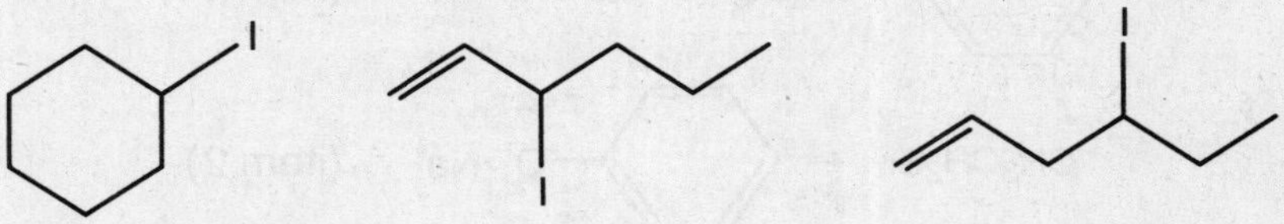

I I I

iodocyclohexane 3-iodo-1-hexene 4-iodo-1-hexene

6.12 Each of these reactions involves displacement of a halogen by a nucleophile. Review Sec. 6.1 and Table 6.1.

a.

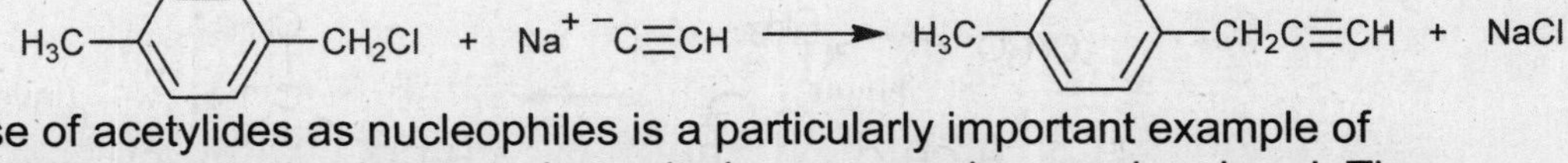

$$H_3C-C_6H_4-CH_2Cl + Na^{+}\,{}^{-}C\equiv CH \longrightarrow H_3C-C_6H_4-CH_2C\equiv CH + NaCl$$

The use of acetylides as nucleophiles is a particularly important example of nucleophilic substitution because it results in a new carbon–carbon bond. Thus, larger organic molecules can be assembled from smaller ones using this method. The same is true for cyanide ion as a nucleophile (part b).

b.

$$CH_3CH_2CH_2Br + Na^{+}\,{}^{-}C\equiv N \longrightarrow CH_3CH_2CH_2-C\equiv N + NaBr$$

c.

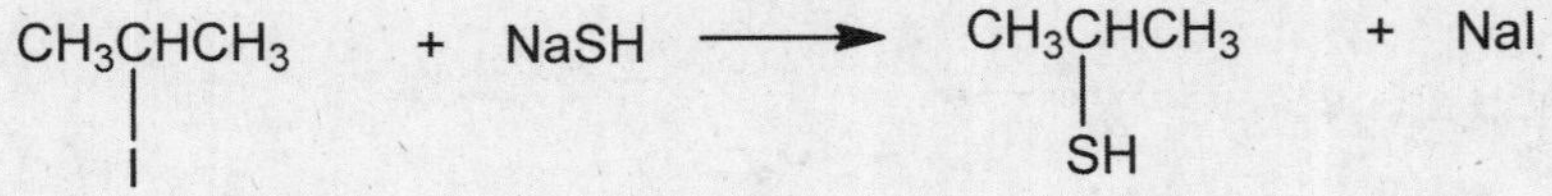

$$CH_3\underset{\displaystyle I}{\underset{|}{C}}HCH_3 + NaSH \longrightarrow CH_3\underset{\displaystyle SH}{\underset{|}{C}}HCH_3 + NaI$$

d. $CH_3CH_2CH_2CH_2CH_2Br$ + NaI ⟶ $CH_3CH_2CH_2CH_2CH_2I$ + NaBr

e. $CH_3CHCH_2CH_3$ (with Cl on C-2) + $NaOCH_2CH_3$ ⟶ $CH_3CHCH_2CH_3$ (with OCH_2CH_3 on C-2) + NaCl

f. $(CH_3)_3CBr$ + CH_3CH_2OH ⟶ $(CH_3)_3COCH_2CH_3$ + HBr

The mechanism here is S_N1 (most of the other reactions in this problem occur by S_N2 mechanism).

g. $BrCH_2CH_2CH_2CH_2CH_2CH_2Br$ + 2 NaCN ⟶ $NC\text{-}CH_2CH_2CH_2CH_2CH_2CH_2\text{-}CN$ + 2 NaBr

Displacement occurs at both possible positions.

h. $H_2C{=}CHCH_2Cl$ + $2NH_3$ ⟶ $H_2C{=}CHCH_2NH_2$ + NH_4Cl

i. 1-chloro-1-methylcyclohexane (CH_3, Cl) + H_2O ⟶ 1-methylcyclohexanol (CH_3, OH) + HCl

The starting halide is tertiary, and the mechanism is S_N1.

6.13 Use the equations in Table 6.1 as a guide.

a. $CH_3CH_2CH_2CH_2Br$ + $NaOCH_3$ (item 2)

b. $CH_3CH_2CH_2Br$ + $HC{\equiv}C^- \, Na^+$ (item 15)

c. $CH_3CH_2CH_2Br$ + NH_3 (item 6)

d. $CH_3CH_2CH_2Br$ + $CH_3CH_2CH_2SNa$ (item 11)

e. C_6H_5–CH_2Br + $Na^+ \, ^-C{\equiv}N$ (item 14)

f. CH_3CH_2Br + C_6H_5–$O^- \, Na^+$ (item 2)

6.14 The configuration inverts if the reaction occurs by an S_N2 mechanism, but if the S_N1 mechanism prevails, considerable racemization occurs.

a. The nucleophile is methoxide ion, CH_3O^-. The alkyl halide is secondary, and the mechanism is S_N2.

CH_3O^- + (R)-2-bromobutane (CH_3, H, H_3CH_2C, Br on C) ⟶ (S)-2-methoxybutane (CH_3, H, H_3CO, CH_2CH_3 on C) (inversion)

(*R*)-2-bromobutane (*S*)-2-methoxybutane

b. The alkyl halide is tertiary and the nucleophile is methanol (a weaker nucleophile than methoxide ion). The mechanism is S_N1, and the product is a mixture of *R* and *S* isomers.

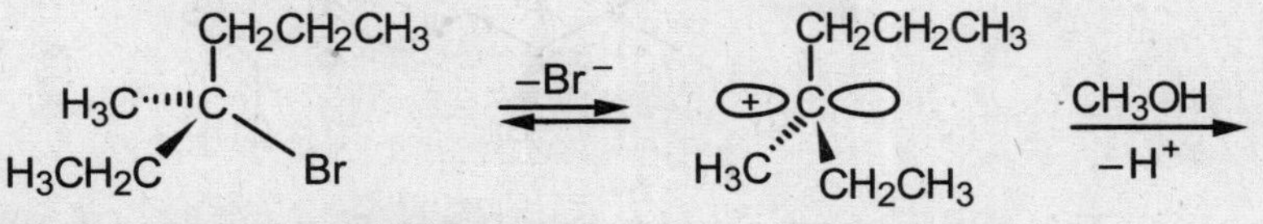

(*S*)-3-bromo-3-methylhexane

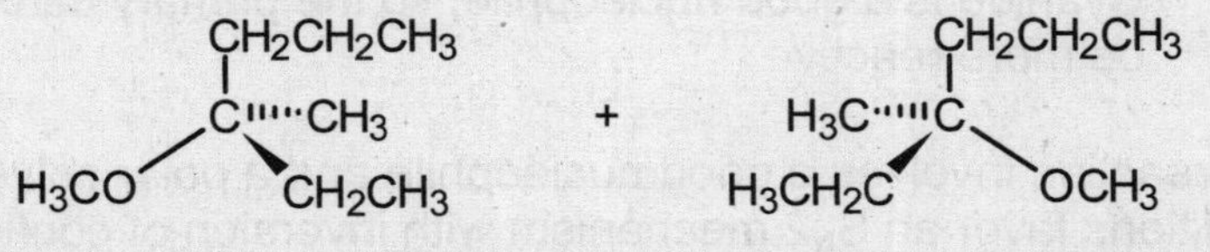

(*R*)-3-methoxy-3-methylhexane (*S*)-3-methoxy-3-methylhexane

c. The alkyl halide is secondary, and the HS^- ion is a strong nucleophile. The mechanism is S_N2.

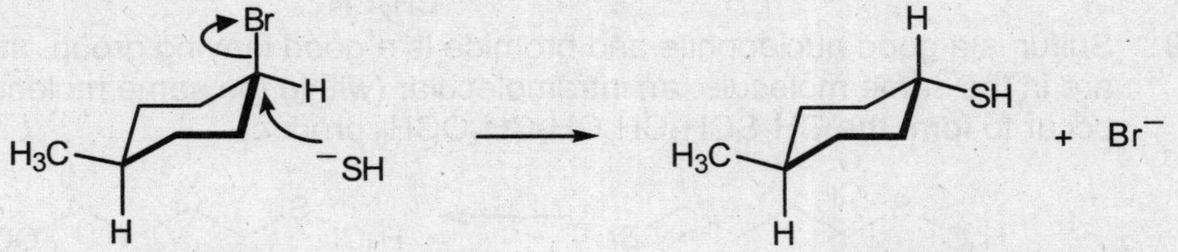

(*cis*) (*trans*)

6.15 The group priorities are $-F > -OCHF_2 > -CF_3 > -H$.

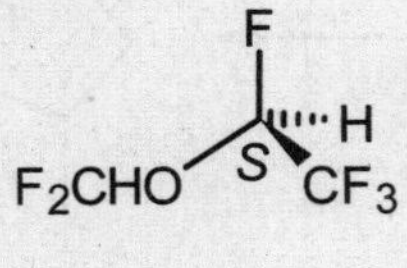

6.16 An S_N2 displacement can occur. Since the leaving group and the nucleophile are identical (iodide ion), there is no change in the gross structure of the product. However, the configuration inverts every time a displacement occurs.

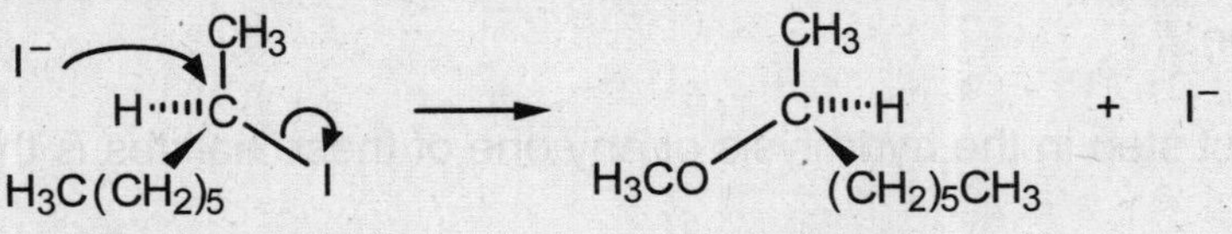

(*R*)-2-iodobutane (*S*)-2-iodobutane

Since the enantiomer is produced, the optical rotation of the solution decreases. As the concentration of the *S* enantiomer builds up, it too reacts with iodide ion to form some *R* isomer. Eventually, an equilibrium (50:50) or racemic mixture is formed, and the solution is optically inactive.

6.17 a. For halomon, the two stereogenic centers are:

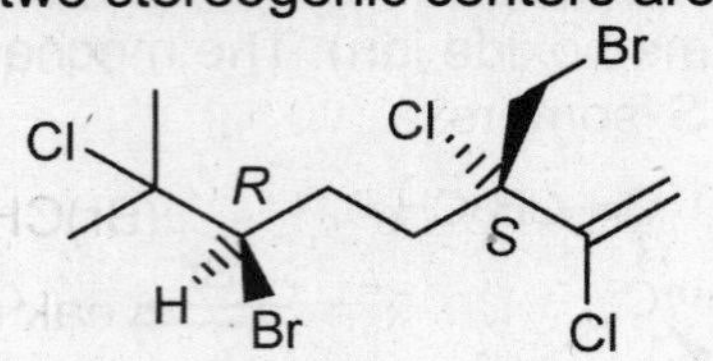

b. The two carbon centers bonded to Br are secondary (left) and primary (right). Cyanide is a good nucleophile, so the primary carbon (right) bearing Br would be more reactive

6.18 The reaction involves a good nucleophile and a polar solvent (acetone). These conditions favor an S_N2 mechanism with inversion of configuration.

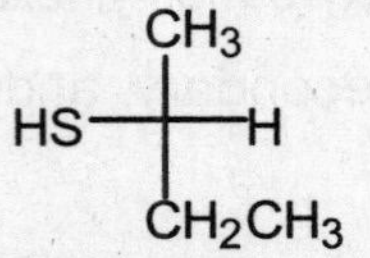

6.19 Sulfur is a good nucleophile and bromide is a good leaving group. If the two units are in the same molecule, an intramolecular (within the same molecule) reaction can occur to form the $CH_3SCH_2CH_2CH_2CH_2OCH_3$ product.

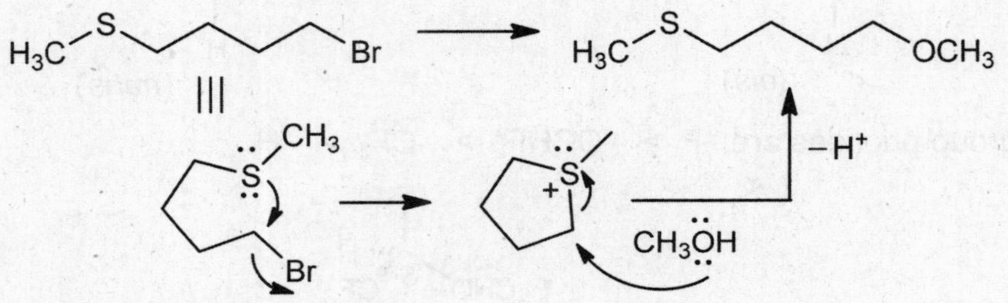

For hexyl bromide, the product would be $CH_3CH_2CH_2CH_2CH_2CH_2–OCH_3$.

Since the sulfur nucleophile is internal to the molecule, the reaction is faster as the two ends can find each other more often, leading to a more efficient, hence faster, reaction.

6.20 The first step in the hydrolysis of any one of these halides is the ionization to a *t*-butyl cation:

$$(CH_3)_3C-X \longrightarrow (CH_3)_3C^+ + X^-$$

(X = Cl, Br, or I)

The product-determining step involves the partitioning of this intermediate between two paths: one is the reaction with water and the other is loss of a proton:

$$(CH_3)_3C-OH \xleftarrow[-H^+]{H_2O} (CH_3)_3C^+ \xrightarrow{-H^+} H_2C=C(CH_3)_2$$

Since the halide ion is, to a first approximation, not involved in these steps, this partition occurs in the same ratio regardless of which alkyl halide is being hydrolyzed. This result provides experimental support for the S_N1 mechanism.

6.21 a. Sodium cyanide is a strong, anionic nucleophile. Thus the mechanism is S_N2 and the reactivity order of the halides is primary > secondary > tertiary. Therefore,

$$(CH_3)_2CHCH_2Br > CH_3CH(Br)CH_2CH_3 >> (CH_3)_3CBr$$

b. With 50% aqueous acetone, there is a weak nucleophile (H_2O) and a highly polar reaction medium favoring ionization, or the S_N1 mechanism. In this mechanism, the reactivity order of alkyl halides is tertiary > secondary > primary. Therefore,

$$(CH_3)_3CBr >> CH_3CH(Br)CH_2CH_3 >> (CH_3)_2CHCH_2Br$$

6.22 a. The halide is tertiary, and the nucleophile is a relatively weak base. Hence the predominant mechanism is S_N1:

1-chloro-1-methylcyclohexane (H_3C, Cl) + CH_3CH_2OH → 1-ethoxy-1-methylcyclohexane (H_3C, OCH_2CH_3) + HCl

Some E1 reaction may occur in competition with S_N1, giving mainly the product with the double bond in the ring:

1-chloro-1-methylcyclohexane (H_3C, Cl) + CH_3CH_2OH —E1→ 1-methylcyclohexene (CH_3) + methylenecyclohexane (CH_2) + HCl

major minor

However, the main product will be the ether (S_N1).

b. The nucleophile in this case is stronger, but the S_N2 process is not possible because the alkyl halide is tertiary. This nucleophile is also a strong base. Therefore, an E2 reaction will be preferred.

1-chloro-1-methylcyclohexane (H_3C, Cl, ring H) + $CH_3CH_2O^-$ —E2→ 1-methylcyclohexene (CH_3) + CH_3CH_2OH + Cl^-

or

1-chloro-1-methylcyclohexane (Cl, CH_2—H) + $CH_3CH_2O^-$ —E2→ methylenecyclohexane (CH_2) + CH_3CH_2OH + Cl^-

The predominant product is 1-methylcyclohexene, the more stable of the two possible alkenes.

6.23

$$(CH_3)_2C(Br)CH_2CH_2CH_2CH_3 \xrightleftharpoons{\text{first step}} (CH_3)_2\overset{+}{C}CH_2CH_2CH_2CH_3 + Br^-$$

$$(CH_3)_2\overset{+}{C}CH_2CH_2CH_2CH_3 \xrightarrow[\text{step}]{\text{second}} H_2C{=}C(CH_3)CH_2CH_2CH_2CH_3 + H^+$$

or

$$(CH_3)_2C{=}CHCH_2CH_2CH_3 + H^+$$

In the second step, the proton may be lost from either of the methyl carbons *or* from the methylene carbon, giving the two alkenes shown.

6.24 The first reaction involves a strong nucleophile (CH_3O^-), and the S_N2 mechanism is favored. Therefore, only one product is obtained.

$$CH_3O^- + H_2C{=}CH{-}\underset{Br}{CH}{-}CH_3 \longrightarrow H_2C{=}CH{-}\overset{OCH_3}{CH}{-}CH_3 + Br^-$$

The second reaction involves a weak nucleophile (CH_3OH) that is also a fairly polar solvent, favoring the S_N1 mechanism:

$$H_2C{=}CH{-}\underset{Br}{CH}{-}CH_3 \rightleftharpoons H_2C{=}CH{-}\overset{+}{C}H{-}CH_3 + Br^-$$

The carbocation is a resonance hybrid:

$$H_2C{=}CH{-}\overset{+}{C}H{-}CH_3 \longleftrightarrow H_2\overset{+}{C}{-}CH{=}CH{-}CH_3$$

It can react with methanol at either positively charged carbon, giving the two observed products.

6.25

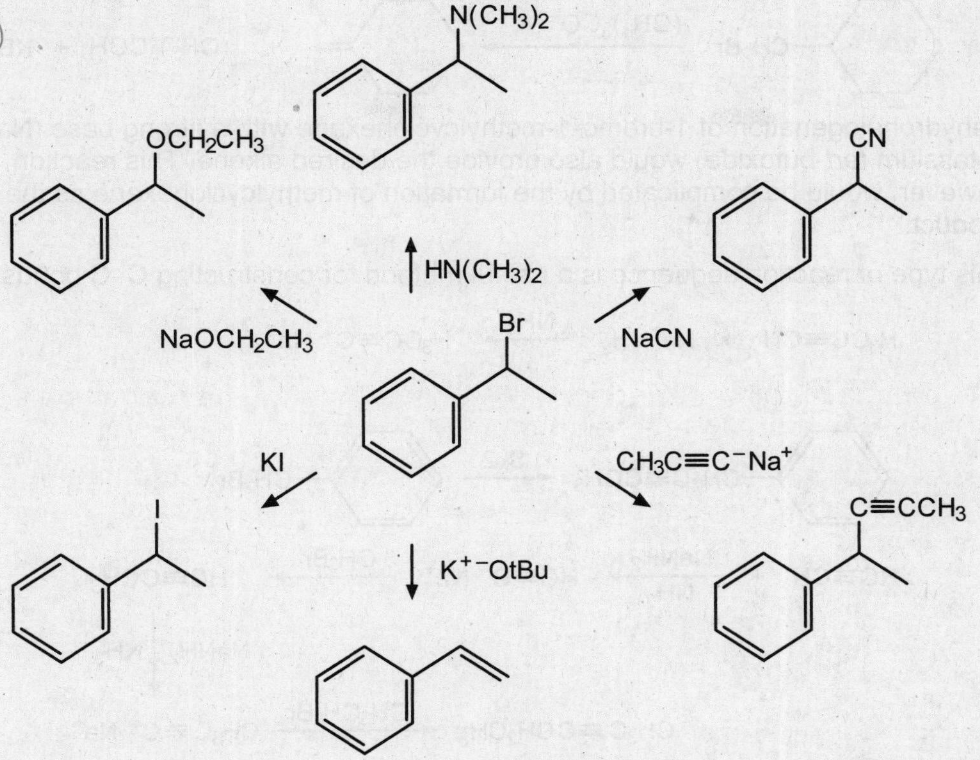

6.26 a.

$$CH_3CH_2CH_2O^- \; Na^+ \; + \; CH_3Br \xrightarrow{S_N2} CH_3CH_2CH_2OCH_3 \; + \; NaBr$$

or

$$CH_3O^- \; Na^+ \; + \; CH_3CH_2CH_2Br \xrightarrow{S_N2} CH_3OCH_2CH_2CH_3 \; + \; NaBr$$

Both reactions would be successful. This synthesis of ethers is called the Williamson synthesis (see Sec. 8.5).

b.

$$(CH_3)_3CBr \; + \; CH_3OH \xrightarrow{S_N1} (CH_3)_3COCH_3 \; + \; HBr$$

We select this combination of reagents, *not* $(CH_3)_3COH + CH_3Br$, because methyl bromide, being primary, will not react by an S_N1 mechanism, and $(CH_3)_3COH$ is too weak a nucleophile to displace Br^- from CH_3Br in an S_N2 process. [$(CH_3)_3CO^-K^+$ would provide a strong enough nucleophile, and the reaction

$$(CH_3)_3CO^- \; K^+ \; + \; CH_3Br \xrightarrow{S_N2} (CH_3)_3COCH_3 \; + \; K^+ \, Br^-$$

would provide an alternative synthesis of the desired product, but it uses an alkoxide rather than an alcohol as the problem specifies.]

6.27 Dehydrohalogenation of (bromomethyl)cyclohexane with potassium *tert*-butoxide, a sterically hindered and strong base, would give the desired alkene as the major product. The substitution derived from S_N2 displacement of the halide by the alkoxide might also be obtained as a minor product.

$$\text{C}_6\text{H}_{11}\text{–CH}_2\text{Br} \xrightarrow{(CH_3)_3CO^- \ K^+} \text{methylenecyclohexane} + (CH_3)_3COH + KBr$$

Dehydrohalogenation of 1-bromo-1-methylcyclohexane with a strong base (NaOH or potassium *tert*-butoxide) would also provide the desired alkene. This reaction, however, would be complicated by the formation of methylcyclohexene as the major product.

6.28 This type of reaction sequence is a useful method for constructing C–C bonds.

a.

$$H_3CC{\equiv}CH + Na^+NH_2^- \xrightarrow{NH_3} H_3CC{\equiv}C^- \ Na^+$$

$$+ \ C_6H_5\text{–}CH_2Br \xrightarrow{S_N2} C_6H_5\text{–}CH_2C{\equiv}CCH_3$$

b.

$$HC{\equiv}CH + \xrightarrow[NH_3]{NaNH_2} HC{\equiv}C^- \ Na^+ \xrightarrow{CH_3Br} HC{\equiv}CCH_3$$

$$HC{\equiv}CCH_3 \xrightarrow{NaNH_2 \ | \ NH_3} CH_3C{\equiv}C^- \ Na^+ \xrightarrow{CH_3CH_2Br} CH_3C{\equiv}CCH_2CH_3$$

The order in which the alkyl halides were used could be reversed, with the same overall result.

6.29 a.

$$CH_3CH_2\text{—}C{\equiv}CH \xrightarrow[NH_3]{NaNH_2} CH_3CH_2\text{—}C{\equiv}C^- \ Na^+ \xrightarrow{CH_3CH_2CH_2Br} CH_3CH_2\text{—}C{\equiv}C\text{—}CH_2CH_2CH_3$$

$$\xrightarrow{H_2,\ \text{Lindlar's catalyst}} \text{cis-}CH_3CH_2(H)C{=}C(H)CH_2CH_2CH_3$$

b.

$$H_2C{=}CHCH_2Br \xrightarrow{Na^+ \ ^-OH} H_2C{=}CHCH_2OH \xrightarrow[Pt]{H_2} CH_3CH_2CH_2OH$$

6.30 This base-induced elimination would proceed by an anti-periplanar E2 elimination, so a H would be attacked by the base and then a leaving group would be lost.

HCFC-123

$$CF_3\text{–}CHCl_2 \xrightarrow{B^-} F_2C{=}CCl_2 + B\text{-}H + F^-$$

HFC-134a

$$CF_3\text{–}CH_2F \xrightarrow{B^-} F_2C{=}CHF + B\text{-}H + F^-$$

7

Alcohols, Phenols, and Thiols

Chapter Summary

The functional group of alcohols and phenols is the **hydroxyl group**. In alcohols, this group is connected to an aliphatic carbon, whereas in phenols, it is attached to an aromatic ring.

In the IUPAC system of nomenclature, the suffix for alcohols is *-ol*. Alcohols are classified as **primary**, **secondary**, or **tertiary** depending on whether one, two, or three organic groups are attached to the hydroxyl-bearing carbon. The nomenclature of alcohols and phenols is summarized in Secs. 7.1–7.3.

Alcohols and phenols form **hydrogen bonds**. These bonds account for the relatively high boiling points of these substances and the water solubility of lower members of the series.

Brønsted–Lowry and Lewis definitions of acids and bases are reviewed in Sec. 7.6. Alcohols are comparable in acidity to water, but phenols are much more acidic. This increased acidity is due to charge delocalization (resonance) in phenoxide ions. Electron-withdrawing groups, such as –F and $-NO_2$, increase acidity, through either an **inductive** or a **resonance** effect, or both.

Alkoxides, the conjugate bases of alcohols, are prepared from alcohols by reaction with reactive metals or metal hydrides. They are used as organic bases. Because of the greater acidity of phenols, phenoxides can be obtained from phenols and aqueous base.

Alcohols and phenols are also weak bases. They can be protonated on the oxygen by strong acids. This reaction is the first step in the acid-catalyzed dehydration of alcohols to alkenes and in the conversion of alcohols to alkyl halides by reaction with hydrogen halides. Alkyl halides can also be prepared from alcohols to alkyl halides by reaction with hydrogen halides. Alkyl halides can also be prepared from alcohols by reaction with **thionyl chloride** or **phosphorus halides**.

Primary alcohols can be oxidized to **carboxylic acids** using **Jones' reagent**, whereas secondary alcohols give **ketones**. Primary alcohols can be oxidized to aldehydes using **pyridinium chlorochromate** (PCC).

Glycols have two or more hydroxyl groups on adjacent carbons. **Ethylene glycol**, **glycerol**, and **sorbitol** are examples of glycols that are commercially important. Three important industrial alcohols are **methanol**, **ethanol**, and **2-propanol**.

Phenols readily undergo aromatic substitution since the hydroxyl group is ring-activating and *ortho*,*para*-directing. Phenols are easily oxidized to **quinones**. Phenols with bulky *ortho* substituents are commercial antioxidants.

Examples of biologically important alcohols are **geraniol**, **farnesol**, and **cholesterol**.

The functional group of **thiols** is the **sulfhydryl group**, –SH. Thiols are also called **mercaptans** because of their reaction with mercury salts to form **mercaptides**. Thiols have intense, disagreeable odors. They are more acidic than alcohols and are easily oxidized to **disulfides**.

Reaction Summary

Alkoxides from Alcohols

$$2\ RO{-}H + 2\ Na \longrightarrow 2\ RO^{-}\ Na^{+} + H_2$$

$$RO{-}H + NaH \longrightarrow RO^{-}\ Na^{+} + H_2$$

Phenoxides from Phenols

$$Ar{-}OH + Na^{+}\ {}^{-}OH \longrightarrow ArO^{-}\ Na^{+} + H_2O$$

Dehydration of Alcohols

$$\text{H–C–C–OH} \xrightarrow{H^+} \text{C=C} + H_2O$$

Alkyl Halides from Alcohols

$$R{-}OH + HX \longrightarrow R{-}X + H_2O$$

$$R{-}OH + SOCl_2 \longrightarrow R{-}Cl + HCl + SO_2$$

$$3\ R{-}OH + PX_3 \longrightarrow 3\ R{-}X + H_3PO_3 \quad (X = Cl \text{ or } Br)$$

Oxidation of Alcohols

$$\underset{\text{primary}}{RCH_2OH} \xrightarrow{PCC} \underset{\text{aldehyde}}{RCH{=}O}$$

$$\underset{\text{primary}}{RCH_2OH} \xrightarrow[H^+]{CrO_3} \underset{\text{carboxylic acid}}{RCO_2H}$$

$$\underset{\text{secondary}}{R_2CHOH} \xrightarrow[H^+]{CrO_3} \underset{\text{ketone}}{R_2C{=}O}$$

Aromatic Substitution in Phenols

$$C_6H_5OH \xrightarrow{HONO_2} O_2N\text{-}C_6H_4\text{-}OH$$

$$C_6H_5OH \xrightarrow{3\ Br_2} 2,4,6\text{-}Br_3C_6H_2OH$$

Oxidation of Phenols to Quinones

$$HO\text{-}C_6H_4\text{-}OH \xrightarrow[H_2SO_4]{Na_2Cr_2O_7} O{=}C_6H_4{=}O$$

Thiols

$$RX + NaSH \longrightarrow RSH + Na^+X^- \quad \text{(preparation)}$$

$$RSH + NaOH \longrightarrow RS^-Na^+ + H_2O \quad \text{(acidity)}$$

$$2\ RSH \longrightarrow RS\text{—}SR \quad \text{(oxidation)}$$

Learning Objectives

1. Know the meaning of: alcohol, phenol, thiol, hydroxyl group, primary, secondary, and tertiary alcohol.
2. Know the meaning of: alkoxide, phenoxide, oxonium ion, alkyloxonium ion.
3. Know the meaning of: dehydration, thionyl chloride, phosphorus trichloride, phosphorus tribromide.
4. Know the meaning of: chromic anhydride, PCC, aldehyde, ketone, carboxylic acid, glycol, glycerol, sorbitol, glyceryl trinitrate (nitroglycerine), quinone, antioxidant.
5. Be familiar with: geraniol, farnesol, isoprene unit, squalene, cholesterol.
6. Know the meaning of: thiol, mercaptan, mercaptide, sulfhydryl group, disulfide.
7. Given the structure of an alcohol, tell whether it is a primary, secondary, or tertiary.
8. Given the IUPAC name of an alcohol or phenol, draw its structure.
9. Given the structure of an alcohol or phenol, assign it a correct name.
10. Explain the significance of hydrogen bonding of an alcohol or phenol with regard to solubility in water and boiling point.

11. Given a small group of compounds, including alcohols, phenols, and hydrocarbons, arrange them in order of water solubility, and construct a scheme for separating them based on acidity differences.

12. Given a group of compounds with similar molecular weights but differing potential for hydrogen bonding, arrange them in order of boiling point.

13. Draw the resonance contributors to phenoxide or substituted phenoxide ions, and discuss the acidity of the corresponding phenols.

14. Account for the acidity difference between alcohols and phenols.

15. Write equations for the reaction of a specific alcohol or phenol with sodium or sodium hydride or with an aqueous base (NaOH, KOH).

16. Write the structures for all possible dehydration products of a given alcohol, and predict which product should predominate.

17. Write the steps in the mechanism for the dehydration of a given alcohol. Given alcohols of different classes, tell which dehydration mechanism is most likely, and what the relative dehydration rates will be.

18. Write equations for the reaction of a given alcohol with HCl, HBr, or HI, with cold, concentrated H_2SO_4 or HNO_3, with PCl_3 or PBr_3, with thionyl chloride ($SOCl_2$), or with an oxidant such as chromic anhydride or PCC.

19. Write the steps in the mechanism for conversion of an alcohol to an alkyl halide upon reaction with a hydrogen halide.

20. Write equations for the reaction of phenol with dilute aqueous nitric acid and with bromine water.

21. Contrast the acidity of alcohols and thiols. Also contrast their reactivity toward oxidizing agents.

22. Write equations for the reaction of a given thiol with Hg^{2+}, base, or an oxidizing agent such as H_2O_2.

ANSWERS TO PROBLEMS

Problems Within the Chapter

7.1 a. Number from the hydroxyl-bearing carbon:

$\overset{2}{ClCH_2}\overset{1}{CH_2}OH$ 2-chloro-1-ethanol or just 2-chloroethanol

b. cyclopentanol

c. The alcohol takes precedence over the double bond.

$\overset{4}{H_2C}=\overset{3}{C}H\overset{2}{C}H_2\overset{1}{C}H_2OH$ 3-buten-1-ol

7.2 a. [structure: cyclohexane ring with H_3C at C3 and OH at C1; numbering 1, 2, 3]

b. [structure: benzene ring attached to C2 of a chain H_3C(3)–CH(2)–CH_2(1)–OH]

c. $\overset{1}{CH_3}\overset{2}{C}H(OH)\overset{3}{C}\equiv\overset{4}{C}\overset{5}{CH_3}$ (written $CH_3CHC\equiv CCH_3$ with OH on C2)

7.3 Methanol is usually grouped with the primary alcohols:

Primary: CH_3OH, CH_3CH_2OH, $CH_3CH_2CH_2OH$, $CH_3CH_2CH_2CH_2OH$, $(CH_3)_2CHCH_2OH$, $CH_2{=}CHCH_2OH$, $C_6H_5CH_2OH$ (all have a $-CH_2OH$ group)

Secondary: $(CH_3)_2CHOH$, $CH_3CH(OH)CH_2CH_3$, cyclohexanol (all have a CHOH group)

Tertiary: $(CH_3)_3COH$

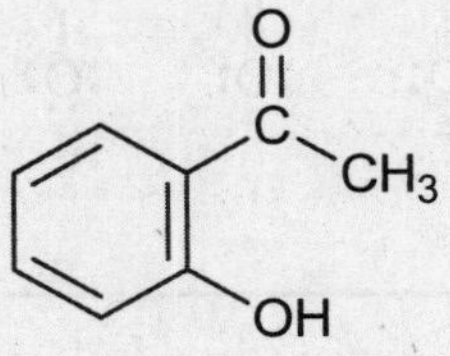

7.4 a. OH, Cl, Cl, Cl, Cl, Cl (pentachlorophenol) b. OH, CH_3, CH_3

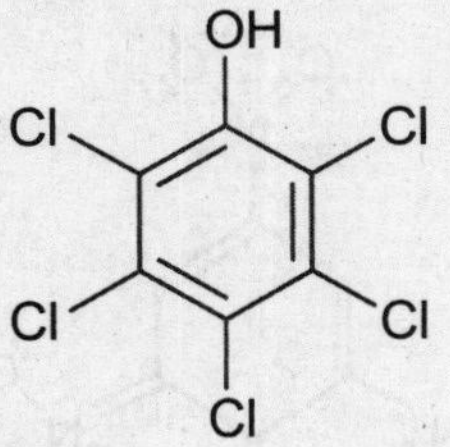

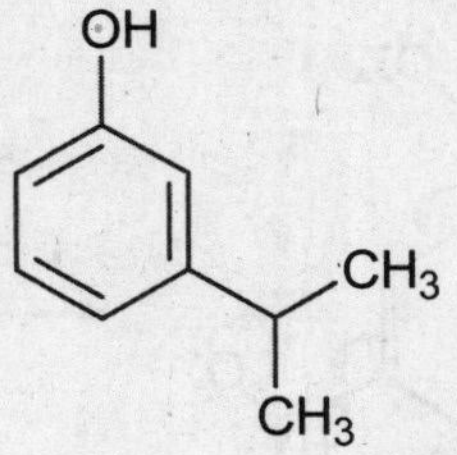

c. O, C, CH_3, OH d. O, O_2N, C, H, OH

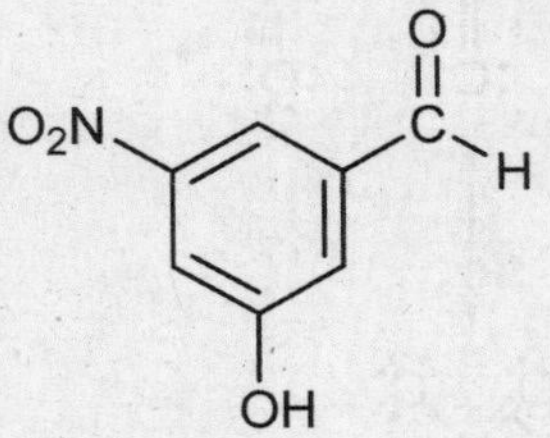

7.5 Another way to write eq. 7.6 is:

$K_a[H_2O] = [H_3O^+][HO^-]$

For the ionization of water (eq. 7.5), $[H_3O^+] = [HO^-]$.

Therefore, $K_a[H_2O] = [H_3O^+]^2$ (or $[HO^-]^2$).
Substituting, $(1.8 \times 10^{-16})(55.5) = 10^{-14} = [H_3O^+]^2$ or $[H_3O^+] = 10^{-7}$.

7.6 Use eq. 7.7.

$$pK_a = -\log(1.0 \times 10^{-16}) = 16.0$$

7.7 Acetic acid has the lower pK_a, and, hence, is the stronger acid.

7.8 a. Lewis acid, can accept electron pairs to neutralize some of the positive charge on Mg^{2+}.

b. Lewis base; can donate its electron pair to an acid; this carbanion is isoelectronic with $(CH_3)_3N$: (part h).

c. Lewis base, because of the unshared electron pair on the nitrogen.

d. Lewis acid, can accept electron pairs to neutralize the positive charge.

e. Lewis base, because of the unshared electron pairs on the oxygen:

$$CH_3CH_2\ddot{\underset{..}{O}}CH_2CH_3$$

f. Lewis acid; this carbocation is isoelectronic with $(CH_3)_3B$ (part g).

g. Lewis acid; the boron has only six valence electrons around it and can accept two more:

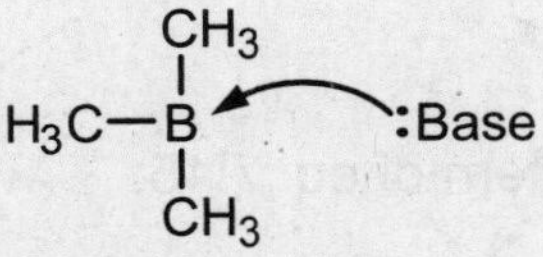

CH_3, H_3C-B, CH_3 ← :Base

h. Lewis base, again because of the unshared electron pair on the nitrogen.

i. Lewis base; hydride can donate its electron pair to an acid.

7.9 The amide anion functions as a Brønsted–Lowry base. It accepts a proton from a terminal acetylene, which functions as a Brønsted–Lowry acid.

7.10 Start pushing electrons with the structure in the box.

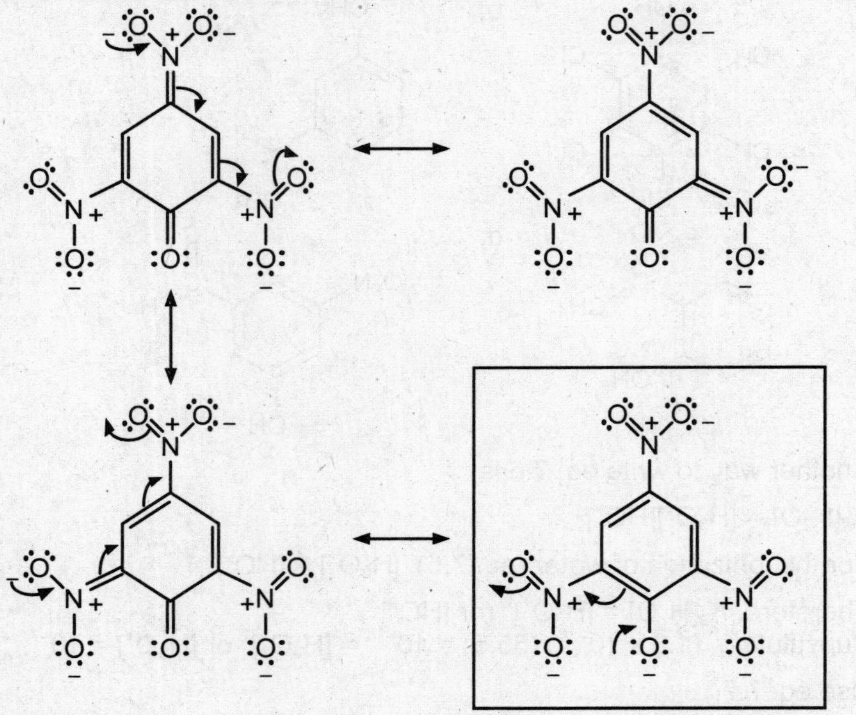

7.11 Both alcohols are weaker acids than the three phenols. Of the alcohols, 2-chloroethanol is the stronger acid because of the electronegativity of the chlorine substituent. Among the three phenols, acidity increases with increasing electronegativity of the *para* substituent: $CH_3 < H < Cl$.

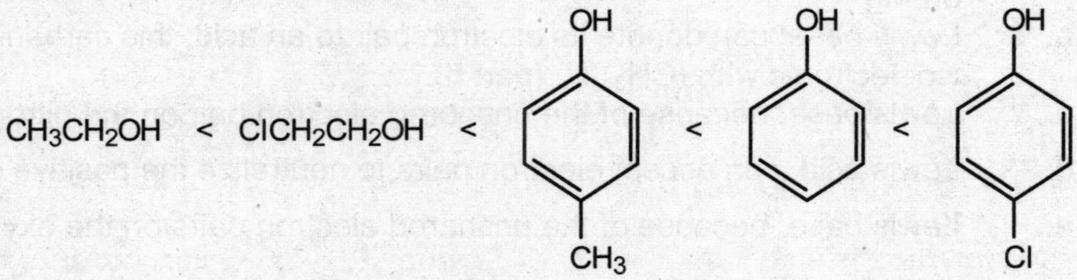

CH_3CH_2OH < $ClCH_2CH_2OH$ < (4-methylphenol) < (phenol) < (4-chlorophenol)

7.12 Follow the pattern of eq. 7.12:

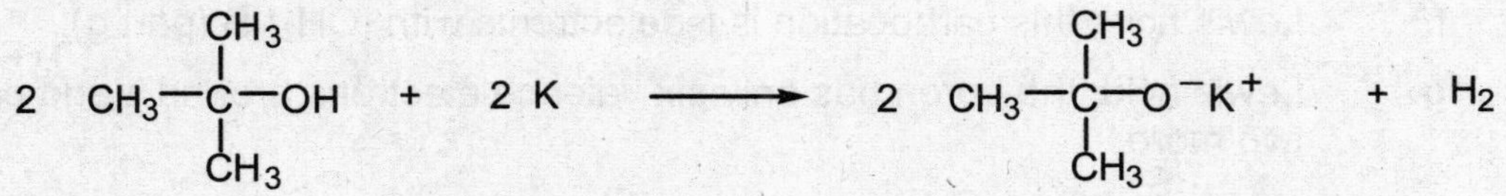

$$2\ (CH_3)_3C{-}OH + 2\ K \longrightarrow 2\ (CH_3)_3C{-}O^-K^+ + H_2$$

potassium *t*-butoxide

7.13 a. Follow the pattern of eq. 7.15:

$$O_2N-C_6H_4-OH + KOH \longrightarrow O_2N-C_6H_4-O^-\ K^+ + H_2O$$

b. Alcohols do not react with aqueous base. The equilibrium favors the starting material because hydroxide ion is a weaker base than the alkoxide ion:

$$C_6H_{11}OH + KOH \rightleftharpoons C_6H_{11}O^-K^+ + H_2O$$

7.14 Write the structure of the alcohol, and consider products with a double bond between the hydroxyl-bearing carbon and each adjacent carbon that also have at least one hydrogen attached.

a. $CH_3CH_2-C(CH_3)(OH)-CH_2CH_3 \xrightarrow[H^+]{H_2O}$ $(H_3C)HC{=}C(CH_3)(CH_2CH_3)$ + $(H_3C)HC{=}C(CH_3)(CH_2CH_3)$ + $CH_3CH_2C(=CH_2)CH_2CH_3$

b. 1-methylcyclohexanol $\xrightarrow[H^+]{H_2O}$ 1-methylcyclohexene (CH_3) + methylenecyclohexane ($=CH_2$)

The predominant product is generally the alkene with the most substituted double bond. In part a, the products with trisubstituted double bonds should predominate over the disubstituted product. To predict more precisely is not possible. In part b, the product with the double bond in the ring (1-methylcyclohexene) is trisubstituted and will predominate.

7.15 Unlike the alcohol in Example 7.2, the alcohol in this case is primary instead of tertiary. The rate-determining step is the S_N2 reaction:

$$CH_3(CH_2)_3OH + H^+ \rightleftharpoons CH_3CH_2CH_2CH_2-\ddot{O}^+(H)-H \xrightarrow[\text{slow step}]{S_N2,\ X^-} CH_3(CH_2)_3X + H_2O$$

The rate of this step varies with the nucleophilicity of X^-; this order of nucleophilicity is $I^- > Br^- > Cl^-$.

7.16 a. 1-methylcyclohexanol (CH_3, OH) + HBr $\longrightarrow$ 1-bromo-1-methylcyclohexane (CH_3, Br) + H_2O

b. $(CH_3)_2CHCH_2CH_2CH_2OH$ + HBr $\longrightarrow$ $(CH_3)_2CHCH_2CH_2CH_2Br$ + H_2O

7.17 a. $3\ C_6H_{11}CH_2OH + PBr_3 \longrightarrow 3\ C_6H_{11}CH_2Br + H_3PO_3$

b. $C_6H_{11}OH + SOCl_2 \longrightarrow C_6H_{11}Cl + SO_2 + HCl$

$3\ C_6H_{11}OH + PCl_3 \longrightarrow 3\ C_6H_{11}Cl + H_3PO_3$

7.18 a. cyclohexanol + HBr ⟶ product + H_2O (see eq. 7.25)

Phenol does not react.

b. cyclohexanol + $H_2SO_4 \xrightarrow{\text{heat}}$ cyclohexene + H_2O

Phenol may undergo electrophilic aromatic substitution:

$C_6H_5OH + H_2SO_4 \longrightarrow HO_3S{-}C_6H_4{-}OH + H_2O$

7.19 a. The alcohol is primary and gives a carboxylic acid.

H_3C–chain(–CH_3)–OH $\xrightarrow[H^+]{CrO_3}$ H_3C–chain(–CH_3)–C(=O)OH

b. The alcohol is primary and gives an aldehyde.

H_3C–chain(–CH_3)–OH $\xrightarrow[CH_2Cl_2]{PCC}$ H_3C–chain(–CH_3)–C(=O)H

c. The alcohol is secondary and gives a ketone.

$C_6H_5CH_2CH_2CH(OH)CH_3 \xrightarrow[H^+]{CrO_3} C_6H_5CH_2CH_2\overset{O}{\overset{\|}{C}}CH_3$

d. The alcohol is secondary and gives a ketone, regardless of the oxidizing agent. The product is the same as in part c.

7.20 The phenoxide ion is negatively charged, and that charge can be delocalized to the *ortho* and *para* ring carbons. Therefore, attack by an electrophile at these positions is facilitated.

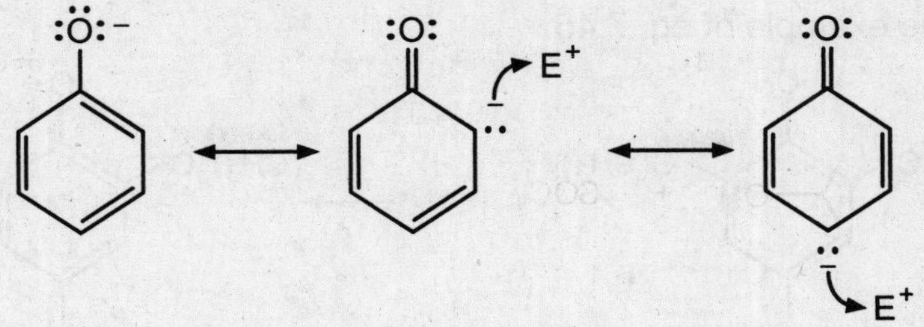

:Ö:⁻ ⟷ :O: E⁺ ⟷ :O: E⁺

7.21 a. The hydroxyl group is more ring-activating than the methyl group. Thus, substitution *ortho* to the hydroxyl group is preferred.

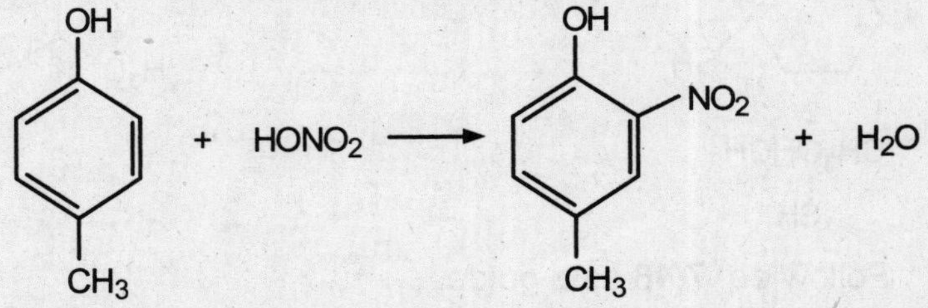

OH, CH_3 + $HONO_2$ ⟶ OH, NO_2, CH_3 + H_2O

b. The hydroxyl group is ring-activating, whereas the chlorine is a ring-deactivating substituent. Therefore, substitution occurs *para* to the hydroxyl group. Substitution *ortho* to the hydroxyl group is less likely since the product would have three adjacent substituents, which would be quite sterically crowded.

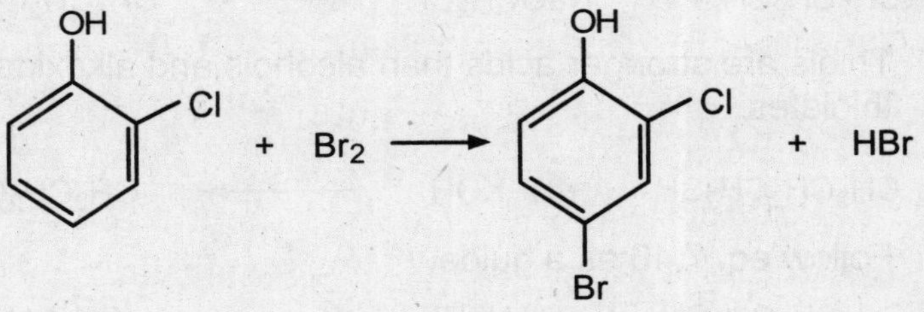

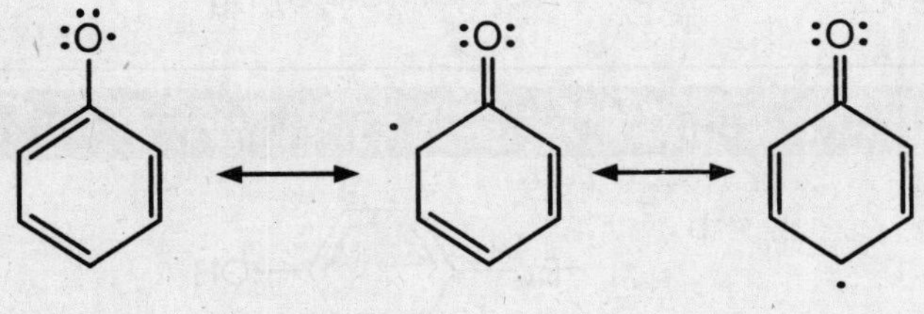

OH, Cl + Br_2 ⟶ OH, Cl, Br + HBr

7.22 :Ö· ⟷ :O: ⟷ :O:

7.23 Follow the example of eq. 7.45.

$$\text{2,6-di-}t\text{-butyl-4-methylphenol (OH, }(CH_3)_3C\text{, }C(CH_3)_3\text{, }CH_3) + HO\cdot \longrightarrow \text{phenoxy radical (O}\cdot\text{, }(CH_3)_3C\text{, }C(CH_3)_3\text{, }CH_3) + H_2O$$

7.24 a. cyclohexane ring bearing H and SH on the same carbon

b. SH on C-2 of a six-carbon chain: H_3C (1), 2 (SH), 3, 4, 5, CH_3 (6)

c. 1 2 3
CH_3CHCH_3 with SH on C-2

7.25 a. Follow eq. 7.48 as a guide.

$$CH_3CH_2CH_2SH \quad + \quad NaH \longrightarrow CH_3CH_2CH_2S^- \; Na^+ \quad + \quad H_2$$

b. Thiols are stronger acids than alcohols and alkoxides are stronger bases than thiolates.

$$CH_3CH_2CH_2SH \quad + \quad NaOCH_2CH_3 \longrightarrow CH_3CH_2CH_2S^- \; Na^+ \quad + \quad CH_3CH_2OH$$

b. Thiols are stronger acids than alcohols and alkoxides are stronger bases than thiolates.

$$CH_3CH_2CH_2SH \quad + \quad KOH \longrightarrow CH_3CH_2CH_2S^- \; K^+ \quad + \quad H_2O$$

c. Follow eq. 7.46 as a guide.

$$2\ CH_3CH_2CH_2SH \quad + \quad HgCl_2 \longrightarrow (CH_3CH_2CH_2S)_2Hg \quad + \quad 2\ HCl$$

ADDITIONAL PROBLEMS

7.26 a. cyclohexane ring bearing CH_3 and OH on the same carbon

b. Br–(benzene ring, para)–OH

c. H_3C–CH(OH)–CH(OH)–CH_3 (OH groups drawn on opposite sides)

d. $HOCH_2-C(CH_3)_2-CH_2CH_3$

e. $CH_3CH_2O^- \; Na^+$

f. cyclopentane ring with H and CH_3 on one carbon and H and OH on the adjacent carbon (CH_3 and OH both hashed)

g. H, SH, C, CH_3CH_2, CH_3

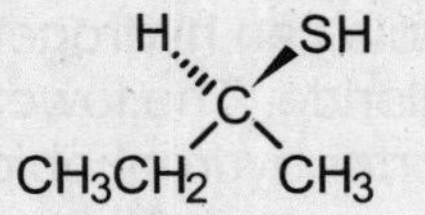

h. OH, H_3C

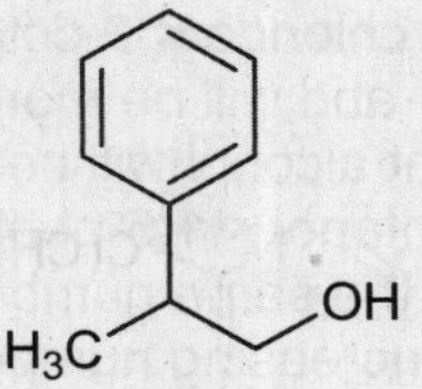

i. 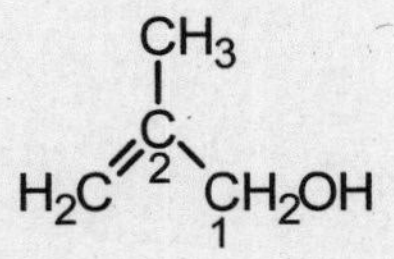 CH_3, C, H_2C, 2, CH_2OH, 1

j. 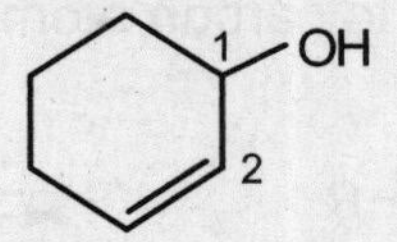 1, OH, 2

7.27 a. 2-chloro-3-pentanol
b. 2-butanol
c. 2-chloro-2-methyl-3-pentanol
d. 4-chloro-2-pentanol; the alcohol gets a lower number than the chloro group

7.28
a. 1,2,3,4-butanetetraol
b. potassium *t*-butoxide
c. 2,4,6-trichlorophenol
d. 1-methylcyclopropanol
e. *o*-bromophenol
f. *trans*-2-methylcyclobutanol
g. 2-buten-1-ol
h. 2-butanethiol (or *sec*-butyl mercaptan)
i. 2-butanol
j. 3-bromo-2-methyl-2-butanol

7.29 a. 2-Methyl-1-butanol; the longest chain was not selected.
b. 3,3-Dimethyl-2-butanol; the hydroxyl should get the lower number.
c. 2-Propen-1-ol (or allyl alcohol); the hydroxyl group should get the lower number.
d. 4-Chloro-2-pentanol; the hydroxyl group should get the lower number.
e. 2,5-Dibromophenol; give substituents the lowest possible numbers.

7.30 There are 4 stereogenic centers are position 3, 7, 11 and 15. The (3*R*,7*R*,11*R*,15*R*) isomer is:

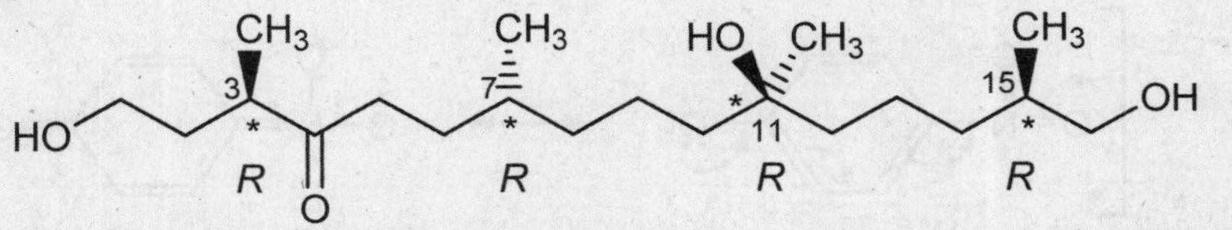

CH_3, CH_3, HO, CH_3, CH_3, 3, 7, 15, 11, HO, OH, O, *R*, *R*, *R*, *R*

There are 16 stereoisomers for all combinations of *R* and *S* at the 4 positions. The base name of the compound is 3,7,11,15-tetramethyl-4-oxo-1,11,16-hexadecanetriol.

7.31 The structure of (*S*)-3-methyl-3-mercapto-1-hexanol is:

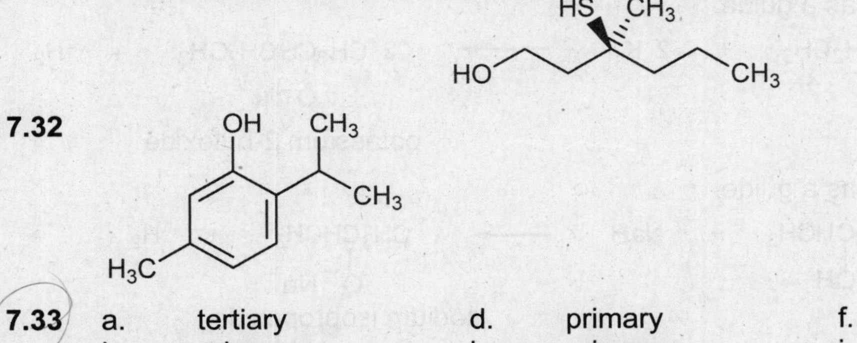

7.32

7.33
a. tertiary
d. primary
f. secondary
h. primary
i. primary
j. secondary

7.34 a. Ethyl chloride < 1-octanol < ethanol. Both alcohols can hydrogen bond with water and will be more soluble than the alkyl chloride. The lower molecular weight alcohol will be more soluble (it has a shorter hydrophobic carbon chain).

b. 1-Pentanol < 1,5-pentanediol < 1,2,3,4,5-pentanepentol. All three compounds have the same number of carbon atoms. Water solubility will therefore increase with increasing numbers of hydroxyl groups (that is, as the ratio of hydroxyl groups to carbon atoms increases).

7.35 a.

$$R{-}\ddot{N}(R){-}R + H^+ \rightleftharpoons R{-}\overset{+}{N}H(R){-}R$$

b.

$$R{-}\ddot{\underset{..}{O}}{-}R + H^+ \rightleftharpoons R{-}\overset{+}{\ddot{O}}H{-}R$$

c.

$$R_2C{=}\ddot{O}{:} + H^+ \rightleftharpoons R_2C{=}\overset{+}{\ddot{O}}{-}H$$

7.36

cyclopentanol < 2-chlorocyclopentanol < phenol < O_2N–C_6H_4–OH (*p*-nitrophenol)

The two alcohols are less acidic than the two phenols. The electron-withdrawing chlorine substituent makes 2-chlorocyclopentanol a stronger acid than cyclopentanol. The electron-withdrawing nitro substituent makes *p*-nitrophenol a stronger acid than phenol. The negative charge in the *p*-nitrophenoxide ion can be delocalized to the nitrogen:

$$\left[O{=}\overset{+}{N}(O^-){-}C_6H_4{-}\ddot{\underset{..}{O}}{:}^- \longleftrightarrow {}^-O{-}\overset{+}{N}(O^-){=}C_6H_4{=}\ddot{O}{:} \right]$$

7.37 Since *t*-butyl alcohol is a weaker acid than ethanol (pK_a = 18 and pK_a = approximately 16, respectively), it follows that if we consider the conjugate bases, *t*-butoxide ion is a stronger base than ethoxide ion.

7.38 a. Use eq. 7.12 as a guide.

$$2\ CH_3CH(OH)CH_2CH_3 + 2\ K \longrightarrow 2\ CH_3CH(O^-\,K^+)CH_2CH_3 + H_2$$

potassium 2-butoxide

b. Use eq. 7.13 as a guide.

$$CH_3CH(OH)CH_3 + NaH \longrightarrow CH_3CH(O^-\,Na^+)CH_3 + H_2$$

sodium isopropoxide (or 2-propoxide)

c. Use eq. 7.15 as a guide.

$$Cl\text{-}C_6H_4\text{-}OH + NaOH \longrightarrow Cl\text{-}C_6H_4\text{-}O^-Na^+ + H_2O$$

sodium *p*-chlorophenoxide

d. Use eq. 7.14 as a guide. The equilibrium lies on the side of the weakest acid (cyclopentanol) and weakest base (sodium hydroxide).

$$C_5H_9\text{-}OH + NaOH \rightleftharpoons C_5H_9\text{-}O^-Na^+ + H_2O$$

e. Use eq. 7.48 as a guide. The equilibrium lies on the side of the weakest acid (the alcohol) and the weakest base (the thiolate salt).

$$CH_3CH{=}CHCH_2SH + NaOH \longrightarrow CH_3CH{=}CHCH_2S^-Na^+ + H_2O$$

7.39 Consider the problem in Problem 7.38 part c. The acid on the left-hand side of the equation is a phenol with a pK_a of approximately 10 (see Table 7.2). The acid on the right-hand side of the equation is water with a pK_a of about 16. Thus, the equilibrium lies to the side of water, the weaker acid. In part d, we need to compare the pK_as of cyclopentanol and water. Table 7.2 suggests that the pK_a of cyclopentanol might be very close to that of water, so one might expect a mixture at equilibrium. It turns out that cyclopentanol, a secondary alcohol, is slightly less acidic than water and the equilibrium lies slightly to the left. The pK_a of a thiol is about 10 (see Sec. 7.16), so, in part e, the equilibrium clearly will lie to the right.

7.40 If you have any difficulty with this problem, review Sec. 7.8.

a. 3-methylcyclopentanol $\xrightarrow[\text{heat}]{H^+}$ 3-methylcyclopentene + 4-methylcyclopentene + H_2O

Both disubstituted alkenes will be formed in equal amounts.

b. 1-methylcyclopentanol $\xrightarrow{H^+}$ methylenecyclopentane ($=CH_2$) and 1-methylcyclopentene ($-CH_3$) + H_2O

The major product is 1-methylcyclopentene as it is the more substituted (more stable) double bond.

c.

$$C_6H_5\text{-}CH_2CH_2OH \xrightarrow[\text{heat}]{H^+} C_6H_5\text{-}CH{=}CH_2 + H_2O$$

d. 2-hexanol (H_3C, OH, CH_3) $\xrightarrow[\text{heat}]{H^+}$ 1-hexene (CH_3) + cis-2-hexene (H_3C, CH_3) and *trans* + H_2O

Of the three possible alkenes, the *cis* and *trans*-2-hexenes are the most stable and will predominate.

7.41 In the reaction

$$R-\overset{+}{O}H_2 \longrightarrow R^+ + H-\ddot{O}-H$$

electrons flow toward the positive oxygen, and positive charge passes from the oxygen to carbon (R group).

In the reaction

$$R-\ddot{O}-H \longrightarrow R^+ + :\ddot{O}-H^-$$

the oxygen is not charged and, therefore, is less electron-demanding and two oppositely charged species, R^+ and OH^-, must be separated. The second reaction thus requires much more energy than the first.

7.42 The second step, formation of a carbocation, is the rate-determining step.

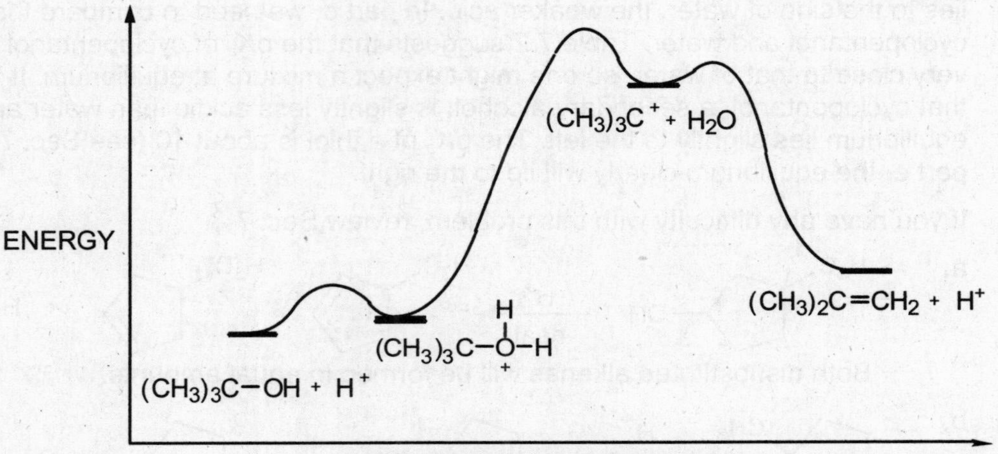

7.43 To begin, protonation of the oxygen and loss of water yield a tertiary carbocation:

$$CH_3-C(OH)(CH_3)-CH_2CH_3 \overset{H^+}{\rightleftharpoons} CH_3-C(\overset{+}{O}H_2)(CH_3)-CH_2CH_3$$

$$\xrightarrow[\text{rate-determining step}\ (S_N1)]{\text{slow}} CH_3-\overset{+}{C}(CH_3)-CH_2CH_3 + H_2O$$

The carbocation can then lose a proton from a carbon adjacent to the one that bears the positive charge, to give either product:

$$CH_3-\overset{+}{C}(CH_3)-CH_2CH_3 \xrightarrow{-H^+} H_2C{=}C(CH_3)-CH_2CH_3 \text{ and } CH_3-C(CH_3){=}CHCH_3$$

7.44 Acid-catalyzed dehydration of nerolidol would yield two alkenes in similar yields:

7.45 The acid-catalyzed hydration of propene proceeds through formation of the secondary carbocation and then attack by water to form propanol.

$$CH_3CH{=}CH_2 \xrightarrow{H^+} H_3C\overset{+}{C}HCH_3 \xrightarrow{H_2\ddot{O}} H_3C{-}CH(\overset{+}{O}H_2){-}CH_3 \xrightarrow{-H^+} H_3C{-}CH(OH){-}CH_3$$

7.46 The by-product in eq. 7.26 is isobutylene, formed by an E1 process that competes with the main S_N1 reaction:

$$(CH_3)_3C{-}OH \xrightarrow{H^+} (CH_3)_3C{-}\overset{+}{O}H_2 \xrightarrow[\text{step}]{\text{slow}} (CH_3)_3C^+ + H_2O$$

$$(CH_3)_3C^+ \xrightarrow[\text{step}]{\text{fast}} \begin{cases} \xrightarrow[S_N1]{Cl^-} (CH_3)_3C{-}Cl & 80\% \\ \xrightarrow[E1]{-H^+} H_2C{=}C(CH_3)_2 & 20\% \end{cases}$$

The reaction in eq. 7.28, on the other hand, involves a primary alcohol and proceeds by an S_N2 process:

$$CH_3CH_2CH_2CH_2{-}\overset{+}{O}H_2 + Cl^- \xrightarrow{S_N2} CH_3CH_2CH_2CH_2Cl + H_2O$$

Chloride ion is a *very* weak base; thus, the E2 process cannot compete with the S_N2 reaction, and the yield of the substitution product is nearly 100%.

7.47 The mechanism involves protonation of the hydroxyl group and loss of water to form a carbocation:

$$H_2C{=}CH{-}CH(OH){-}CH_3 \xrightarrow[-H_2O]{H^+} H_2C{=}CH{-}\overset{+}{C}H{-}CH_3$$

The carbocation is allylic and stabilized by resonance. This allylic ion can react with the nucleophile Cl^- at either end, giving the observed products:

$$H_2C{=}CH{-}\overset{+}{C}H{-}CH_3 \leftrightarrow \overset{+}{C}H_2{-}CH{=}CH{-}CH_3 \xrightarrow{Cl^-} H_2C{=}CH{-}CH(Cl){-}CH_3 \text{ (3-chloro-1-butene)} + CH_2(Cl){-}CH{=}CH{-}CH_3 \text{ (1-chloro-2-butene)}$$

7.48 a. $H_3C{-}C(CH_3)(OH){-}CH_2CH_3 + HCl \longrightarrow H_3C{-}C(CH_3)(Cl){-}CH_2CH_3 + H_2O$

b. $2\ CH_3(CH_2)_3CH_2OH + 2\ Na \longrightarrow 2\ CH_3(CH_2)_3CH_2O^-Na^+ + H_2$

sodium pentoxide

c. 3 cyclopentanol (H, OH) + $PBr_3 \longrightarrow$ 3 bromocyclopentane (H, Br) + H_3PO_3

d. $C_6H_5{-}CH_2CH_2OH + SOCl_2 \longrightarrow C_6H_5{-}CH_2CH_2Cl + SO_2 + HCl$

e. 1-methylcyclopentanol (CH_3, OH) + $HOSO_3H \xrightarrow[-H_2O]{heat}$ 1-methylcyclopentene ($-CH_3$) (major) + methylenecyclopentane ($=CH_2$) (minor)

f. $HO{-}CH_2CH_2{-}OH + 2\ HONO_2 \longrightarrow O_2NO{-}CH_2CH_2{-}ONO_2 + 2\ H_2O$

g. $CH_3(CH_2)_6CH_2OH + HBr \xrightarrow{ZnBr_2} CH_3(CH_2)_6CH_2Br + H_2O$

h. $CH_3CH_2CH_2CH_2CH_2OH + NaOH \longrightarrow$ no reaction

i. $CH_3CH_2CH_2CH_2CH_2OH \xrightarrow[H^+]{CrO_3} CH_3CH_2CH_2CH_2CO_2H$

j. $C_6H_{11}{-}CH_2CH_2OH \xrightarrow{PCC} C_6H_{11}{-}CH_2CH{=}O$

7.49 In problems of this type, working backward from the final product sometimes helps. Ask yourself: Which reactions that I have studied give this type of product? If there is more than one, see which one requires a precursor which is easily obtained from the compound given as the starting point for the synthesis.

a. cyclohexene $\xrightarrow[H^+]{H_2O}$ cyclohexanol (H, OH) $\xrightarrow[H^+]{CrO_3}$ cyclohexanone (O)

b.

$$CH_3CH_2CH_2CH_2Cl \xrightarrow[H_2O]{NaOH} CH_3CH_2CH_2CH_2OH \xrightarrow[CH_2Cl_2,\ 25^{\circ}C]{PCC} CH_3CH_2CH_2CH{=}O$$

c.

$$CH_3CH_2CH_2CH_2OH \xrightarrow[ZnBr_2]{HBr} CH_3CH_2CH_2CH_2Br \xrightarrow{Na^+\ ^-SH} CH_3CH_2CH_2CH_2SH$$

7.50 Oxidation of a primary alcohol with PCC gives an aldehyde:

CH_2OH PCC O H

7.51 The secondary alcohol should be oxidized to a ketone:

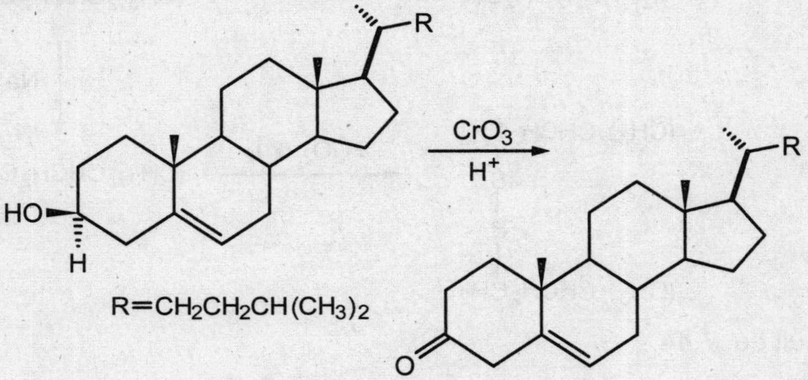

R HO H CrO_3 H^+ R O

$R{=}CH_2CH_2CH(CH_3)_2$

7.52 The structure of (2*E*)-4-hydroxy-2-nonenal is:

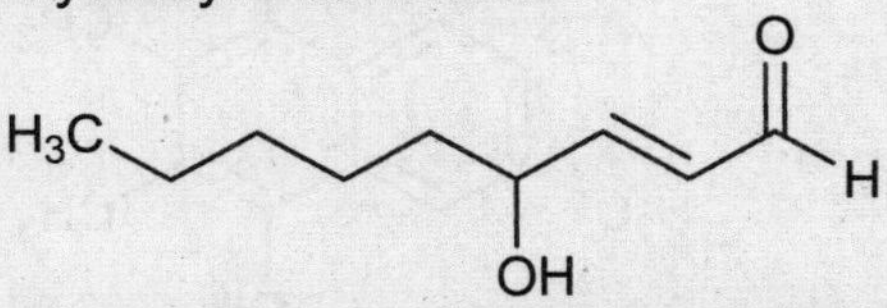

H_3C O H OH

Oxidation by PCC or by Jones' reagent would lead to the dicarbonyl product:

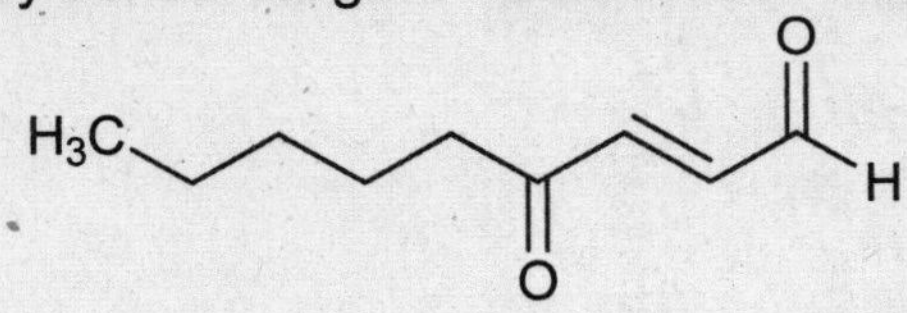

H_3C O H O

7.53 The oxygens in the quinones are located on the same carbons as the hydroxyl groups in the starting phenols.

a. O O

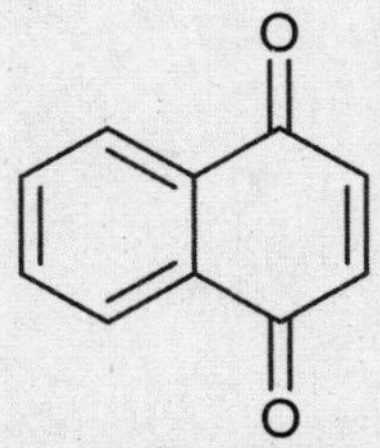

1,4-naphthoquinone

b. O O

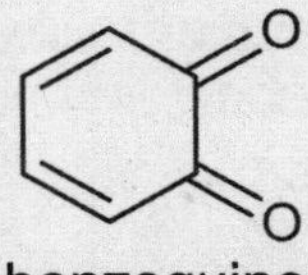

o-benzoquinone

c.

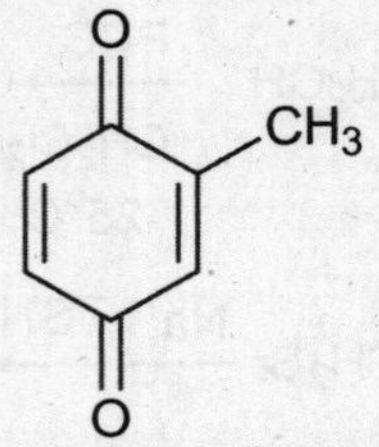

2-methyl-1,4-benzoquinone

7.54 $2\ CH_3SH \xrightarrow{H_2O_2 \text{ or } I_2} CH_3S{-}SCH_3$ (compare with eq. 7.49)

7.55 Work backward: from the disulfide to the thiol (eq. 7.49), to the alkyl halide (eq. 7.47), to the alcohol (eq. 7.30 or eq. 7.31 or eq. 7.32).

$$(CH_3)_2CHCH_2CH_2OH \xrightarrow{HBr \text{ or } PBr_3} (CH_3)_2CHCH_2CH_2Br$$

$$(CH_3)_2CHCH_2CH_2Br \xrightarrow{NaSH} (CH_3)_2CHCH_2CH_2SH$$

$$(CH_3)_2CHCH_2CH_2SH \xrightarrow{H_2O_2 \text{ or } I_2} (CH_3)_2CHCH_2CH_2{-}S{-}S{-}CH_2CH_2CH(CH_3)_2$$

7.56 Consult eq. 7.44.

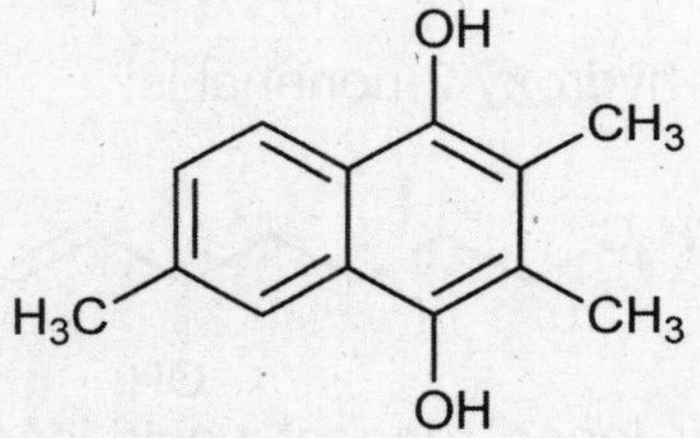

OH, CH_3, CH_3, H_3C, OH

8

Ethers and Epoxides

Chapter Summary

Ethers have two organic groups, either alkyl or aryl, connected to a single oxygen atom (R–O–R'). In common names, the two organic groups are named and followed by the word *ether*, as in *ethyl methyl ether*, $CH_3CH_2OCH_3$. In the IUPAC system, the smaller **alkoxy group** is named as a substituent on the longer carbon chain. For the preceding example, the IUPAC name is *methoxyethane*.

Ethers have much lower boiling points than the corresponding isomeric alcohols because ethers cannot form intermolecular hydrogen bonds with themselves. They do, however, act as Lewis bases to form hydrogen bonds with compounds containing an –OH group (alcohols or water).

Ethers are excellent solvents for organic compounds. Their relative inertness makes them good solvents in which to carry out organic reactions.

Alkyl or aryl halides react with magnesium metal in **diethyl ether** or **tetrahydrofuran** (**THF**) to form **Grignard reagents, R–MgX**. Ethers stabilize these reagents by coordinating with the magnesium. Grignard reagents react with water, and the –MgX is replaced by –H, or if D_2O is used, by –D. Alkyl halides also react with lithium to form organolithium reagents.

Diethyl ether is prepared commercially by intermolecular dehydration of ethanol with sulfuric acid. The **Williamson ether synthesis**, another route to ethers, involves preparation of an alkoxide from an alcohol and a reactive metal, followed by an S_N2 displacement between the alkoxide and an alkyl halide.

Ethers can be cleaved at the C–O bond by strong protonic (HBr) or Lewis (BBr_3) acids. The products are alcohols and/or alkyl halides.

Epoxides (oxiranes) are three-membered cyclic ethers. The simplest and commercially most important example is **ethylene oxide**, manufactured from ethylene, air, and a silver catalyst. In the laboratory, epoxides are most commonly prepared from alkenes and organic peroxy acids.

Epoxides react with nucleophiles to give products in which the ring has opened. For example, acid-catalyzed hydration of ethylene oxide gives **ethylene glycol**. Other nucleophiles (such as alcohols) add similarly to epoxides, as do Grignard reagents and organolithium reagents. The reactions with **organometallic reagents** (RMgX and RLi) are useful two-carbon chain-lengthening reactions.

Cyclic ethers with larger rings than epoxides include **tetrahydrofuran (THF)**, **tetrahydropyran (THP)**, and **dioxane**. Large-ring cyclic polyethers, called **crown ethers**, can selectively bind metal ions, depending on the ring size.

Reaction Summary

Grignard Reagents

$$R{-}X + Mg \xrightarrow{\text{ether or THF}} R{-}MgX \quad \text{(preparation)}$$

$$R{-}MgX + H{-}OH \longrightarrow RH + Mg(OH)X$$

$$R{-}MgX + D{-}OD \longrightarrow RD + Mg(OD)X$$

(hydrolysis)

Lithium Reagents

$$R{-}X + 2\,Li \xrightarrow{\text{hexane}} R{-}Li + LiX \quad \text{(preparation)}$$

$$R{-}Li + H{-}OH \longrightarrow RH + LiOH \quad \text{(hydrolysis)}$$

Ether Preparation

$$2\,ROH \xrightarrow[140^{\circ}C]{H_2SO_4} ROR + H_2O$$

(best for primary alcohols; gives symmetric ethers)

$$2\,ROH + 2\,Na \longrightarrow 2\,RO^- Na^+ + H_2$$

$$RO^- Na^+ + R'X \longrightarrow ROR' + Na^+ X^-$$

(Williamson synthesis)

(best for R' = primary)

Ether Cleavage

$$R{-}O{-}R + HBr \longrightarrow RBr + ROH \xrightarrow{HBr} RBr + H_2O$$

$$R{-}O{-}R + BBr_3 \longrightarrow RBr + RO{-}BBr_2 \xrightarrow{H_2O} ROH + H_3BO_3$$

Ethylene Oxide

$$2\,H_2C{=}CH_2 + O_2 \xrightarrow[250^{\circ}C,\ \text{pressure}]{Ag} H_2C{-}CH_2 \text{ (bridged by O)}$$

Other Epoxides

$$\text{>C=C<} + RCO_3H \longrightarrow \text{>C}{-}\text{C<} \text{ (bridged by O)} + RCO_2H$$

alkene + peroxy acid → epoxide + acid

Epoxide Ring Openings

$$\text{epoxide} + H{-}OH \xrightarrow{H^+} HO{-}C{-}C{-}OH$$

glycol

$$\text{epoxide} + R{-}OH \xrightarrow{H^+} HO{-}C{-}C{-}OR$$

2-alkoxyalcohol

$$\text{epoxide} + R{-}MgX \longrightarrow R{-}C{-}C{-}OMgX \xrightarrow[H^+]{H_2O} R{-}C{-}C{-}OH$$

alcohol with two more carbons than the Grignard reagent

Learning Objectives

1. Know the meaning of: ether, alkoxy group, Grignard reagent, organolithium reagents, organometallic compound, ether cleavage.
2. Know the meaning of: epoxide, oxirane, organic peroxy acid, nucleophilic addition to epoxides, diethylene glycol.
3. Know the meaning of: cyclic ether, tetrahydrofuran, furan, tetrahydropyran, dioxane, crown ethers.
4. Given the name of an ether or epoxide, write its structure, and vice versa.
5. Given the molecular formula, draw the structures of isomeric ethers and alcohols.
6. Compare the boiling points and solubilities in water of isomeric ethers and alcohols.
7. Write an equation for the preparation of a given Grignard reagent, and be able to name it.
8. Write an equation for the reaction of a given Grignard reagent with H_2O or D_2O.
9. Write an equation, using the appropriate Grignard reagent, for the preparation of a specific deuterium-labeled hydrocarbon.
10. Write an equation for the preparation of a given organolithium reagent.
11. Write equations for the reaction of organolithium reagents with water or epoxides.
12. Write equations for the preparation of a symmetrical and an unsymmetrical ether.
13. Write equations for the preparation of an ether using a Williamson synthesis.
14. Write the equation for the cleavage of an ether by a strong acid (HBr, HI, H_2SO_4) or a Lewis acid (BBr_3).

15. Write the steps in the mechanism for cleavage of an ether.
16. Write an equation for the preparation of an epoxide from the corresponding alkene.
17. Write equations for the reaction of ethylene oxide or other epoxides with nucleophiles such as H^+, and H_2O, H^+ and alcohols, or a Grignard reagent.
18. Write the steps in the mechanism for ring-opening reactions of ethylene oxide and other epoxides.

ANSWERS TO PROBLEMS

Problems Within the Chapter

8.1
a. ethyl isobutyl methyl ether or 1-ethoxy-2-methyl-propane
b. phenyl *n*-propyl ether or 1-propoxybenzene or 1-phenoxypropane
c. 1-ethoxy-1-methylcyclohexane

8.2
a. (cyclopropyl)–O–(cyclopropyl)
b. $C_6H_5-CH_2-O-CH_2-C_6H_5$
c. H_3C, CH_3, OCH_3

8.3
$HOCH_2CH_2CH_2CH_2OH$ 1,4-butanediol
$CH_3OCH_2CH_2CH_2OH$ 3-methoxy-1-propanol
$CH_3OCH_2CH_2OCH_3$ 1,2-dimethoxyethane

The compounds are listed in order of decreasing boiling point. The fewer the hydroxyl groups, the fewer the possibilities there are for intermolecular hydrogen bonding, and the lower the boiling point.

8.4
a.
$$CH_3CH_2MgI + H_2O \longrightarrow CH_3CH_3 + HO^{-}\,{}^{+}MgI$$
b.
$$C_6H_5-MgBr + CH_3CH_2OH \longrightarrow C_6H_5-H + CH_3CH_2OMgBr$$

8.5 Yes. There are no acidic protons in the starting alkyl bromide.

8.6 We must first convert the alcohol to an alkyl halide before we can make the Grignard reagent.

$$(CH_3)_2CHOH \xrightarrow{\text{HBr or }PBr_3} (CH_3)_2CHBr$$
$$(CH_3)_2CHBr \xrightarrow{\text{Mg, ether (dry)}} (CH_3)_2CHMgBr$$
$$(CH_3)_2CHMgBr \xrightarrow{D_2O} (CH_3)_2CHD$$

8.7 a. Follow eq. 8.7 and then eq. 8.6 as guides.

$$CH_3CH_2CH_2Br \xrightarrow[\text{ether}]{2\,Li} CH_3CH_2CH_2Li + LiBr \xrightarrow{D_2O} CH_3CH_2CH_2D$$

The resulting propane is labeled with one deuterium atom on one of the terminal carbon atoms.

b. Follow eq. 3.53, then eq. 8.6 as guides.

$$CH_3(CH_2)_3C{\equiv}C{-}H \xrightarrow{NaNH_2} CH_3(CH_2)_3C{\equiv}C^-\ Na^+ \xrightarrow{D_2O} CH_3(CH_2)_3C{\equiv}C{-}D$$

8.8 The reaction occurs by an S_N2 mechanism. First a molecule of ethanol is protonated by the acid catalyst:

$$CH_3CH_2{-}O{-}H \xrightarrow{+H^+} CH_3CH_2{-}\overset{+}{O}H_2$$

$$CH_3CH_2{-}\ddot{\underset{..}{O}}{-}H + CH_3CH_2{-}\overset{+}{O}H_2 \rightarrow CH_3CH_2{-}\overset{+}{O}(H){-}CH_2CH_3 + H_2O$$

$$CH_3CH_2{-}\overset{+}{O}(H){-}CH_2CH_3 \xrightarrow{-H^+} CH_3CH_2{-}O{-}CH_2CH_3$$

Then a second molecule of ethanol displaces water, followed by loss of a proton to regenerate the acid catalyst. The reaction does not go by an S_N1 mechanism because the ethyl cation, which would be an intermediate, is primary and not easily formed.

8.9 Use eq. 8.8 as a model.

$$2\ CH_3CH_2CH_2OH \xrightarrow{H_2SO_4} (CH_3CH_2CH_2)_2O + H_2O$$

8.10 First the double bond is protonated to give a *t*-butyl cation, which then reacts with methanol, a nucleophile, to give an oxonium ion. Loss of a proton gives *t*-butyl methyl ether. Notice that the acid is a catalyst; it is needed for a reaction to occur, but it is not consumed.

$$H_2C{=}C(CH_3)_2 + H^+ \rightleftharpoons (CH_3)_3C^+$$

$$(CH_3)_3C^+ + CH_3\ddot{\underset{..}{O}}H \rightleftharpoons CH_3{-}\overset{+}{O}(H){-}C(CH_3)_3$$

$$CH_3{-}\overset{+}{O}(H){-}C(CH_3)_3 \underset{}{\overset{-H^+}{\rightleftharpoons}} CH_3{-}O{-}C(CH_3)_3$$

8.11 a. C_6H_5OH + NaH ⟶ $C_6H_5O^-\ Na^+$ + H_2

$C_6H_5O^-\ Na^+$ + CH_3Br ⟶ $C_6H_5OCH_3$ + Na^+Br^-

We cannot use the alternative combination because the second step would fail; aryl halides do *not* undergo S_N2 reactions.

b. 2 $(CH_3)_3COH$ + 2 K ⟶ 2 $(CH_3)_3CO^-\ K^+$ + H_2

$(CH_3)_3CO^-\ K^+$ + CH_3Br ⟶ $(CH_3)_3COCH_3$ + $K^+\ Br^-$

We cannot use the alternative combination because the second step would fail; tertiary halides do *not* undergo S_N2 reactions:

$CH_3O^-\ K^+$ + $(CH_3)_3CBr$ ⟶ (does not occur) $(CH_3)_3COCH_3$ + $K^+\ Br^-$

c. $(CH_3)_3COH$ + 2K ⟶ $(CH_3)_3CO^-\ K^+$ + H_2

$(CH_3)_3CO^-\ K^+$ + $C_6H_5CH_2Br$ ⟶ $C_6H_5CH_2–O–C(CH_3)_3$ + KBr

We cannot use the alternative combination because the second step would fail; tertiary halides do *not* undergo S_N2 reactions.

8.12 The reaction occurs by an S_N1 mechanism (review Sec. 6.6). The C–O bond to the *t*-butyl group cleaves to give an intermediate *t*-butyl cation that is much more stable than a phenyl cation.

$C_6H_5–O–C(CH_3)_3$ $\overset{+H^+}{\rightleftharpoons}$ $C_6H_5–\overset{+}{O}(H)–C(CH_3)_3$ ⇌

C_6H_5OH + H–Ö–H + $^+C(CH_3)_3$ ⇌

$H–\overset{+}{O}(H)–C(CH_3)_3$ $\overset{-H^+}{\rightleftharpoons}$ H–O–$C(CH_3)_3$

8.13 Follow eq. 8.18 as a guide.

cyclohexene + 3-Cl-C_6H_4–C(=O)–O–O–H ⟶ cyclohexene oxide (cis H, H) + 3-Cl-C_6H_4–C(=O)–O–H

8.14 In the first step, the epoxide oxygen is protonated by the acid catalyst:

$$\text{cyclohexene oxide} \underset{}{\overset{+H^+}{\rightleftharpoons}} \text{protonated epoxide } (\overset{+}{O}{-}H)$$

In the second step, water acts as a nucleophile in an S_N2 displacement:

$$H{-}\ddot{O}{-}H + \text{protonated epoxide} \xrightarrow{S_N2} \text{protonated diol} \xrightarrow{-H^+} \text{diol}$$

The product is *trans*-1,2-cyclohexanediol:

$$\text{cyclohexene oxide} \xrightarrow[H^+]{H_2O} \textit{trans}\text{-1,2-cyclohexanediol}$$

8.15 a. Follow eq. 8.21.

$$CH_3CH_2CH_2CH_2MgCl + \text{ethylene oxide} \longrightarrow CH_3CH_2CH_2CH_2CH_2CH_2OMgCl \xrightarrow{H_2O} CH_3CH_2CH_2CH_2CH_2CH_2OH$$

b. Follow eq. 8.21.

$$C_6H_5CH_2MgBr + \text{ethylene oxide} \longrightarrow C_6H_5CH_2CH_2CH_2OMgBr \xrightarrow{H_2O} C_6H_5CH_2CH_2CH_2OH$$

c. Follow eq. 8.22.

$$CH_2{=}CHLi + \text{ethylene oxide} \longrightarrow CH_2{=}CH_2CH_2CH_2O^-\ Li^+ \xrightarrow{H^+} CH_2{=}CHCH_2CH_2OH$$

d. Acetylides (see eq. 3.53) are organometallic reagents. They react with epoxides as would a Grignard reagent or an organolithium reagent.

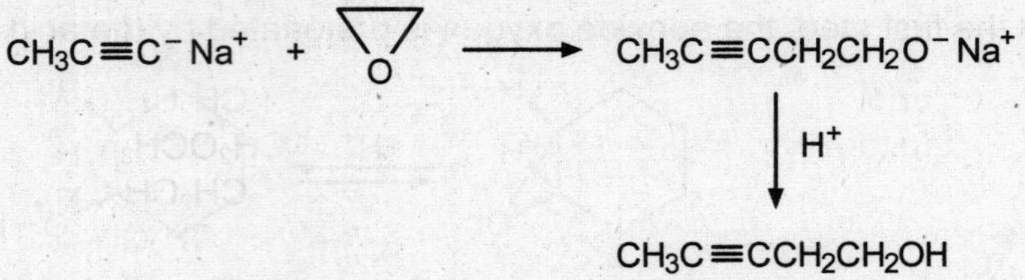

ADDITIONAL PROBLEMS

8.16

a. $CH_3CH_2–O–CH_2CH_3$

b. 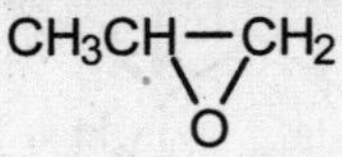

c. 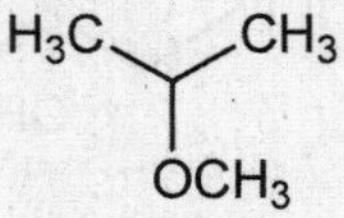

d. 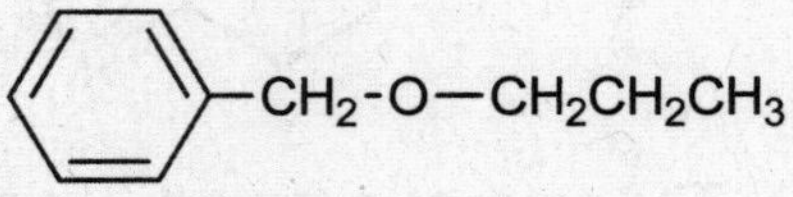

e. 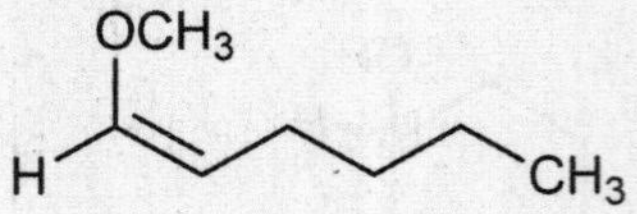

f. 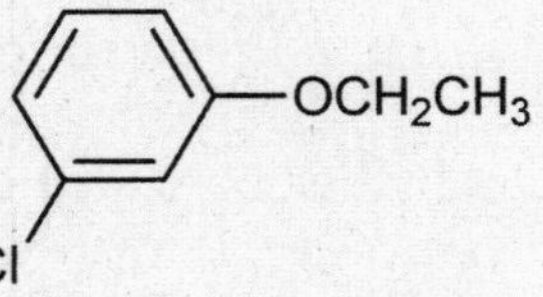

g. 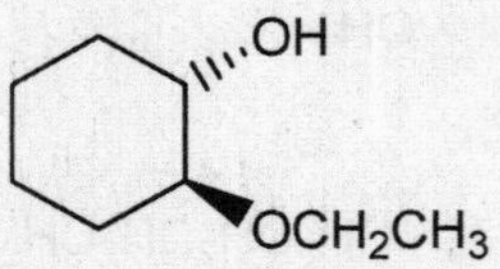

h. $CH_3CH_2–O–CH(CH_3)_2$

i. 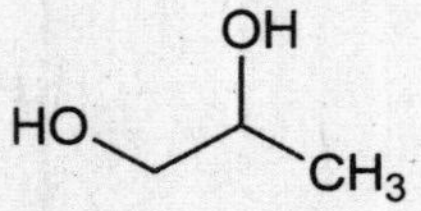

j. 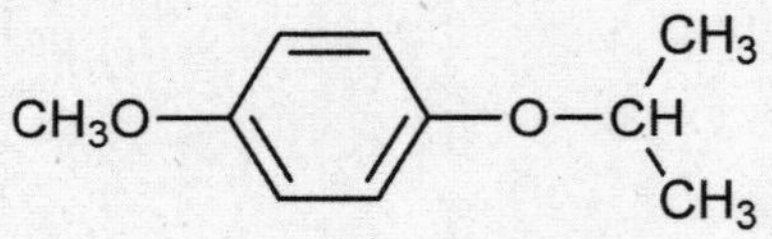

8.17

a. isobutyl propyl ether

b. isobutyl methyl ether

c. propylene oxide (or methyloxirane)

d. *p*-chloroanisole (or *p*-chlorophenyl methyl ether)

e. 4-methoxyhexanol

f. 2-ethoxyethanol

g. 2-methyltetrahydrofuran

h. 4-methoxy-1-butyne

8.18 Be systematic.

$CH_3CH_2CH_2CH_2OH$ 1-butanol	$CH_3OCH_2CH_2CH_3$ methyl *n*-propyl ether
$CH_3CH_2CH(OH)CH_3$ 2-butanol	$CH_3OCH(CH_3)_2$ methyl isopropyl ether
$(CH_3)_2CHCH_2OH$ 2-methyl-1-propanol	$CH_3CH_2OCH_2CH_3$ diethyl ether
$(CH_3)_3COH$ 2-methyl-2-propanol	

8.19 The structures and actual boiling points are:

1-pentanol	$CH_3CH_2CH_2CH_2CH_2OH$	137°C
1,2-dimethoxyethane	$CH_3OCH_2CH_2OCH_3$	83°C
hexane	$CH_3CH_2CH_2CH_2CH_2CH_3$	69°C
ethyl *n*-propyl ether	$CH_3CH_2OCH_2CH_2CH_3$	64°C

1-Pentanol is the only one of these compounds capable of forming hydrogen bonds with itself. Thus it has the highest boiling point. Judging from the table in Sec. 8.2, we might expect hexane to have a slightly higher boiling point than a corresponding monoether, although the boiling points should be quite close. 1,2-Dimethoxyethane, with four polar C–O bonds, is expected to associate more than the monoether (with only two C–O bonds). Therefore, we expect it to have a boiling point that is appreciably higher than that of ethyl *n*-propyl ether.

Regarding water solubility, 1,2-dimethoxyethane has two oxygens that can have hydrogen-bonding interactions with water. Pentanol and the other ether have only one oxygen, and the hexane has none. We expect 1,2-dimethoxyethane to be the most soluble in water of these four compounds, and it is. In fact, dimethoxyethane is completely soluble in water; 1-pentanol and ethyl *n*-propyl ether are only slightly soluble in water, and hexane is essentially insoluble in water.

8.20 One possible name would be: 2-methoxy-4-methylphenol.

8.21 a.

$$CH_3CH_2CH_2CH_2I \xrightarrow[Et_2O]{Mg} CH_3CH_2CH_2CH_2MgI \xrightarrow{D_2O} CH_3CH_2CH_2CH_2D$$

b.

$$CH_3CH_2OCH_2CH_2CH_2Br \xrightarrow[ether]{Mg} CH_3CH_2OCH_2CH_2CH_2MgBr$$

$$\xrightarrow{D_2O} CH_3CH_2OCH_2CH_2CH_2D$$

c.

$$C_6H_5CH_2Br \xrightarrow[Et_2O]{Mg} C_6H_5CH_2MgBr \xrightarrow{D_2O} C_6H_5CH_2D$$

d.

$$H_2C{=}CHCH_2Br \xrightarrow[Et_2O]{Mg} H_2C{=}CHCH_2MgBr \xrightarrow{D_2O} H_2C{=}CHCH_2D$$

Note that the ether functional group can be tolerated in making a Grignard reagent (part b).

8.22

$$C_6H_5OCH_3 \xrightarrow{Br_2} p\text{-}Br\text{-}C_6H_4\text{-}OCH_3$$

$$\downarrow (CH_3)_2C{=}CH_2,\ HF$$

$$\text{2-}C(CH_3)_3\text{-}C_6H_4\text{-}OCH_3 \xleftarrow[\text{2. } H_2O]{\text{1. Mg, ether}} Br\text{-}C_6H_3(C(CH_3)_3)\text{-}OCH_3$$

Direct Friedel–Crafts alkylation of anisole would give *p-t*-butylanisole, the sterically less hindered product. The bromine is therefore used to block the *para* position, then is removed when no longer needed. (For an activated substrate like anisole, a $FeBr_3$ catalyst is not required for step 1.)

8.23 a. Since both alkyl groups are identical and primary, the dehydration route using sulfuric acid is preferred because it is least expensive and gives a good yield (see eq. 8.8).

$$2\ CH_3CH_2CH_2OH \xrightarrow[\text{ether}]{H_2SO_4} CH_3CH_2CH_2OCH_2CH_2CH_3 + H_2O$$

b. The Williamson method is preferred. The sodium phenoxide can be prepared using NaOH instead of Na, because of the acidity of phenols:

$$C_6H_5O^-\ Na^+ + CH_3CH_2Br \longrightarrow C_6H_5OCH_2CH_3 + Na^+Br^-$$

c. The following Williamson ether synthesis is preferred. The alternate Williamson ether synthesis (the reaction between sodium ethoxide and *tert*-butyl iodide) would fail because dehydrohalogenation (that is, E2 elimination) would be faster than substitution.

$$(CH_3)_3COH \xrightarrow{Na} (CH_3)_3CO^-\ Na^+ + 1/2\ H_2 \xrightarrow{CH_3CH_2I} (CH_3)_3COCH_2CH_3 + NaI$$

Another good method for the preparation of this ether is the acid-catalyzed addition of ethanol to 2-methylpropene:

$$(H_3C)_2C{=}CH_2 \xrightarrow[CH_3CH_2OH]{H^+\text{ (catalyst)}} (CH_3)_3C{-}OCH_2CH_3$$

8.24 The second step fails because S_N2 displacements cannot be carried out on aryl halides.

$$C_6H_5O^- Na^+ + C_6H_5Br \not\longrightarrow C_6H_5{-}O{-}C_6H_5 + Na^+Br^-$$

8.25 The product (benzyl *t*-butyl ether) is primary on the benzyl side and tertiary on the *t*-butyl side. So, an S_N2 type displacement by potassium *t*-butoxide of benzyl bromide is one approach, and an acid-catalyzed (S_N1) process of benzyl alcohol with isobutylene (2-methylpropene) is another method:

$$(CH_3)_3COH + 2K \longrightarrow (CH_3)_3CO^- K^+ + H_2$$

$$(CH_3)_3CO^- K^+ + C_6H_5{-}CH_2Br \longrightarrow C_6H_5{-}CH_2{-}O{-}C(CH_3)_3 + KBr$$

and

$$C_6H_5{-}CH_2OH + CH_2{=}C(CH_3)_2 \xrightarrow{H^+} C_6H_5{-}CH_2{-}O{-}C(CH_3)_3$$

8.26 Acid-catalyzed dehydration of methanol would give dimethyl ether (follow eq. 8.8):

$$2\ CH_3OH \xrightarrow{H^+\text{ (catalyst)}} CH_3OCH_3 + H_2O$$

8.27 a. $CH_3OCH_2CH_2CH_3 + 2HBr \longrightarrow CH_3Br + CH_3CH_2CH_2Br + H_2O$

b. No reaction; ethers (except for epoxides) are inert toward base.

c. $CH_3CH_2{-}\ddot{O}{-}CH_2CH_3 + H_2SO_4 \longrightarrow CH_3CH_2{-}\overset{H}{\underset{..}{O^+}}{-}CH_2CH_3 + HSO_4^-$

The ether acts as a base and dissolves in the strong acid.

d. No reaction; ethers can be distinguished from alcohols by their inertness toward sodium metal.

e. $C_6H_5{-}OCH_2CH_3 \xrightarrow[2.\ H_2O]{1.\ BBr_3} C_6H_5{-}OH + CH_3CH_2Br$

Compare with eq. 8.15.

8.28 The oxygen of the ether can be protonated, and the resulting highly polar dialkyloxonium ion is soluble in sulfuric acid:

$$R-\ddot{O}-R + H_2SO_4 \longrightarrow R-\overset{H}{\underset{..}{\overset{|}{O^+}}}-R + HSO_4^-$$

diakyloxonium ion

Alkanes have no unshared electron pairs and are not protonated by sulfuric acid and thus remain insoluble in it.

8.29 (tetrahydrofuran) $\xrightarrow{2\ HBr}$ $BrCH_2CH_2CH_2CH_2Br$

8.30 $H_2C{=}CHCH_2CH_2 \xrightarrow{CH_3CO_3H} H_2C\overset{O}{—}CHCH_2CH_3 \xrightarrow{H_2O,\ H^+} H_2C(HO)-CH(OH)CH_2CH_2$

(refer to eq. 8.18) (refer to eq. 8.19)

8.31 a. $H_2C\overset{O}{—}CH_2 + HCl \longrightarrow H_2C(HO)-CH_2(Cl)$

b. $H_2C\overset{O}{—}CH_2 + 2HCl \longrightarrow H_2C(Cl)-CH_2(Cl) + H_2O$

The 2-chloroethanol formed in part a reacts as an alcohol with the second mole of HCl to produce the dichloride.

c. $H_2C\overset{O}{—}CH_2 + HO-C_6H_5 \xrightarrow{H^+} HOCH_2CH_2O-C_6H_5$

2-phenoxyethanol

d. $H_2C\overset{O}{—}CH_2 + C_6H_5MgBr \longrightarrow C_6H_5CH_2CH_2OMgBr \xrightarrow{H_2O} C_6H_5CH_2CH_2OH$

8.32 See eq. 8.20 for comparison.

$$H_2C\overset{O}{-}CH_2 + CH_3CH_2OH \xrightarrow{H^+} HOCH_2CH_2OCH_2CH_3$$

ethyl cellosolve

$$H_2C\overset{O}{-}CH_2 + HOCH_2CH_2OCH_2CH_3 \xrightarrow{H^+} HOCH_2CH_2OCH_2CH_2OCH_2CH_3$$

ethyl carbitol

8.33

$$C_6H_5Br \xrightarrow[\text{ether}]{Mg} C_6H_5MgBr \xrightarrow{H_2C\overset{O}{-}CH_2} C_6H_5CH_2CH_2OMgBr \xrightarrow[H_2O]{H^+} C_6H_5CH_2CH_2OH$$

2-phenylethanol (oil of roses)

Compare with eqs. 8.4 and 8.21.

8.34 After protonation of the oxygen, the epoxide ring opens in an S_N1 manner to give the tertiary carbocation. This ion then reacts with a molecule of methanol (a weak nucleophile).

$$(CH_3)_2C\overset{O}{-}CH_2 \underset{}{\overset{H^+}{\rightleftharpoons}} (CH_3)_2C\overset{O^+H}{-}CH_2 \xrightarrow{S_N1} (CH_3)_2\overset{+}{C}-CH_2OH \xrightarrow{CH_3OH} (CH_3)_2C(\overset{+}{O}HCH_3)-CH_2OH \underset{}{\overset{-H^+}{\rightleftharpoons}} (CH_3)_2C(OCH_3)-CH_2OH$$

The regioisomer $(CH_3)_2C(OH)CH_2OCH_3$ is *not* formed because methanol is a weak nucleophile and S_N2 attack on the protonated oxirane at the primary carbon cannot compete with the fast S_N1 process.

8.35 $H_2C{-}CH_2$ (epoxide, O bridging) + NH_3 $\longrightarrow$ $H_2NCH_2CH_2OH$

8.36 $CH_3C{\equiv}CH + NaNH_2 \xrightarrow{\text{(see eq. 3.53)}} CH_3C{\equiv}C^-\,Na^+ + NH_3$

(see eq. 8.22) $\downarrow$ $H_2C{-}CH_2$ (epoxide, O bridging)

$CH_3C{\equiv}CCH_2CH_2OH \xleftarrow[H_2O]{H^+} CH_3C{\equiv}CCH_2CH_2O^-\,Na^+$

8.37 First the ethylene oxide is protonated by the acid catalyst:

$$H_2C{-}CH_2\ (\text{ring with } {:}\ddot{O}{:}) \underset{}{\overset{H^+}{\rightleftharpoons}} H_2C{-}CH_3\ (\text{ring with } O^+{-}H)$$

The alcohol or glycol then acts as a nucleophile in an S_N2 displacement, which occurs quite easily because the epoxide ring opens in the process, thus relieving the strain associated with the small ring:

$$H_2C{-}CH_2\ (\text{ring with } {+}O{-}H) + H{-}\ddot{O}{-}R \longrightarrow H_2C(OH){-}CH_2{-}\overset{+}{\ddot{O}}(H){-}R$$

$$\rightleftharpoons H_2C(OH){-}CH_2{-}\ddot{O}R + H^+$$

For eq. 8.20, R = $-CH_3$ or $-CH_2CH_2OH$

8.38 As described, a synthesis of disparlure is:

$$(CH_3)_2CH(CH_2)_4-C{\equiv}CH \xrightarrow{NaNH_2} (CH_3)_2CH(CH_2)_4-C{\equiv}C^-\ Na^+$$

$$\downarrow CH_3(CH_2)_8CH_2Br$$

$$(CH_3)_2CH(CH_2)_4-C{\equiv}C-CH_2(CH_2)_8CH_3$$

$$\downarrow H_2\text{, Lindlar's catalyst}$$

cis-alkene: $(CH_3)_2CH(CH_2)_4$ and $CH_2(CH_2)_8CH_3$ on the same side of C=C, with H and H

$$\downarrow CH_3CO_3H$$

cis-epoxide: $(CH_3)_2CH(CH_2)_4$ and $CH_2(CH_2)_8CH_3$ on the epoxide ring carbons, H O H

8.39 a. By analogy with crown ethers, it is the ether oxygens of nonactin and monensin that are expected to bind to the metal cations. The carboxylate and alcohol oxygens in monensin, and the ester oxygens in nonactin, might also be involved.

b. Nonactin is cyclic and probably has a cavity size that is only conducive to complexation of sodium cation. Monensin is acyclic, more flexible than nonactin, and probably can fold in a manner that accommodates larger potassium ions.

8.40 Excess HBr would convert 1,4-dioxane to:

$$\text{1,4-dioxane} \xrightarrow[HBr]{excess} 2\ BrCH_2CH_2Br$$

8.41 Since the product has only two carbons and the starting material has four carbons, two groups of two carbons must be separated by an ether oxygen:

$$-C-C-O-C-C-$$

The remaining two oxygens ($C_4H_{10}O_3$) must be at the ends of the chain. The desired structure is

$$HO-CH_2-CH_2-O-CH_2-CH_2-OH$$

and the equation for the reaction with HBr is

$$HOCH_2CH_2OCH_2CH_2OH \quad + \quad 4\ HBr \longrightarrow 2\ BrCH_2CH_2Br \quad + \quad 3\ H_2O$$

In this step, the ether is cleaved, and the alcohol functions are also converted to alkyl halides.

8.42 a. Add each compound to a small amount of 10% aqueous sodium hydroxide. The phenol dissolves, whereas the ether is inert toward the base.

$$C_6H_5{-}OH + Na^{+}\,{}^{-}OH \longrightarrow C_6H_5{-}O^{-}Na^{+} + H_2O$$

$$C_6H_5{-}OCH_3 + Na^{+}\,{}^{-}OH \longrightarrow \text{no reaction}$$

b. Add a small piece of sodium to each compound. The alcohol liberates a gas (hydrogen), whereas no gas bubbles are apparent in the ether.

$$2\ CH_3CH_2CH_2CH_2OH + 2\ Na \longrightarrow 2\ CH_3CH_2CH_2CH_2O^{-}Na^{+} + H_2$$

$$CH_3OCH_2CH_2CH_3 + Na \longrightarrow \text{no reaction}$$

c. In separate test tubes, add a little of each compound to concentrated sulfuric acid. The ether is protonated and dissolves, whereas the hydrocarbon, being inert and less dense than sulfuric acid, simply floats on top.

d. Add a little bromine in carbon tetrachloride to each ether. The allyl phenyl ether, being unsaturated, quickly decolorizes the bromine, but the ethyl phenyl ether does not.

$$H_2C{=}CHCH_2OC_6H_5 + Br_2 \longrightarrow H_2\underset{Br}{C}{-}\underset{Br}{C}HCH_2OC_6H_5$$

$$CH_3CH_2OC_6H_5 + Br_2 \longrightarrow \text{no reaction}$$

9

Aldehydes and Ketones

Chapter Summary

The **carbonyl group**, **C=O**, is present in both **aldehydes** (RCH=O) and **ketones** (R_2C=O). The IUPAC ending for naming aldehydes is *-al*, and numbering begins with the carbonyl carbon. The ending for the names of ketones is *-one*, and the longest chain is numbered as usual. Common names are also widely used. Nomenclature is outlined in Sec. 9.1.

Formaldehyde, **acetaldehyde**, and **acetone** are important commercial chemicals, synthesized by special methods. In the laboratory, aldehydes and ketones are most commonly prepared by oxidizing alcohols, but they can also be prepared by hydrating alkynes and by Friedel–Crafts acylation of arenes. Aldehydes and ketones occur widely in nature (see Figure 9.1).

The carbonyl group is planar, with the sp^2 carbon trigonal planar. The C=O bond is polarized, with C partially positive and O partially negative. Many carbonyl reactions are initiated by nucleophilic addition to the partially positive carbon and completed by addition of a proton to the oxygen.

With acid catalysis, alcohols add to the carbonyl group of aldehydes to give **hemiacetals** [RCH(OH)OR']. Further reaction with excess alcohol gives **acetals** [RCH(OR')$_2$]. Ketones react similarly. These reactions are reversible; that is, acetals can be readily hydrolyzed by aqueous acid to their alcohol and carbonyl components. Water adds similarly to the carbonyl group of certain aldehydes (for example, formaldehyde and chloral) to give hydrates. Hydrogen cyanide adds to carbonyl compounds as a carbon nucleophile to give **cyanohydrins** [R_2C(OH)CN].

Grignard reagents add to carbonyl compounds. The products, after hydrolysis, are alcohols whose structures depend on that of the starting carbonyl compound. Formaldehyde gives *primary* alcohols, other aldehydes give *secondary* alcohols, and ketones give *tertiary* alcohols.

Nitrogen nucleophiles add to the carbonyl group. Often, addition is followed by elimination of water to give a product with a R_2C=NR group in place of the R_2C=O group. For example, primary amines (R'NH_2) give **imines** (R_2C=NR'); **hydroxylamine** (NH_2OH) gives **oximes** (R_2C=NOH); and **hydrazine** (NH_2NH_2) gives **hydrazones** (R_2C=NNH_2).

Aldehydes and ketones are easily reduced to primary or secondary alcohols, respectively. Useful reagents for this purpose are various metal hydrides such as **lithium aluminum hydride** ($LiAlH_4$) or **sodium borohydride** ($NaBH_4$).

Aldehydes are more easily oxidized than ketones. The **Tollens' silver mirror test** is positive for aldehydes and negative for ketones.

Aldehydes or ketones with an α-hydrogen exist as an equilibrium mixture of **keto** (H–C_α–C=O) and **enol** (C_α=C–OH) **tautomers**. The keto form usually predominates. An α-hydrogen is weakly acidic and can be removed by a base to produce a resonance-stabilized **enolate anion**. Deuterium exchange of α-hydrogens provides experimental evidence for enols as reaction intermediates.

In the **aldol condensation**, an enolate anion acts as a carbon nucleophile and adds to a carbonyl group to form a new carbon–carbon bond. Thus, the α-carbon of one aldehyde molecule becomes bonded to the carbonyl carbon of another aldehyde molecule to form an aldol (a 3-hydroxyaldehyde). In the **mixed aldol condensation**, the reactant with an α-hydrogen supplies the enolate anion, and the other reactant, usually without an α-hydrogen, supplies the carbonyl group to which the enolate ion adds. The aldol reaction is used commercially and also occurs in nature.

Quinones are cyclic conjugated diketones. They are colored compounds used as dyes. They also play important roles in reversible biological oxidation–reduction (electron-transfer) reactions.

Reaction Summary

Preparation of Aldehydes and Ketones

$$R-CH_2-OH \xrightarrow{PCC} R-C(=O)-H \quad \text{(from primary alcohols)}$$

$$R-CH(R)-OH \xrightarrow{Cr^{6+}} R-C(=O)-R \quad \text{(from secondary alcohols)}$$

$$C_6H_6 + R-C(=O)-Cl \xrightarrow{AlCl_3} C_6H_5-C(=O)-R \quad \text{(from arenas)}$$

$$R-C\equiv C-H \xrightarrow[Hg^{2+}]{H_3O^+} R-C(=O)-CH_3 \quad \text{(from alkynes)}$$

Hemiacetals and Acetals

$$ROH + R'-C(=O)-R'' \underset{}{\overset{H^+}{\rightleftharpoons}} R'-C(OH)(OR)-R'' \underset{H^+}{\overset{ROH}{\rightleftharpoons}} R'-C(OR)(OR)-R'' + H_2O$$

Grignard Reagents

$$RMgX + H_2C{=}O \longrightarrow RCH_2OMgX \xrightarrow{H_3O^+} RCH_2OH$$

formaldehyde → primary alcohol

$$RMgX + R'HC{=}O \longrightarrow R'{-}C(H)(R){-}OMgX \xrightarrow{H_3O^+} R'{-}C(H)(R){-}OH$$

other aldehydes → secondary alcohol

$$RMgX + R'R''C{=}O \longrightarrow R''{-}C(R')(R){-}OMgX \xrightarrow{H_3O^+} R''{-}C(R')(R){-}OH$$

ketone → tertiary alcohol

Hydrogen Cyanide

$$HC{\equiv}N + R_2C{=}O \rightleftharpoons R{-}C(OH)(CN){-}R$$

R = H, alkyl; product: cyanohydrin

Nitrogen Nucleophiles

$$R'{-}\ddot{N}H_2 + R_2C{=}O \rightleftharpoons R{-}C(NHR')(OH){-}R \xrightarrow{-H_2O} R_2C{=}NR'$$

R = H, alkyl; product: imine

Reduction

$$R_2C{=}O \xrightarrow{LiAlH_4 \text{ or } NaBH_4} R{-}C(OH)(H){-}R$$

Oxidation

$$RCH{=}O + 2\,Ag(NH_3)_2^+ + 3\,HO^- \xrightarrow[\text{test}]{\text{Tollens'}} RCO_2^- + 2\,Ag^\circ + 4\,NH_3 + 2\,H_2O$$

aldehyde → silver mirror

$$2\,RCH{=}O \xrightarrow{CrO_3,\ H^+} 2\,RCO_2H$$

Tautomerism

$$-\overset{H}{\underset{|}{C}}-\overset{O}{\overset{\|}{C}}- \rightleftharpoons -\underset{|}{C}=\overset{OH}{\overset{|}{C}}-$$

keto form — enol form

base ↓

$$\left[-\overset{\cdot\cdot}{\overset{-}{\underset{|}{C}}}-\overset{:O:}{\overset{\|}{C}}- \longleftrightarrow -\underset{|}{C}=\overset{:\ddot{O}:^-}{\overset{|}{C}}- \right]$$

enolate anion

Deuterium Exchange

$$RCH_2\overset{O}{\overset{\|}{C}}CH_2R' \xrightarrow[\text{or } D_2O,\ D^+]{D_2O,\ DO^-} RCD_2\overset{O}{\overset{\|}{C}}CD_2R'$$

(only α-hydrogens exchange)

Aldol Condensation

$$2\ RCH_2CH{=}O \xrightarrow{\text{base}} RCH_2\overset{OH}{\overset{|}{C}H}\underset{R}{\underset{|}{C}H}CH{=}O$$

Mixed Aldol Condensation

$$C_6H_5-CH{=}O + RCH_2CH{=}O \xrightarrow{\text{base}} C_6H_5-\overset{OH}{\overset{|}{C}H}-\underset{R}{\underset{|}{C}H}-CH{=}O$$

(no α-hydrogen)

$$\xrightarrow{-H_2O} C_6H_5-CH{=}\underset{R}{\underset{|}{C}}-CH{=}O$$

Mechanism Summary

Nucleophilic Addition

$$\overset{-}{Nu}: + R_2C{=}\ddot{O}: \longrightarrow R_2C(Nu)-\ddot{O}:^- \xrightarrow[\text{or ROH}]{H_2O} R_2C(Nu)-\ddot{O}-H$$

Learning Objectives

1. Know the meaning of: aldehyde, ketone, carbonyl group, formaldehyde, acetaldehyde, benzaldehyde, acetone, salicylaldehyde, acetophenone, benzophenone, carbaldehyde group.
2. Know the meaning of: nucleophilic addition, hemiacetal and acetal, aldehyde hydrate, cyanohydrin.
3. Know the meaning of: imine, hydroxylamine, oxime, hydrazine, hydrazone, phenylhydrazine, phenylhydrazone, semicarbazone.
4. Know the meaning of: lithium aluminum hydride, sodium borohydride, Tollens' reagent, silver mirror test.
5. Know the meaning of: keto form, enol form, tautomers, tautomerism, enolate anion, α-hydrogen and α-carbon, aldol condensation, mixed aldol condensation.
6. Given the structure of an aldehyde or ketone, state its IUPAC name.
7. Given the IUPAC name of an aldehyde or ketone, write its structure.
8. Write the resonance contributors to the carbonyl group.
9. Given the structure or name of an aldehyde or ketone, write an equation for its reaction with the following nucleophiles: alcohol, cyanide ion, Grignard reagent or acetylide, hydroxylamine, hydrazine, phenylhydrazine, 2,4-dinitrophenylhydrazine, primary amine, lithium aluminum hydride, and sodium borohydride.
10. Explain the mechanism of acid catalysis of nucleophilic additions to the carbonyl group.
11. Write the steps in the mechanism of acetal formation and hydrolysis. Draw the structures of resonance contributors to intermediates in the mechanism.
12. Given a carbonyl compound and a Grignard reagent, write the structure of the alcohol that is formed when they react.
13. Given the structure of a primary, secondary, or tertiary alcohol, deduce what combination of aldehyde or ketone and Grignard reagent can be used for its synthesis.
14. Given the structure of an aldehyde or ketone, write the formula of the alcohol that is obtained from it by reduction.
15. Given the structure of an aldehyde, write the structure of the acid that is formed from it by oxidation.
16. Know which tests can distinguish an aldehyde from a ketone.
17. Given the structure of an aldehyde or ketone, write the structure of the corresponding enol and enolate anion.
18. Identify the α-hydrogens in an aldehyde or ketone, and be able to recognize that these hydrogens can be exchanged readily for deuterium.
19. Write the structure of the aldol product formed by the self-condensation of an aldehyde of given structure.

20. Given two reacting carbonyl compounds, write the structure of the mixed aldol product obtained from them.

21. Write the steps in the mechanism of the aldol condensation.

ANSWERS TO PROBLEMS

Problems Within the Chapter

9.1 a. H_3C–$CH_2CH_2CH_2CH_2$–C(=O)–H

b. (2-chlorobenzaldehyde: benzene ring with C(=O)H and Cl)

c. H_3C–CH_2CH_2–C(=O)–CH_2CH_2–CH_3

d. $H_3C-\overset{O}{\overset{\|}{C}}-CH(CH_3)_2$

e. (cyclobutyl)–C(=O)–H

f. $CH_3CH_2CH_2CH_2-C{\equiv}C-\overset{O}{\overset{\|}{C}}-CH_3$

9.2 a. 3-bromo-3-methylbutanal (no number is necessary for the aldehyde function)

b. 2-butenal (the number locates the double bond between C-2 and C-3)

c. cyclobutanone d. 3-heptanone

9.3 a. cyclohexanol $\xrightarrow[H_2SO_4]{CrO_3}$ cyclohexanone

b. 3-methylcyclopentanol (H_3C, OH) $\xrightarrow{PCC}$ 3-methylcyclopentanone (H_3C, =O)

b. H_3C, CH_3, OH $\xrightarrow[H_2SO_4]{PCC}$ H_3C, CH_3, $\overset{O}{\overset{\|}{C}}$–OH

d. H_3C, CH_3, OH $\xrightarrow{PCC}$ H_3C, CH_3, $\overset{O}{\overset{\|}{C}}$–H

9.4 a. cyclopentyl–CH_2OH $\xrightarrow[25°C]{PCC, CH_2Cl_2}$ cyclopentyl–CHO
(see eq. 7.38)

b. cyclopentanol (–OH) $\xrightarrow[H^+]{CrO_3}$ cyclopentanone (=O)
(see eq. 7.36)

9.5

$$C_6H_6 + CH_3-\overset{O}{\overset{\|}{C}}-Cl \xrightarrow{AlCl_3} C_6H_5-\overset{O}{\overset{\|}{C}}-CH_3$$

acetophenone

9.6

$$HC{\equiv}CCH_2CH_2CH_2CH_2CH_3 \xrightarrow[Hg^{2+}]{H_3O^+} H_3C-\underset{\underset{O}{\|}}{C}-CH_2CH_2CH_2CH_2CH_3$$

(see eq. 3.52)

9.7 a. The carbonyl compound is more polar than the hydrocarbon but cannot form hydrogen bonds with itself.

p-xylene (bp 138°C) < benzaldehyde (bp 179°C) < benzyl alcohol (bp 205°C)

b. Water solubility increases in the same order because of increasing possibilities for hydrogen bonding with water.

9.8

$$CH_3OH + CH_3CH_2-\overset{O}{\overset{\|}{C}}-H \rightleftharpoons CH_3CH_2-\underset{H}{\overset{OCH_3}{C}}-OH$$

hemiacetal

Mechanism:

$$CH_3CH_2-\overset{:O:}{\overset{\|}{C}}-H + H^+ \rightleftharpoons \left[CH_3CH_2-\overset{\overset{+}{O}-H}{\overset{\|}{C}}-H \longleftrightarrow CH_3CH_2-\overset{O-H}{\overset{|}{\overset{+}{C}}}-H \right]$$

$$\xrightarrow{CH_3\ddot{O}H} CH_3CH_2-\underset{H}{\overset{H-\overset{+}{O}CH_3}{C}}-OH \xrightleftharpoons{-H^+} CH_3CH_2-\underset{H}{\overset{OCH_3}{C}}-OH$$

9.9

$$H_3C-\underset{H}{\overset{OCH_2CH_3}{C}}-OH + CH_3CH_2OH \xrightarrow{H^+} H_3C-\underset{H}{\overset{OCH_2CH_3}{C}}-OCH_2CH_3 + H_2O$$

<u>Mechanism</u>:

$$H_3C-\underset{H}{\overset{OCH_2CH_3}{C}}-O-H \overset{H^+}{\rightleftharpoons} H_3C-\underset{H}{\overset{OCH_2CH_3}{C}}-\overset{+}{\underset{H}{O}}-H \overset{-H_2O}{\rightleftharpoons}$$

$$\left[H_3C-\underset{H}{\overset{+}{C}}-OCH_2CH_3 \longleftrightarrow H_3C-\underset{H}{C}=\overset{+}{O}CH_2CH_3 \right] \rightleftharpoons$$

$$CH_3CH_2\ddot{O}H$$

$$H_3C-\underset{H}{\overset{H-\overset{+}{O}CH_2CH_3}{C}}-OCH_2CH_3 \overset{-H^+}{\rightleftharpoons} H_3C-\underset{H}{\overset{O-CH_2CH_3}{C}}-OCH_2CH_3$$

9.10 Follow eqs. 9.11 and 9.13. Protonation of benzaldehyde is followed by addition of methanol to the protonated carbonyl group. Loss of a proton gives the intermediate hemiacetal.

$$PhCH{=}O + H^+ \rightleftharpoons \left[PhCH{=}\overset{+}{O}H \longleftrightarrow Ph\overset{+}{C}H-OH \right]$$

$$CH_3\ddot{O}H \quad Ph\overset{+}{C}H-OH \rightleftharpoons Ph-\underset{H}{\overset{H-\overset{+}{O}-CH_3}{C}}-OH$$

$$\Big\downarrow\uparrow -H^+$$

$$Ph-\underset{H}{\overset{:\ddot{O}-CH_3}{C}}-OH$$

Ph = (phenyl group)

Protonation of the hydroxyl group and loss of water gives a resonance-stabilized carbocation. Nucleophilic attack of methanol on the carbocation followed by loss of a proton gives the acetal.

$$\text{Ph-CH(OCH}_3\text{)-OH} \overset{H^+}{\rightleftharpoons} \text{Ph-CH(OCH}_3\text{)-}\overset{+}{\text{O}}\text{H}_2 \overset{-H_2O}{\rightleftharpoons}$$

$$\left[\text{Ph-}\overset{+}{\text{C}}\text{H-OCH}_3 \longleftrightarrow \text{Ph-CH=}\overset{+}{\text{O}}\text{CH}_3\right] \rightleftharpoons$$

$$CH_3OH$$

$$\text{Ph-CH(}\overset{+}{\text{O}}\text{H-CH}_3\text{)-OCH}_3 \overset{-H^+}{\rightleftharpoons} \text{Ph-CH(OCH}_3\text{)-OCH}_3$$

Notice that each step is reversible. The reaction can be driven to the right by removing water as it is formed.

9.11 a. Follow eq. 9.17:

$$\text{cyclohexanone} \xrightarrow[H^+]{CH_3CH_2OH\ (2\ \text{equiv})} \text{1,1-(OCH}_2\text{CH}_3)_2\text{-cyclohexane} + H_2O$$

b. Follow eq. 9.16:

$$\text{cyclohexanone} \xrightarrow[H^+]{HOCH_2CH_2OH} \text{spiro cyclic acetal (1,3-dioxolane)} + H_2O$$

9.12 The proton first adds to one of the oxygens:

$$C_6H_5\text{-CH(OCH}_3\text{)-OCH}_3 \overset{H^+}{\rightleftharpoons} C_6H_5\text{-CH(OCH}_3\text{)-}\overset{+}{\text{O}}\text{H-CH}_3$$

Loss of methanol gives a resonance-stabilized carbocation:

$$C_6H_5\text{-CH(OCH}_3\text{)-}\overset{+}{\text{O}}\text{H-CH}_3 \overset{-CH_3OH}{\rightleftharpoons} \left[C_6H_5\text{-}\overset{+}{\text{C}}\text{H-OCH}_3 \longleftrightarrow C_6H_5\text{-CH=}\overset{+}{\text{O}}\text{CH}_3\right]$$

The carbocation reacts with water, which is a nucleophile and is present in large excess:

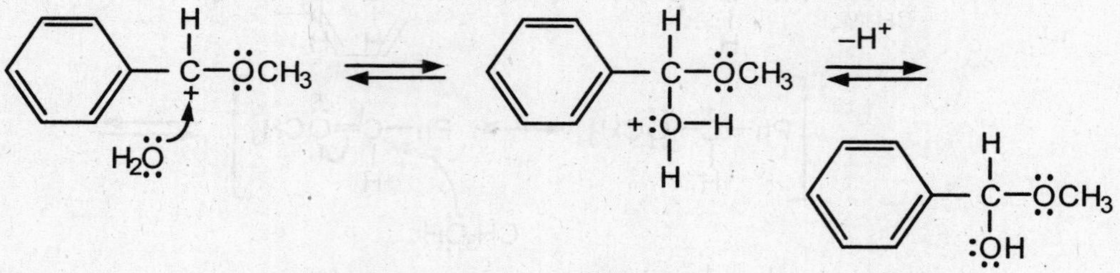

The sequence is then repeated, beginning with protonation of the methoxy oxygen of the hemiacetal:

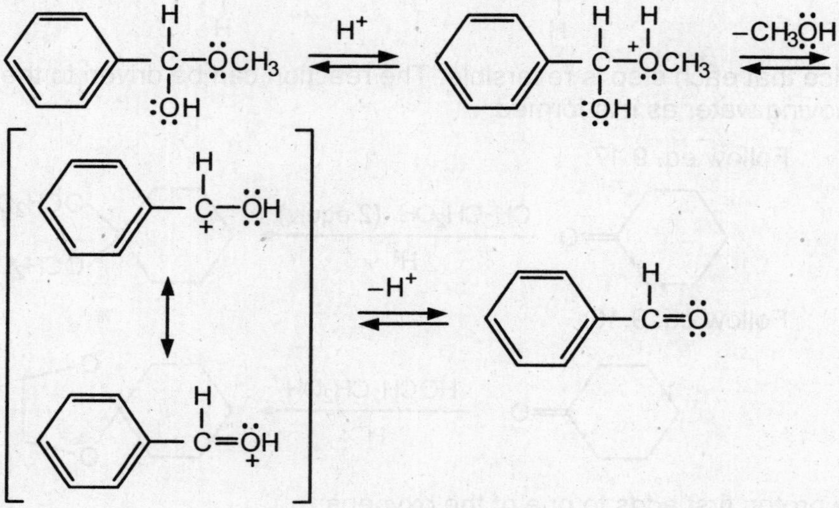

The whole process is driven forward because water is present in excess.

9.13 In the first step, one bromine is replaced by a hydroxyl group:

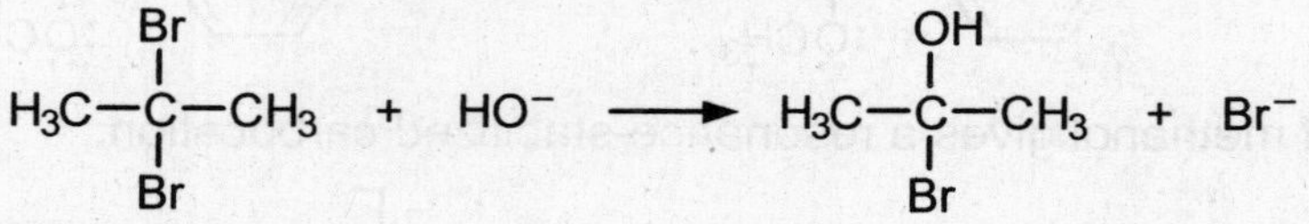

Loss of HBr gives acetone:

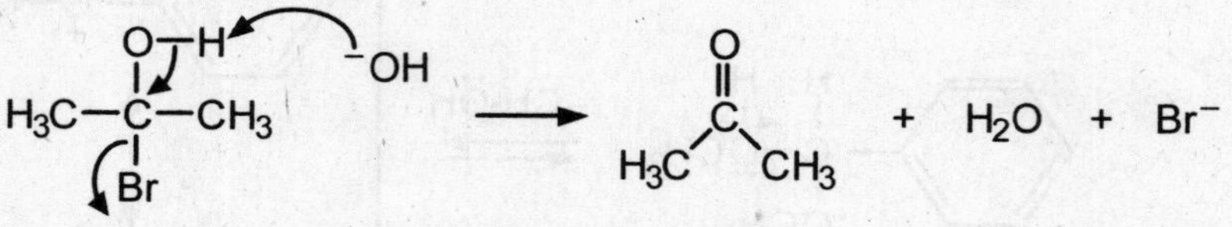

Even if both bromines were replaced by hydroxyl groups, the resulting diol (acetone hydrate) would lose water since acetone does not form a stable hydrate.

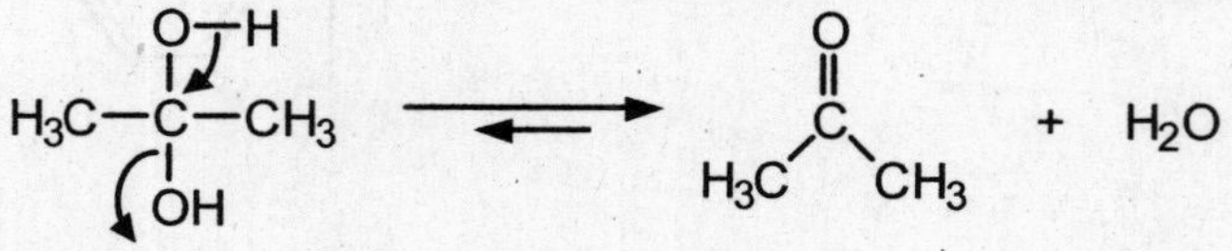

9.14 a. Follow eq. 9.22:

$C_6H_5CH_2MgBr + H_2C{=}O \longrightarrow C_6H_5CH_2CH_2OMgBr \xrightarrow{H_2O} C_6H_5CH_2CH_2OH + Mg(OH)Br$

b. Follow eq. 9.23:

$C_6H_5MgBr + CH_3CH_2CH_2CH_2CH{=}O \longrightarrow C_6H_5CH(OMgBr){-}CH_2CH_2CH_2CH_3 \xrightarrow{H_2O} C_6H_5CH(OH){-}CH_2CH_2CH_2CH_3$

c. Follow eq. 9.23:

$CH_3CH_2MgBr + C_6H_5CHO \longrightarrow C_6H_5C(H)(OMgBr)CH_2CH_3 \xrightarrow{H_2O} C_6H_5C(H)(OH)CH_2CH_3$

d. Follow eq. 9.24:

CH_3MgBr + cyclohexanone $\longrightarrow$ 1-methylcyclohexyl (H_3C, $OMgBr$) $\xrightarrow{H_2O}$ 1-methylcyclohexanol (H_3C, OH)

9.15 a. The alcohol is primary, so formaldehyde must be used as the carbonyl component:

$$CH_3O\text{-}C_6H_4\text{-}MgBr \xrightarrow{CH_2=O} CH_3O\text{-}C_6H_4\text{-}CH_2OMgBr \xrightarrow{H_2O,\ H^+} CH_3O\text{-}C_6H_4\text{-}CH_2OH$$

b. Only one R group attached to –C–OH comes from the Grignard reagent. The alcohol is tertiary, so the carbonyl component must be a ketone.

Two possibilities:

$$H_3C\text{-}C(=O)\text{-}CH_3 \xrightarrow{CH_3\text{-}C_6H_4\text{-}MgBr} CH_3\text{-}C_6H_4\text{-}C(CH_3)_2\text{-}OMgBr$$

$$CH_3\text{-}C_6H_4\text{-}C(=O)\text{-}CH_3 \xrightarrow{CH_3MgBr} CH_3\text{-}C_6H_4\text{-}C(CH_3)_2\text{-}OMgBr$$

$$CH_3\text{-}C_6H_4\text{-}C(CH_3)_2\text{-}OMgBr \xrightarrow{H_2O,\ H^+} CH_3\text{-}C_6H_4\text{-}C(CH_3)_2\text{-}OH$$

9.16 Follow eq. 9.26:

$$CH_3CH_2C(=O)CH_2CH_3 + H_3C\text{-}C\equiv C^- Na^+ \longrightarrow (CH_3CH_2)_2C(ONa)\text{-}C\equiv C\text{-}CH_3 \xrightarrow{H_2O} (CH_3CH_2)_2C(OH)\text{-}C\equiv C\text{-}CH_3$$

9.17 a.

$$CH_3CH_2CH=O + HCN \longrightarrow CH_3CH_2\text{-}CH(OH)\text{-}CN$$

b.

$$C_5H_9\text{-}CH=O + HCN \longrightarrow C_5H_9\text{-}CH(OH)\text{-}CN$$

c. Benzophenone + HCN → $(C_6H_5)_2C(OH)CN$

HO CN

9.18 C_6H_5–CH=O + H_2N–C_6H_5 → C_6H_5–CH=N–C_6H_5 + H_2O

9.19

a. $CH_3CH_2CH_2CH_2CH{=}O$ + NH_2OH → $CH_3CH_2CH_2CH_2CH{=}NOH$

b. $CH_3CH_2CH_2CH_2CH{=}O$ + $CH_3CH_2NH_2$ → $CH_3CH_2CH_2CH_2CH{=}NCH_2CH_3$

c. $CH_3CH_2CH_2CH_2CH{=}O$ + NH_2NH_2 → $CH_3CH_2CH_2CH_2CH{=}NNH_2$

d. $CH_3CH_2CH_2CH_2CH{=}O$ + $NH_2NHC_6H_5$ → $CH_3CH_2CH_2CH_2CH{=}NNHC_6H_5$

9.20

a. Follow eq. 9.33:

$$C_6H_5\text{–}C({=}O)\text{–}CH_3 \xrightarrow[2.\ H_2O]{1.\ LiAlH_4} C_6H_5\text{–}CH(OH)\text{–}CH_3$$

b. Follow eq. 9.33:

$$CH_3CH_2CH_2CH_2\text{–}C({=}O)\text{–}H \xrightarrow[2.\ H_2O]{1.\ LiAlH_4} CH_3CH_2CH_2CH_2\text{–}CH(OH)\text{–}H$$

9.21 A metal hydride will reduce the C=O bond but not the aromatic ring.

$$C_6H_5\text{–}C({=}O)\text{–}CH_3 \xrightarrow[H_2O]{NaBH_4} C_6H_5\text{–}C(HO)(H)\text{–}CH_3$$

9.22 Follow eq. 9.39 as a guide, replacing R with H:

$$H_2C{=}O + 2\ Ag(NH_3)_2^+ + 3\ HO^- \longrightarrow HCO_2^- + 2\ Ag\downarrow + 4\ NH_3\uparrow + 2\ H_2O$$

9.23 Remove an α-hydrogen and place it on the oxygen; make the carbon–oxygen bond single, and make the bond between the α-carbon and what was the carbonyl carbon double.

a.

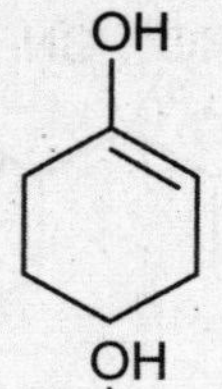

b.

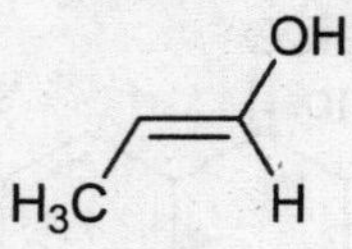

cis and *trans*

c.

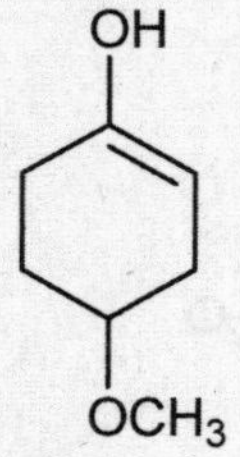

9.24 a.

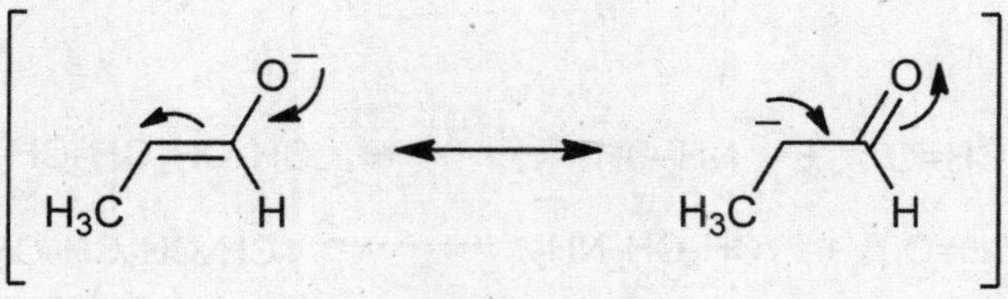

b.

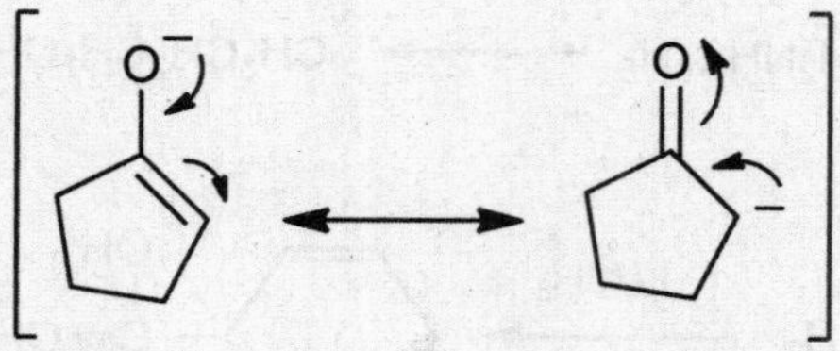

9.25 In each case, only the α-hydrogens can be readily exchanged.

a. There are three exchangeable hydrogens, indicated by the arrows.

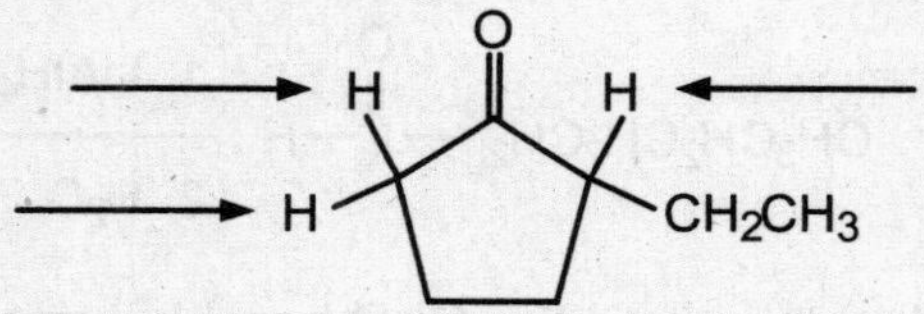

b. Only the three methyl hydrogens indicated by the arrow can be readily exchanged. The remaining methyl hydrogens are β, not α, with respect to the carbonyl group.

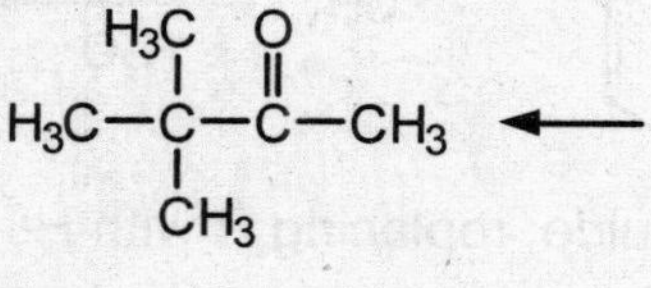

9.26 Follow eqs. 9.51–9.53 as a guide.

$$CH_3CH_2CH{=}O \;+\; HO^- \rightleftharpoons CH_3\ddot{C}HCH{=}O \;+\; H_2O$$

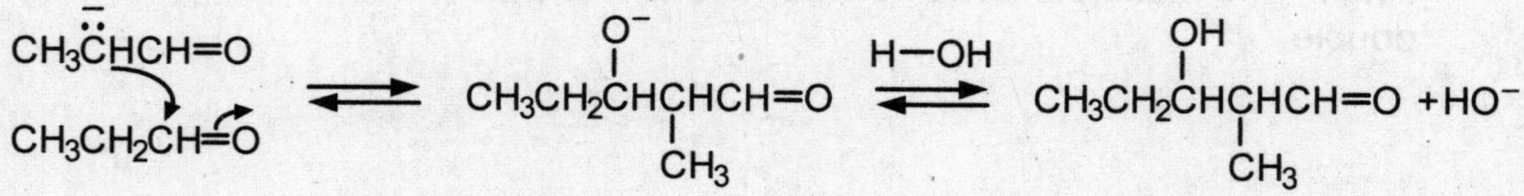

9.27 Only the acetaldehyde has an α-hydrogen, so it reacts with the basic catalyst to produce an enolate anion:

$$CH_3CH{=}\ddot{O} + HO^- \rightleftharpoons :\overset{-}{C}H_2CH{=}\ddot{O} + H_2O$$

The enolate anion then attacks the carbonyl group of benzaldehyde:

$$\ddot{O}{=}CHCH_2^- + C_6H_5CH{=}\ddot{O} \longrightarrow C_6H_5\underset{}{C}H(\ddot{O}:^-)CH_2CH{=}\ddot{O} \xrightleftharpoons{H_2O} C_6H_5CH(\ddot{O}H)CH_2CH{=}\ddot{O} + HO^-$$

Dehydration occurs by an elimination mechanism:

$$C_6H_5CH(OH){-}CH(H){-}CH{=}O \longrightarrow C_6H_5{-}CH{=}CH{-}CH{=}O + H_2O$$

9.28 Propanal has the α-hydrogens. The aldol condensation product is

$$C_6H_5{-}CH(OH)CH(CH_3)CH{=}O$$

and its dehydration product is

$$C_6H_5{-}CH{=}C(CH_3){-}CH{=}O$$

9.29 An aldol condensation-dehydration sequence can be used to construct the carbon skeleton.

$$2\ CH_3CH_2CH_2CH{=}O \xrightarrow{NaOH} CH_3CH_2CH_2CH(OH)CH(CH_2CH_3)CH{=}O$$

$$\xrightarrow{-H_2O\quad H^+} CH_3CH_2CH_2CH{=}C(CH_2CH_3)CH{=}O \xrightarrow[Ni]{H_2} CH_3CH_2CH_2CH_2CH(CH_2CH_3)CH_2OH$$

Catalytic hydrogenation of the double bonds completes the synthesis.

ADDITIONAL PROBLEMS

9.30 a. 3-pentanone b. heptanal
c. benzophenone (diphenyl ketone) d. *o*-bromobenzaldehyde
e. cyclobutanone f. 2,2-dimethylpropanal
g. dicyclopentylketone h. 3-penten-2-one
i. bromoacetone (or bromopropanone)

9.31 a.

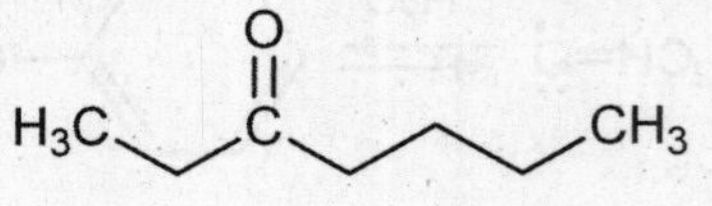

b.

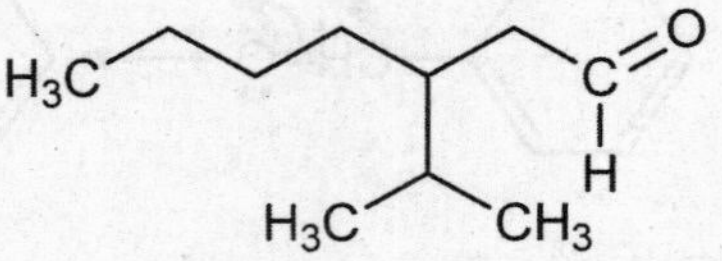

c.

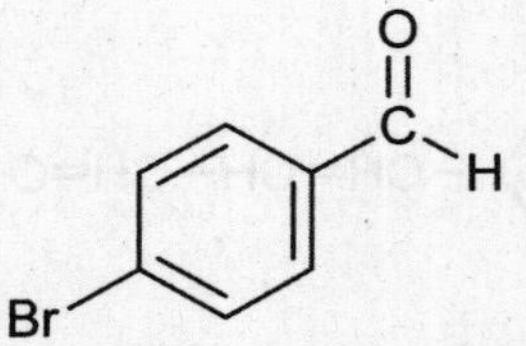

d.

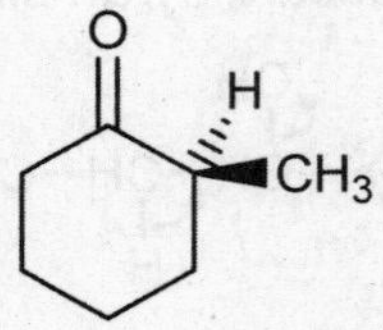

e. $CH_3CH_2CH{=}CHCH_2CHO$

f.

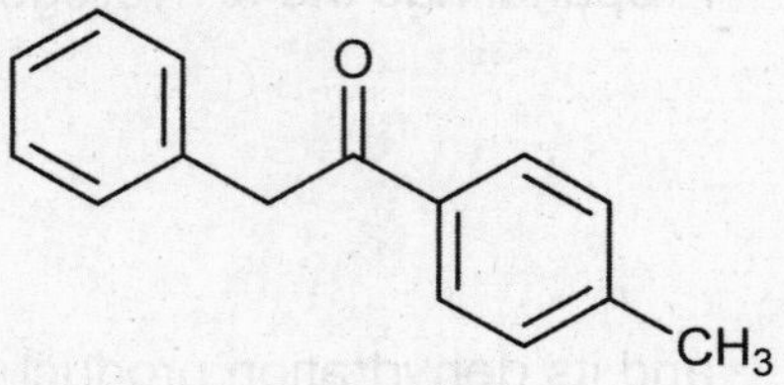

g.

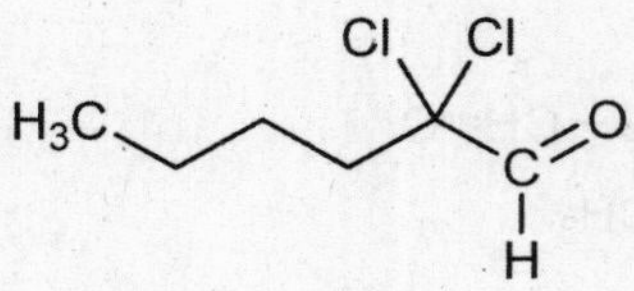

h.

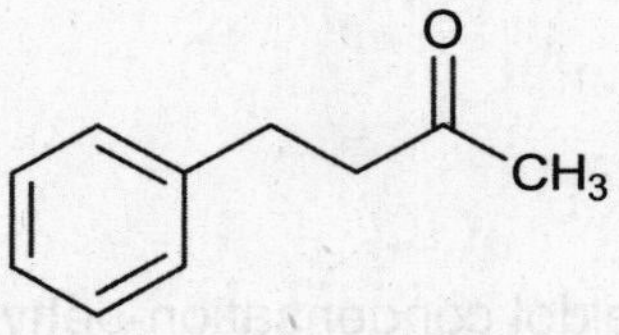

i.

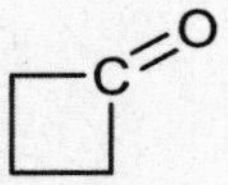

j.

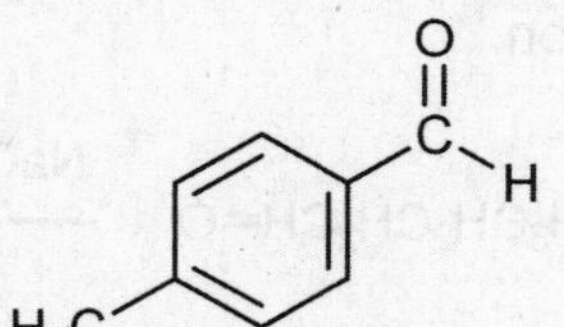

9.32 In each case, see the indicated section of the text for typical examples.

a. Sec. 9.7 b. Sec. 9.7 c. Sec. 9.10
d. Sec. 9.11 e. Sec. 9.11 f. Sec. 9.11
g. Sec. 9.14 h. Secs. 9.14 and 9.18
i. Sec. 9.15 j. Sec. 9.11

9.33

$CH_3CH_2CH_2CH_2CH_2CH_2CH{=}O$ bp 155°C

$CH_3CH_2CH_2\underset{\underset{O}{\|}}{C}CH_2CH_2CH_3$ bp 144°C

$CH_3\overset{\overset{H_3C}{|}}{CH}\underset{\underset{O}{\|}}{C}\overset{\overset{CH_3}{|}}{CH}CH_3$ bp 124°C

The compounds are isomers and have identical molecular weights. Each has a carbonyl group which, because of its polarity, can associate as follows as a consequence of intermolecular attraction between opposite charges:

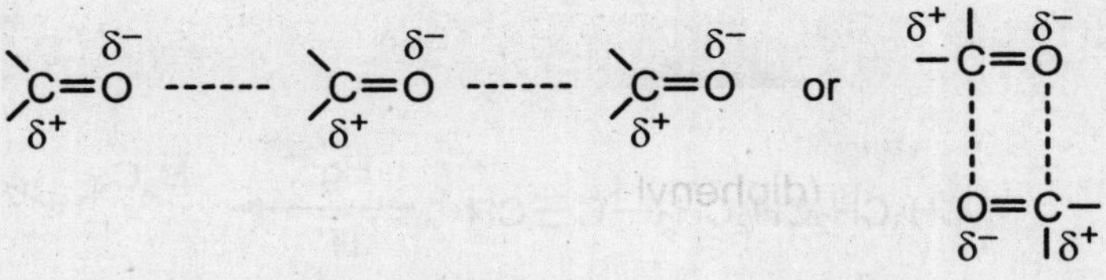

As we go from left to right in the series (as shown above), the carbonyl group is more and more hindered, or buried in the structure. Thus, association is more difficult, and the boiling point decreases.

9.34 a. CrO_3, H^+ (see eq. 7.36)

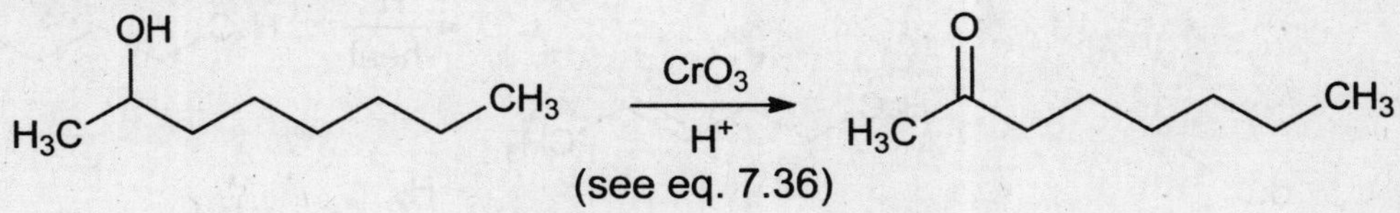

b. Hg^{2+}, H^+ (see eq. 3.52)

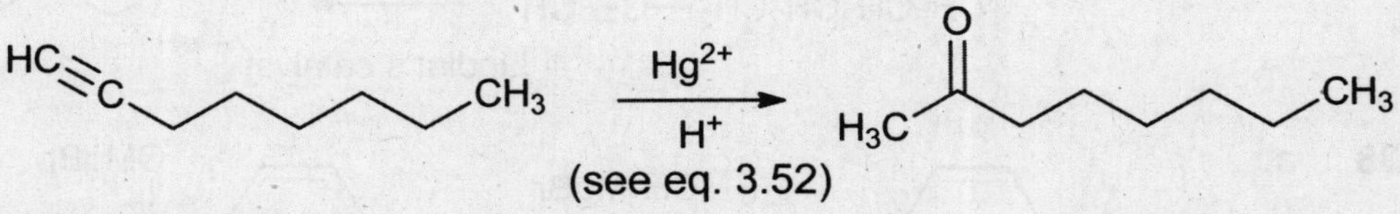

9.35 Use eq. 7.38 as a guide:

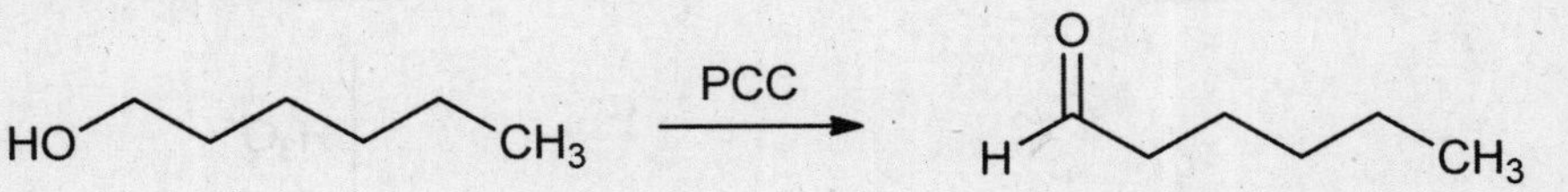

PCC

9.36

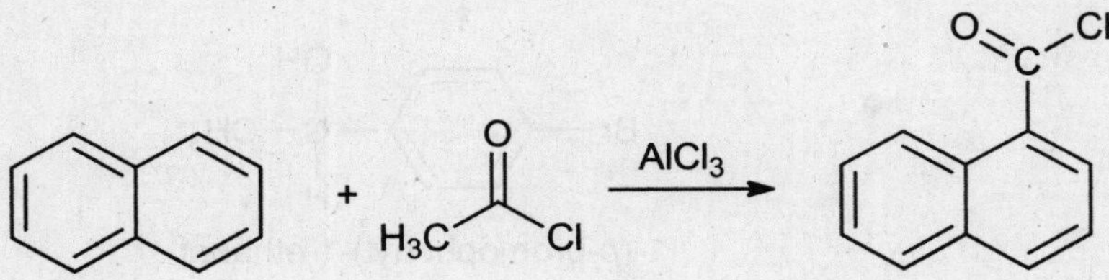

$AlCl_3$

9.37 a.

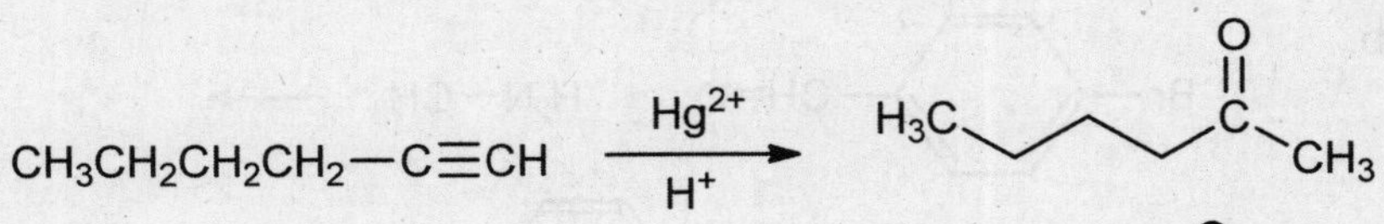

$CH_3CH_2CH_2CH_2-C\equiv CH \xrightarrow[H^+]{Hg^{2+}} H_3C\text{–}CH_2CH_2CH_2\text{–}C(=O)\text{–}CH_3$

b.

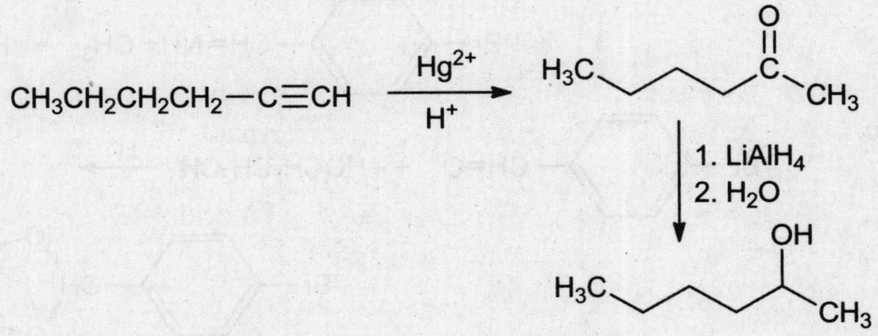

$CH_3CH_2CH_2CH_2-C\equiv CH \xrightarrow[H^+]{Hg^{2+}} H_3C\text{–}CH_2CH_2CH_2\text{–}C(=O)\text{–}CH_3$

1. $LiAlH_4$
2. H_2O

$H_3C\text{–}CH_2CH_2CH_2\text{–}CH(OH)\text{–}CH_3$

c.

$$CH_3CH_2CH_2CH_2-C\equiv CH \xrightarrow[H^+]{Hg^{2+}} H_3C\text{–CH}_2\text{–CH}_2\text{–CH}_2\text{–C(=O)–}CH_3$$

1. $LiAlH_4$
2. H_2O

$$H_3C\text{–CH}_2\text{–CH}_2\text{–CH}_2\text{–CH(OH)–}CH_3 \xrightarrow[heat]{H^+} H_3C\text{–CH}_2\text{–CH}_2\text{–CH}_2\text{–CH=CH}_2 + H_3C\text{–CH}_2\text{–CH}_2\text{–CH=CH–}CH_3$$

d.

$$CH_3CH_2CH_2CH_2-C\equiv CH \xrightarrow[\text{Lindlar's catalyst}]{H_2} H_3C\text{–CH}_2\text{–CH}_2\text{–CH}_2\text{–CH=CH}_2$$

9.38 a.

$$Br\text{–}C_6H_4\text{–}CH{=}O \xrightarrow{CH_3MgBr} Br\text{–}C_6H_4\text{–}C(OMgBr)(H)\text{–}CH_3$$

H_3O^+

$$Br\text{–}C_6H_4\text{–}C(OH)(H)\text{–}CH_3$$

1-(*p*-bromophenyl)-1-ethanol

b.

$$Br\text{–}C_6H_4\text{–}CH{=}O + H_2N-CH_3 \longrightarrow$$

$$Br\text{–}C_6H_4\text{–}CH{=}NH-CH_3 + H_2O$$

c.

$$Br\text{–}C_6H_4\text{–}CH{=}O + HOCH_2CH_2OH \xrightarrow{H^+}$$

$$Br\text{–}C_6H_4\text{–}CH(O\text{–}CH_2CH_2\text{–}O) + H_2O$$

2-*p*-bromophenyl-1,3-dioxolane
(*p*-bromobenzaldehyde ethylene glycol acetal)

d.

$$Br-C_6H_4-CH{=}O + H_2N-NH-C_6H_5 \longrightarrow Br-C_6H_4-CH{=}N-NH-C_6H_5 + H_2O$$

p-bromobenzaldehyde phenylhydrazone

e.

$$Br-C_6H_4-CH{=}O + HCN \rightleftharpoons Br-C_6H_4-CH(OH)(CN)$$

p-bromobenzaldehyde cyanohydrin

f.

$$Br-C_6H_4-CH{=}O + 2Ag(NH_3)^{+2} + 3HO^- \longrightarrow Br-C_6H_4-CO_2^- + 2\,Ag + 4\,NH_3 + 2\,H_2O$$

p-bromobenzoate ion

g.

$$Br-C_6H_4-CH{=}O \xrightarrow[H^+]{CrO_3} Br-C_6H_4-CO_2H$$

p-bromobenzoic acid

h.

$$Br-C_6H_4-CH{=}O \xrightarrow{C_6H_5MgBr} Br-C_6H_4-CH(OMgBr)-C_6H_5 \xrightarrow{H_3O^+} Br-C_6H_4-CH(OH)-C_6H_5$$

1-(*p*-bromophenyl)-1-(phenyl)-1-methanol

i.

$$Br-C_6H_4-CH{=}O + 2\;CH_3OH \xrightarrow{H^+} Br-C_6H_4-CH(OCH_3)_2 + H_2O$$

p-bromobenzaldehyde dimethylacetal

j. $Br-C_6H_4-CH{=}O + H_2N-OH \longrightarrow Br-C_6H_4-CH{=}NOH + H_2O$

p-bromobenzaldoxime

k. $Br-C_6H_4-CH{=}O \xrightarrow{NaBH_4} Br-C_6H_4-CH_2OH$

9.39 a. Use Tollens' reagent (the silver mirror test). The pentanal (an aldehyde) will react, whereas 2-pentanone (a ketone) will not.

b. Again use Tollens' reagent. Alcohols (such as benzyl alcohol) do not react.

c. Both compounds are ketones, but 2-cyclohexenone has a carbon–carbon double bond and will be easily oxidized by potassium permanganate (the purple color of the $KMnO_4$ will turn to the brown color of MnO_2). The saturated ketone, cyclohexanone, will not react.

9.40 a. $C_6H_5-CH{=}O \xrightarrow[H^+]{CrO_3} C_6H_5-CO_2H$

b. $C_6H_5-CH{=}CHCH{=}O + 2Ag(NH_3)_2^+ + 3HO^- \longrightarrow C_6H_5-CH{=}CHCO_2^- + 2\,Ag + 4\,NH_3 + 2\,H_2O$

c. $(H_3CO)(HO)C_6H_3-CH{=}O \xrightarrow{NH_2NH_2} (H_3CO)(HO)C_6H_3-CH{=}N-NH_2$

d. 2-methyl-3-($CH_2CH{=}CHCH_2CH_3$)cyclopent-2-en-1-one (O, $CH_2CH{=}CHCH_2CH_3$, CH_3) $\xrightarrow{NaBH_4}$ corresponding cyclopentenol (HO, $CH_2CH{=}CHCH_2CH_3$, CH_3)

e. cyclohexenone (O) $\xrightarrow[2.\ H_2O]{1.\ LiAlH_4}$ cyclohexenol (OH)

f.

$$\text{camphor} \xrightarrow[\text{2. } H_3O^+]{\text{1. } CH_3MgBr} \text{tertiary alcohol (OH, } CH_3)$$

g.

$$HO\text{–}C_6H_3(OCH_3)\text{–}CH{=}O \xrightarrow[HCl]{CH_3OH} HO\text{–}C_6H_3(OCH_3)\text{–}CH(OCH_3)_2$$

9.41 All parts of this problem involve the preparation or hydrolysis of hemiacetals or acetals (or the corresponding ketone derivatives). See Sec. 9.7.

a.

$$CH_3CH_2CH_2CH{=}O + 2\ CH_3OH \xrightarrow{H^+} CH_3CH_2CH_2CH(OCH_3)_2 + H_2O$$

b.

$$CH_3CH(OCH_3)_2 \xrightarrow[H_2O]{H^+} CH_3CH{=}O + 2\ CH_3OH$$

c. In this case, the acetal is cyclic, and the product is a hydroxy aldehyde, which may exist in its cyclic hemiacetal form.

$$\text{2-methoxytetrahydropyran} \xrightarrow[H_2O]{H^+} HOCH_2CH_2CH_2CH_2CH{=}O + CH_3OH$$

$$HOCH_2CH_2CH_2CH_2CH{=}O \rightleftharpoons \text{2-hydroxytetrahydropyran}$$

d. In this reaction, a hemiacetal is converted to an acetal.

$$\text{2-hydroxytetrahydropyran} + CH_3CH_2OH \xrightarrow{H^+} \text{2-ethoxytetrahydropyran } (OCH_2CH_3) + H_2O$$

9.42 To prepare ethynylestradiol, we can treat estrone with 2 equivalents of an alkynyl anion:

$$\text{estrone} \xrightarrow[\text{2. } H_2O]{\text{1. } 2\ HC{\equiv}C^-\ Na^+} \text{ethynylestradiol}$$

Two equivalents are needed as the first equivalent would react with the phenol in an acid-base reaction. The second equivalent would react with the ketone moiety.

9.43 Protonation of acetone makes a carbocation that can react as an electrophile in sequential electrophilic aromatic substitution reactions with phenol.

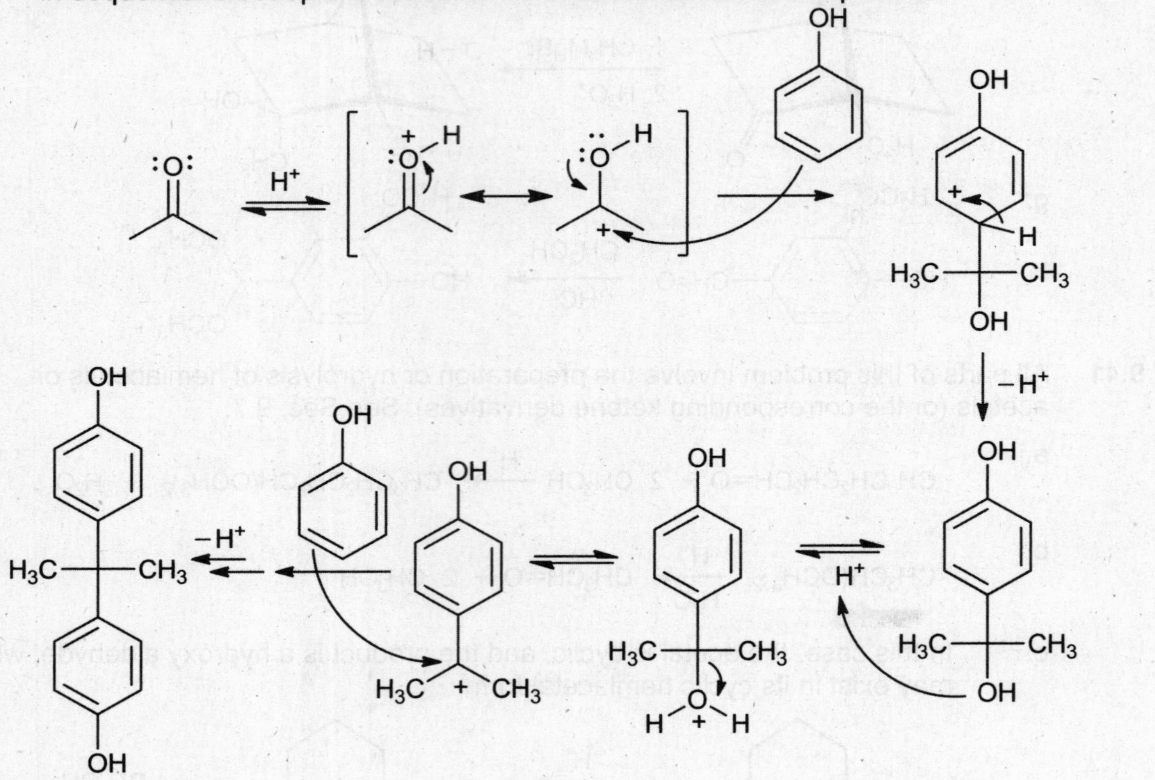

H^+

H_3C CH_3 OH OH

$-H^+$

OH OH OH OH

$-H^+$

H_3C CH_3 H^+

H_3C CH_3 H_3C + CH_3 H_3C CH_3 H_3C CH_3 OH

H + H

OH

9.44 For guidance, review Sec. 9.9.

a. $H_2C{=}O$ $\xrightarrow[\text{2. } H_2O,\ H^+]{\text{1. } CH_3MgBr}$ CH_3CH_2OH

b. H_3C(CH₂)₄CHO $\xrightarrow[\text{2. } H_2O]{\text{1. } CH_3CH_2MgBr}$ H_3C(CH₂)₄CH(OH)CH₂CH_3

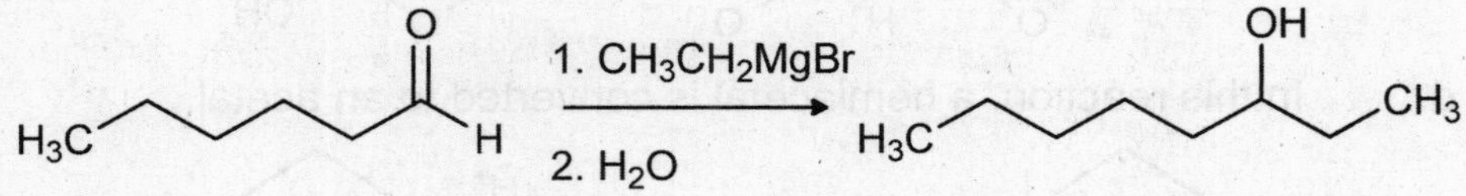

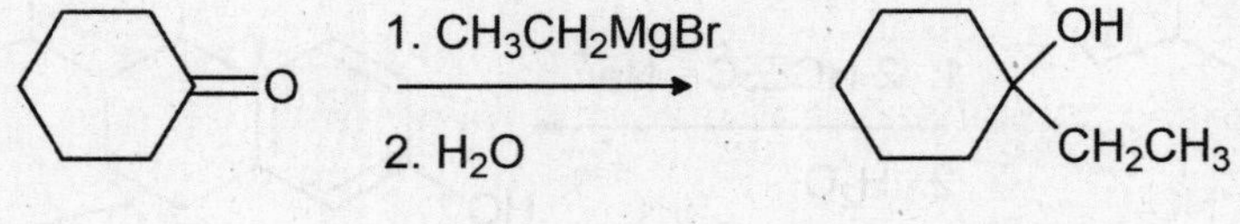

c. $C_6H_5C(=O)CH_3$ $\xrightarrow[\text{2. } H_2O,\ H^+]{\text{1. } CH_3MgBr}$ $C_6H_5C(CH_3)_2OH$

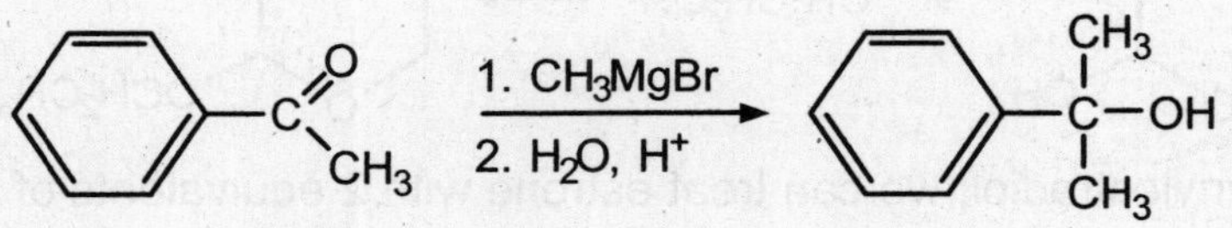

d. ethylene oxide $\xrightarrow[\text{2. } H_2O]{\text{1. } CH_3CH_2MgBr}$ $HOCH_2CH_2CH_2CH_3$

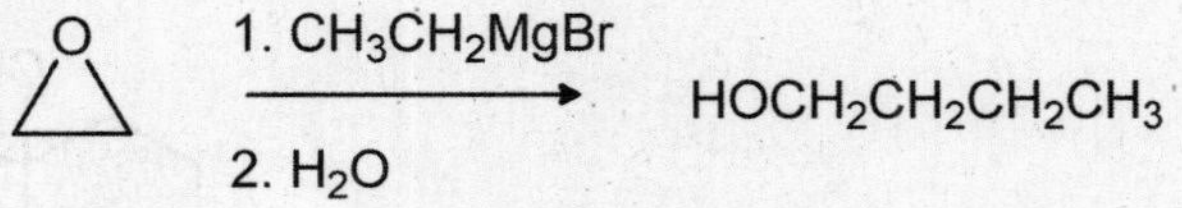

e. cyclohexanone $\xrightarrow[\text{2. } H_2O]{\text{1. } CH_3CH_2MgBr}$ 1-ethylcyclohexanol (OH, CH_2CH_3)

9.45 In each case, write the structure of the alcohol:

$$R_1-\overset{R_2}{\underset{R_3}{C}}-O-H$$

One of the R groups comes from the Grignard reagent. The rest of the molecule comes from the carbonyl compound. For example, If we select R_1 as the alkyl group to come from the Grignard reagent, then the carbonyl compound is:

$$\overset{R_2}{\underset{R_3}{C}}=O$$

a. 1-methylcyclohexanol (CH_3, OH on the same ring carbon)

CH_3MgX + cyclohexanone ⟶ 1-methylcyclohexyl O^-MgX $\xrightarrow{H_3O^+}$ 1-methylcyclohexanol (CH_3, OH)

For the remaining cases, we will not write the equations, but simply show how the initial reactants are derived.

b. $CH_3CH_2-\overset{H}{\underset{CH_2CH_2CH_2CH_2CH_3}{C}}-OH$ from CH_3CH_2MgX + $CH_3CH_2CH_2CH_2CH_2CH{=}O$

c. $CH_3-\overset{CH_3}{\underset{CH_2CH_2CH_3}{C}}-OH$ from CH_3MgX + $CH_3\underset{O}{\underset{\|}{C}}CH_2CH_2CH_3$

or

$CH_3CH_2CH_2-\overset{CH_3}{\underset{CH_3}{C}}-OH$ from $CH_3CH_2CH_2MgX$ + $CH_3\underset{O}{\underset{\|}{C}}CH_3$

Either of these combinations of reagents will work.

d. 1-cyclopentylcyclopentanol (HO) from cyclopentyl–MgX + O=cyclopentanone

In this case, the "free-standing" R group is selected to come from the Grignard reagent.

e. $C_6H_5-CH(OH)-CH_2CH_3$ from $C_6H_5-MgX + CH_3CH_2CH{=}O$

or

$C_6H_5-CH(OH)-CH_2CH_3$ from $CH_3CH_2MgX + O{=}CH-C_6H_5$

Either of these combinations of reagents will work.

f. $H_2C{=}CH-CH(OH)-CH_3$ from $H_2C{=}CHMgX + CH_3CH{=}O$

or

$H_2C{=}CH-CH(OH)-CH_3$ from $CH_3MgX + H_2C{=}CH-CH{=}O$

Either of these combinations of reagents will work.

Vinyl Grignard reagents are known, although they are a bit more difficult to prepare than simple alkyl Grignard reagents. Either pair of reagents will work.

9.46 a. The reaction is similar to that of a Grignard reagent with a ketone (see eq. 9.24). Also see eq. 9.26.

$$\text{cyclohexanone} \xrightarrow{Na^{+}\,{}^{-}C{\equiv}CH} \text{1-ethynylcyclohexanolate } (Na^{+}\,{}^{-}O) \xrightarrow[H^{+}]{H_2O} \text{1-ethynylcyclohexanol (HO)}$$

b. See eq. 9.28 for guidance.

$$\text{cyclopentanone} + HCN \rightleftharpoons \text{(HO, CN)}$$

cyclopentanone cyanohydrin

c. See Table 9.1 for guidance.

$$CH_3\underset{\displaystyle O}{\overset{\|}{C}}CH_2CH_3 + NH_2OH \rightleftharpoons CH_3\underset{\displaystyle NOH}{\overset{\|}{C}}CH_2CH_3 + H_2O$$

d. See Sec. 9.11 for examples.

$$C_6H_5-CH{=}O + H_2N-CH_2-C_6H_5 \longrightarrow C_6H_5-CH{=}N-CH_2-C_6H_5 + H_2O$$

e. See Table 9.1.

$$CH_3CH_2CH{=}O + H_2N-NH-C_6H_5 \longrightarrow CH_3CH_2CH{=}N-NH-C_6H_5 + H_2O$$

9.47 Chloral has three electronegative chlorine units attached to the α-carbon (Cl_3C–) to the aldehydes. The carbonyl carbon bears a partial positive charge so such electronegative elements destabilizes the carbonyl and favors the equilibrium towards significant formation of the hydrate product. For acetaldehyde, on the other hand, the equilibrium constant is 1.

9.48 a. See Sec. 9.12.

$$C_6H_5-C(=O)-CH_3 \xrightarrow[\text{2. } H_2O,\ H^+]{\text{1. } LiAlH_4} C_6H_5-CH(OH)-CH_3$$

b. See Sec. 9.12. Usually the aromatic ring will not be reduced, although under certain reaction conditions even this is possible.

$$C_6H_5-CH{=}CH-CH{=}O \xrightarrow[Ni]{\text{excess } H_2} C_6H_5-CH_2CH_2CH_2OH$$

c. The carbonyl group is reduced, but the carbon–carbon double bond is not reduced.

$$\text{2-methylcyclohex-2-enone} \xrightarrow[\text{2. } H_2O,H^+]{\text{1. } NaBH_4} \text{2-methylcyclohex-2-enol}$$

d. The carbonyl group is oxidized, but the aromatic ring is not.

$$C_6H_5-CH_2CH_2CH{=}O \xrightarrow[\text{reagent}]{\text{Jones'}} C_6H_5-CH_2CH_2CO_2H$$

e. The carbonyl group is oxidized, while the carbon–carbon double bond is not.

$$CH_3CH{=}CHCH{=}O \xrightarrow{Ag_2O} CH_3CH{=}CHCO_2H$$

9.49 Review Sec. 9.14.

a.

$$H_2C{=}\overset{OH}{\overset{|}{C}}{-}CH_2CH_3 \quad \text{and} \quad H_2C{-}\overset{OH}{\overset{|}{C}}{=}CHCH_3 \qquad (\textit{cis} \text{ and } \textit{trans})$$

There are two types of α-hydrogens in 2-butanone, and either may enolize:

$$\overset{\alpha}{CH_3}\overset{O}{\overset{\|}{C}}\overset{\alpha}{CH_2}CH_3$$

b.

$$C_6H_5{-}CH_2{-}CH{=}O \; : \; \text{The enol is} \; C_6H_5{-}CH{=}CH{-}OH$$

(*cis* and *trans*)

c. All the hydrogens are α to a carbonyl group. The CH_2 hydrogens that are α to *two* carbonyl groups are most likely to enolize.

$$\overset{\alpha}{CH_3}{-}\overset{O}{\overset{\|}{C}}{-}\overset{\alpha}{CH_2}{-}\overset{O}{\overset{\|}{C}}{-}\overset{\alpha}{CH_3}$$

$$CH_3{-}\overset{O}{\overset{\|}{C}}{-}CH{=}\overset{OH}{\overset{|}{C}}{-}CH_3 \longleftarrow \text{favored}$$

$$CH_2{=}\overset{OH}{\overset{|}{C}}{-}CH_2{-}\overset{O}{\overset{\|}{C}}{-}CH_3 \longleftarrow \text{not favored}$$

9.50 The α-hydrogens are exchanged for deuteriums:

$$CH_3CH_2{-}\overset{O}{\overset{\|}{C}}{-}C_6H_5 \xrightarrow[CH_3OD\ (\text{excess})]{CH_3ONa} CH_3CD_2{-}\overset{O}{\overset{\|}{C}}{-}C_6H_5$$

9.51 Review Sec. 9.16. Only hydrogens α to a carbonyl group will be replaced by deuterium.

a. 3-methylcyclopentanone (H, H on each α-carbon) $\xrightarrow[D_2O]{NaOD}$ 3-methylcyclopentanone with D, D on each α-carbon

Four hydrogens are replaced by deuterium.

b.

$$CH_3CH_2\underset{CH_3}{\underset{|}{C}}HCH{=}O \xrightarrow[D_2O]{NaOD} CH_3CH_2\underset{CH_3}{\underset{|}{C}}DCH{=}O$$

One hydrogen is replaced by deuterium.

9.52 Treatment of acetone with base will yield a small amount of the enolate anion. In the presence of Cl_2 as a halogen source, α chlorination can occur. Repeating this process twice would then lead to trichloroacetone. Note that after the first chlorination, the enolate will always be formed on the same side as the existing Cl since that enolate will be more stable (due to the presence of the electron-withdrawing Cl).

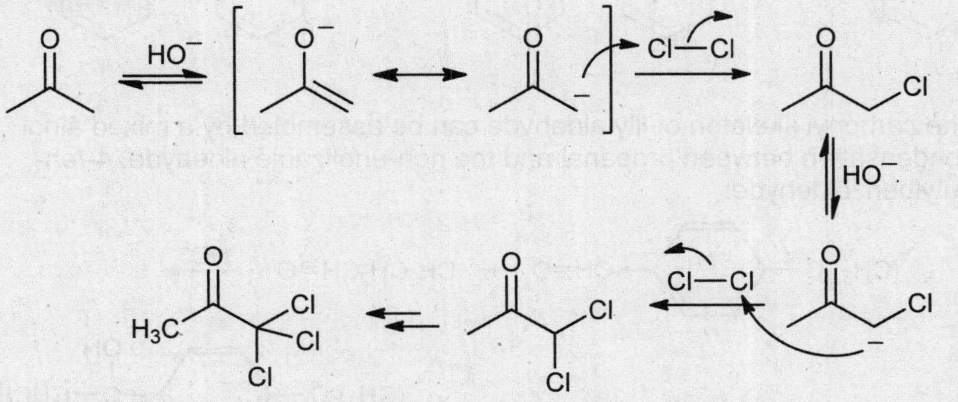

HO^- Cl—Cl HO^- Cl—Cl H$_3$C Cl Cl Cl Cl Cl Cl

9.53 Follow eqs. 9.51–9.53. The steps in the mechanism are as follows:

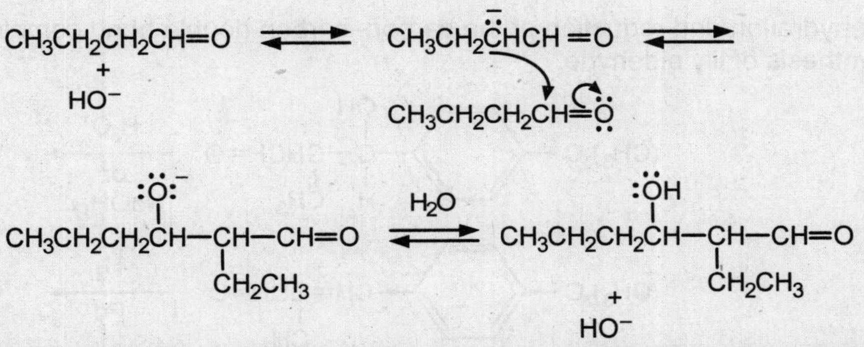

$$CH_3CH_2CH_2CH{=}O + HO^- \rightleftharpoons CH_3CH_2\ddot{C}^{-}HCH{=}O \quad CH_3CH_2CH_2CH{=}\ddot{O}$$

$$CH_3CH_2CH_2CH(O^-)\text{—}CH(CH_2CH_3)\text{—}CH{=}O \xrightleftharpoons{H_2O} CH_3CH_2CH_2CH(OH)\text{—}CH(CH_2CH_3)\text{—}CH{=}O + HO^-$$

9.54 The enol form of phenol benefits from the resonance stabilization and aromaticity of the benzene ring, but the keto form, with the single sp^3 center in the ring, cannot.

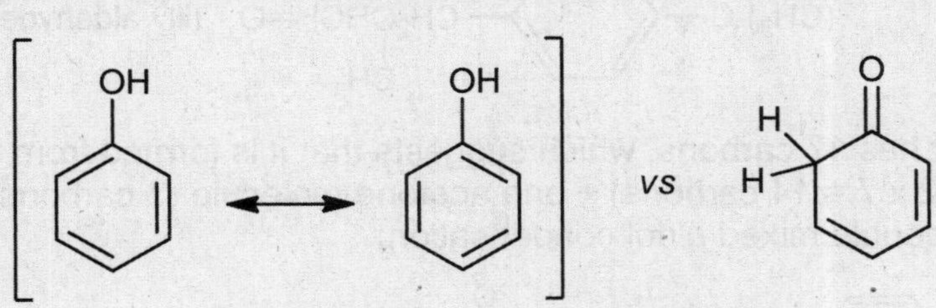

OH OH *vs* H H O

9.55 There are two enol forms (center and right) that are in equilibrium with each other. Both enol forms benefit from intramolecular hydrogen-bonding interactions.

9.56 The carbonyl skeleton of lily aldehyde can be assembled by a mixed aldol condensation between propanal and the non-enolizable aldehyde, 4-*tert*-butylbenzaldehyde:

$$(CH_3)_3C\text{—}C_6H_4\text{—}CH{=}O \; + \; CH_3CH_2CH{=}O \xrightarrow{HO^-} (CH_3)_3C\text{—}C_6H_4\text{—}CH(OH)\text{—}CH(CH_3)CH{=}O$$

Dehydration and reduction of the carbon–carbon double bond completes the synthesis of lily aldehyde.

$$(CH_3)_3C\text{—}C_6H_4\text{—}CH(OH)\text{—}CH(CH_3)CH{=}O \xrightarrow[\text{NaOH, }\Delta]{H_3O^+ \text{ or}} (CH_3)_3C\text{—}C_6H_4\text{—}CH{=}C(CH_3)CH{=}O$$

$$(CH_3)_3C\text{—}C_6H_4\text{—}CH{=}C(CH_3)CH{=}O \xrightarrow[Pd]{H_2} (CH_3)_3C\text{—}C_6H_4\text{—}CH_2CH(CH_3)CH{=}O \text{ (lily aldehyde)}$$

9.57 The product has 17 carbons, which suggests that it is formed from two benzaldehyde molecules (2 x 7 = 14 carbons) + one acetone molecule (3 carbons). The product forms by a double mixed aldol condensation:

$$2\, C_6H_5\text{—}CH{=}O \; + \; CH_3\overset{O}{\overset{\|}{C}}CH_3 \xrightarrow[\text{(see eq. 9.55)}]{HO^-} C_6H_5\text{—}CH{=}CH\text{—}\overset{O}{\overset{\|}{C}}\text{—}CH{=}CH\text{—}C_6H_5$$

dibenzalacetone

The product is yellow because of the extended conjugated system of double bonds.

9.58 The phenol-imine (pale yellow) structure retains a fully conjugated aromatic ring on the left side of the molecule, while the keto-enamine (red) compound has a quinoidal structure. Due to the benefits of aromaticity, the more stable phenol-imine compound requires a higher energy of excitation and appears pale yellow.

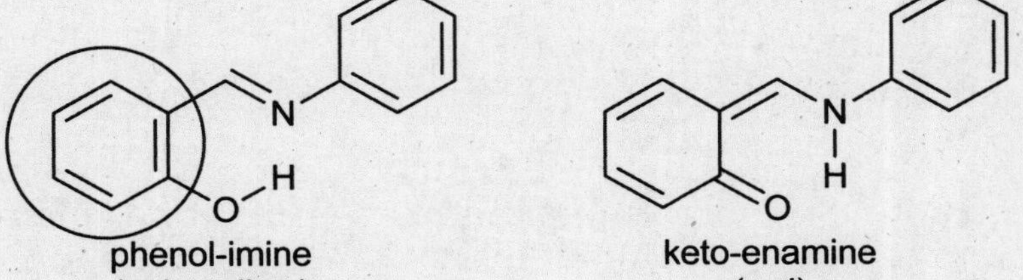

9.59 a. The vitamin B_6-enzyme complex is an imine (see Sec. 9.11):

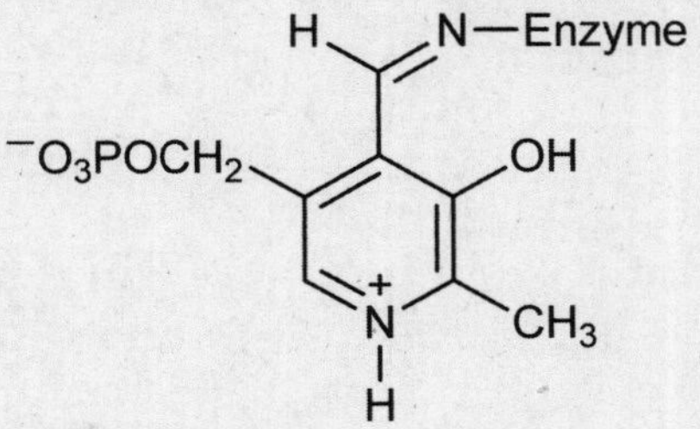

b. The amino group is replaced by a carbonyl group:

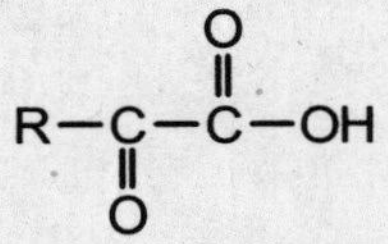

an α-keto acid

10

Carboxylic Acids and Their Derivatives

Chapter Summary

Carboxylic acids, the most important class of organic acids, contain the **carboxyl group**, –C(=O)-OH. The IUPAC ending for the names of these compounds is *-oic acid* but many common names (such as formic acid and acetic acid) are also used. An **acyl group**, R–C(=O)–, is named by changing the *-ic* ending of the corresponding acid to *-yl* [$CH_3C(=O)$– is acetyl].

The carboxyl group is polar and readily forms hydrogen bonds. A carboxylic acid dissociates to a **carboxylate anion** and a proton. In the carboxylate anion, the negative charge is delocalized equally over both oxygens. The pK_as of simple carboxylic acids are about 4–5, but the acidity can be increased by electron-withdrawing substituents (such as chlorine) close to the carboxyl group.

Carboxylic acids react with bases to give salts. These are named by naming the cation first and then the carboxylate anion. The name of the anion is obtained by changing the *-ic* ending of the acid name to *-ate* (acetic becomes acetate).

Carboxylic acids are prepared by at least four methods: (1) by oxidation of primary alcohols or aldehydes, (2) by oxidation of an aromatic side chain, (3) from a Grignard reagent and carbon dioxide, or (4) by hydrolysis of a nitrile, $RC{\equiv}N$.

Carboxylic acid derivatives are compounds in which the carboxyl –OH group is replaced by other groups. Examples include **esters, acyl halides, anhydrides,** and **amides.**

Esters, RCO_2R', are named as salts are; the R' group is named first, followed by the name of the carboxylate group (for example, $CH_3CO_2CH_2CH_3$ is ethyl acetate). Esters can be prepared from an acid and an alcohol, with a mineral acid catalyst (**Fischer esterification**). The key step of the mechanism is nucleophilic attack by the alcohol on the protonated carbonyl group of the acid. Many esters are used as flavors and perfumes.

Saponification is the base-mediated hydrolysis of an ester, yielding its component carboxylate salt and alcohol. **Ammonolysis** of esters gives amides. Esters react with Grignard reagents to give tertiary alcohols. With lithium aluminum hydride, on the other hand, they are reduced to primary alcohols.

Acid derivatives undergo nucleophilic substitution. The mechanism is as follows: the nucleophile adds to the trigonal carbonyl carbon to form a **tetrahedral intermediate**, which, through loss of a leaving group, becomes the trigonal product. The reaction can be regarded as an **acyl transfer**, the transfer of an acyl group from one nucleophile to another. The reactivity order of acid derivatives toward nucleophiles is acyl halides > anhydrides > esters > amides.

Acyl chlorides are prepared from acids and either $SOCl_2$ or PCl_5. They react rapidly with water to give acids, with alcohols to give esters, and with ammonia to give amides. **Acid anhydrides** react similarly but less rapidly. **Thioesters** are nature's acylating agents. They react with nucleophiles less rapidly than anhydrides but more rapidly than ordinary esters.

Amides can be prepared from ammonia and other acid derivatives. They can also be prepared by heating ammonium salts. They are named by replacing the *-ic* or *-oic* acid ending with *amide*.

Because of resonance, the C–N bond in amides has considerable C=N character. Rotation about that bond is restricted, and the amide group is planar. Amides are polar, form hydrogen bonds, and have high boiling points considering their molecular weights.

Amides react slowly with nucleophiles (such as water and alcohols). They are reduced to amines by $LiAlH_4$. **Urea**, made from CO_2 and NH_3, is an important fertilizer.

Some reactions of acid derivatives are summarized in Table 10.5. β-**Keto esters** can be prepared by the **Claisen condensation**, a reaction analogous to the aldol condensation but involving **ester enolates** as the reactive intermediates.

Reaction Summary

Acids

$$RCO_2H \rightleftharpoons RCO_2^- + H^+ \quad \text{(ionization)}$$

$$RCO_2H + NaOH \longrightarrow RCO_2^-\ Na^+ + H_2O \quad \text{(salt formation)}$$

Preparation of Acids

$$RCH_2OH \xrightarrow{Cr^{6+}} RCH{=}O \xrightarrow[\text{or } Ag^+]{Cr^{6+}} RCO_2H$$

$$ArCH_3 \xrightarrow[\text{or } O_2,\ Co^{3+}]{KMnO_4} ArCO_2H$$

$$RMgX + CO_2 \longrightarrow RCO_2MgX \xrightarrow{H_3O^+} RCO_2H$$

$$RC{\equiv}N + 2H_2O \xrightarrow{H^+ \text{ or } HO^-} RCO_2H + NH_3$$

Fischer Esterification

$$RCO_2H + R'OH \xrightarrow{H^+} RCO_2R' + H_2O$$

Saponification

$$RCO_2R' + NaOH \longrightarrow RCO_2^-\ Na^+ + R'OH$$

Ammonolysis of Esters

$$RCO_2R' + NH_3 \longrightarrow RCONH_2 + R'OH$$

Esters and Grignard Reagents

$$RCO_2R' \xrightarrow[2.\ H_3O^+]{1.\ 2\ R''MgX} R\text{—}C(R'')_2\text{—}OH + R'OH$$

Reduction of Esters

$$RCO_2R' \xrightarrow[2.\ H_3O^+]{1.\ LiAlH_4} RCH_2OH + R'OH$$

Preparation of Acyl Chlorides

$$RCO_2H + SOCl_2 \longrightarrow RCOCl + HCl + SO_2$$

$$RCO_2H + PCl_5 \longrightarrow RCOCl + HCl + POCl_3$$

Preparation of Anhydrides

$$R\text{—}C(=O)\text{—}Cl + Na^+\ {}^-O\text{—}C(=O)\text{—}R' \longrightarrow R\text{—}C(=O)\text{—}O\text{—}C(=O)\text{—}R' + NaCl$$

Reactions of Acyl Halides (or Anhydrides)

$R\text{—}C(=O)\text{—}Cl$ or $R\text{—}C(=O)\text{—}O\text{—}C(=O)\text{—}R$

$$\xrightarrow{H_2O} RCO_2H + HCl\ \text{(or } RCO_2H)$$

$$\xrightarrow{R'OH} RCO_2R' + HCl\ \text{(or } RCO_2H)$$

$$\xrightarrow{NH_3} RCONH_2 + NH_4Cl\ \text{(or } RCO_2H)$$

Amides from Ammonium Salts

$$RCO_2^-\ NH_4^+ \xrightarrow{heat} RCONH_2 + H_2O$$

Reactions of Amides

$$RCONH_2 + H_2O \xrightarrow{H^+ \text{ or } HO^-} RCO_2H + NH_3$$

$$RCONH_2 \xrightarrow{LiAlH_4} RCH_2NH_2$$

Urea

$$CO_2 + 2NH_3 \xrightarrow[pressure]{heat} H_2N\text{—}C(=O)\text{—}NH_2 + H_2O$$

Mechanism Summary

Nucleophilic Addition-Elimination (Nucleophilic Acyl Substitution)

$$Nu:^- + R(L)C{=}O \longrightarrow \left[R(L)(Nu)C{-}\ddot{O}:^- \right] \longrightarrow R(Nu)C{=}O + L:^-$$

Learning Objectives

1. Know the meaning of: carboxylic acid, carboxyl group, acyl group, carboxylate anion, acidity or ionization constant, inductive effect.
2. Know the meaning of: carboxylate salt, ester, acyl halide, acid anhydride, primary amide.
3. Know the meaning of: Fischer esterification, nucleophilic addition-elimination (nucleophilic acyl substitution), tetrahedral intermediate, saponification, ammonolysis, acyl transfer.
4. Given the IUPAC name of a carboxylic acid, salt, ester, amide, acyl halide, or anhydride, write its structural formula, and given the structure, write the name.
5. Know the common names of the monocarboxylic acids listed in Table 10.1 and the dicarboxylic acids listed in Table 10.2.
6. Know the systems for designating carbons in a carboxylic acid chain by numbers (IUPAC) or by Greek letters (common).
7. Know how to name acyl groups and how to write a formula given a name that includes acyl group nomenclature.
8. Given the formula of a carboxylic acid, write an expression for its ionization constant, K_a.
9. Write the resonance structures of a carboxylate anion.
10. Given two or more carboxylic acids with closely related structures, rank them in order of increasing (or decreasing) acidities (pK_as).
11. Tell whether a particular substituent will increase or decrease the acidity of a carboxylic acid.
12. Given a carboxylic acid and a base, write the equation for salt formation.
13. Given a carboxylic acid, tell which aldehyde or primary alcohol is needed for its preparation by oxidation.
14. Given an aromatic compound with alkyl substituents, tell which aromatic acid would be obtained from its oxidation.

15. Given a carboxylic acid, write an equation for its synthesis by hydrolysis of a nitrile (cyanide) or by the Grignard method.

16. Given an alcohol and an acid, write the equation for formation of the corresponding ester.

17. Write the steps in the mechanism for the acid-catalyzed (Fischer) esterification of a given carboxylic acid with a given alcohol.

18. Given the name or the structure of an ester, write the structure of the alcohol and acid from which it is derived.

19. Write an equation for the reaction of a given ester with aqueous base (saponification).

20. Write an equation for the reaction of a given ester with ammonia, a Grignard reagent, or lithium aluminum hydride.

21. Given a particular acid halide, write an equation for its preparation from an acid.

22. Given a particular anhydride, write an equation for its preparation from an acid chloride and a carboxylic acid salt.

23. Write the equation for the reaction of a given acyl halide or anhydride with a given nucleophile (especially with water, an alcohol, or ammonia).

24. Write equations for the preparation of a given amide from an acyl halide, acid anhydride, or ammonium salt.

25. Given a particular amide, write equations for its hydrolysis and reduction with lithium aluminum hydride.

26. Given a particular product that can be prepared by any of the reactions in this chapter, deduce the structures of the reactants required for its preparation, and write the equation for the reaction.

27. Write the Claisen condensation product from the reaction of an ester with an alkoxide or sodium hydride.

28. Write the steps in the mechanism of the Claisen condensation.

ANSWERS TO PROBLEMS

Problems Within the Chapter

10.1 a. OH O H_3C OH

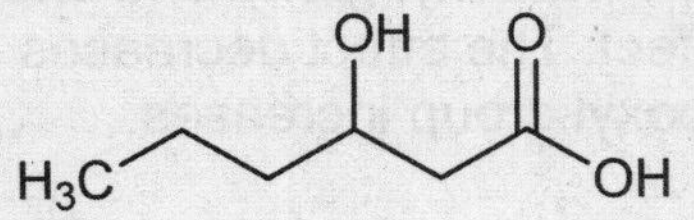

b. O H_3C OH I CH_3

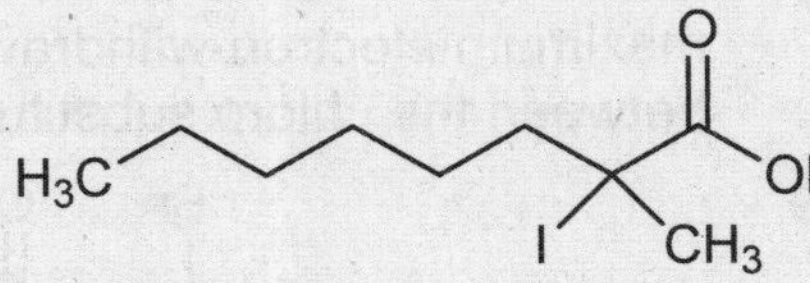

c. $CH_3C{\equiv}CCO_2H$

d. CH_3 O H_3C OH O

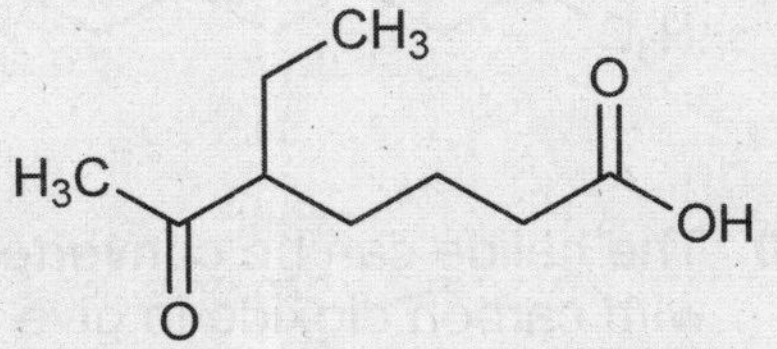

10.2 a. phenylethanoic acid b. 3,3-dibromopropanoic acid

c. 2-butenoic acid d. 5,5-dimethylpentanoic acid

10.3 a. b.

H$_3$C CH$_3$ OH O CO$_2$H NO$_2$

10.4 a. cyclopropanecarboxylic acid b. *p*-methoxybenzoic acid

10.5 a. b.

O C H CO$_2$H O C Br

c. d.

H$_3$C O Br O CH$_3$

10.6 K_a is 1.8 x 10^{-5} for acetic acid and 5.0 x 10^{-2} for dichloroacetic acid. K_a is larger for dichloroacetic acid; it is the stronger acid. The ratio is:

$$\frac{5.0 \times 10^{-2}}{1.8 \times 10^{-5}} = 2.8 \times 10^{3}$$

In other words, dichloroacetic acid is 2800 times stronger than acetic acid.

10.7 The negative charge in the benzoate anion cannot be delocalized into the aromatic, but can be distributed over the two oxygen atoms.

10.8 K_a for benzoic acid is 6.6 x 10^{-5} or 0.66 x 10^{-4}. For *o*-, *m*-, and *p*-chlorobenzoic acids, K_a is 12.5, 1.6, and 1.0 x 10^{-4}, respectively. All three chloro acids are stronger than benzoic acid. However, the difference is greatest for the ortho isomer since, in this isomer, the chloro substituent is closest to the carboxyl group and exerts the maximum electron-withdrawing inductive effect. The effect decreases as the distance between the chloro substituent and the carboxyl group increases.

10.9

H$_3$C Br O OH $\xrightarrow{KOH}$ H$_3$C Br O O$^-$ K$^+$ + H$_2$O

10.10 The halide can be converted to the corresponding Grignard reagent, which can react with carbon dioxide to give the carboxylic acid.

Cyclohexyl-Cl $\xrightarrow[\text{ether}]{\text{Mg}}$ cyclohexyl-MgCl $\xrightarrow[\text{2. } H^+]{\text{1. } CO_2}$ cyclohexyl-CO_2H

10.11 There are several possible approaches. First, the alcohol must be converted to a halide. The halide can then be converted to the corresponding Grignard reagent, which can react with carbon dioxide to provide the carboxylic acid.

$$CH_3CH_2CH_2OH \xrightarrow{PBr_3} CH_3CH_2CH_2Br \xrightarrow{\text{Mg, ether}} CH_3CH_2CH_2MgBr \xrightarrow[\text{2. } H_3O^+]{\text{1. } CO_2} CH_3CH_2CH_2CO_2H$$

Conversion of the halide to a nitrile followed by hydrolysis would also provide the carboxylic acid. This reaction is introduced in the next section (10.7d), so do not be concerned if it did not occur to you when initially working this problem.

$$CH_3CH_2CH_2Br \xrightarrow{NaCN} CH_3CH_2CH_2CN \xrightarrow[\text{heat}]{H_3O^+} CH_3CH_2CH_2CO_2H$$

10.12 It is not possible to conduct S_N2 displacements at sp^2-hybridized (aryl, vinyl) carbon atoms. The conversion can be accomplished via the Grignard reagent:

$$C_6H_5\text{-Br} \xrightarrow[\text{ether}]{\text{Mg}} C_6H_5\text{-MgBr} \xrightarrow[\text{2. } H_3O^+]{\text{1. } CO_2} C_6H_5\text{-}CO_2H$$

10.13

$$C_6H_5\text{-}CH_2Br \xrightarrow{\text{Mg, ether}} C_6H_5\text{-}CH_2MgBr \xrightarrow[\text{2. } H_3O^+]{\text{1. } CO_2} C_6H_5\text{-}CH_2CO_2H$$

$$C_6H_5\text{-}CH_2Br \xrightarrow{NaCN} C_6H_5\text{-}CH_2CN \xrightarrow[\text{heat}]{H_2O,\ H^+} C_6H_5\text{-}CH_2CO_2H$$

10.14 a. methyl methanoate b. cyclopropyl propanoate

10.15 a.

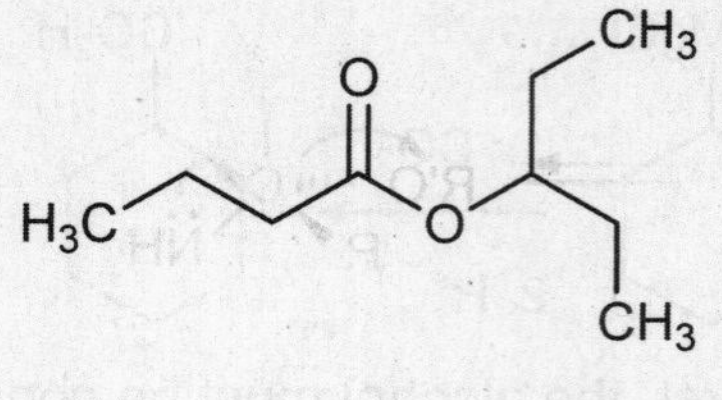

b.

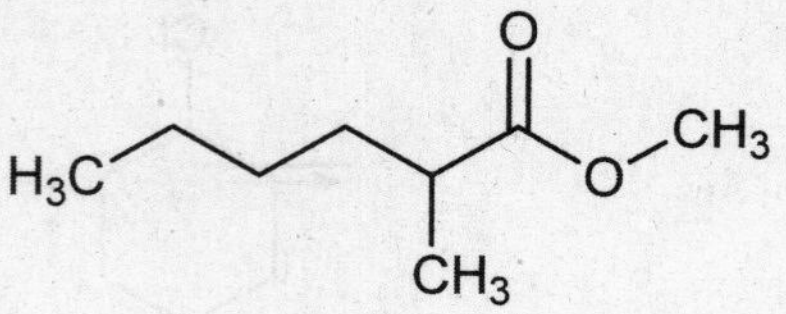

10.16

$$CH_3CH_2CH_2CH_2CO_2H + CH_3CH_2OH \overset{H^+}{\rightleftharpoons} CH_3CH_2CH_2CH_2CO_2CH_2CH_3$$

pentanoic acid — ethanol — ethyl pentanoate

10.17

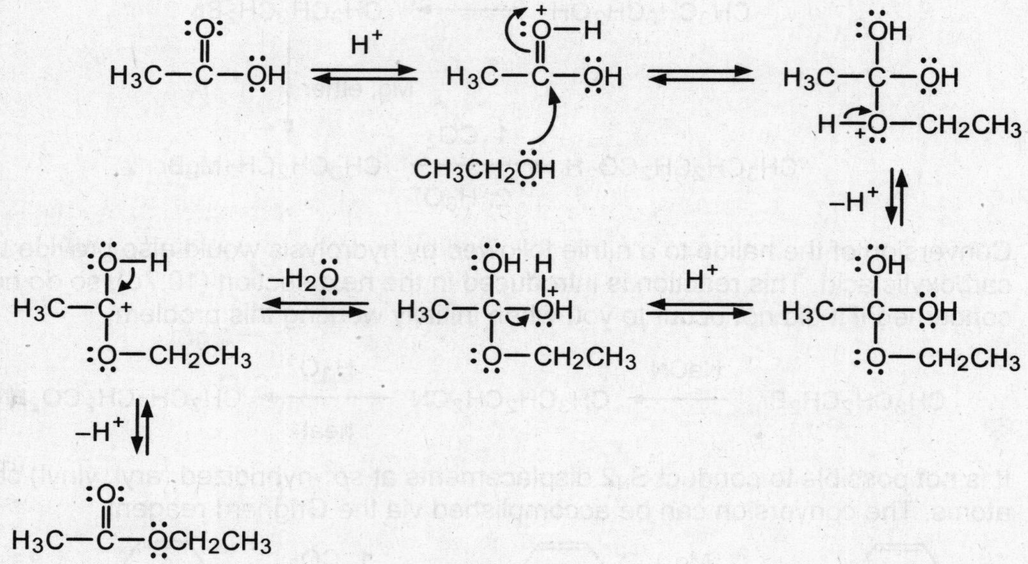

10.18

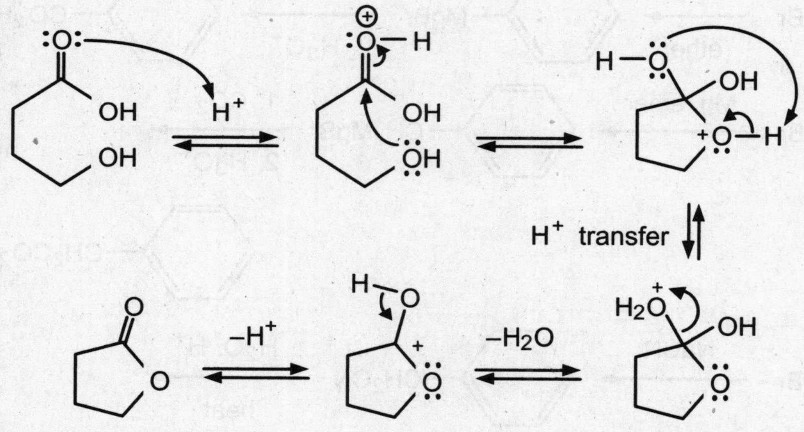

10.19

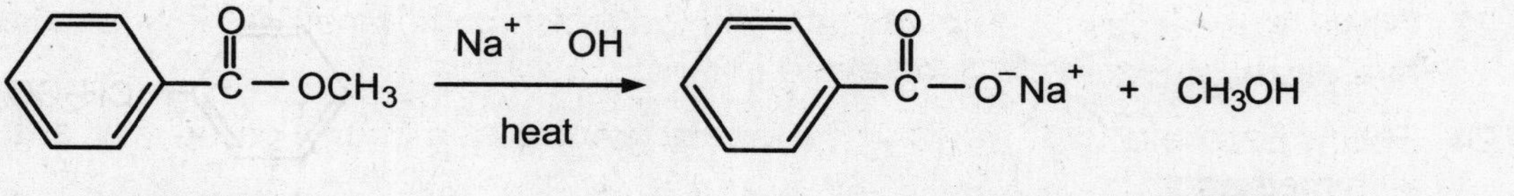

methyl benzoate — sodium benzoate — methanol

10.20

$$R'O-C(=O)-R \; + \; :NH_3 \rightleftharpoons R'O-C(O^-)(R)-NH_2^+-H \rightleftharpoons R'O-C(OH)(R)-\ddot{N}H_2 \rightleftharpoons R'O^- \; + \; H-O-C(R)=N^+H_2 \rightleftharpoons R'OH + R-C(=O)-NH_2$$

10.21 The Grignard reagent provides two of the three R groups attached to the hydroxyl-bearing carbon of the tertiary alcohol. The ester provides the third R group. So, from

$$\text{cyclopropyl}-\overset{O}{\overset{\|}{C}}-OCH_3 \; + \; \text{excess} \; C_6H_5MgBr$$

we get the tertiary alcohol

$$\text{cyclopropyl}-C(C_6H_5)_2-OH$$

10.22

$$C_6H_5-\overset{O}{\overset{\|}{C}}-OH \; + \; SOCl_2 \longrightarrow C_6H_5-\overset{O}{\overset{\|}{C}}-Cl \; + \; HCl \; + \; SO_2$$

benzoic acid → benzoyl chloride

10.23 When acyl halides come in contact with the moist membranes of the nose, they hydrolyze, producing HCl, a severe irritant.

10.24 Follow Example 10.7. Replace the methyl group with a phenyl group (Ph) and water with methanol:

$$Ph-\overset{O}{\overset{\|}{C}}-Cl + CH_3\ddot{O}H \rightleftharpoons Ph-\overset{O^-}{\overset{|}{C}}(-Cl)-\overset{+}{O}(H)CH_3 \xrightleftharpoons{-Cl^-} Ph-\overset{O}{\overset{\|}{C}}-\overset{+}{O}(H)CH_3$$

$$\xrightleftharpoons{-H^+} Ph-\overset{O}{\overset{\|}{C}}-OCH_3 + HCl$$

10.25 First prepare butanoyl chloride:

$$CH_3CH_2CH_2CO_2H \xrightarrow{SOCl_2} CH_3CH_3CH_2COCl$$

Then perform a Friedel–Crafts acylation:

$$H_3C-C_6H_5 \xrightarrow[AlCl_3]{CH_3CH_2CH_2COCl} H_3C-C_6H_4-\overset{O}{\overset{\|}{C}}-CH_2CH_2CH_3$$

10.26 a. $H_3C\text{–}CH_2CH_2\text{–}C(=O)\text{–}O\text{–}C(=O)\text{–}CH_2CH_2\text{–}CH_3$ b. $C_6H_5-\overset{O}{\overset{\|}{C}}-O-\overset{O}{\overset{\|}{C}}-C_6H_5$

10.27

$$C_6H_4(CO_2H)_2 \xrightarrow{heat} \text{phthalic anhydride} + H_2O$$

phthalic acid phthalic anhydride

10.28 No. The two carboxyl groups are *trans* to one another and cannot interact in an intramolecular fashion.

10.29 Use the middle part of eq. 10.39 as a guide:

$$H_3C\text{–}C(=O)\text{–}O\text{–}C(=O)\text{–}CH_3 + HOCH_2CH_2CH_2CH_2CH_3 \longrightarrow H_3C\text{–}C(=O)\text{–}OCH_2CH_2CH_2CH_2CH_3 + CH_3CO_2H$$

10.30 a. Use the top part of eq. 10.39 as a guide:

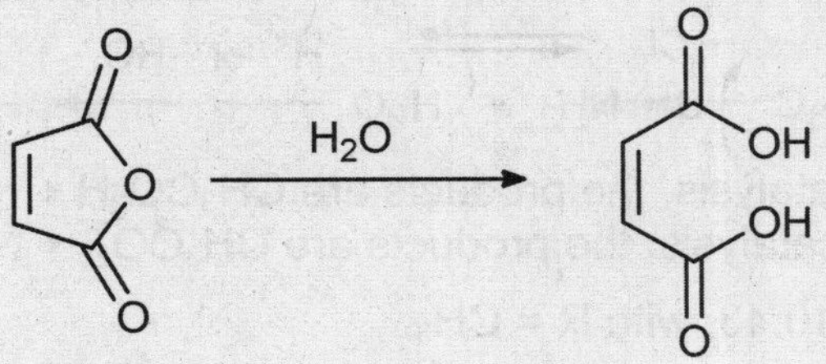
O
O
O
H_2O
O
OH
OH
O

b. Use the middle of eq. 10.39 as a guide:

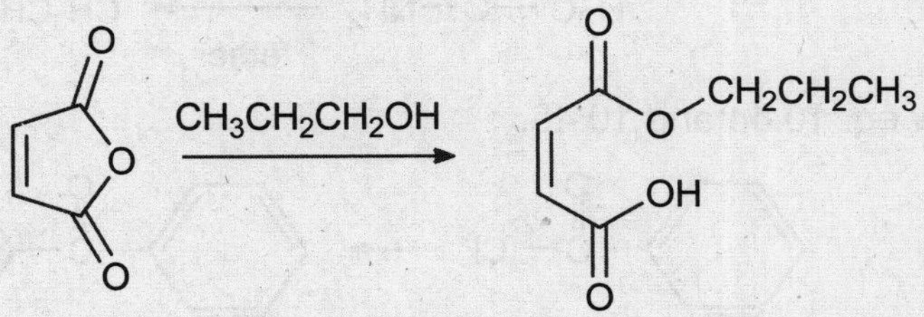
O
O
O
$CH_3CH_2CH_2OH$
O
O
$CH_2CH_2CH_3$
OH
O

c. Use the bottom of eq. 10.39 as a guide:

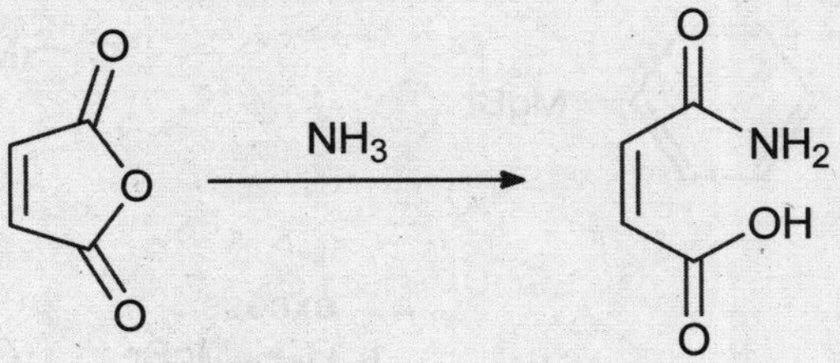
O
O
O
NH_3
O
NH_2
OH
O

10.31 a. 3-methylbutanamide b.

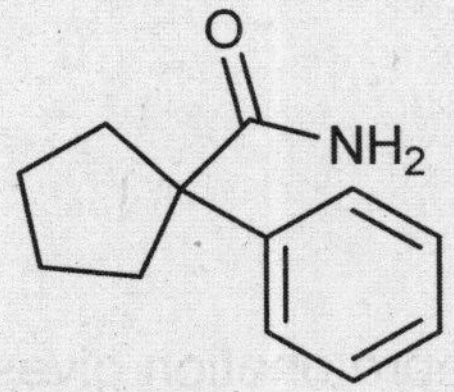
O
NH_2

10.32 See Sec. 10.20:

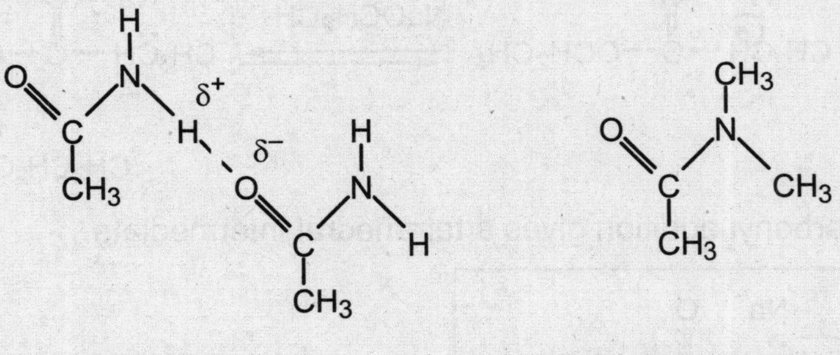
H
O
N
δ+
H
δ−
C
CH3
O
H
N
C
H
CH3
CH3
O
N
C
CH3
CH3

acetamide *N,N*-dimethylacetamide

The oxygen behaves as a Lewis base and the hydrogen behaves as a Lewis acid. *N,N*-dimethylacetamide has no acidic hydrogen, and hydrogen bonding cannot occur.

10.33 Follow eq. 10.42 with R = CH_3.

$$H_3C-\overset{O}{\overset{\|}{C}}-NH_2 + H_2O \xrightarrow{H^+ \text{ or } HO^-} H_3C-\overset{O}{\overset{\|}{C}}-OH + NH_3$$

With acid catalysis, the products are $CH_3CO_2H + NH_4^+$.
With base catalysis, the products are $CH_3CO_2^- + NH_3$.

10.34 Follow eq. 10.43, with R = CH_3.

$$H_3C-\overset{O}{\overset{\|}{C}}-NH_2 \xrightarrow[\text{ether}]{LiAlH_4} CH_3CH_2NH_2$$

10.35 Follow eq. 10.44 and 10.45.

$$C_6H_5-\overset{O}{\overset{\|}{C}}-Cl + C_6H_5MgBr \longrightarrow C_6H_5-\overset{O}{\overset{\|}{C}}-C_6H_5 + Cl-Mg-Br$$

$$\xrightarrow[\text{2. } H_3O^+]{\text{excess} \atop \text{1. } C_6H_5MgBr} (C_6H_5)_3C-OH$$

10.36 Step 1: Deprotonation gives an ester enolate.

$$CH_3\overset{H}{\overset{|}{C}}H-\overset{O}{\overset{\|}{C}}-OCH_2CH_3 \underset{}{\overset{NaOCH_2CH_3}{\rightleftharpoons}} CH_3\ddot{C}H^{-}\,Na^+-\overset{O}{\overset{\|}{C}}-OCH_2CH_3 + CH_3CH_2OH$$

Step 2: Carbonyl addition gives a tetrahedral intermediate.

$$\boxed{CH_3\ddot{C}H^{-}\,Na^+-\overset{O}{\overset{\|}{C}}-OCH_2CH_3} + CH_3CH_2-\overset{O}{\overset{\|}{C}}-OCH_2CH_3 \rightleftharpoons CH_3CH_2-\overset{Na^+\,^-O}{\overset{|}{\underset{CH_3CH_2O}{\underset{|}{C}}}}-\boxed{\underset{CH_3}{\underset{|}{C}H}-\overset{O}{\overset{\|}{C}}-OCH_2CH_3}$$

Step 3: The tetrahedral intermediate gives an ester and ethoxide.

$$CH_3CH_2-\overset{O^-Na^+}{\underset{CH_3CH_2O}{C}}-\underset{CH_3}{C}HCO_2CH_2CH_3 \rightleftharpoons CH_3CH_2-\overset{O}{\overset{\|}{C}}-\underset{CH_3}{CH}-\overset{O}{\overset{\|}{C}}-OCH_2CH_3 + CH_3CH_2O^-Na^+$$

Step 4: An acid-base reaction drives the equilibria to the right.

$$CH_3CH_2-\overset{O}{\overset{\|}{C}}-\underset{CH_3}{CH}-\overset{O}{\overset{\|}{C}}-OCH_2CH_3 + NaOCH_2CH_3 \longrightarrow CH_3CH_2-\overset{O}{\overset{\|}{C}}-\overset{Na^+}{\underset{CH_3}{\ddot{C}}}-\overset{O}{\overset{\|}{C}}-OCH_2CH_3 + CH_3CH_2OH$$

ADDITIONAL PROBLEMS

10.37 a. (structure: H_3C, H_3C, CO_2H) b. (structure: H_3C, Br, CO_2H)

c. (structure: H_3C, Cl, CO_2H) d. (structure: cyclopentane–CO_2H)

e. (structure: benzene ring with CO_2H and $CH(CH_3)_2$) f. (structure: H_3C, O, CO_2H)

g. $CH_3-C_6H_4-CO_2H$ h. (structure: H_3C, H_3C, CO_2H)

i. $CH_3O-C_6H_4-CH_2CO_2H$ j. (structure: naphthalene with CO_2H)

k. $H_2C{=}C(CH_3)CH(CH_3)CO_2H$

10.38 a. 4-bromo-4-methylpentanoic acid b. 3-methoxy-2-methylbutanoic acid
c. *p*-nitrobenzoic acid d. cyclohexanecarboxylic acid
e. propenoic acid (or acrylic acid) f. 2-phenylpropanoic acid
g. 2,2-difluoropropanoic acid h. 3-butynoic acid

10.39 a. The molecular weights are identical (74), but acids hydrogen-bond more effectively than alcohols do.

$$CH_3CH_2CO_2H > CH_3CH_2CH_2CH_2OH$$

bp 141°C bp 118°C

b. Chain branching generally lowers the boiling point. Thus, for these isomeric acids, the order is:

$$CH_3CH_2CH_2CH_2CO_2H > (CH_3)_3CCO_2H$$

bp 187°C bp 164°C

10.40 The factors that affect acidity of carboxylic acids are discussed in Sec. 10.5.

a. $ClCH_2CO_2H$; both substituents, chlorine and bromine, are approximately the same distance from the carboxyl group, but chlorine is more electronegative than bromine.

b. *o*-Bromobenzoic acid; the bromine is closer to the carboxyl group and is an electron-withdrawing substituent. Compare the pK_as of the corresponding chloro acids, given in Table 10.4.

c. CF_3CO_2H; fluorine is more electronegative than chlorine.

d. Benzoic acid; the methoxy group is an electron-releasing substituent when in the *para* position and may destabilize the anion because of the presence of resonance structures such as

$$\left[CH_3\ddot{O}-C_6H_4-C(=O)O^- \longleftrightarrow CH_3\overset{+}{\ddot{O}}=C_6H_4=C(O^-)O^- \right]$$

which bring two negative charges near one another.

e. $CH_3CHClCO_2H$; the chlorine, which is electron-withdrawing, is closer to the carboxyl group.

10.41 See Sec. 10.6 if you have any difficulty.

a.

$$ClCH_2CH_2CO_2H \ + \ K^{+}\,{}^{-}OH \longrightarrow ClCH_2CH_2CO_2^{-}\ K^{+} \ + \ H_2O$$

Salt formation occurs at room temperature. If the reagents are heated for some time, an S_N2 displacement on the primary chloride may also occur, giving the salt of hydroxypropanoic acid, $HOCH_2CH_2CO_2^-K^+$.

b.

$$2\ CH_3(CH_2)_8CO_2H + Ca(OH)_2 \longrightarrow [CH_3(CH_2)_8CO_2^-]\ Ca^{2+} + 2\ H_2O$$

10.42 a.

$$CH_3CH_2CH_2CH_2OH \xrightarrow{Na_2Cr_2O_7} CH_3CH_2CH_2CO_2H$$

b.

$$CH_3CH_2CH_2OH \xrightarrow{HBr} CH_3CH_2CH_2Br$$

$$CH_3CH_2CH_2Br \xrightarrow{NaCN} CH_3CH_2CH_2CN \xrightarrow[\text{heat}]{H_3O^+} CH_3CH_2CH_2CO_2H$$

$$CH_3CH_2CH_2Br \xrightarrow[\text{ether}]{Mg} CH_3CH_2CH_2MgBr \xrightarrow[\text{2. } H_2O,\ H^+]{\text{1. } CO_2} CH_3CH_2CH_2CO_2H$$

c.

$$Cl-C_6H_4-CH_3 \xrightarrow[\text{2. } H_3O^+]{\text{1. } KMnO_4\text{, heat}} Cl-C_6H_4-CO_2H$$

d. cyclohexyl–CN $\xrightarrow[\text{heat}]{H_3O^+}$ cyclohexyl–CO_2H

e. ethylene oxide $\xrightarrow[H^+]{CH_3OH}$ $CH_3OCH_2CH_2OH$ $\xrightarrow{Na_2Cr_2O_7}$ $CH_3OCH_2CO_2H$

f. C_6H_5–Br $\xrightarrow[\text{ether}]{Mg}$ C_6H_5–MgBr $\xrightarrow[2.\ H_2O,\ H^+]{1.\ CO_2}$ C_6H_5–CO_2H

10.43 The nitrile route would involve an S_N2 displacement of bromide by cyanide, a highly unlikely step when the alkyl halide is tertiary. The Grignard route, on the other hand, works well for all alkyl halides—primary, secondary, and tertiary.

10.44 a. H_3C–C(=O)–O–$CH_2CH(CH_3)CH_3$

b. $H-\overset{O}{\overset{\|}{C}}-OCH(CH_3)_2$

c. $CH_3CH_2\underset{Cl}{CH}CO_2^-Na^+$

d. $(CH_3CO_2^-)_2Ca^{+2}$

e. C_6H_5–C(=O)–O–C_6H_5

f. benzene ring bearing C(=O)NH_2 and, ortho to it, CH_3

g. H_3C–CH_2–CH(OCH_3)–C(=O)–Cl

h. C_6H_5–C≡N

i. $CH_3CH_2-\overset{O}{\overset{\|}{C}}-O-\overset{O}{\overset{\|}{C}}-CH_2CH_3$

j. cyclohexane ring bearing CO_2H and, on the adjacent carbon, C(=O)CH_3

k. five-membered lactone ring (C=O next to the ring O) with H_3C on the carbon next to the carbonyl

10.45 a. ammonium *p*-bromobenzoate
b. calcium butanoate
c. phenyl 4-methylpentanoate
d. methyl trifluoroacetate
e. formamide
f. butanoic anhydride

10.46 The first word of the name refers to the alcohol portion and the second to the acid portion:

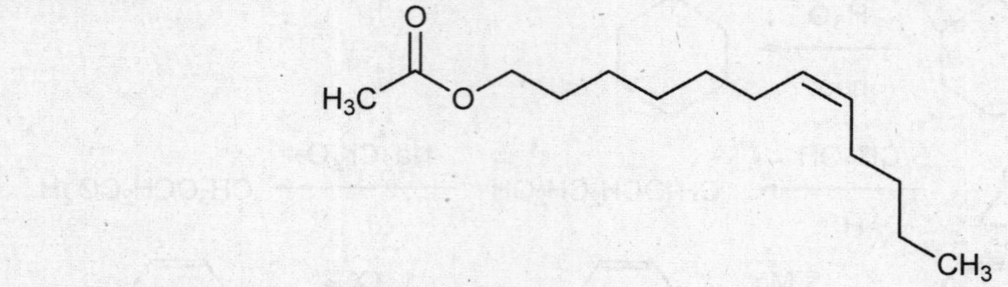

10.47 Carboxylic acids form dimmers as good hydrogen bonds can form from one carboxylic acid to another. Similar structures can be formed between mixed acids, such as carboxylic acids and sulfuric acid.

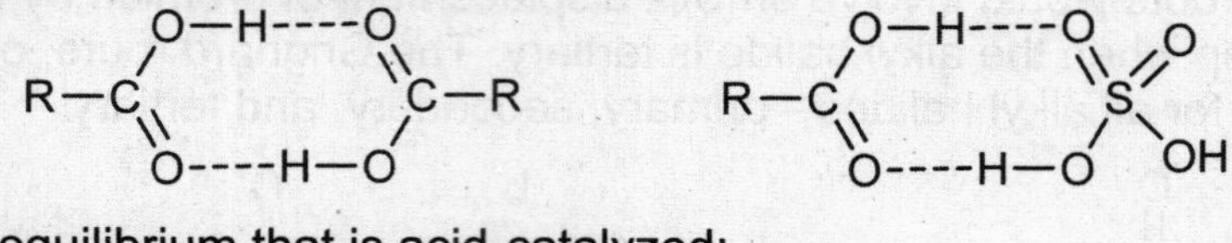

10.48 This is an equilibrium that is acid-catalyzed:

$$CH_3CH_2CH_2CO_2H \quad + \quad CH_3CH_2OH \overset{H^+}{\rightleftharpoons} CH_3CH_2CH_2CO_2CH_2CH_3 \quad + \quad H_2O$$

10.49 Compare with the answer to Problem 10.17.

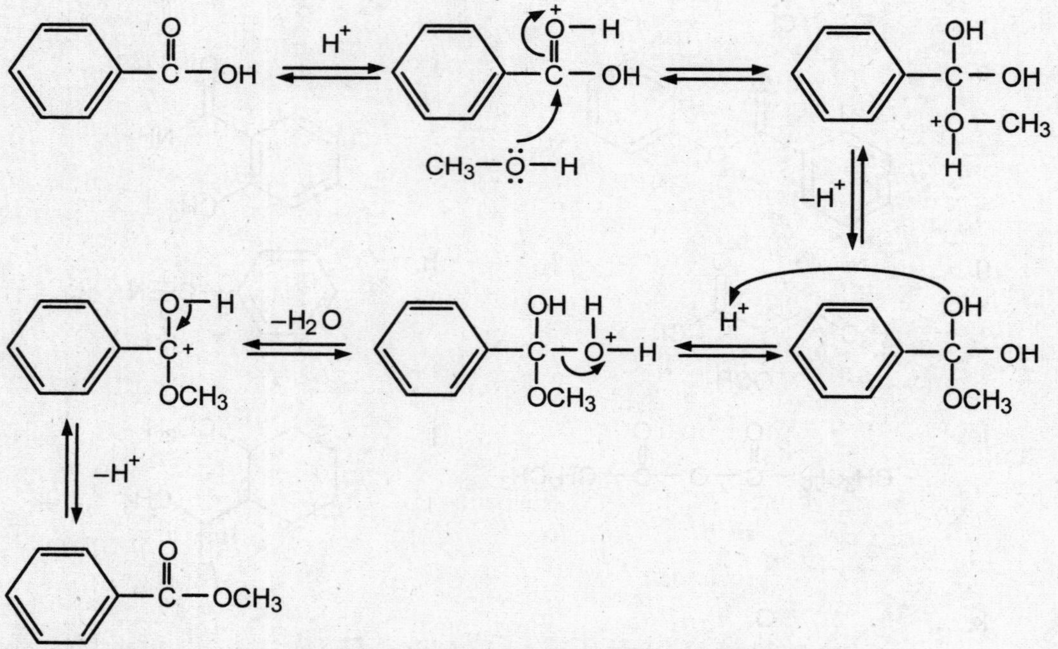

10.50 First, we need to add the carboxyl unit and then form the ester.

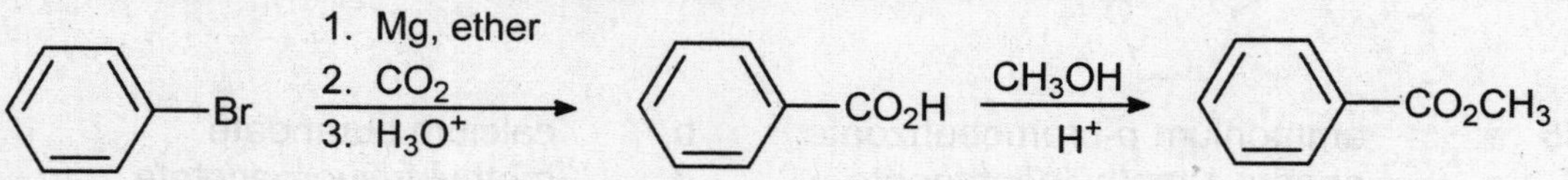

1. Mg, ether
2. CO_2
3. H_3O^+

CH_3OH, H^+

$-CO_2H$ → $-CO_2CH_3$

10.51 a. See eq. 10.22, where R = phenyl and R' = propyl.

$$C_6H_5\text{–}CO_2CH_2CH_2CH_3 + Na^+\ {}^-OH \xrightarrow{\text{heat}} C_6H_5\text{–}CO_2^-\ {}^+Na + CH_3CH_2CH_2OH$$

b. Compare with eq. 10.25.

$$C_6H_5\text{–}CO_2CH_2CH_2CH_3 + NH_3 \xrightarrow{\text{heat}} C_6H_5\text{–}\overset{O}{\overset{\|}{C}}\text{–}NH_2 + CH_3CH_2CH_2OH$$

c. See Sec. 10.15.

$$C_6H_5\text{–}CO_2CH_2CH_2CH_3 \xrightarrow[\text{2. } H_3O^+]{\text{1. 2 } C_6H_5MgI} (C_6H_5)_3C\text{–}OH + CH_3CH_2CH_2OH$$

d. See Sec. 10.16.

$$C_6H_5\text{–}CO_2CH_2CH_2CH_3 \xrightarrow[\text{2. } H_3O^+]{\text{1. 2 } LiAlH_4} C_6H_5\text{–}CH_2OH + CH_3CH_2CH_2OH$$

10.52 a.

$$CH_3CH_2\text{–}\overset{O}{\overset{\|}{C}}\text{–}OCH_2CH_3 + HO^- \rightleftharpoons CH_3CH_2\text{–}\overset{O^-}{\overset{|}{C}}(OH)\text{–}OCH_2CH_3$$

$$\rightleftharpoons CH_3CH_2\text{–}\overset{O}{\overset{\|}{C}}\text{–}O\text{–}H + CH_3CH_2O^- \rightleftharpoons CH_3CH_2\text{–}\overset{O}{\overset{\|}{C}}\text{–}O^- + CH_3CH_2OH$$

b.

$$CH_3CH_2C(=O)OCH_2CH_3 + H_3N \rightleftharpoons CH_3CH_2-C(O^-)(OCH_2CH_3)-{}^+NH_3$$

H^+ transfer

$$CH_3CH_2-C(OH)(NH_2)-OCH_2CH_3 \rightleftharpoons CH_3CH_2C(=O)NH_2 + CH_3CH_2OH$$

10.53 In each case, the two identical organic groups attached to the hydroxyl-bearing carbon come from the Grignard reagent, and the third group comes from the acid part of the ester.

a. CH_3CH_2MgBr + $C_6H_5-C(=O)-OR$

b. C_6H_5-MgBr + $CH_3CH_2CH_2-C(=O)-OR$

The identity of the R group in the ester does not affect the product structure. R is usually CH_3 or CH_3CH_2.

10.54 a. The carbonyl group in esters is less reactive than the carbonyl group of ketones because of the possibility of resonance in esters that delocalizes, to the "ether" oxygen, some of the positive charge associated with the carbonyl carbon atom.

$$\left[R-C(=\ddot{O}:)-\ddot{O}-R \longleftrightarrow R-\overset{+}{C}(-\ddot{O}:^-)-\ddot{O}-R \longleftrightarrow R-C(-\ddot{O}:^-)=\overset{+}{O}-R \right]$$

Such a resonance structure is not possible in a ketone.

$$\left[R_2C=\ddot{O}: \longleftrightarrow R_2\overset{+}{C}-\ddot{O}:^- \right]$$

Therefore, the carbonyl carbon in esters is less positively charged and less susceptible to nucleophilic attack than is the carbonyl carbon of ketones.

b. In benzoyl chloride, the positive charge on the carbonyl carbon can be delocalized in the aromatic ring.

$$\left[C_6H_5-\overset{O}{\overset{\|}{C}}-Cl \longleftrightarrow C_6H_5-\overset{O^-}{\overset{|}{\underset{+}{C}}}-Cl \longleftrightarrow {}^{+}C_6H_5{=}\overset{O^-}{\overset{|}{C}}-Cl \right]$$

Such delocalization is not possible in cyclohexanecarbonyl chloride or any other aliphatic acid chloride. For this reason, aryl acid chlorides are usually less reactive toward nucleophiles than are aliphatic acid chlorides.

10.55 a.

$$CH_3CH_2CH_2-\overset{O}{\overset{\|}{C}}-Cl + H_2O \longrightarrow CH_3CH_2CH_2-\overset{O}{\overset{\|}{C}}-OH + HCl$$

b.

$$CH_3CH_2CH_2-\overset{O}{\overset{\|}{C}}-Br + NH_3 \longrightarrow CH_3CH_2CH_2-\overset{O}{\overset{\|}{C}}-NH_2 + HBr$$

c.

$$CH_3CH_2-C_6H_5 + (CH_3)_2CH-\overset{O}{\overset{\|}{C}}-Cl \xrightarrow{AlCl_3} CH_3CH_2-C_6H_4-\overset{O}{\overset{\|}{C}}-CH(CH_3)_2$$

d.

$$HO_2C-CH_2CH_2-CO_2H \xrightarrow{235°C} \text{succinic anhydride} + H_2O$$

e.

$$C_6H_5-\overset{O}{\overset{\|}{C}}-Cl + CH_3CH_2OH \longrightarrow C_6H_5-\overset{O}{\overset{\|}{C}}-OCH_2CH_3 + HCl$$

f.

$$H_3C-CH_2CH_2CH_2CH_2-OH \xrightarrow{Ac_2O} H_3C-CH_2CH_2CH_2CH_2-O-\overset{O}{\overset{\|}{C}}-CH_3 + CH_3CO_2H$$

g.

$$\text{phthalic anhydride} \xrightarrow[H^+]{CH_3CH_2OH \text{ (1 equiv)}} C_6H_4(CO_2CH_2CH_3)(CO_2H)$$

h.

$$\text{phthalic anhydride} \xrightarrow[H^+]{CH_3CH_2OH\ (excess)} C_6H_4(CO_2CH_2CH_3)_2$$

i.

$$Cl-\overset{O}{\overset{\|}{C}}-(CH_2)_4-\overset{O}{\overset{\|}{C}}-Cl \xrightarrow{NH_3\ (excess)} H_2N-\overset{O}{\overset{\|}{C}}-(CH_2)_4-\overset{O}{\overset{\|}{C}}-NH_2 + 2NH_4^+\ Cl^-$$

10.56 a. See eq.10.33.

$$CH_3CH_2CH_2-\overset{O}{\overset{\|}{C}}-OH + PCl_5 \longrightarrow CH_3CH_2CH_2-\overset{O}{\overset{\|}{C}}-Cl + HCl + POCl_3$$

b. See eq. 10.32.

$$CH_3(CH_2)_6CO_2H + SOCl_2 \longrightarrow CH_3(CH_2)_6COCl + HCl + SO_2$$

c. See eq. 10.8.

$$C_6H_4(CH_3)_2\ (ortho) \xrightarrow[heat]{KMnO_4} C_6H_4(CO_2H)_2\ (ortho)$$

d. See eq. 10.41.

$$C_6H_5-\overset{O}{\overset{\|}{C}}-O^-\ NH_4^+ \xrightarrow{heat} C_6H_5-\overset{O}{\overset{\|}{C}}-NH_2 + H_2O$$

e. See eq. 10.43.

$$CH_3(CH_2)_5-\overset{O}{\overset{\|}{C}}-NH_2 \xrightarrow{LiAlH_4} CH_3(CH_2)_5CH_2NH_2$$

f. See eq. 10.28.

$$C_5H_9-CO_2CH_2CH_3 \xrightarrow[2.\ H_3O^+]{1.\ LiAlH_4} C_5H_9-CH_2OH + CH_3CH_2OH$$

10.57 Ketones are more reactive toward nucleophiles than esters. Reduction therefore occurs at the ketone carbonyl group, to give

$$CH_3CH(OH)CH_2CH_2CO_2CH_3$$

10.58 The method combines the formation of a cyanohydrin (Sec. 9.10) with the hydrolysis of a cyanide to an acid (Sec. 10.7d).

$$C_6H_5-CH{=}O \xrightarrow{HCN} C_6H_5-\underset{H}{\overset{OH}{C}}-CN \xrightarrow[heat]{H_3O^+} C_6H_5-\underset{H}{\overset{OH}{C}}-CO_2H$$

10.59 Use eqs. 10.48–10.50 as a guide. The overall equation is

2 [ester] $\xrightarrow[\text{2. } H^+]{\text{1. } NaOCH_3}$ [β-keto ester] + CH_3OH

The steps are as follows:

Step 1:

CH_3O^- ⇌ + CH_3OH

Step 2:

⇌

+ CH_3OH

Step 3:

$\overset{CH_3O^-}{\rightleftharpoons}$ + CH_3OH

The reaction is neutralized with acid to obtain the final product.

10.60

$$CH_3CH_2O\overset{O}{\overset{\|}{C}}CH(CH_2)_3\overset{O}{\overset{\|}{C}}OCH_2CH_3 + CH_3CH_2O^- \rightleftharpoons CH_3CH_2O\overset{O}{\overset{\|}{C}}\overset{-}{C}H(CH_2)_3\overset{O}{\overset{\|}{C}}OCH_2CH_3$$

$CH_3CH_2O^-$ + $CH_3CH_2O_2C$–(2-oxocyclopentyl) ⇌ $CH_3CH_2O_2C$–(cyclopentyl bearing O^- and OCH_2CH_3)

↓

$CH_3CH_2O\overset{O}{\overset{\|}{C}}$–(carbanion of 2-oxocyclopentyl) + H^+ + CH_3CH_2OH ⟶ $CH_3CH_2O\overset{O}{\overset{\|}{C}}$–(2-oxocyclopentyl, H)

Deprotonation of the β-ketoester drives the equilibrium to the right. Neutralization of the reaction mixture gives ethyl 2-oxocyclopentanecarboxylate.

10.61 The enolate of ethyl acetate behaves as a nucleophile and the nonenolizable ester, ethyl benzoate, behaves as an electrophile.

$$C_6H_5CO_2CH_2CH_3 + CH_3CH_2O\overset{O}{\overset{\|}{C}}CH_3 \xrightarrow[2.\ H_3O^+]{1.\ NaOCH_2CH_3} C_6H_5\overset{O}{\overset{\|}{C}}CH_2\overset{O}{\overset{\|}{C}}OCH_2CH_3$$

10.62 Review Sec. 10.20.

$$\left[H_3C-\overset{:\ddot{O}}{\overset{\|}{C}}-\underset{H}{\underset{|}{\ddot{N}}}-H \longleftrightarrow H_3C-\overset{:\ddot{O}:^-}{\overset{|}{C}}=\underset{H}{\underset{|}{\overset{+}{N}}}-H \right]$$

The six atoms that lie in one plane are

$$C-\overset{:\ddot{O}:^-}{\overset{|}{C}}=\underset{H}{\underset{|}{\overset{+}{N}}}-H$$

10.63 Acetaminophen can be prepared as follows:

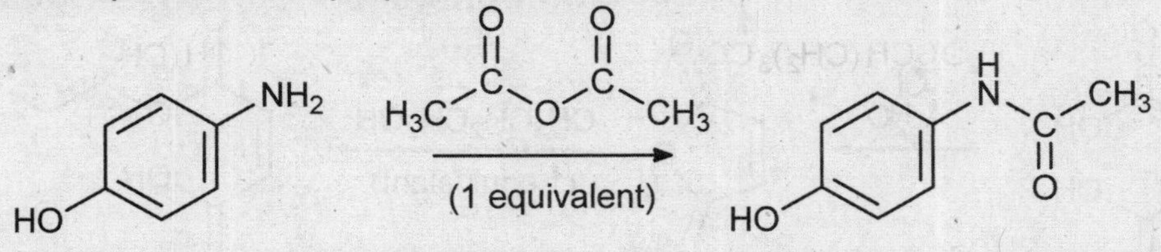

(1 equivalent)

Note that with only 1 equivalent of acetic anhydride, the more reactive amino group would react first.

10.64 Addition of catalytic amounts of NaOH would form the six-membered ring lactone:

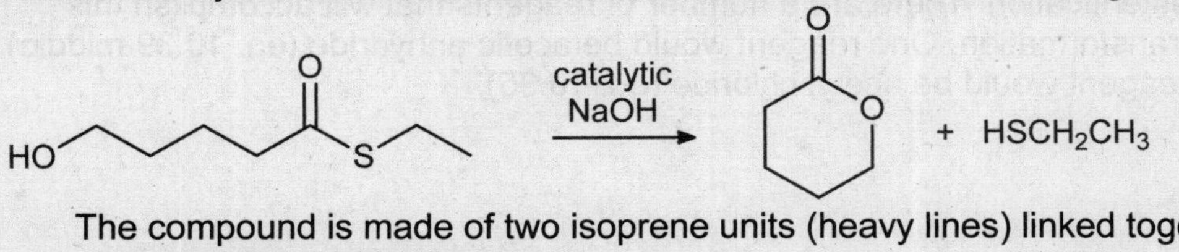

catalytic NaOH

+ $HSCH_2CH_3$

10.65 a. The compound is made of two isoprene units (heavy lines) linked together as shown by the dashed lines.

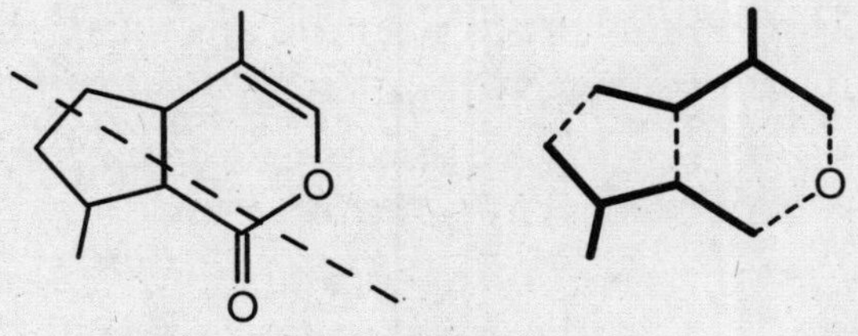

b. Nepetalactone has three stereogenic carbons:

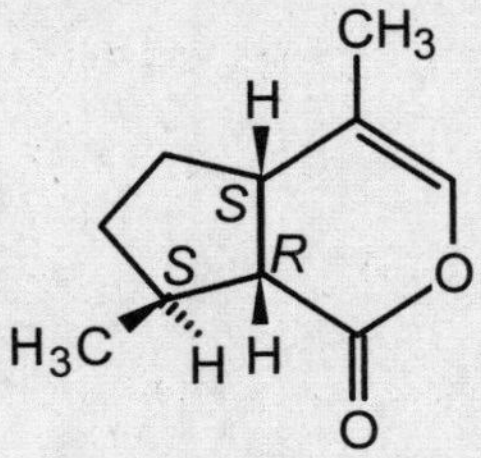

10.66 There are three stereogenic centers with the following absolute configurations:

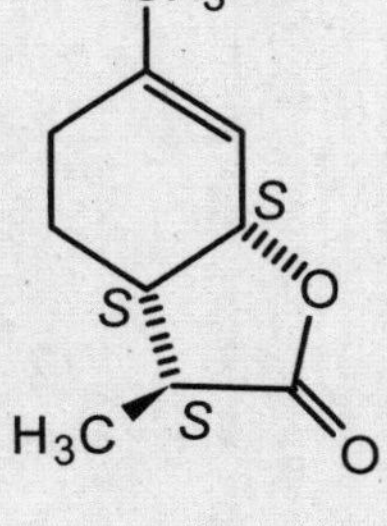

10.67 The anhydride is critical for this sequence:

$$\text{maleic acid (HOOC–CH=CH–COOH)} \xrightarrow{Ac_2O} \text{maleic anhydride} \xrightarrow[\text{(1 equivalent)}]{CH_3CH_2CH_2OH} \text{HOOC–CH=CH–CO–O–CH}_2\text{CH}_2\text{CH}_3$$

10.68 The first step involves formation of a lactone (Sec. 10.12) and can be accomplished by heating with H^+ (H_2SO_4) as shown in eq. 10.21. The second step is an esterification. There are a number of reagents that will accomplish this transformation. One reagent would be acetic anhydride (eq. 10.39 middle). Another reagent would be acetyl chloride (eq. 10.35).

11

Amines and Related Nitrogen Compounds

Chapter Summary

Amines are organic derivatives of ammonia. They may be **primary**, **secondary**, or **tertiary**, depending on whether one, two, or three organic groups are attached to the nitrogen. The nitrogen is sp^3-hybridized and pyramidal, nearly tetrahedral.

The **amino group** is $-NH_2$. Amines are named according to the Chemical Abstracts (CA) system by adding the suffix *-amine* to the names of the alkyl groups attached to the nitrogen. Amines can also be named using the IUPAC system in which the amino group is named as a substituent. Aromatic amines are named as derivatives of aniline or of the aromatic ring system.

Primary and secondary amines form intermolecular N–H···N bonds. Their boiling points are higher than those of alkanes but lower than those of alcohols with comparable molecular weights. Lower members of the series are water-soluble because of N···H–O bonding.

Amines can be prepared by S_N2 alkylation of ammonia or 1° and 2° amines. Aromatic amines are made by reduction of the corresponding nitro compounds. Amides, nitriles, and imines can also be reduced to amines.

Amines are weak bases. Alkylamines and ammonia are of comparable basicity, but aromatic amines are much weaker as a result of delocalization of the unshared electron pair on nitrogen to the *ortho* and *para* carbons of the aromatic ring. **Amides** are much weaker bases than amines because of delocalization of the unshared electron pair on nitrogen to the adjacent carbonyl oxygen. Amides are stronger Brønsted acids than amines because of the partial positive charge on the amide nitrogen and resonance in the **amidate anion**.

Amines react with strong acids to form **amine salts**. The pK_as of amine salts are related to the base strength of the corresponding amines. Alkylammonium salts have pK_as of 9–10 while arylammonium salts have pK_as of 4–5. The fact that these salts are usually water-soluble can be exploited in separating amines from neutral or acidic contaminants. Chiral amines can be used to resolve enantiomeric acids, through the formation of diastereomeric salts.

Primary and secondary amines react with acid derivatives to form amides. Amides made commercially this way include **acetanilide** and ***N,N*-diethyl-*m*-toluamide** (the insect repellent Off®).

Tertiary amines react with alkyl halides to form **quaternary ammonium salts**. An example of this type of salt with important biological properties is **choline** (2-hydroxyethyltrimethylammonium ion).

Primary aromatic amines react with nitrous acid to give **aryldiazonium ions**, ArN_2^+, which are useful intermediates in synthesis of aromatic compounds. The process by which they are formed is called **diazotization**. The nitrogen in these ions can readily be replaced by various nucleophiles (OH, Cl, Br, I, CN). Diazonium ions couple with reactive aromatics, such as amines or phenols, to form **azo compounds**, which are useful as dyes.

Reaction Summary

Alkylation of Ammonia and Amines

$$R-X + 2NH_3 \longrightarrow R-NH_2 + NH_4^+ X^-$$

$$C_6H_5NH_2 \xrightarrow{R-X} C_6H_5NHR \xrightarrow{R-X} C_6H_5NR_2$$

Reduction Routes to Amines

$$C_6H_5NO_2 \xrightarrow[\text{or } SnCl_2,\ HCl]{H_2,\ \text{Ni catalyst}} C_6H_5NH_2$$

$$R-C(=O)-NR'R'' \xrightarrow{LiAlH_4} R-CH_2-NR'R''$$

$$R-C{\equiv}N: \xrightarrow{LiAlH_4} R-CH_2-NH_2$$

$$R-C(=O)-R' \xrightarrow[NaBH_3CN]{R''NH_2} R-CH(NHR'')-R'$$

Amine Basicity

$$R-NH_2 + H-Cl \longrightarrow R-\overset{+}{N}H_3\ Cl^-$$

Acylation of Primary and Secondary Amines

$R-NH_2$ (primary amine) + $R'C(=O)OC(=O)R'$ or $R'C(=O)Cl$ → $R'C(=O)NHR$ (secondary amide)

R_2NH (secondary amine) + $R'C(=O)OC(=O)R'$ or $R'C(=O)Cl$ → $R'C(=O)NR_2$ (tertiary amide)

Quaternary Ammonium Salts

$$R_3N + R'X \longrightarrow R_3\overset{+}{N}{-}R' \quad X^-$$

Aryldiazonium Salts

$$ArNH_2 + HONO \xrightarrow{HX} ArN_2^+ \quad X^-$$

(aryldiazonium ion)

$$ArN_2^+ + H_2O \xrightarrow{heat} ArOH + N_2 + H^+$$

(phenols)

$$ArN_2^+ + HX \xrightarrow{Cu_2X_2} ArX \quad (X = Cl, Br)$$

$$ArN_2^+ + KI \longrightarrow ArI$$

$$ArN_2^+ + KCN \xrightarrow{Cu_2(CN)_2} ArCN$$

$$ArN_2^+ + HBF_4 \longrightarrow ArF$$

$$ArN_2^+ + H_3PO_2 \longrightarrow ArH$$

Diazo Coupling

$$ArN_2^+ X^- + C_6H_5OH \longrightarrow p\text{-}HO{-}C_6H_4{-}N{=}N{-}Ar$$

(azo compound)

Mechanism Summary

Diazotization

$$\left[Ar(H)\ddot{N}{-}\ddot{N}{=}\ddot{O}: \rightleftharpoons Ar{-}\ddot{N}{=}\ddot{N}{-}\ddot{O}{-}H\right] \xrightarrow{H^+}$$

$$Ar{-}\ddot{N}{=}\ddot{N}{-}\overset{+}{O}H_2 \longrightarrow Ar{-}\overset{+}{N}{\equiv}N: + H_2O$$

Learning Objectives

1. Know the meaning of: primary, secondary, and tertiary amine, amino group, aniline, amine salt, quaternary ammonium salt.
2. Know the meaning of: nitrous acid, diazonium ion, diazotization, azo coupling.
3. Given the structure of an amine, identify it as primary, secondary, or tertiary.
4. Given the structure of an amine, name it. Also, given the name of an amine, write its structural formula.
5. Explain the effect of hydrogen bonding on the boiling points of amines and their solubility in water.
6. Write an equation for the reaction between ammonia or an amine of any class and an alkyl halide.
7. Write an equation for the preparation of a given aromatic amine from the corresponding nitro compound.
8. Write an equation for the preparation of a given amine of the type RCH_2NH_2 or $ArCH_2NH_2$ by reduction of the appropriate nitrile.
9. Write an equation for the preparation of a secondary amine from a ketone, primary amine, and sodium cyanoborohydride.
10. Write an equation for the dissociation of an amine in water.
11. Write an expression for K_a of any amine salt.
12. Draw the important contributors to the resonance hybrid for an aromatic amine.
13. Given the structures of several amines, rank them in order of relative basicity.
14. Account for the difference in basicity between an aliphatic and an aromatic amine.
15. Write an equation for the reaction of a given amine of any class with a strong acid. Also, write an equation for the reaction of an amine salt with a strong base.
16. Account for the basicity and acidity difference between amines and amides.
17. Explain, with the aid of equations, how you can separate an amine from a mixture containing neutral and/or acidic compounds.
18. Explain how chiral amines can be used to resolve a mixture of enantiomeric acids.
19. Write an equation for the reaction of a given primary or secondary amine with an acid anhydride or acyl halide.
20. Write the steps in the mechanism for acylation of a primary or secondary amine.
21. Write an equation for the diazotization of a given primary aromatic amine.
22. Write the equations for the reaction of an aromatic diazonium salt with: aqueous base; $HX + Cu_2X_2$ (X = Cl, Br); $KCN + Cu_2(CN)_2$; HBF_4; and H_3PO_2.
23. Write an equation for the coupling of an aromatic diazonium salt with a phenol or aromatic amine.

ANSWERS TO PROBLEMS

Problems Within the Chapter

11.1 a. primary b. secondary c. primary d. tertiary

11.2 *N,N*-dimethyl-3-pentanamine

11.3 a. neopentylamine or 2,2-dimethylpropanamine
b. 3-aminopropanol c. *p*-nitroaniline

11.4 a. $(CH_3CH_2CH_2CH_2)_2NH$
b. (benzene ring with CH_3, H_3C, NH_2, CH_3: 2,4,6-trimethylaniline)
c. H_3C–CH_2–$CH(NH_2)$–CH_2–CH_3
d. H_3C–$CH(N(CH_3)_2)$–$CH_2CH_2CH_2$–CH_3 (N with H_3C and CH_3)

11.5 Trimethylamine has no hydrogens on the nitrogen: $(CH_3)_3N$. Thus, intermolecular hydrogen bonding is not possible. In contrast, intermolecular hydrogen bonding is possible for $CH_3CH_2CH_2NH_2$ and this raises its boiling point considerably above that of its tertiary isomer.

11.6 a. $CH_3CH_2CH_2CH_2Br + 2\ NH_3 \longrightarrow CH_3CH_2CH_2CH_2NH_2 + NH_4^+\ Br^-$

b. $CH_3CH_2I + 2\ (CH_3CH_2)_2NH \longrightarrow (CH_3CH_2)_3N + (CH_3CH_2)_2\overset{+}{N}H_2\ I^-$

c. $(CH_3)_3N + CH_3I \longrightarrow (CH_3)_4N^+\ I^-$

d. $CH_3CH_2CH_2NH_2$ + (benzene ring)–CH_2Br $\longrightarrow$ (benzene ring)–$CH_2\overset{+}{N}H_2CH_2CH_2CH_3\ Br^-$

$\xrightarrow{CH_3CH_2CH_2NH_2}$ (benzene ring)–$CH_2NHCH_2CH_2CH_3$ + $CH_3CH_2CH_2\overset{+}{N}H_3\ Br^-$

11.7 2 (benzene ring)–NH_2 + $CH_3CH_2Br \longrightarrow$ (benzene ring)–$NHCH_2CH_3$ + (benzene ring)–$\overset{+}{N}H_3\ Br^-$

11.8 Nitration of toluene twice gives mainly the 2,4-dinitro product. Reduction of the nitro groups completes the synthesis. The NaOH converts the amine hydrochloride salt to the free amine.

$$\text{toluene} \xrightarrow[H_2SO_4]{HONO_2} \text{2,4-dinitrotoluene} \xrightarrow[\text{1. }SnCl_2,\ HCl;\ \text{2. NaOH (see eq. 11.10)}]{H_2,\ Ni\ \text{or}} \text{2,4-diaminotoluene}$$

11.9 See eq. 11.11

$$H_3C{-}C({=}O){-}N(CH_3)_2 \xrightarrow{LiAlH_4} CH_3CH_2N(CH_3)_2$$

11.10 See eq. 11.12

$$C_6H_5CH_2Br \xrightarrow[\text{(see eq. 10.16)}]{Na^+CN^-} C_6H_5CH_2CN \xrightarrow[\text{(see eq. 11.12)}]{LiAlH_4} C_6H_5CH_2CH_2NH_2$$

11.11 See eq. 11.13

$$CH_3\overset{O}{\overset{\|}{C}}CH_2CH_2CH_3 + NH_3 \xrightarrow{NaBH_3CN} CH_3\overset{NH_2}{\overset{|}{C}}HCH_2CH_2CH_3$$

11.12 See eq. 11.14

$$(CH_3)_3N + H_2O \rightleftharpoons (CH_3)_3\overset{+}{N}H + {}^-OH$$

11.13 $ClCH_2CH_2NH_2$ is a weaker base than $CH_3CH_2NH_2$. The chlorine substituent is electron-withdrawing compared to hydrogen and will destabilize the protonated ammonium ion because of the repulsion between the positive charge on nitrogen and the partial positive charge on C-2 due to the C–Cl bond moment:

$$\overset{\delta^-}{Cl}{-}\overset{\delta^+}{C}H_2CH_2\overset{+}{N}H_3 \quad \text{compared to} \quad CH_3CH_2\overset{+}{N}H_3$$

Ethylamine is therefore easier to protonate (more basic) than 2-chloroethylamine.

11.14 Alkyl groups are electron-donating and stabilize the positively charged ammonium ion relative to the amine. Therefore, *N,N*-dimethylaniline is a stronger base than *N*-methylaniline, which is a stronger base than aniline. The electron-withdrawing chlorine in *p*-chloroaniline destabilizes the positively charged ammonium ion. Therefore *p*-chloroaniline is a weaker base than aniline:

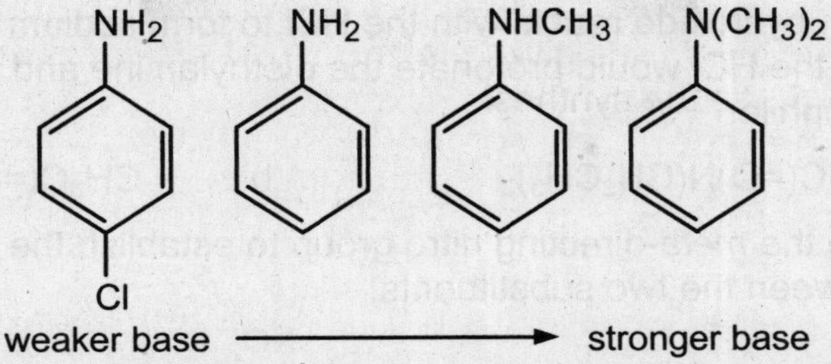

NH2 NH2 NHCH3 N(CH3)2

Cl

weaker base ⟶ stronger base

11.15 The order of the substituents by increasing electron-donating ability is $-NO_2 < -H < -CH_3$. Therefore, the basicities will increase in that order:

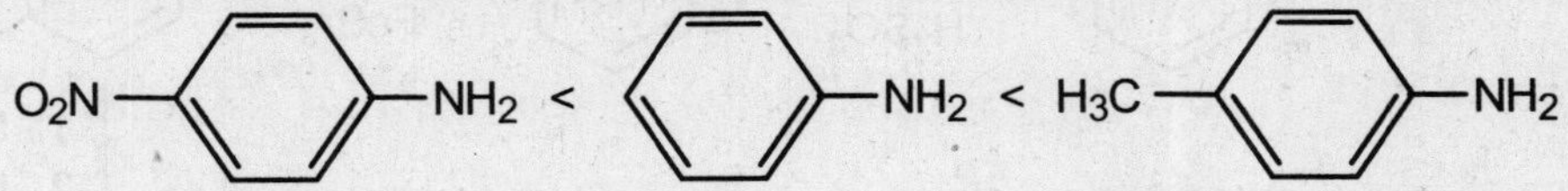

$O_2N-C_6H_4-NH_2$ < $C_6H_5-NH_2$ < $H_3C-C_6H_4-NH_2$

11.16 Amides are less basic than amines, and aromatic amines are less basic than aliphatic amines. Therefore, the order is:

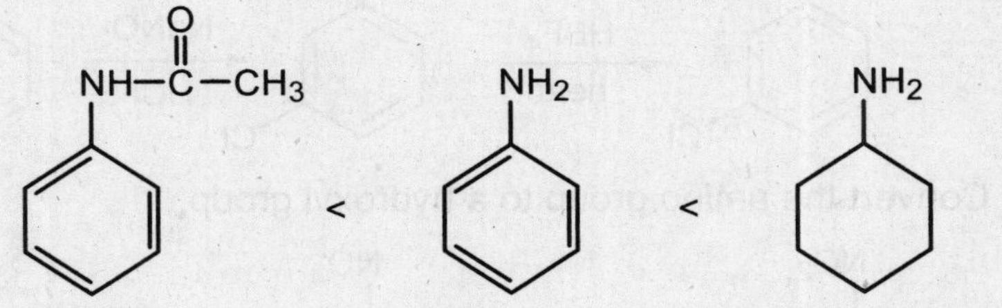

acetanilide < aniline < cyclohexylamine

The acidity increases in the reverse direction.

11.17

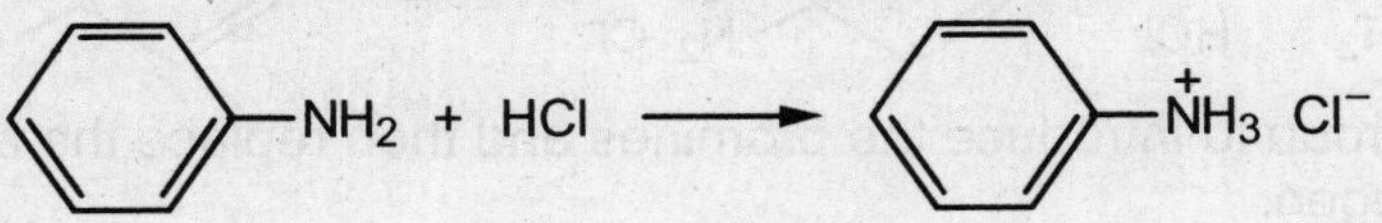

$$C_6H_5NH_2 + HCl \longrightarrow C_6H_5\overset{+}{N}H_3\ Cl^-$$

anilinium chloride

11.18

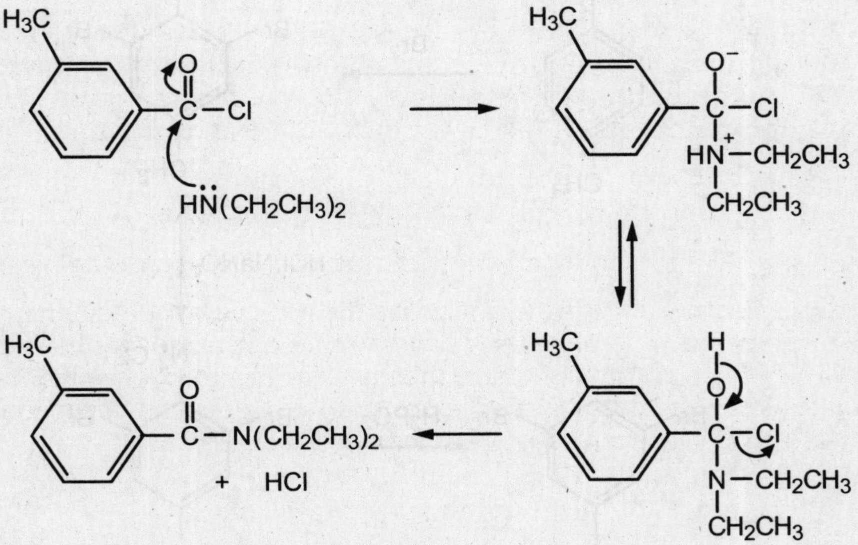

$H_3C-C_6H_4-C(=O)Cl$ + $H\ddot{N}(CH_2CH_3)_2$ ⟶ $H_3C-C_6H_4-C(O^-)(Cl)-\overset{+}{N}H(CH_2CH_3)_2$ ⇌ $H_3C-C_6H_4-C(OH)(Cl)-N(CH_2CH_3)_2$ ⟶ $H_3C-C_6H_4-C(=O)-N(CH_2CH_3)_2$ + HCl

The sodium hydroxide reacts with the HCl to form sodium chloride and water. Otherwise, the HCl would protonate the diethylamine and prevent it from functioning as a nucleophile.

11.19 a. $CH_3C(=O)N(CH_2CH_3)_2$ b. $CH_3C(=O)NHCH_2CH_3$

11.20 a. Use the *meta*-directing nitro group to establish the proper relationship between the two substituents:

benzene → (HNO_3, H_2SO_4) → nitrobenzene (NO_2) → (Br_2, $FeBr_3$) → 3-chloronitrobenzene (NO_2, Cl) → (1. $SnCl_2$, HCl; 2. NaOH) → 3-chloroaniline (NH_2, Cl) → ($NaNO_2$, HCl) → $\overset{+}{N_2}$ Cl^- (Cl) → (HBF_4, heat) → 1-chloro-3-fluorobenzene (F, Cl)

b. Convert the amino group to a hydroxyl group.

3-nitroaniline (NO_2, NH_2) → ($NaNO_2$, HCl) → (NO_2, $\overset{+}{N_2}$ Cl^-) → (H_2O) → 3-nitrophenol (NO_2, OH)

c. Use the amino group to introduce the bromines and then replace the amino group by a hydrogen.

4-methylaniline (NH_2, CH_3) → (Br_2) → 2,6-dibromo-4-methylaniline (NH_2, Br, Br, CH_3) → (HCl, $NaNO_2$) → (N_2^+ Cl^-, Br, Br, CH_3) → (H_3PO_2) → 3,5-dibromotoluene (Br, Br, CH_3)

d. Introduce the fluoro groups as follows:

CH3, HNO3 / H2SO4 → CH3, NO2, NO2

1. SnCl2, HCl 2. NaOH or H2, Ni

CH3, F, F ← 1. NaNO2 HBF4 2. Δ ← CH3, NH2, NH2

The use of HBF_4 in place of HCl gives an intermediate diazonium tetrafluoroborate.

11.21

:N: N+ ⟷ :N: N+ + ⟷ :N: N+ +

11.22

NH2, SO3H sulfanilic acid → HNO2 / HCl → N2+ Cl−, SO3H → N(CH3)2 → N(CH3)2, N=N, SO3H

NaOH

Na+ −O3S–⟨⟩–N=N–⟨⟩–N(CH3)2

methyl orange

ADDITIONAL PROBLEMS

11.23 Many correct answers are possible, but only one example is given in each case.

a. CH_3NH_2 methylamine

b. *N*-methylpyrrolidine

c. *N*-methylaniline

d. tetramethylammonium chloride

e. $C_6H_5\overset{+}{N}{\equiv}N\colon\ Cl^-$

f. azobenzene

g. acetamide

11.24 a. $(CH_3CH_2CH_2)N(CH_2CH_3)_2$

b. $C_6H_5CH_2NHCH_3$

c. $CH_3CHCH_2CH_3$ with NH_2 on C-2

d. $H_2N(CH_2)_6NH_2$

e. $C_5H_9N(CH_3)_2$

f. $CH_3CH_2CHCH_2CH_2CH_3$ with $N(CH_3)_2$ on C-3

g. $(CH_3CH_2)_4\overset{+}{N}\ \overset{-}{Br}$

h. H_2N–C_6H_4–NO_2 (para)

i. $CH_3CH(NH_2)CH_2CH_2CH_2CH_3$

j. C_6H_5–NH–C_6H_5

k. *o*-methylaniline (NH_2, CH_3)

l. $CH_3CH_2CH(CH_3)CH(NH_2)CH_3$

11.25 a. *p*-chloroaniline
b. methylpropylamine (or *N*-methylpropanamine)
c. diethylmethylamine
d. tetramethylammonium chloride
e. 3-amino-2-butanol
f. 2-aminocyclohexanone

g. *p*-chlorobenzenediazonium chloride
h. *N*-methyl-*p*-methoxyaniline
i. *trans*-1,3-diaminocyclobutane
j. 1,6-diaminohexane

11.26

$CH_3CH_2CH_2CH_2NH_2$	*n*-butylamine or 1-butanamine (primary)
$CH_3CH_2CH(NH_2)CH_3$	2-butylamine or 2-butanamine (primary)
$(CH_3)_2CHCH_2NH_2$	2-methylpropanamine (primary)
$(CH_3)_3CNH_2$	2-methyl-2-propanamine or *t*-butylamine (primary)
$CH_3CH_2CH_2NHCH_3$	*N*-methylpropanamine (secondary)
$(CH_3)_2CHNHCH_3$	*N*-methyl-2-propanamine (secondary)
$(CH_3CH_2)_2NH$	diethylamine or *N*-ethylethanamine (secondary)
$(CH_3)_2NCH_2CH_3$	ethyldimethylamine or *N,N*-dimethylethanamine (tertiary)

11.27 a. Aniline is the stronger base. The *p*-cyano group is electron-withdrawing and therefore decreases the basicity of aniline. Note that the possibilities for delocalization of the unshared electron pair are greater in *p*-cyanoaniline than in aniline.

$$\left[:N{\equiv}C\text{-}C_6H_4\text{-}\ddot{N}H_2 \longleftrightarrow :\ddot{N}^{-}{=}C{=}C_6H_4{=}N^{+}H_2 \right]$$

Resonance stabilizes the free base relative to its protonated form, and the effect is greater with *p*-cyanoaniline than with aniline.

b. The possibilities for delocalization of an electron pair are greater in diphenylamine than in aniline (two phenyl groups versus one phenyl group). Thus, aniline is the stronger base.

11.28 The mixture is first dissolved in an inert, low-boiling solvent such as ether. The following scheme describes a separation procedure:

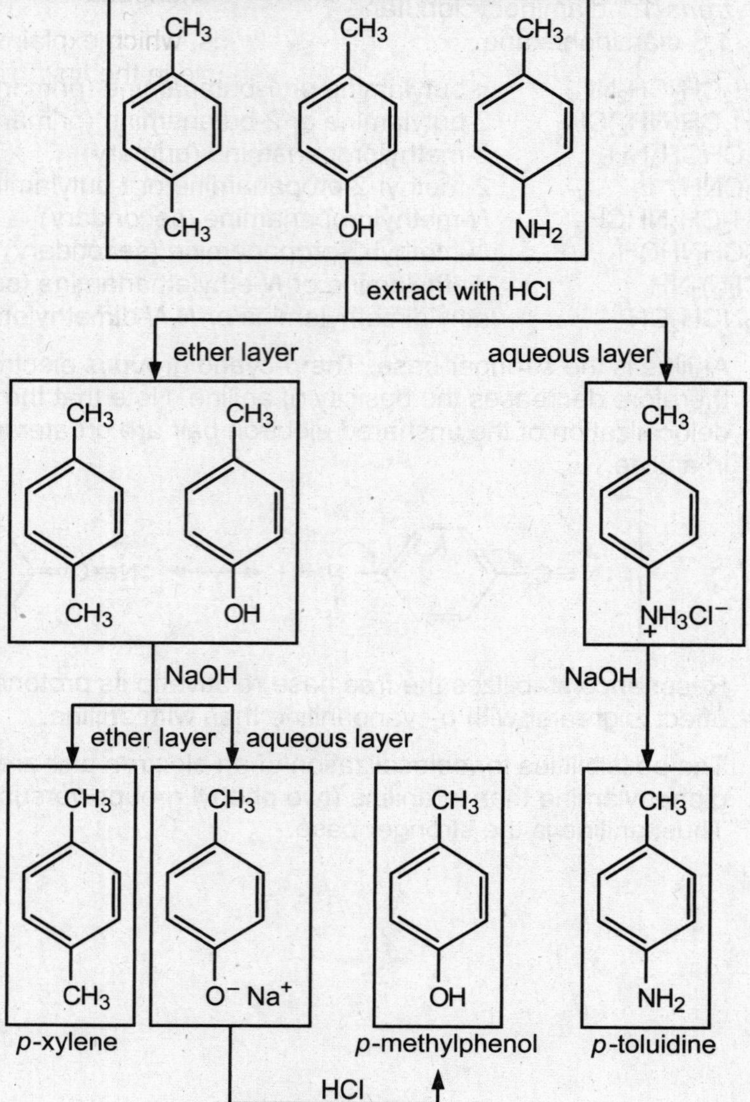

To recover the *p*-xylene, the ether is evaporated and the *p*-xylene distilled. In the case of *p*-toluidine and *p*-methylphenol, once the product is liberated from the corresponding salt, it is extracted from the water by ether. The ether is then evaporated and the desired product is distilled. The order of extraction—acid first, then base—can be reversed.

11.29 The unshared electron pair on the amino group can be delocalized not only to the *ortho* and *para* carbons of the ring as with aniline (see Sec.11.6), but also to the oxygen of the nitro group.

11.30 The boiling point order is

pentane < methyl propyl ether < 1-aminobutane < 1-butanol

O–H···O bonds are stronger than N–H···N bonds, which explains the order of the last two compounds. No hydrogen bonding is possible in the first two compounds, but C–O bonds are polar, giving the ether a higher boiling point than the alkane. The actual boiling points are pentane, 36°C; methyl propyl ether, 39°C; 1-aminobutane, 78°C; and 1-butanol, 118°C.

11.31 The *R* and *S* enantiomers of ammonium salt **A** can interconvert only by processes that involve breaking a carbon–nitrogen bond. This does not occur easily, and thus, the enantiomers can be separated by formation of diastereomeric salts.

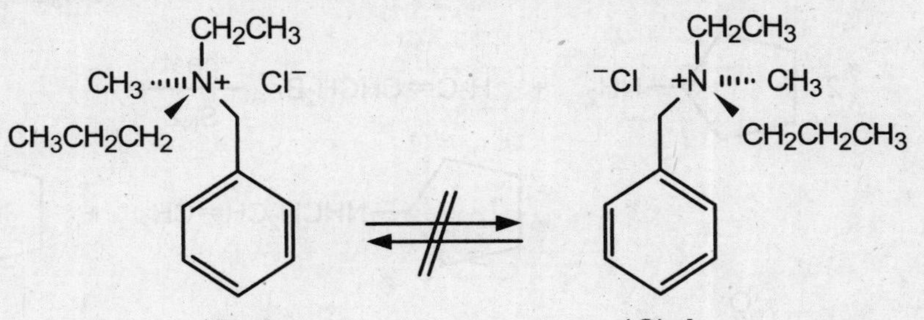

CH_2CH_3, CH_3, N^+, Cl^-, $CH_3CH_2CH_2$ — ^-Cl, ^+N, CH_3, CH_2CH_3, $CH_2CH_2CH_3$

(*R*)-**A** (*S*)-**A**

The enantiomers of amine **B** can easily interconvert by "inversion" of the nitrogen lone pair (see eq. 11.1) and thus cannot be separated.

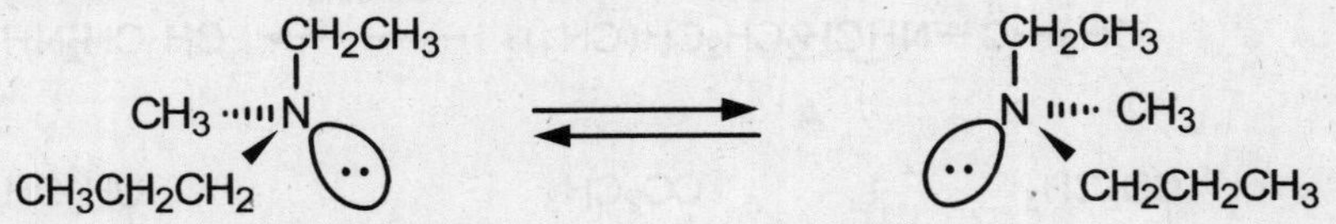

CH_2CH_3, CH_3, N, $CH_3CH_2CH_2$ ⇌ CH_2CH_3, N, CH_3, $CH_2CH_2CH_3$

11.32 The priority order is $CH_2C_6H_5 > CH_2CH_2CH_3 > CH_2CH_3 > CH_3$

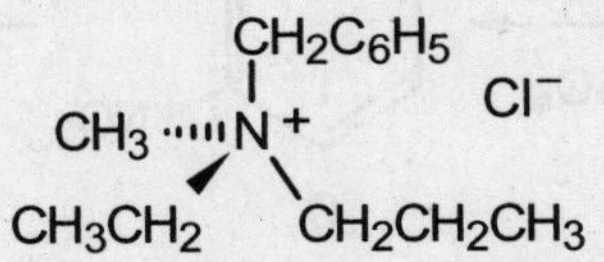

$CH_2C_6H_5$, CH_3, N^+, Cl^-, CH_3CH_2, $CH_2CH_2CH_3$

11.33 a. Alkylate aniline twice with ethyl bromide or ethyl iodide.

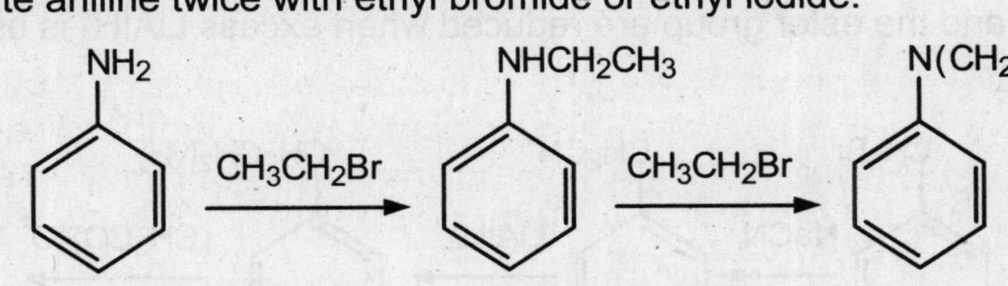

NH_2 (aniline) $\xrightarrow{CH_3CH_2Br}$ $NHCH_2CH_3$ $\xrightarrow{CH_3CH_2Br}$ $N(CH_2CH_3)_2$

b. First nitrate, then brominate, to obtain the *meta* orientation.

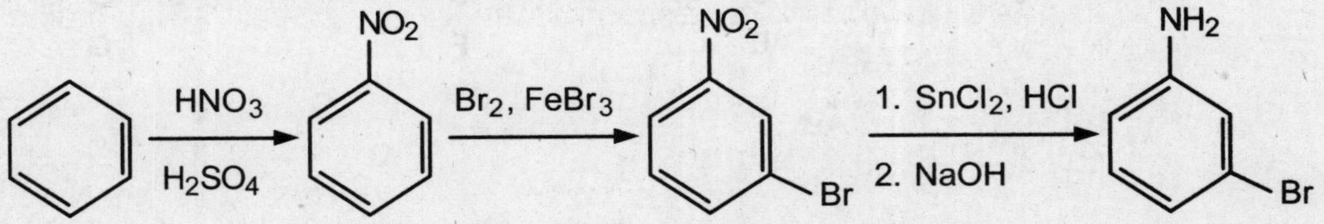

benzene $\xrightarrow[H_2SO_4]{HNO_3}$ NO_2 $\xrightarrow{Br_2,\ FeBr_3}$ NO_2, Br $\xrightarrow[2.\ NaOH]{1.\ SnCl_2,\ HCl}$ NH_2, Br

c. The reverse of the sequence in part b gives mainly *para* orientation.

$$C_6H_6 \xrightarrow{Br_2,\ FeBr_3} C_6H_5Br \xrightarrow[H_2SO_4]{HNO_3} p\text{-}BrC_6H_4NO_2 \xrightarrow[2.\ NaOH]{1.\ SnCl_2,\ HCl} p\text{-}BrC_6H_4NH_2$$

d. Displace the bromide and then reduce the nitrile.

$$CH_3CH_2CH_2CH_2CH_2Br \xrightarrow{NaCN} CH_3CH_2CH_2CH_2CH_2C{\equiv}N \xrightarrow{LiAlH_4} CH_3CH_2CH_2CH_2CH_2CH_2NH_2$$

11.34 a.

$$2\ C_5H_9\text{-}NH_2 + H_2C{=}CHCH_2Br \xrightarrow[S_N2]{heat} C_5H_9\text{-}NHCH_2CH{=}CH_2 + C_5H_9\text{-}\overset{+}{N}H_3\ Br^-$$

b.

$$CH_3\text{-}\overset{O}{\overset{\|}{C}}\text{-}Cl + H_2NCH_2CH_2CH(CH_3)_2 \longrightarrow$$

$$\underset{\mathbf{A}}{CH_3\text{-}\overset{O}{\overset{\|}{C}}\text{-}NHCH_2CH_2CH(CH_3)_2} \xrightarrow{LiAlH_4} \underset{\mathbf{B}}{CH_3CH_2NHCH_2CH_2CH(CH_3)_2}$$

c.

$$C_6H_5CO_2CH_3 \xrightarrow[H_2SO_4]{HONO_2} \underset{\mathbf{C}}{m\text{-}O_2NC_6H_4CO_2CH_3} \xrightarrow{LiAlH_4} \underset{\mathbf{D}}{m\text{-}H_2NC_6H_4CH_2OH} + CH_3OH$$

In the first step, the ester group is *meta*-directing. In the second step, both the nitro group and the ester group are reduced when excess $LiAlH_4$ is used.

d.

$$C_6H_5CH_2Br \xrightarrow{NaCN} \underset{\mathbf{E}}{C_6H_5CH_2CN} \xrightarrow{LiAlH_4} \underset{\mathbf{F}}{C_6H_5CH_2CH_2NH_2} \xrightarrow{(CH_3CO)_2O} \underset{\mathbf{G}}{C_6H_5CH_2CH_2NH\overset{O}{\overset{\|}{C}}CH_3}$$

e. Cyclohexanone + $H_2NCH(CH_3)_2$ $\xrightarrow{NaBH_3CN}$ cyclohexyl–$NHCH(CH_3)_2$

H

11.35 a. Compare with eq. 11.18.

$$CH_3-C_6H_4-NH_2 + HCl \longrightarrow CH_3-C_6H_4-\overset{+}{N}H_3\ Cl^-$$

b.

$$(CH_3CH_2)_3N + H{-}OSO_3H \longrightarrow (CH_3CH_2)_3\overset{+}{N}H\ {}^-OSO_3H$$

c.

$$(CH_3)_2N\overset{+}{H_2}\,Cl^- + Na^+\,{}^-OH \longrightarrow (CH_3)_2NH + Na^+\,Cl^- + H_2O$$

d.

$$C_6H_5-N(CH_3)_2 + CH_3I \longrightarrow C_6H_5-\overset{+}{N}(CH_3)_3\ I^-$$

e.

$$C_6H_{11}-NH_2 + CH_3-\overset{O}{\overset{\|}{C}}-O-\overset{O}{\overset{\|}{C}}-CH_3 \longrightarrow C_6H_{11}-NH\overset{O}{\overset{\|}{C}}CH_3 + CH_3CO_2H$$

11.36 The reaction begins with nucleophilic attack by the amine on the carbonyl group of the anhydride.

$$CH_3-\overset{O}{\overset{\|}{C}}-O-\overset{O}{\overset{\|}{C}}-CH_3 + H_2\ddot{N}CH_2CH_3 \longrightarrow CH_3-\underset{+NH_2CH_2CH_3}{\overset{O^-}{C}}-O-\overset{O}{\overset{\|}{C}}-CH_3$$

$\downarrow$ H^+ transfer

$$H_3C-\overset{O}{\overset{\|}{C}}-\ddot{N}HCH_2CH_3 \xleftarrow{-CH_3CO_2H} CH_3-\underset{\ddot{N}HCH_2CH_3}{\overset{H-O}{C}}-O-\overset{O}{\overset{\|}{C}}-CH_3$$

Even though the resulting amide has an unshared electron pair on nitrogen, it does not react with a second mole of acetic anhydride to become diacylated:

$$CH_3-\overset{O}{\overset{\|}{C}}-O-\overset{O}{\overset{\|}{C}}-CH_3 + H_3C-\overset{O}{\overset{\|}{C}}-\ddot{N}HCH_2CH_3 \not\longrightarrow \left(H_3C-\overset{O}{\overset{\|}{C}}\right)_2\ddot{N}CH_2CH_3$$

The reason is that amides are poor nucleophiles because the unshared electron pair on the nitrogen is delocalized through resonance:

$$\left[H_3C-\overset{O}{\overset{\|}{C}}-\ddot{N}HCH_2CH_3 \longleftrightarrow H_3C-\overset{O^-}{\overset{|}{C}}=\overset{+}{N}HCH_2CH_3 \right]$$

The amide is ineffective with respect to nucleophilic attack on the carbonyl group of acetic anhydride.

11.37 a. There are several ways to accomplish this synthesis. One approach would follow the example described in eq. 11.13. Reaction of 1-phenyl-2-propanone with methylamine would give an imine. Reduction of the imine would afford methamphetamine, which would give methamphetamine hydrochloride upon reaction with HCl.

1-phenyl-2-propanone $\xrightarrow{CH_3NH_2}$ imine ($=NCH_3$) $\xrightarrow{NaCNBH_3}$ methamphetamine ($NHCH_3$, *) $\xrightarrow{HCl}$ $^+NH_2CH_3$ Cl^-

imine

methamphetamine

b. The imine reduction generates a stereogenic center (marked with an asterisk). Because both of the reactants are achiral, this reaction would afford a 50:50 mixture (racemic mixture) of the *R* and *S* enantiomers (see Sec. 5.11).

11.38 Each of these alkaloids have 4 stereogenic centers.

11.39 Use eq. 11.26 as a guide.

$$Br(CH_2)_{12}Br \xrightarrow{2\,(CH_3)_3N} Br^-\ (CH_3)_3\overset{+}{N}(CH_2)_{12}\overset{+}{N}(CH_3)_3\ Br^-$$

11.40 The reaction involves an S_N2 ring opening of ethylene oxide.

$$(CH_3)_3N: + H_2C\!-\!CH_2\ (\text{O bridge}) \xrightarrow{\text{base}} (CH_3)_3\overset{+}{N}CH_2CH_2O^- \underset{}{\overset{H_2O}{\rightleftharpoons}} [(CH_3)_3\overset{+}{N}CH_2CH_2OH]\ ^-OH$$

choline

11.41

$$CH_3-\overset{O}{\overset{\|}{C}}-S-CoA + [(CH_3)_3\overset{+}{N}CH_2CH_2OH]\ ^-OH \longrightarrow$$

$$CH_3-\overset{O}{\overset{\|}{C}}-O-CH_2CH_2\overset{+}{N}(CH_3)_3\ ^-OH + HS-CoA$$

11.42 Alkyl diazonium salts can lose nitrogen to give 1°, 2°, or 3° carbocations, depending on the nature of the alkyl group.

$$R-N_2^+ \longrightarrow R^+ + N_2$$

Aryl cations are less stable than 1°, 2°, or 3° carbocations, so aryl diazonium salts are more stable than alkyl diazonium salts.

11.43 These equations illustrate the reactions in Secs. 11.12 and 11.13.

a.

$$H_3C-C_6H_4-N_2^+\ HSO_4^- \xrightarrow[\text{2. heat}]{\text{1. } HBF_4} H_3C-C_6H_4-F + N_2 + H_2SO_4$$

b.

$$H_3C-C_6H_4-N_2^+\ HSO_4^- + H-OH \xrightarrow[\text{(see eq. 11.33)}]{\text{heat}} H_3C-C_6H_4-OH + N_2 + H_2SO_4$$

c.

$$H_3C-C_6H_4-N_2^+\ HSO_4^- + KCN \xrightarrow[\text{(see eq. 11.33)}]{Cu_2(CN)_2} H_3C-C_6H_4-CN + N_2 + KHSO_4$$

d. Since *para* coupling is blocked by the methyl substituent, *ortho* coupling occurs:

$$H_3C-C_6H_4-N_2^+\ HSO_4^- + HO-C_6H_4-OCH_3 \xrightarrow[\text{(see eq. 11.34)}]{HO^-} HSO_4^- + H_3C-C_6H_4-N{=}N-C_6H_3(OH)(OCH_3) + H_2O$$

e $$H_3C-C_6H_4-N_2^+\ HSO_4^- + HCl \xrightarrow[\text{(see eq. 11.33)}]{Cu_2Cl_2} H_3C-C_6H_4-Cl + N_2 + H_2SO_4$$

f. $$H_3C-C_6H_4-N_2^+\ HSO_4^- + C_6H_5-N(CH_2CH_3)_2 \xrightarrow{HO^-} H_2O + HSO_4^- + H_3C-C_6H_4-N{=}N-C_6H_4-N(CH_2CH_3)_2$$

g. $$H_3C-C_6H_4-N_2^+\ HSO_4^- \xrightarrow{H_3PO_2} H_3C-C_6H_4-H + N_2 + H_2SO_4$$

h. $$H_3C-C_6H_4-N_2^+\ HSO_4^- + KI \xrightarrow{\text{(see eq. 11.33)}} H_3C-C_6H_4-I + N_2 + KHSO_4$$

11.44 a. $$ClC_6H_4NH_2 \xrightarrow[H^+,\ 0°C]{HONO} ClC_6H_4N_2^+ \xrightarrow[Cu_2(CN)_2]{KCN} ClC_6H_4CN + N_2 \xrightarrow{H_3O^+} ClC_6H_4CO_2H$$

COOH

b.

NO_2 NO_2 NH_2 N_2^+ I

HNO_3, H_2SO_4; 1. Cl_2, $FeCl_3$; Cl; 1. $SnCl_2$, HCl 2. NaOH; $NaNO_2$, H^+; KI, heat; Cl

Note that the order of each step in the sequence is important. The benzene must be nitrated first and then chlorinated to attain *meta* orientation. Chlorination of iodobenzene would not give *meta* product, so this indirect route must be used.

c.

CH_3COCl, $AlCl_3$; HNO_3, H_2SO_4; NO_2; H_2, Ni; NH_2; $NaNO_2$, H^+; N_2^+; KI, heat; I

d.

$$\text{toluene} \xrightarrow[SO_3]{H_2SO_4} \text{p-}CH_3C_6H_4SO_3H \xrightarrow[H_2SO_4]{HNO_3} \text{(}CH_3\text{, }NO_2\text{, }SO_3H\text{)} \xrightarrow{H_2,\ Ni} \text{(}CH_3\text{, }NH_2\text{, }SO_3H\text{)} \xrightarrow[H^+]{NaNO_2} \text{(}CH_3\text{, }N_2^+\text{, }SO_3H\text{)} \xrightarrow[Cu_2(CN)_2]{KCN} \text{(}CH_3\text{, }CN\text{, }SO_3H\text{)}$$

11.45 Benzidine can be diazotized at each amino group. It can then couple with two equivalents of the aminosulfonic acid. Coupling occurs *ortho* to the amino group since the *para* position is blocked by the sulfonic acid group.

$$H_2N\text{–}C_6H_4\text{–}C_6H_4\text{–}NH_2 \text{ (benzidine)} \xrightarrow[H^+]{HONO}$$

$$N_2^+\text{–}C_6H_4\text{–}C_6H_4\text{–}N_2^+ + \text{4-amino-1-naphthalenesulfonic acid (}NH_2\text{, }SO_3H\text{)} \longrightarrow$$

Congo Red (NH_2, SO_3H naphthalene–N=N–biphenyl–N=N–naphthalene NH_2, SO_3H)

11.46

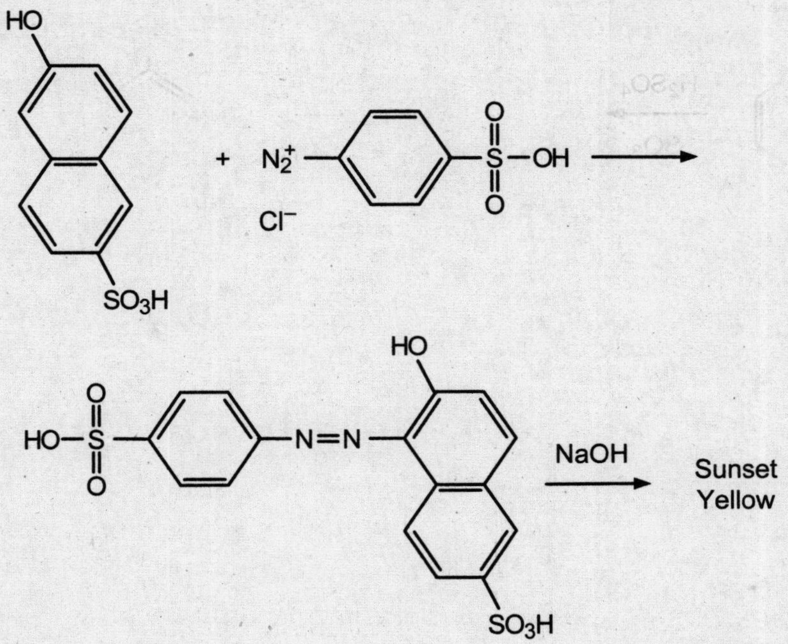
HO
+
N_2^+
Cl^-
O
S
OH
O
SO_3H
HO
O
HO
S
O
N=N
NaOH
Sunset
Yellow
SO_3H

12

Spectroscopy and Structure Determination

Chapter Summary

Spectroscopic methods provide rapid, nondestructive ways to determine molecular structures. One of the most powerful of these methods is **nuclear magnetic resonance (NMR) spectroscopy**, which involves the excitation of nuclei from lower to higher energy spin states while they are placed between the poles of a powerful magnet. In organic chemistry, the most important nuclei measured are ^{1}H and ^{13}C.

Protons in different chemical environments have different **chemical shifts**, measured in δ (delta) units from the reference peak of tetramethylsilane [TMS, $(CH_3)_4Si$]. **Peak areas** are proportional to the number of protons. Peaks may be split (**spin-spin coupling**) depending on the number of protons. Proton NMR gives at least three types of structural information: (1) the number of signals and their chemical shifts can be used to identify the kinds of chemically different protons in the molecules; (2) peak areas tell how many protons of each kind are present; (3) spin-spin coupling patterns identify the number of near-neighbor protons.

^{13}C NMR spectroscopy can tell how many different "kinds" of carbon atoms are present, and ^{1}H–^{13}C splitting can be used to determine the number of hydrogens on a given carbon.

NMR spectroscopy can be used to study problems of biological and medicinal importance, including analysis of bodily fluids for metabolites and monitoring of muscle tissue function. **Magnetic resonance imaging** (MRI) relies on the ^{1}H NMR spectrum of water and has been used in diagnostic medicine and to address problems in food science, agriculture, and the building industry.

Infrared spectroscopy is mainly used to tell what types of bonds are present in a molecule (using the **functional group region**, 1500–5000 cm^{-1}) and whether two substances are identical or different (using the **fingerprint region**, 700–1500 cm^{-1}).

Visible and ultraviolet spectroscopy employs radiation with wavelengths of 200–800 nm. This radiation corresponds to energies that are associated with **electronic transitions**, in which an electron "jumps" from a filled orbital to a vacant orbital with higher energy. Visible-ultraviolet spectra are most commonly used to detect conjugation. In general, the greater the degree of the conjugation, the longer the wavelength of energy absorbed.

Mass spectra are used to determine molecular weights and molecular composition (from the **parent** or **molecular ion**) and to obtain structural information from the fragmentation of the molecular ion into **daughter ions**. Electrospray ionization (ESI-MS) and matrix-assisted laser desorption ionization (MALDI-MS) mass spectroscopy can be used to obtain structural information about macromolecules, including proteins, polymers, and drug-DNA complexes.

Learning Objectives

1. Know the meaning of: NMR, applied magnetic field, spin state, chemical shift, δ value, TMS (tetramethylsilane), peak area, spin-spin coupling, $n + 1$ rule, singlet, doublet, triplet, quartet, MRI.
2. Given a structure, tell how many different "kinds" of protons are present.
3. Given a simple structure, use the data in Table 12.2 to predict the appearance of its ^1H NMR spectrum.
4. Given a ^1H NMR spectrum and other data such as molecular formula, use Table 12.2 to deduce a possible structure.
5. Use spin-spin splitting patterns on a spectrum to help assign a structure.
6. Use ^1H and ^{13}C NMR spectra with other data or spectra to deduce a structure.
7. Know the meaning of the functional group region and fingerprint region of an infrared spectrum, and tell what kind of information can be obtained from each.
8. Use the infrared stretching frequencies in Table 12.4 to distinguish between different classes of organic compounds.
9. In connection with visible-ultraviolet spectroscopy, know the meaning of: nanometer, electronic transition, Beer's law, molar absorptivity or extinction coefficient.
10. Know the relationship between conjugation and visible-ultraviolet absorption, and use this relationship to distinguish between closely related structures.
11. Know the meaning of: mass spectrum, *m/z* ratio, molecular ion, parent ion, fragmentation, daughter ion.
12. Given the molecular formula of a compound, deduce the *m/z* ratio of the parent ion and predict the relative intensity of the parent + 1 and parent ions.
13. Given the structure of a simple compound and its mass spectrum, deduce possible structures for the daughter ions.
14. Use all spectroscopic methods in conjunction to deduce a structure.

ANSWERS TO PROBLEMS

Problems Within the Chapter

12.1 All of the protons in the structures in part b are equivalent and appear as a sharp, single peak. One way to test whether protons are equivalent is to replace any one of them by some group X. If the same product is obtained regardless of which proton is replaced, then the protons must be equivalent. Try this test with the compounds in parts a and c. Note that diethyl ether (part a) and propanoyl chloride (part c), has two sets of equivalent protons (CH_2 and CH_3).

12.2 a. CH_3CH_2OH 3:2:1 b. $CH_3CO_2CH_3$ 1:1 (or 3:3)
c. $CH_3CH_2–O–CH_2CH_3$ 2:3 (or 4:6)

12.3 All of the hydrogens in 1,2-dichloroethane are equivalent, and it will show only one peak in its NMR spectrum. The NMR spectrum of 1,1-dichloroethane will show two peaks in a 3:1 ratio.

12.4 a. CH_3CO_2H

There will be two peaks, at approximately δ 2.1–2.6 and δ 10–13, with relative areas 3:1.

b. $CH_3C{\equiv}CH$

There will be two peaks, at approximately δ 1.6–1.9 and δ 2.4–2.7, with relative areas 3:1.

12.5 The compound is

$$(CH_3)_3C{-}\overset{O}{\overset{\|}{C}}{-}OCH_3$$

δ 0.9 (9 H) δ 3.6 (3 H)

The expected spectrum for its isomer is

$$H_3C{-}\overset{O}{\overset{\|}{C}}{-}OC(CH_3)_3$$

δ 2.1–2.6 (3 H) δ 0.9 (9 H)

The only difference is the chemical shift of the methyl protons.

12.6 The greater the electronegativity of the atom attached to the methyl group (the more electron-withdrawing the atom), the further downfield the chemical shift.

CH_3–H	CH_3–I	CH_3–Br	CH_3–Cl
δ 0.23	δ 2.16	δ 2.68	δ 3.05
E_N(H) = 2.1	E_N(I) = 2.5	E_N(Br) = 2.8	E_N(Cl) = 3.0

E_N is a measure of electronegativity.

12.7 One would expect peaks at approximately δ 1.0 (for the methyl groups) and δ 5.2–5.7 (for the vinylic protons), with relative areas of 9:1 (or 18:2). The experimental values are actually δ 0.97 (18 H) and δ 5.30 (2 H).

$$\overset{H}{\underset{(CH_3)_3C}{\ }}\!\!>C{=}C<\!\!\overset{C(CH_3)_3}{\underset{H}{\ }}$$

12.8 CH_3CHCl_2 (underlined CH_3) doublet, δ 0.85–0.95, area = 3

CH_3CHCl_2 (underlined CH) quartet, δ 5.8–5.9, area = 1

12.9 a. $BrCH_2CH_2Cl$ (underlined first CH_2) triplet, δ 3.1–3.3, area = 1

$BrCH_2CH_2Cl$ (underlined second CH_2) triplet, δ 3.6–3.8, area = 1

Actually, the spectrum will be more complex than this because the chemical shifts of the two types of protons are close in value.

b. $ClCH_2CH_2Cl$. The protons are all equivalent, and the spectrum will consist of a sharp singlet at δ 3.4–3.5.

12.10 $CH_3CH_2CH_2OH$

The spectrum *without* ^1H coupling will appear as three peaks:

C-1 δ ~ 65
C-2 δ ~ 32

C-3 $\delta \sim 11$

In the [1]H-coupled spectrum, the peaks for C-1 and C-2 would be triplets, and the C-3 peak would be a quartet.

12.11 a. Two; the methyl carbons are equivalent and the hydroxyl-bearing carbon is unique.

b. Five; the methylene carbons (CH_2) next to the carbonyl group are equivalent as are the remaining two methylene (CH_2) carbons. The methane and the exocyclic methyl group are unique. The carbonyl carbon is unique.

c. Three; the methyl carbons are equivalent.

d. Four; the molecule has an internal plane of symmetry passing through C-2 and the C(4)–C(5) bond.

12.12 The IR spectrum of 1-hexyne will show a band in the C≡C stretching frequency region (2100–2260 cm^{-1}) and in the ≡C–H stretching frequency region (3200–3350 cm^{-1}). These bands will not appear in the IR spectrum of 1,3-hexadiene, which will show bands in the C=C (1600–1680 cm^{-1}) and =C–H (3030–3140 cm^{-1}) regions.

12.13 We use Beer's law: $A = \varepsilon c l$. Rearranging the equation, we get:

$$c = A \div (\varepsilon l) = 2.2 \div (12{,}600 \times 1) = 1.75 \times 10^{-4} \text{ mol/L}$$

Note that ultraviolet spectra are often obtained on very dilute solutions.

12.14 Conjugation is possible between the two rings in biphenyl, but, in diphenylmethane, the $-CH_2-$ group interrupts this conjugation. Thus, for comparable electronic transitions, biphenyl is expected to absorb at longer wavelengths.

12.15 Azulene. The blue color indicates that azulene absorbs light in the visible region of the spectrum (400–800 nm). Naphthalene undergoes electronic transitions in the ultraviolet region of the spectrum (314 nm). The longer the wavelength (nm), the lower the energy of the electronic transition.

12.16 Alkanes have the molecular formula C_nH_{2n+2}. Therefore, for any alkane

$$n(\text{atomic wt. of C}) + (2n+2)(\text{atomic wt. of H}) = m/z \text{ ratio of parent ion } (M^+)$$

In this case, $n(12) + (2n+2)(1) = 114$. Rearranging,

$$14n = 114 \text{ or } n = 8$$

and the molecular formula of the alkane is C_8H_{18}. The intensity of the peak at *m/z* 115 should be (1.1)(8) or 9% of the intensity of the peak at *m/z* 114.

12.17 Using the approach described in Problem 12.16, the molecular formula of the compound is C_3H_3Cl. The molecular formula indicates that the compound has two π bonds or one π bond and one ring. Possible structures are:

$HC{\equiv}CCH_2Cl$ $\qquad H_2C{=}C{=}CHCl \qquad CH_3C{\equiv}CCl$

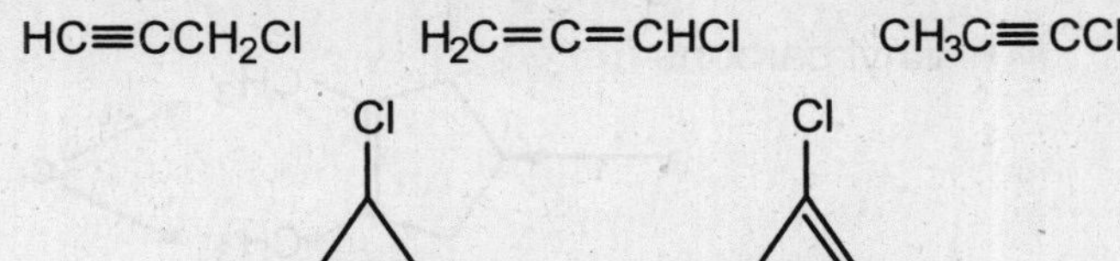

12.18 The molecular ion peak ($C_7H_{14}O^+$) will appear at m/z = 114 (14 mass units, or one –CH_2– group less than for 4-octanone). Since the ketone is symmetrical, it should fragment to yield a $C_3H_7CO^+$ peak (m/z = 71) and a $C_3H_7^+$ peak (m/z = 43). Only one set of daughter ions (instead of the two seen in Figure 12.10) will be observed.

ADDITIONAL PROBLEMS

12.19 Possible solutions are:

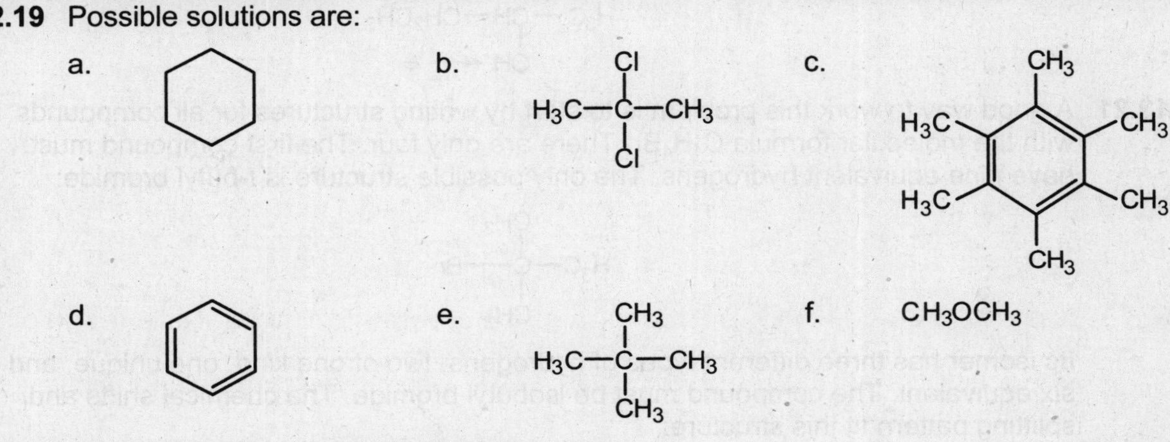

12.20 a. There are four different types of protons:

$$\begin{array}{c} \text{a} \rightarrow CH_3 \\ \text{a} \rightarrow H_3C-\overset{|}{\underset{|}{C}}(b)-CH_2(c)-CH_2-CH_3(d) \\ \text{a} \rightarrow CH_3 \end{array}$$

b. There are five different types of protons:

$$\underset{a}{H_3C}-\underset{b}{CH}(\underset{a}{CH_3})-\underset{}{N}(\underset{c}{H})-\underset{d}{CH_2}-\underset{e}{CH_3}$$

c. There are three different types of protons:

b

CH_3

a → (ring) ← c

CH_3

b

d. There are five different types of protons:

a b c d

$H_3C-CH-CH_2CH_3$

OH ← e

12.21 A good way to work this problem is to start by writing structures for all compounds with the molecular formula C_4H_4Br. There are only four. The first compound must have nine equivalent hydrogens. The only possible structure is *t*-butyl bromide:

CH_3

$H_3C-C-Br$

CH_3

Its isomer has three different types of hydrogens, two of one kind, one unique, and six equivalent. The compound must be isobutyl bromide. The chemical shifts and splitting pattern fit this structure:

δ 1.9, complex, area = 1

δ 0.9, doublet, area = 6 → $(CH_3)_2CHCH_2Br$

δ 3.2, doublet, area = 2

(It is notable that a singlet or doublet that integrates for 6 H is almost always two CH_3's rather than three CH_2's.)

12.22 For the structure of isobutyl acetate, we predict the peaks to appear as:

δ3.5, doublet, 2H

$H_3C-C(=O)-O-CH_2-CH(CH_3)-CH_3$

δ2.3, singlet, 3H

δ1.5 multiplet 1H

δ0.9, doublet, 6H

12.23 The chemical shifts suggest that carbon is more electronegative (more electron-withdrawing) than silicon. Electronegativity generally decreases as one goes down a column in the periodic table. The Pauling electronegativities of carbon and silicon are 2.5 and 1.8, respectively.

12.24 a. CH_3CCl_3 All the protons will appear as a singlet.
$CH_2ClCHCl_2$ There will be two sets of peaks, with an area ratio of 2:1. The former will be a doublet, the latter a triplet.

b. $CH_3CH_2CH_2OH$ In addition to the O–H peak, there will be three sets of proton peaks, with area ratio 3:2:2.
$(CH_3)_2CHOH$ In addition to the O–H peak, there will be only two sets of proton peaks, with area ratio 6:1.

c. $H_3C{-}C({=}O){-}OCH_3$

The spectrum will show a singlet at about δ 3.5–3.8 for the O–CH_3 protons and a singlet at δ 2.0–2.5 for the other methyl group.

$H{-}C({=}O){-}OCH_2CH_3$

The spectrum will show a triplet at about δ 1.0 for the methyl group, a quartet near δ 3.8 for the –CH_2– group, and a singlet near δ 9.0 for the –CHO group.

d. $C_6H_5{-}CH_2CH{=}O$

A one-proton aldehyde peak at δ 9.5–9.7 (a triplet) will easily distinguish this aldehyde from its ketone isomer.

$C_6H_5{-}C({=}O){-}CH_3$

The aliphatic protons will give a sharp three-proton singlet at about δ 2.1–2.6.

12.25 The peaks are assigned as follows:

Methyl 4-methylbenzoate ($CH_3{-}C_6H_4{-}C({=}O){-}O{-}CH_3$): ring H ortho to CH_3: δ 7.15, δ 7.15; ring H ortho to C=O: δ 7.87, δ 7.87; CH_3 (ring): δ 2.35; O–CH_3: δ 3.82

In general, aromatic protons show up at lower field strengths if they are adjacent to electron-withdrawing substituents. This is the basis for distinguishing between the two sets of aromatic protons.

12.26 a. $(CH_3)_2CHCHO$

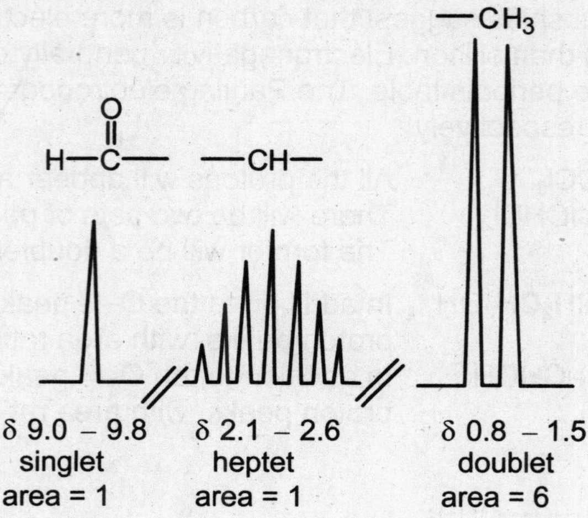

b. $(CH_3)_2CHOCH(CH_3)_2$

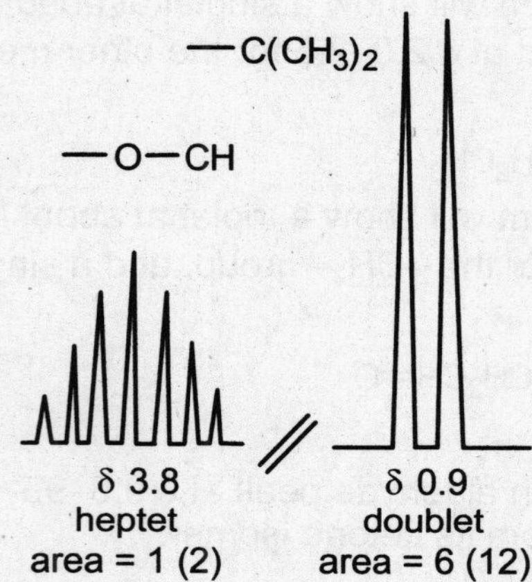

c. $CH_3CH{=}CCl_2$

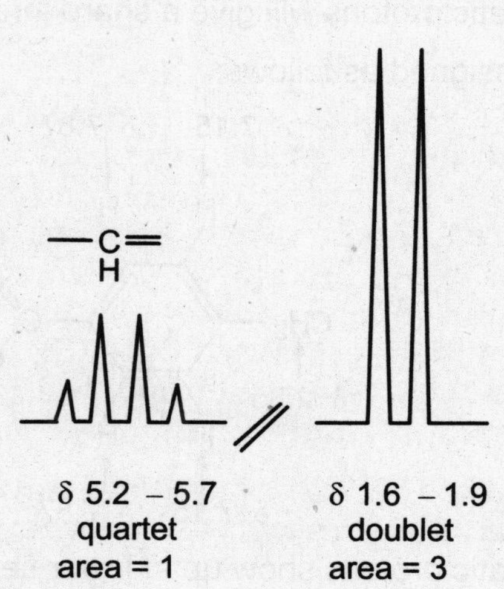

d. 1,4-dimethoxybenzene

CH₃O–C₆H₃(H)–OCH₃

δ 6.5 – 7.5
singlet
area = 2 (4)

δ 3.0 – 4.0
singlet
area = 3 (6)

12.27 There are three possibilities. The number of different types of carbons is shown under each structure.

8 8 8

The ^{1}H NMR spectrum would show a singlet (3H) near δ 2.5 (aromatic CH_3), a singlet (3H) near δ 3.5 (OCH_3), and two doublets (2H each) near δ 7.0.

12.28 The four isomeric butanols are:

$CH_3CH_2CH_2CH_2OH$	$CH_3CH_2CH(OH)CH_3$	$(CH_3)_2CHCH_2OH$	$(CH_3)_3C—OH$
1-butanol	2-butanol	2-methyl-1-propanol (isobutyl alcohol)	2-methyl-2-propanol (*tert*-butyl alcohol)

a. The NMR spectrum is that of 2-methyl-2-propanol (*tert*-butyl alcohol). This is the only isomer of butanol that has only two unique carbon atoms (three equivalent methyl groups and the carbon bonded to the hydroxyl group).

b. 1-Butanol and 2-butanol have four different carbons and would each exhibit four peaks in their decoupled ^{13}C NMR spectra. The $^{13}C–^{1}H$ coupled spectrum of 1-butanol would show one quartet for the methyl group (CH_3) and three triplets for the methylene (CH_2) groups. This is quite different from the coupled spectrum of 2-butanol, which would show two quartets for the methyl groups, a triplet for the methylene, and a doublet for the methine (CH).

12.29 In a proton decoupled ^{13}C NMR spectrum, each of the unique carbons appears as a separate peak, so one needs to know how many unique carbons there are in each structure.

a. 3

$$CH_3CH_2-\overset{O}{\overset{\|}{C}}-OH$$

b. 4

$$C_6H_5-NH_2$$

c. 4

(1,2-dimethylcyclohexene)

12.30 The absence of a band at 3500 cm^{-1} indicates that there is no hydroxyl group. The absence of a band at 1720 cm^{-1} indicates that the compound is not an aldehyde or ketone. This suggests that the oxygen function is probably an ether. Possible structures are:

$CH_3OCH{=}CH_2$ $\quad$ H_2C-O / H_2C-CH_2 (oxetane) $\quad$ $H_2C-CH-CH_3$ with O bridging (epoxide)

An NMR spectrum would readily distinguish between these possibilities. For example, the vinyl ether would show a methyl singlet at about δ 3–4, the middle structure would show a 4H multiplet at δ 3–4, and the epoxide would show a methyl doublet at about δ 1.0–1.5. The presence of a band at about 1650 cm^{-1} in the IR spectrum (C=C stretch) would indicate that the compound is the vinyl ether (the first of the three compounds) and not the cyclic ethers.

12.31 The lower the wavenumber (cm^{-1}), the easier it is to stretch the bond.

$-\overset{\vert}{\underset{\vert}{C}}-H$	$=C\langle^{H}$	$\equiv C-H$
2850–3000 cm^{-1}	3030–3140 cm^{-1}	3200–3350 cm^{-1}
C(sp^3)–H	C(sp^2)–H	C(sp)–H

The longer the bond, the weaker the bond, and the easier it will be to stretch the bond. C–H bond lengths increase as one goes from *sp*-hybridized carbon to sp^3-hybridized carbon, consistent with the observed trend in stretching frequency.

12.32 These infrared data provide direct evidence for hydrogen bonding in alcohols. In dilute solution, the alcohol molecules are isolated, being surrounded by inert solvent molecules. The sharp band at 3580 cm^{-1} is caused by the O–H stretching frequency in an isolated ethanol molecule. As the concentration of ethanol is increased, alcohol molecules can come in contact with one another and form hydrogen bonds. Hydrogen-bonded O–H is less "tight" than an isolated O–H and has a variable length (as the proton is transferred back and forth between oxygen atoms). Consequently, hydrogen-bonded O–H absorbs at a lower frequency and with a broader range (3250–3350 cm^{-1}) than the isolated O–H group.

12.33 Figures 12.15–12.18 represent 3-pentanone, hexanoic acid, 1-pentanol, and cyclohexane, respectively. The C=O stretch typical of ketones is at 1715 cm^{-1} in the spectrum of 3-pentanone. The hexanoic acid shows a C=O stretch at 1710 cm^{-1} and a strong, broad O–H stretch from 2500–3500 cm^{-1}. The O–H stretch typical of alcohols is apparent at 3500 cm^{-1} in the spectrum of 1-pentanol. The cyclohexane shows no functional groups, just a lot of bands in the C–H stretch and fingerprint region of the spectrum.

12.34 Both compounds will be similar in the functional group region of the spectrum with bands at 3500 cm^{-1} for the O–H stretch and 1600–1680 cm^{-1} for the C=C stretch. Their fingerprint regions (700–1500 cm^{-1}) are expected to differ from one another.

12.35 a. The first compound, a ketone, would show a carbonyl stretching band at about 1720 cm^{-1}. The alcohol would not show this band and would show bands for the hydroxyl and alkene groups at 3200–3700 cm^{-1} and 1600–1680 cm^{-1}, respectively.

b. The first compound, an aldehyde, would show a carbonyl stretching band at about 1720 cm^{-1}. This band would be absent in the IR spectrum of the vinyl ether, which would show a C=C stretching band at 1600–1680 cm^{-1}.

c. The secondary amine [$(CH_3CH_2)_2NH$] will show an N–H stretching band at 3200–3600 cm^{-1}. The tertiary amine (triethylamine) has no N–H bonds and this band will be absent.

12.36 You could monitor the disappearance of the O–H stretching band in 1-octanol and the appearance of the carbonyl stretching band in octanal to follow the progress of the reaction and determine the purity of the product.

12.37 Compounds a, c, and d have no unsaturation and will not absorb in the ultraviolet region of the spectrum.

12.38 Since UV-vis excitations are the energy differences between the ground state and the excited state, then the more stable the ground state (and with no change in the excited state), then the energy needed for excitation would be greater. This greater energy would be longer wavelength λ_{max} transition, and then the color would move from the visible portion of the spectrum to the UV region. Hence, the phenol-imine would appear yellow (greater aromaticity in the ground state) and the keto-enamine would appear red.

12.39 See Sec. 12.5 in the text. As the number of double bonds increases, so does the extent of conjugation. Thus the absorption maximum moves to longer and longer wavelengths.

12.40 In *trans*-1,2-diphenylethylene, both phenyl groups can be conjugated with the double bond, and all π bonds can lie in one plane. In *cis*-1,2-diphenylethylene, only one phenyl group at a time can be conjugated with the double bond for steric reasons. The more extended conjugation in the *trans* compound leads to a lower energy (longer wavelength) λ_{max} than in the *cis* compound.

λ_{max} = 295 nm

H H C=C

H H C=C

λ_{max} = 280 nm

12.41 We must use Beer's law (eq. 12.5) to solve this problem.

$$A = \varepsilon cl, \text{ or } c = A/\varepsilon l$$
$$c = 0.65 / 224 \text{ c } 1 = 2.90 \times 10^{-3} \text{ mol/L}$$

To go further, 1 L methylcyclohexane will contain $2.90 \times 10^{-3} \times 98 = 0.284$ g toluene as a contaminant. As you can see, ultraviolet spectroscopy can be a very sensitive tool for detecting impurities.

12.42 Follow the example in eq. 12.7.

$$[CH_3CH_2CH_2CH_2CH_2-\dot{\ddot{O}}-H]^+$$

12.43 The molecular ion of 1-butanol will be:

$$[CH_3CH_2\overset{2}{C}H_2\overset{1}{C}H_2\ddot{O}H]^{+\cdot}$$

Fragmentation between C-1 and C-2 would give a daughter ion with m/z = 31 (compare with eq. 12.8). A possible mechanism for this cleavage is:

$$\left[CH_3CH_2CH_2-\underset{H}{\overset{H}{C}}-\dot{\ddot{O}}-H\right]^+ \longrightarrow CH_3CH_2CH_2\cdot + \left[\underset{H}{\overset{H}{>}}C=\ddot{O}-H\right]^+$$

m/z = 31

12.44 The band at 1725 cm^{-1} in the infrared spectrum is due to a carbonyl group, probably a ketone. The quartet-triplet pattern in the NMR spectrum suggests an ethyl group. The compound is 3-pentanone:

δ 2.7, quartet, area = 2 (or 4H)

$$H_3C-CH_2-\overset{O}{\overset{\|}{C}}-CH_2-CH_3$$

δ 0.9, triplet, area = 3 (or 6H)

12.45 The quartet-triplet pattern suggests that the ten protons are present as two ethyl groups. This gives a partial structure of $(CH_3CH_2)_2CO_3$. The chemical shift of the CH_2 groups (δ 4.15) suggests that they are attached to the oxygen atoms. Finally, the infrared band at 1745 cm^{-1} suggests a carbonyl function. The structure is diethyl carbonate:

$$H_3C-CH_2-O-\overset{O}{\overset{\|}{C}}-O-CH_2-CH_3$$

12.46 The molecular formula, $C_4H_{10}O$, tells us that compound **A** is saturated. The fact that **A** can be oxidized with PCC to an aldehyde (note that **B** gives a positive Tollens' test) tells us that **A** is a primary alcohol. We can draw only two structures for **A** that are consistent with these data:

$CH_3CH_2CH_2CH_2OH$ (1-butanol) and $(CH_3)_2CHCH_2OH$ (2-methyl-1-propanol)

Of these, only 2-methyl-1-propanol is consistent with the ^{1}H NMR spectrum shown in Figure 12.1. The 6H signal at δ 0.9 is due to the two equivalent methyl groups, the heptet at δ 1.7 is due to the CH, the 2H doublet at δ 3.35 is due to the CH_2 adjacent to the hydroxyl group, and the remaining 1H signal is due to the –OH. Compound **B** must therefore be $(CH_3)_2CHCH{=}O$.

12.47 The formula $C_5H_{12}O = C_5H_{11}OH$ tells us that the alcohols are saturated. The peak at $m/z = 59$ cannot be caused by a four-carbon fragment ($C_4 = 4 \times 12 = 48$; this would require 11 hydrogens, too many for four carbons). Thus the peak must contain one oxygen, leaving 59 – 16 = 43 for carbon and hydrogen. A satisfactory composition for the peak at 59 is $C_3H_7O^+$ (or $C_3H_6OH^+$). Similarly, the $m/z = 45$ peak corresponds to $C_2H_5O^+$ (or $C_2H_4OH^+$). Fragmentation of alcohols often occurs between the hydroxyl-bearing carbon and an attached hydrogen or carbon (eq. 12.8). Possible structures for the first alcohol are

H
CH_3CH_2—C—OH
CH_2CH_3
3-pentanol

or

CH_3
H_3C—C—OH
CH_2CH_3
2-methyl-2-butanol

Possible structures of its isomer are

H
H_3C—C—OH
$CH_2CH_2CH_3$
2-pentanol

or

H
CH_3—C—OH
$CH(CH_3)_2$
3-methyl-2-butanol

The correct structure for the first isomer could be deduced by NMR spectroscopy as follows:

H ← δ 4, 1 H, quintet
$H_3C-CH_2-C-CH_2-CH_3$ (C bears OH)
δ 1.3, 4 H, quintet or multiplet
δ 1, 3 H, triplet
three ^{13}C peaks

δ 1.3, 2 H, quartet
CH_3
$CH_3-C-CH_2-CH_3$ (C bears OH)
δ 1, 6 H, singlet
δ 1, 3 H, triplet
four ^{13}C peaks

The correct structure for the second isomer also could be deduced by NMR spectroscopy. However, both proton spectra are likely to be quite complex because of similar chemical shifts and a great deal of spin-spin coupling. The ^{13}C spectra are simpler and diagnostic:

$CH_3CHCH_2CH_2CH_3$ (OH on C-2)
six types of protons
five ^{13}C peaks

$CH_3CHCH(CH_3)_2$ (OH on C-2)
five types of protons
four ^{13}C peaks

12.48 The NMR peak at δ 7.4 with an area of 5 suggests the compound may have a phenyl group, C_6H_5–. If so, this accounts for 77 of the 102 mass units. This leaves only 25 mass units, one of which must be a hydrogen (for the NMR peak at δ 3.08). The other 24 units must be two carbon atoms, since the compound is a hydrocarbon (no other elements are present except C and H). Phenylacetylene fits all the data:

–C≡CH

12.49 The molecular formula is C_8H_8O. The ^{1}H NMR peaks above δ 7.4 ppm suggests that there is an aromatic ring, and apparently 3 unique sets of aromatic hydrogens, often indicative of a single substitution on a benzene ring (i.e. a C_6H_5– group). The ^{1}H NMR peak at δ 2.6 ppm is indicative of an alkyl group next to a carbonyl group. The ^{13}C NMR spectrum confirms that there is a carbonyl group (δ 198 ppm), 4 unique aromatic carbons around δ 130 ppm (consistent with a single substitution on a benzene ring), and an alkyl group (δ 26 ppm). The IR spectrum confirms a C=O group as well as sp^2 and sp^3 C–H bonds. Finally, the mass spectrum suggests that the molecular weight is 120. Prominent fragments in the mass spectrum are 105 (loss of 15, a CH_3 group) and 77 (subsequent loss of 28, a CO unit) to leave 77 (a C_6H_5 unit). A C=O group has a mass of 28, a CH_3 group is 15 and a phenyl (C_6H_5–) group is 77—the three groups add to 120. Therefore, the structure is acetophenone:

O
||
C
CH3

12.50 The natural abundance of ^{13}C is very low (~1%) so the greatest difficulty will be in detecting the desired signal over the noise in the spectrum. However, one way to accomplish the detection of a specific compound would be to intentionally label the compound with ^{13}C by synthetic means. For example if one wanted to monitor the concentration of acetophenone, we could intentionally install a ^{13}C isotope, in nearly 100% abundance, at (say) the carbonyl carbon and then monitor that signal in MRI.

O
||
C
CH3
^{13}C

13

Heterocyclic Compounds

Chapter Summary

Atoms other than carbon and hydrogen that appear in organic compounds are called **heteroatoms**. Cyclic organic compounds that contain one or more heteroatoms are called **heterocycles**. Heterocyclic compounds are the largest class of organic compounds and can be either **aromatic** (such as pyridine, pyrrole, and furan) or **nonaromatic** (such as piperidine, pyrrolidine, and tetrahydrofuran).

Pyridine is a six-membered heterocycle that has a structure **isoelectronic** with the aromatic hydrocarbon benzene. In pyridine, one of the –(CH)= units in benzene is replaced by an sp^2-hybridized nitrogen [–(N:)=], and the nitrogen contributes one electron to the aromatic ring. Pyridine undergoes **electrophilic aromatic substitution** reactions at the 3-position, but at reaction rates much slower than benzene, partly because of the electron-withdrawing and deactivating effect of the nitrogen. Pyridine undergoes **nucleophilic aromatic substitution** upon treatment with strong nucleophiles like sodium amide and sodium methoxide. The reaction mechanism involves addition of the nucleophile to the electron-deficient pyridine followed by elimination of a leaving group. The nitrogen of pyridine is basic, and the nonbonded lone pair is protonated by mineral acids to give **pyridinium salts**. Catalytic hydrogenation of pyridine gives the nonaromatic six-membered heterocycle **piperidine**. **Nicotine** (from tobacco) and **pyridoxine** (vitamin B_6) are two naturally occurring substituted pyridines.

Polycyclic aromatic heterocycles that contain pyridine rings fused with benzene rings include **quinoline** and **isoquinoline**. **Quinine**, used to treat malaria, is an example of a naturally occurring quinoline.

The **diazines (pyridazine**, **pyrimidine**, and **pyrazine)** are six-membered aromatic heterocycles that have two nitrogens in the ring. **Cytosine**, **thymine**, **and uracil** are derivatives of pyrimidine that are important bases in nucleic acids (DNA and RNA). Heterocyclic analogs of the aromatic hydrocarbon naphthalene include **pteridines**, which have four nitrogens in the rings. Naturally occurring pteridine derivatives include **xanthopterin** (a pigment) and **folic acid** (a vitamin). **Methotrexate** is a pteridine used in cancer chemotherapy.

Pyrylium ions are six-membered heterocycles in which a positively charged sp^2-hybridized oxygen replaces the nitrogen in pyridine. The pyrylium ring appears in many naturally occurring flower **pigments**.

Pyrrole, **furan**, and **thiophene** are five-membered aromatic heterocycles with one heteroatom. In pyrrole, the nitrogen is sp^2-hybridized and contributes two electrons to the 6π aromatic ring. Furan and thiophene are isoelectronic with pyrrole, the [–(N:)=] unit being replaced by –(:O:)– and –(:S:)– units, respectively. Pyrrole, furan, and thiophene are electron-rich (there are six π electrons distributed over five atoms) and undergo electrophilic aromatic substitution at the 2-position with reaction rates much faster than benzene. Pyrrole rings form the building blocks of biologically important pigments called **porphyrins**. **Hemoglobin** and **myoglobin** (important in oxygen transport) and **chlorophyll** (important in photosynthesis) are examples of naturally occurring porphyrins. Furans are obtained commercially from **furfural** (furan-2-carbaldehyde), which is produced by heating corn cobs with strong acid.

Polycyclic aromatic heterocycles in which the 2- and 3-positions of a pyrrole are fused to a benzene ring are called **indoles**. The indole ring occurs in many medicinally important natural products, such as the neurotransmitter serotonin.

The **azoles** (**oxazole**, **imidazole**, and **thiazole**) are five-membered aromatic heterocycles that have two heteroatoms in the ring. One of the heteroatoms in each of these heterocycles is an sp^2-hybridized nitrogen that contributes one electron to the 6π aromatic system and has a basic nonbonded lone pair. The other heteroatom (oxygen, nitrogen, or sulfur) contributes two electrons to the 6π system. The imidazole skeleton is present in the amino acid **histidine.** The thiazole ring occurs in **thiamin** (vitamin B_1).

The **purines** are an important class of heterocycles in which an imidazole ring is fused to a pyrimidine ring. **Uric acid** (the main product of nitrogen metabolism in birds and reptiles), **caffeine** (present in coffee), and **adenine** and **guanine** (nitrogen bases present in the nucleic acids DNA and RNA) are examples of naturally occurring purines.

Reaction Summary

Reactions of Pyridine and Related Six-Membered Aromatic Heterocycles

1. Protonation

pyridine + HX → pyridinium (N^+–H) X^-

X = Cl, Br, I, HSO_4, and so forth

2. Electrophilic Aromatic Substitution

pyridine + E^+ → 3-E-pyridine + H^+

3. Nucleophilic Aromatic Substitution

Nu-metal

(for example, H_2N–Na, Ph–Li)

N

N Nu

+ Metal–H

Cl

N

Nu-metal

(for example, CH_3O–Na)

Nu

N

+ Metal–Cl

4. Oxidation

CH_3

N

$KMnO_4$

CO_2H

N

5. Reduction

N

H_2

Pt

N

Electrophilic Aromatic Substitution Reactions of Five-Membered Aromatic Heterocycles

X + E^+ → X E + H^+

X = O:, S:, N–H

Mechanism Summary

Electrophilic Aromatic Substitution of Pyridine

E⁺

+E⁺

H E N

H E + N

H E + N

E N

−H⁺

Nucleophilic Aromatic Substitutions of Pyridines

N X Nu:⁻

X = leaving group

N X Nu

N Nu + X⁻

Electrophilic Aromatic Substitution of Five-Membered Heterocycles

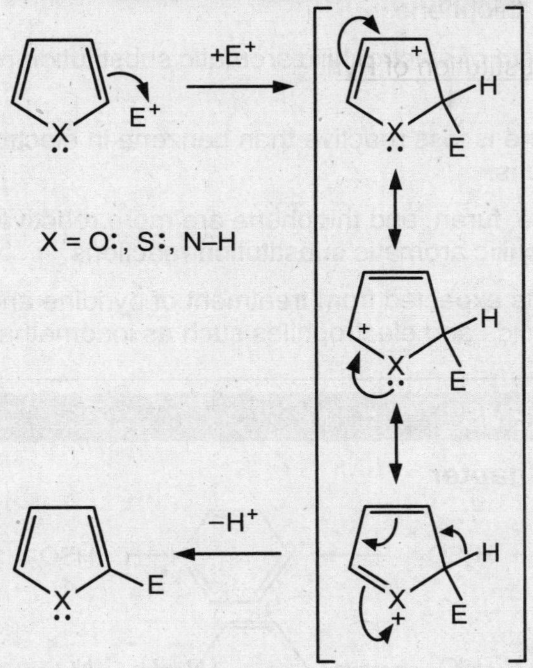

$+E^+$

$-H^+$

X = O:, S:, N–H

Learning Objectives

1. Know the meaning of: heteroatom, aromatic heterocycle, nonaromatic heterocycle, diazine, azole.
2. Draw the structure of: pyridine, pyrrolidine, nicotine, quinoline, isoquinoline, pyrimidine, pteridine, the pyrylium cation.
3. Draw the structure of: pyrrole, furan, thiophene, furfural, indole, oxazole, imidazole, thiazole, purine.
4. Determine how many electrons are contributed to the aromatic π system by each heteroatom in the aforementioned heterocycles.
5. Determine the position of basic nonbonded lone pairs in the aforementioned heterocycles.
6. Predict the product of electrophilic aromatic substitution reactions of pyridine and quinoline.
7. Write the mechanism of electrophilic aromatic substitution reactions of pyridine.
8. Predict the product of nucleophilic aromatic substitution reactions of pyridine and substituted pyridines.
9. Write the mechanism of nucleophilic aromatic substitution reactions of pyridine.
10. Predict the product expected from hydrogenation of aromatic heterocycles.
11. Predict the product expected from potassium permanganate oxidation of a methylated pyridine.

12. Predict the product expected from electrophilic aromatic substitution reactions of pyrrole, furan, and thiophene.
13. Write the mechanism of electrophilic aromatic substitution reactions of pyrrole, furan, and thiophene.
14. Explain why pyridine is less reactive than benzene in electrophilic aromatic substitution reactions.
15. Explain why pyrrole, furan, and thiophene are more reactive than benzene or pyridine in electrophilic aromatic substitution reactions.
16. Predict the products expected from treatment of pyridine and imidazole with Brønsted–Lowry acids and electrophiles such as iodomethane.

ANSWERS TO PROBLEMS

Problems Within the Chapter

13.1 a. N: + H_2SO_4 ⟶ N^+–H HSO_4^-

b. N: + HNO_3 ⟶ N^+–H NO_3^-

13.2 The contributors to the resonance hybrid derived from electrophilic attack at C-2 are:

E ⟷ E ⟷ E
N H N H N H

The contributors to the resonance hybrid derived from attack at C-4 are:

H E ⟷ H E ⟷ H E
N N N

In each case, positive charge is placed on the nitrogen atom in one resonance contributor. This is a destabilizing effect due to the electronegativity of nitrogen. Electrophilic attack at C-3 gives a cation in which the positive charge does not reside on nitrogen (see Example 13.1). Therefore, attack at C-3 is favored relative to attack at C-2 and C-4.

13.3 Methoxide adds to C-4 followed by elimination of chloride ion.

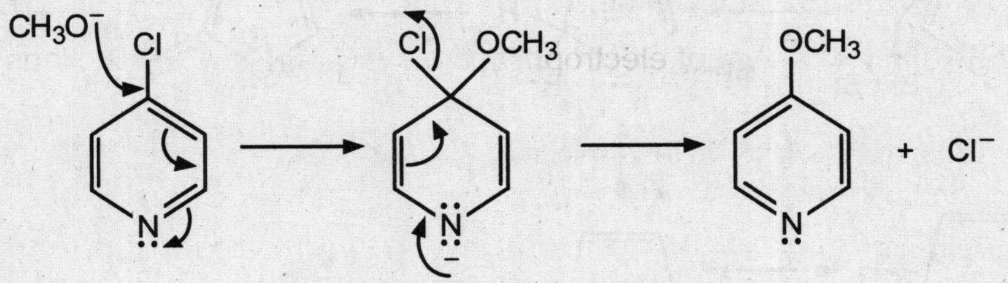

CH_3O^- Cl → Cl OCH$_3$ → OCH$_3$ + Cl^-

13.4 The configuration of (+)-coniine is *R* (review Sec. 5.3).

13.5

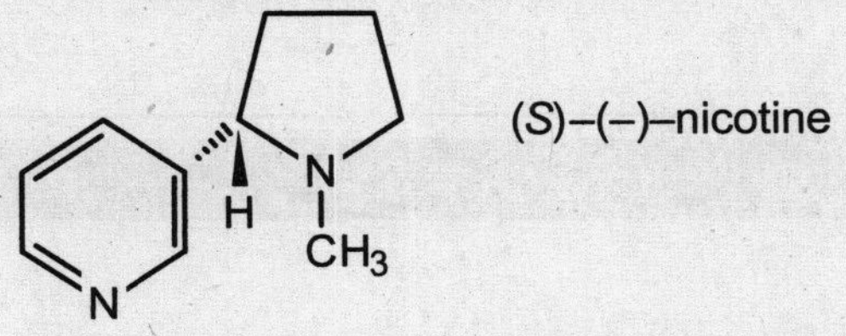

(*S*)–(–)–nicotine

13.6 The first equivalent of HCl reacts with the more basic, sp^3-hybridized nitrogen to give:

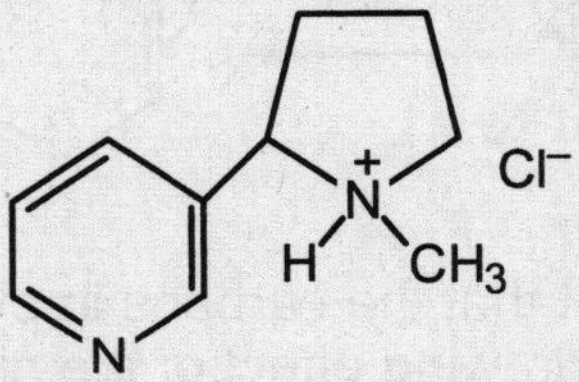

Cl^-

The second equivalent of HCl reacts with the sp^2-hybridized pyridine nitrogen:

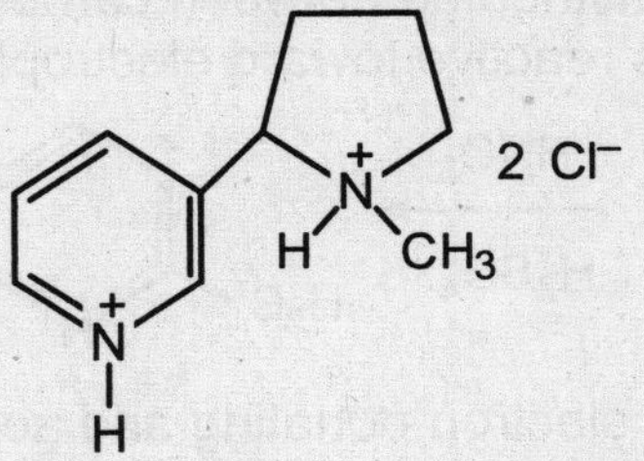

2 Cl^-

13.7 The cation derived from protonation of pyrrole at C-2 is a hybrid of the following resonance structures.

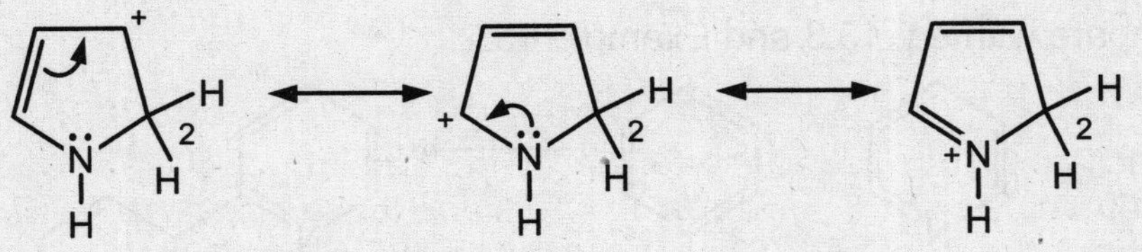

13.8

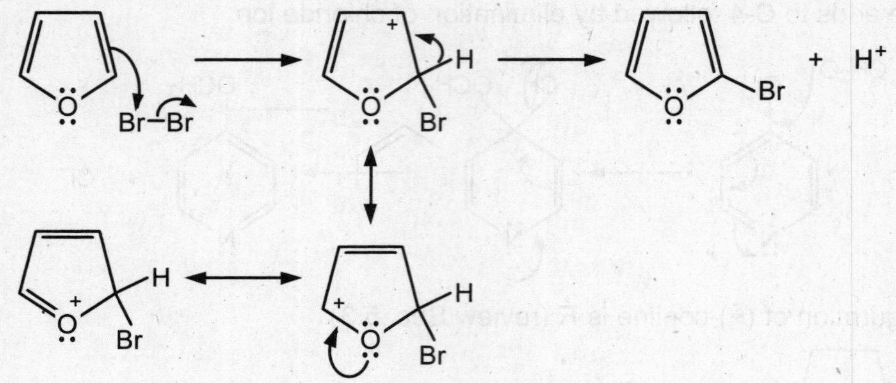

ADDITIONAL PROBLEMS

13.9 Three minor dipolar contributors are:

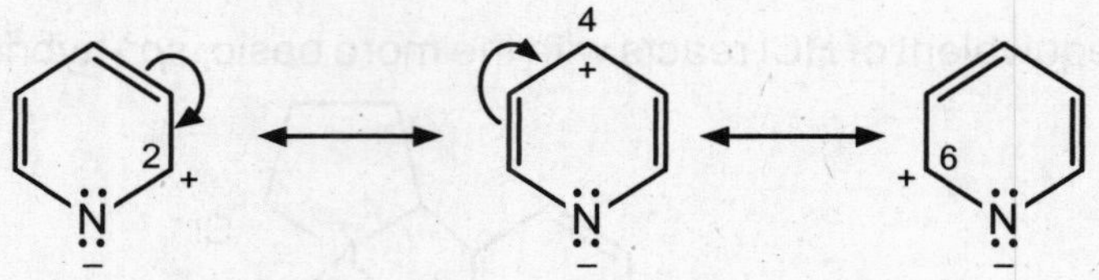

These structures suggest that the carbons in pyridine are partially positively charged (due to the electron-withdrawing effect of the nitrogen) and, therefore, are expected to be deactivated (relative to benzene) toward reaction with electrophiles. Note that the positive charge is distributed between carbons 2, 4, and 6. Therefore, these carbons should be less reactive toward electrophiles than carbon 3 (or 5).

13.10

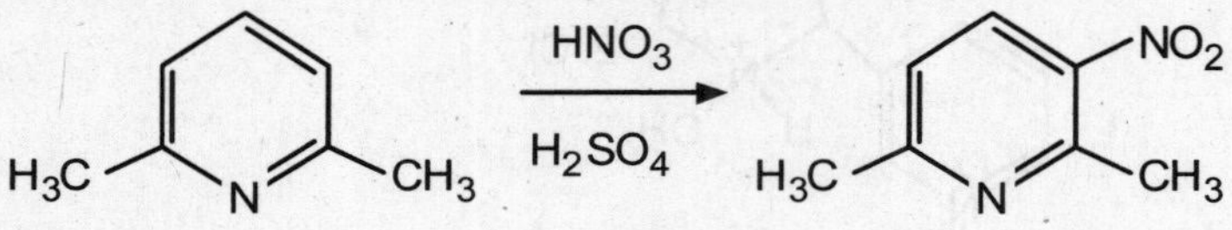

The methyl groups are electron-donating and activate aromatic compounds toward electrophilic aromatic substitution (see Secs. 4.10 and 4.11). Therefore, 2,6-dimethylpyridine undergoes nitration under much milder conditions than required by pyridine. Notice that nitration occurs *ortho* and *para* to the methyl groups.

13.11 Compare with eq. 13.3 and Example 13.2.

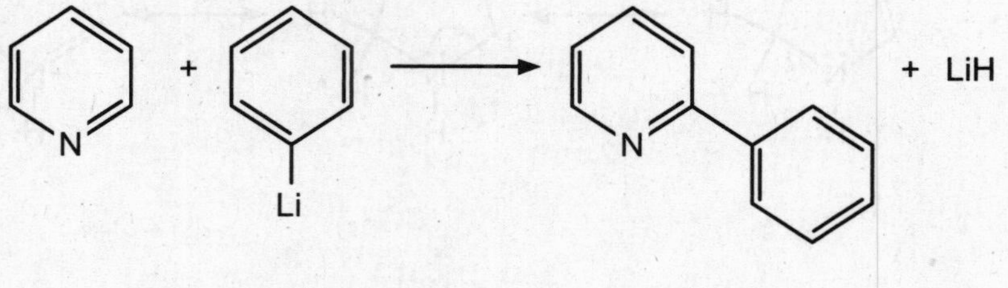

Mechanism

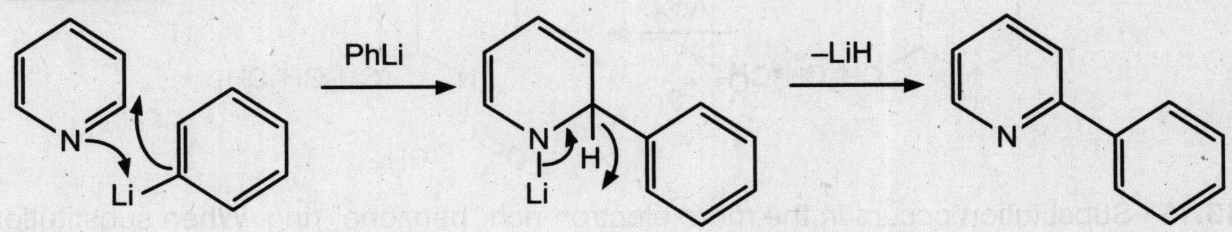

PhLi → −LiH →

13.12

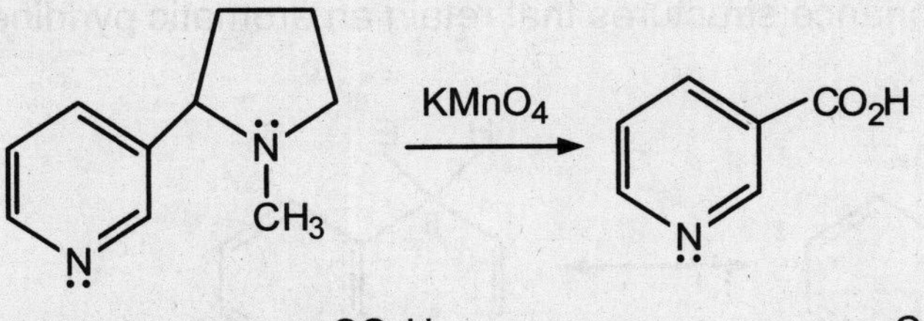

$KMnO_4$ → CO_2H

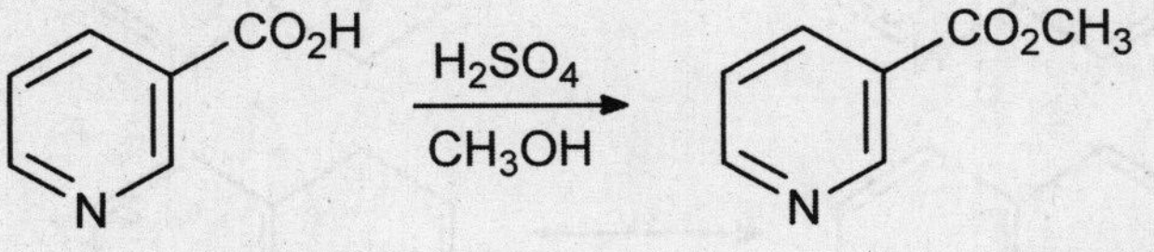

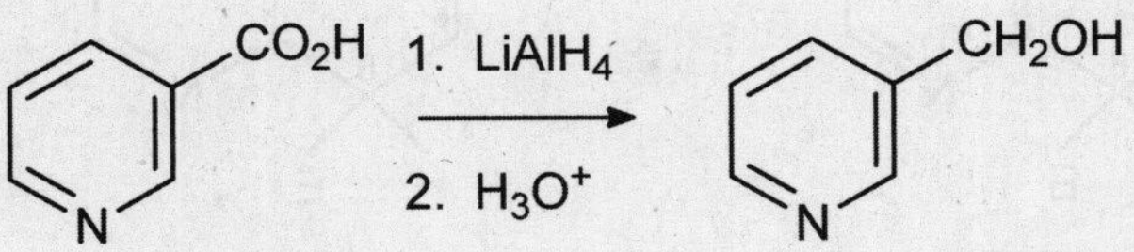

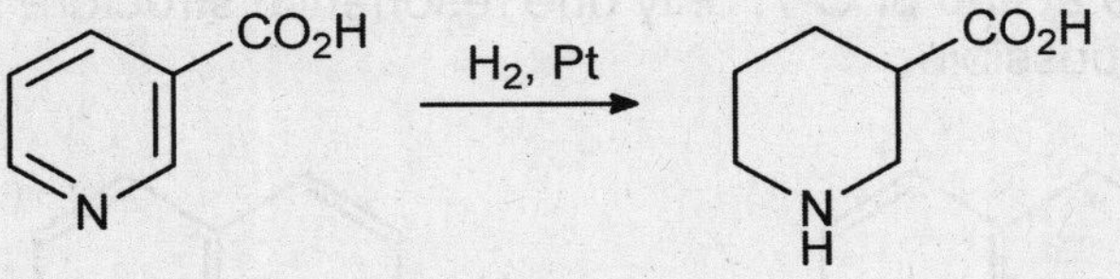

13.13 a.

CO_2H $\xrightarrow[CH_3OH]{H_2SO_4}$ CO_2CH_3

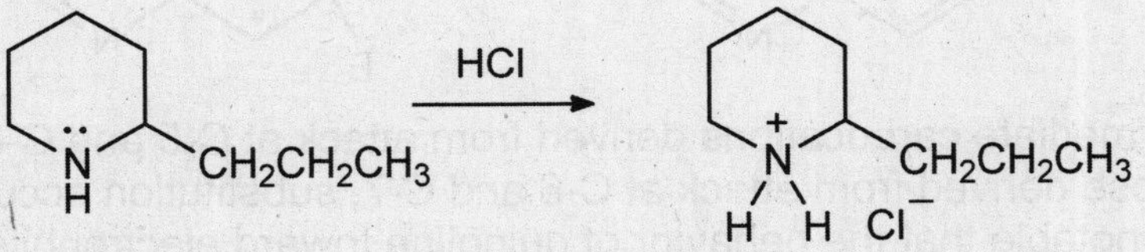

b.

CO_2H $\xrightarrow[2.\ H_3O^+]{1.\ LiAlH_4}$ CH_2OH

c.

CO_2H $\xrightarrow{H_2,\ Pt}$ CO_2H

13.14 a.

$CH_2CH_2CH_3$ $\xrightarrow{HCl}$ $CH_2CH_2CH_3$ Cl^-

b.

$CH_2CH_2CH_3$ $\xrightarrow{C_6H_5CH_2I\ (1\ eq.)}$ $CH_2CH_2CH_3$, $CH_2C_6H_5$ I^-

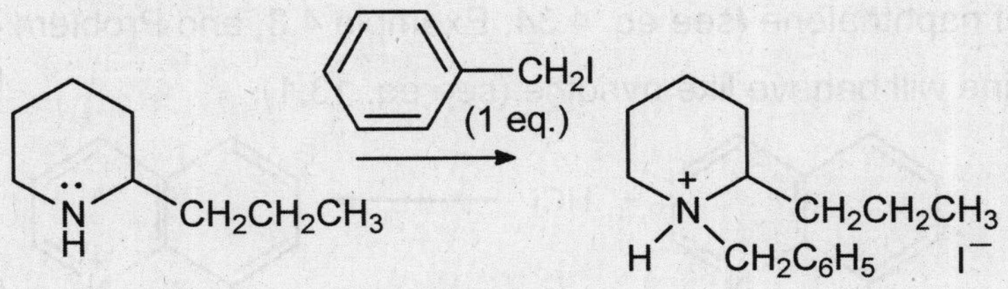

c.

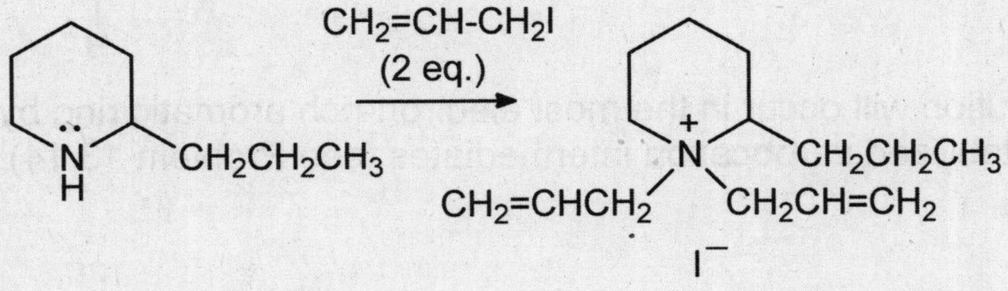

$CH_2CH_2CH_3$ $\xrightarrow{CH_2{=}CH{-}CH_2I\ (2\ eq.)}$ $CH_2CH_2CH_3$, $CH_2{=}CHCH_2$, $CH_2CH{=}CH_2$ I^-

d. 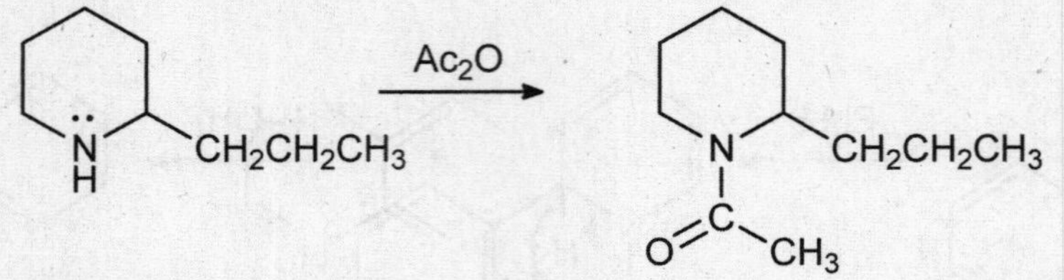 (2-propylpiperidine) $\xrightarrow{Ac_2O}$ N-acetyl-2-propylpiperidine

$CH_2CH_2CH_3$ ... $CH_2CH_2CH_3$, CH_3

13.15 Substitution occurs in the more electron-rich "benzene" ring. When substitution occurs at C-5 or C-8, two resonance structures that retain an aromatic pyridine ring are possible.

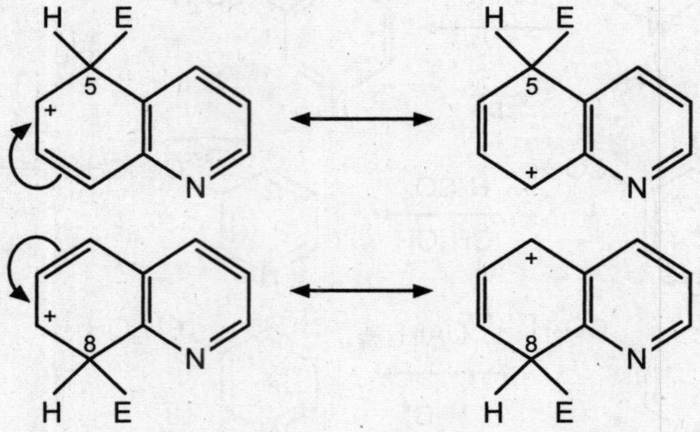

When substitution occurs at C-6 or C-7, only one resonance structure that retains an aromatic pyridine ring is possible.

Since the intermediate carbocations derived from attack at C-5 and C-8 are more stable than those derived from attack at C-6 and C-7, substitution occurs at those positions. It is notable that the behavior of quinoline toward electrophiles is similar to the behavior of naphthalene (see eq. 4.34, Example 4.3, and Problem 4.18).

13.16 a. Quinoline will behave like pyridine (see eq. 13.1).

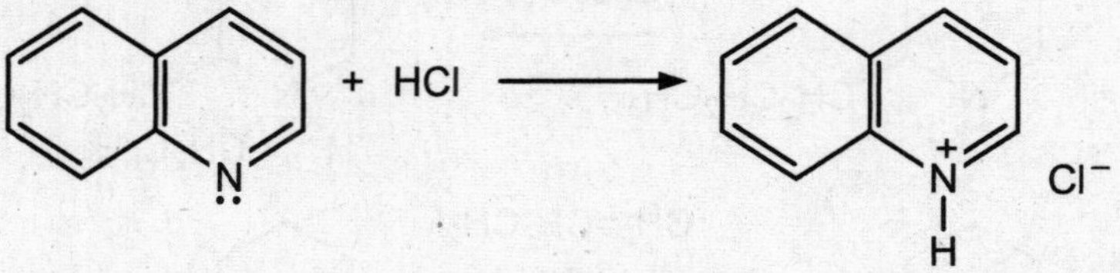

quinoline + HCl ⟶ quinolinium Cl^-

b. Substitution will occur in the most electron-rich aromatic ring by way of the most stabilized carbocation intermediates (see Problem 13.14).

Quinoline $\xrightarrow[H_2SO_4]{HNO_3}$ 5-nitroquinoline + 8-nitroquinoline

c. Quinoline will behave like pyridine (see eq. 13.3).

Quinoline $\xrightarrow[2.\ H_2O]{1.\ NaNH_2}$ 2-aminoquinoline (NH_2)

d. Substitution occurs at C-2 of the electron-deficient pyridine ring.

Quinoline + phenyllithium (Li) $\rightarrow$ 2-phenylquinoline + LiH

e. See eq. 11.6.

Quinoline + CH_3I $\rightarrow$ N-methylquinolinium (CH_3) I^-

13.17 a. See Sec. 13.5.

Furan + Br_2 $\rightarrow$ 2-bromofuran (Br) + HBr

b. See Sec. 13.5

Furan $\xrightarrow{HNO_3}$ 2-nitrofuran (NO_2)

c. See Sec. 13.5.

Furan + $CH_3\overset{O}{\overset{\|}{C}}Cl$ $\xrightarrow{SnCl_4}$ 2-acetylfuran ($C(=O)CH_3$)

13.18 a. 2-methylfuran (CH_3) $\xrightarrow{Br_2}$ 2-bromo-5-methylfuran (Br, CH_3) + HBr

b. [2-methylfuran] $\xrightarrow{HNO_3}$ [O_2N–furan–CH_3]

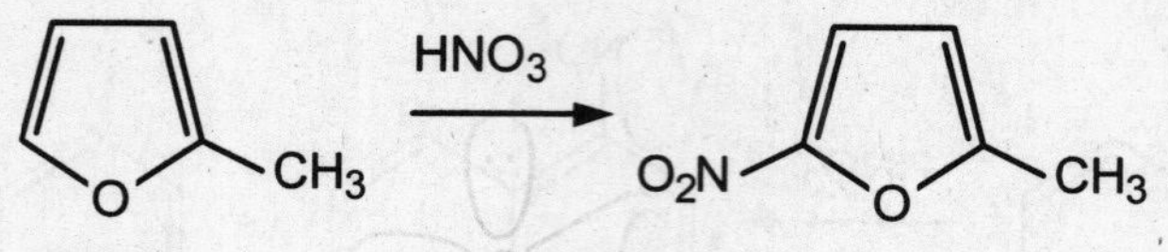

c. [2-methylfuran] $\xrightarrow[SnCl_4]{CH_3COCl}$ [H_3C–C(=O)–furan–CH_3]

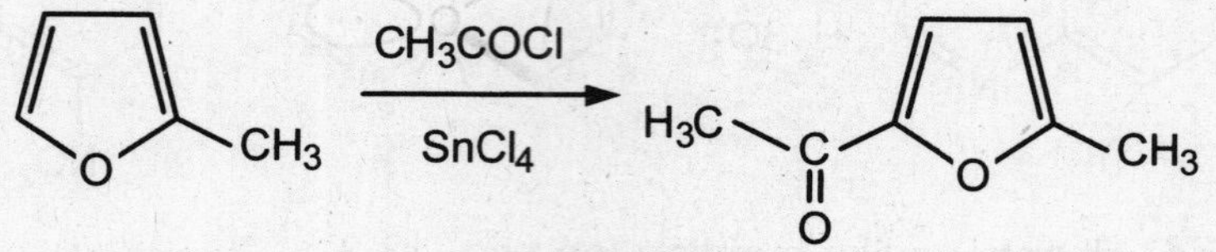

13.19 Examine the intermediates derived from electrophilic attack at each position.

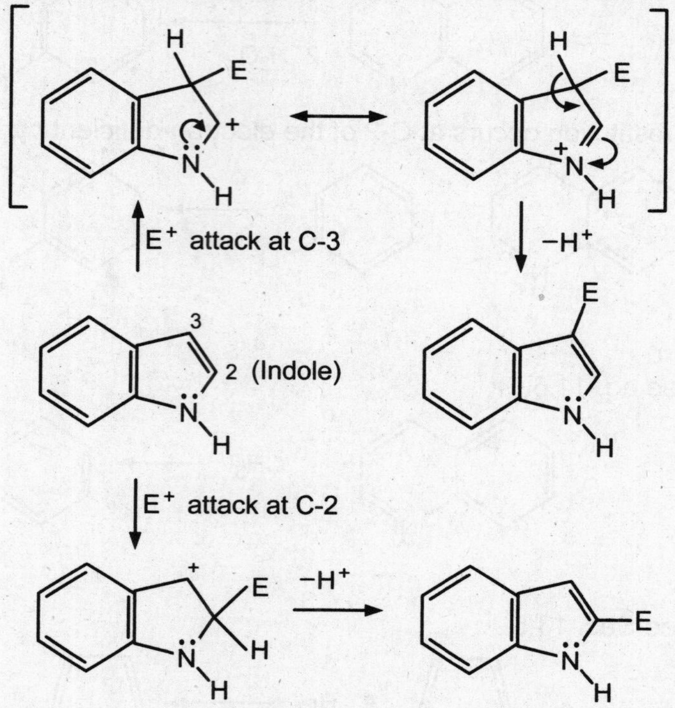

E⁺ attack at C-3

−H⁺

3

2 (Indole)

E⁺ attack at C-2

−H⁺

Attack at C-3 gives a carbocation that is resonance-stabilized by a nonbonded pair of electrons *on the adjacent nitrogen*. Attack at C-2 gives a cation in which the positive charge is not adjacent to an atom bearing nonbonded electrons. Any additional resonance contributors disrupt the benzenoid structure in the "left" ring. Therefore, electrophilic aromatic substitution reactions of indole occur primarily at C-3.

13.20 Imidazole is a good nucleophile and will react with primary alkyl halides, at one nitrogen and then the other:

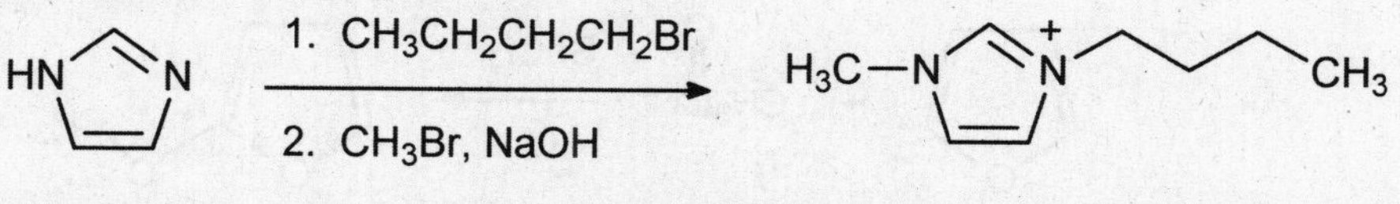

HN–imidazole–N $\xrightarrow[2.\ CH_3Br,\ NaOH]{1.\ CH_3CH_2CH_2CH_2Br}$ H_3C–N–imidazole–N^+–$CH_2CH_2CH_2CH_3$

13.21 The nitrogen has a nonbonded electron-pair in an sp^2 orbital at right angles to the cyclic array of *p* orbitals. Therefore, it will be more basic than pyrrole.

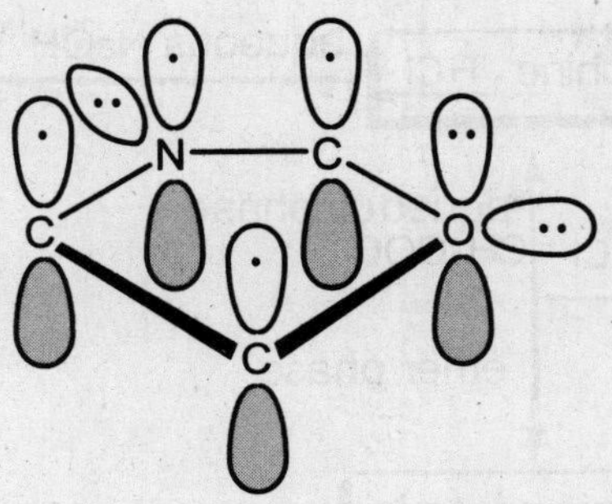

13.22 a. O, F, F

b. S, Br

c. OCH_3, N

d. OH, N

e. H_3C, CH_3, HN, N

f. CO_2H, N

g. H, N, CH_3CH_2, CH_2CH_3

h.

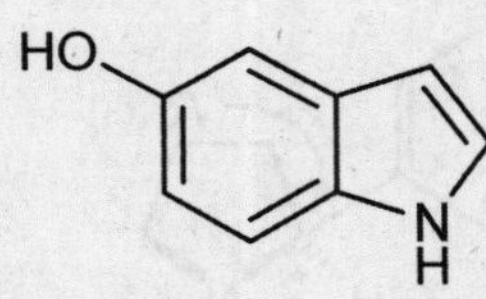

i.

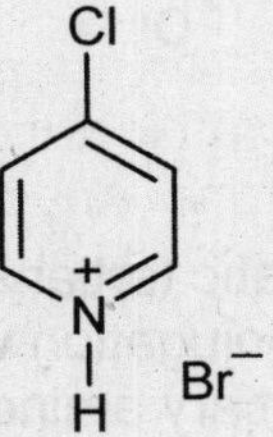

j.

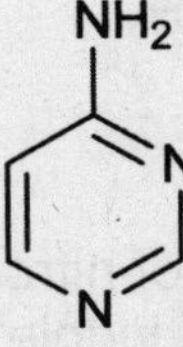

13.23 The configuration of the hydroxyl-bearing carbon atom in codeine is *S* (review Sec. 5.3).

13.24

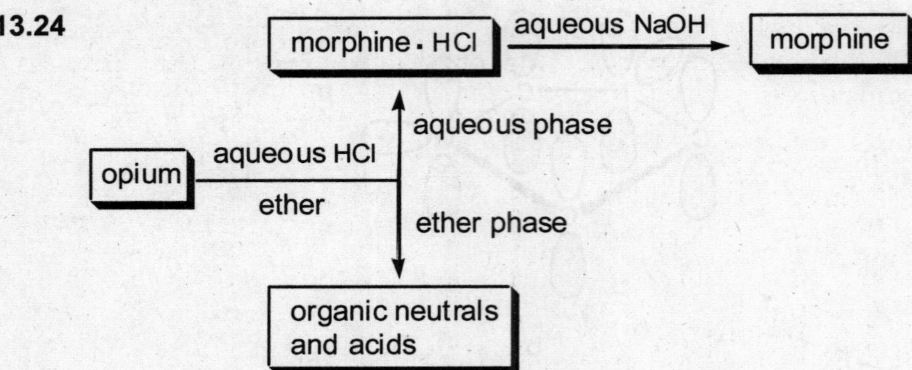

13.25 There are two different kinds of nitrogens in varenicline. The pyridine nitrogens will have a pK_a value for their conjugate acids of about 5, and the aliphatic amine will have a pK_a value for its conjugate acid of about 10.

13.26

Line segment structures of nicotine and cocaine are shown below with each stereogenic center identified by and asterisk and the absolute configuration (*R* or *S*).

nicotine

cocaine

13.27 There are two kinds of nitrogens in lysergic acid. The aliphatic tertiary amine and the indole nitrogen. The indole's nitrogen has its lone pair in conjugation with the aromatic ring system so it is less basic than the aliphatic tertiary amine.

13.28 The *keto* and *enol* forms of phenol and 2-hydroxypyridine are:

phenol

2-hydroxypyridine

aromatic

The *keto* tautomer of phenol does not retain any of the resonance energy associated with the aromatic ring, while the *keto* tautomer of 2-hydroxypyridine does (compare with amide resonance).

13.29 Tautomerization of both amide groups gives:

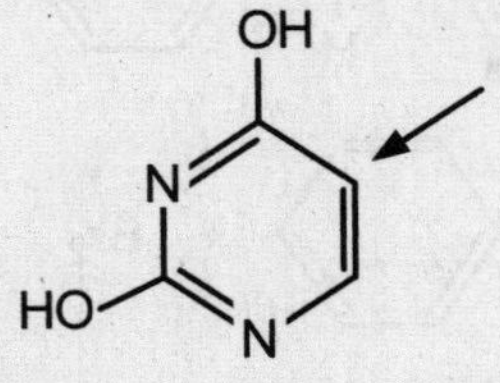

The hydroxyl groups are activating, and thus, nitration occurs at the indicated position, *ortho* to one hydroxyl group and *para* to the other. The product is 5-nitrouracil (the amide form of uracil is more stable than the *enol* tautomer).

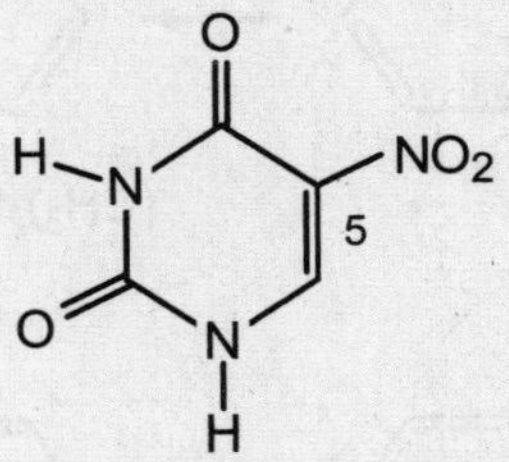

5-nitrouracil

13.30 Two resonance contributors to the structure of uric acid are shown below:

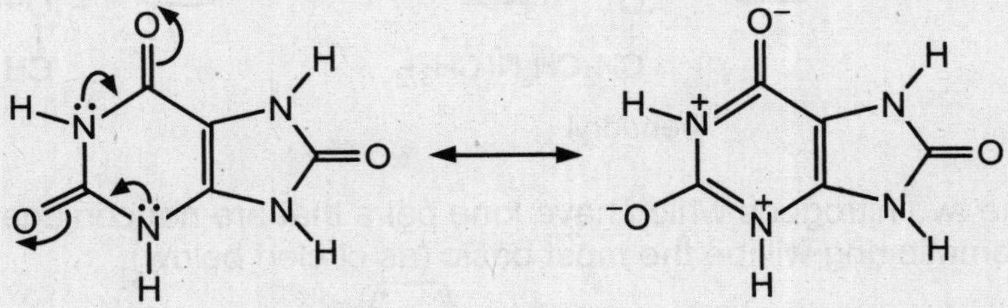

The hydrogens in the pyrimidine ring are expected to be the most acidic due to contribution from an aromatic resonance structure that places positive charge on the pyrimidine ring nitrogens.

13.31 a.

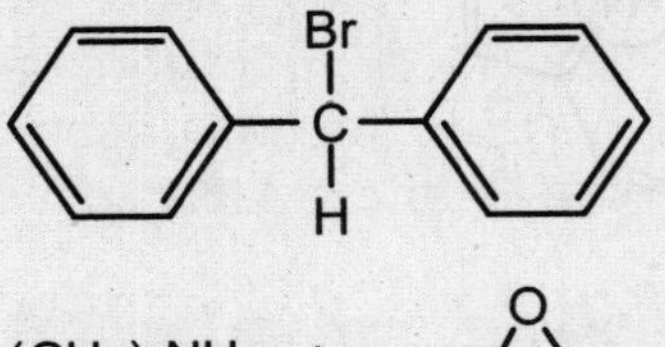

b. $(CH_3)_2NH$ + H_2C–CH_2 (O) ⟶ $(CH_3)_2NCH_2CH_2OH$ (see Sec. 8.8)

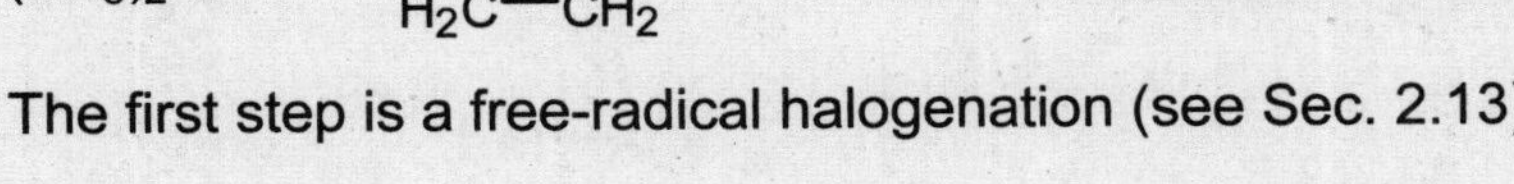

c. The first step is a free-radical halogenation (see Sec. 2.13).

$$Br_2 \xrightleftharpoons{h\nu} 2\ Br\cdot$$

The second step is an S_N1 reaction (see Sec. 6.5).

$(CH_3)_2NCH_2CH_2\ddot{O}H$

$-H^+$

$CH_2CH_2N(CH_3)_2$

benadryl

13.32 The two nitrogens which have lone pairs that are not conjugated with the aromatic ring will be the most basic (as circled below).

14

Synthetic Polymers

Chapter Summary

Polymers are **macromolecules** built of smaller units called **monomers.** The process by which they are formed is called **polymerization.** They may be synthetic (nylon, Teflon, and Plexiglas) or natural (such as the biopolymers starch, cellulose, proteins, DNA, and RNA). **Homopolymers** are made from a single monomer. **Copolymers** are made from two or more monomers. Polymers may be **linear**, **branched**, or **cross-linked**, depending on how the monomer units are arranged. These details of structure affect polymers' properties.

Chain-growth, or **addition**, **polymers** are made by adding one monomer unit at a time to the growing polymer chain. The reaction requires initiation to produce some sort of reactive intermediate, which may be a **free radical**, a **cation**, or an **anion**. The intermediate adds to the monomer, giving a new intermediate, and the process continues until the chain is terminated in some way. **Polystyrene** is a typical free-radical chain-growth polymer.

Chiral centers can be generated when a substituted vinyl compound is polymerized. The resulting polymers are classified as **atactic**, **isotactic**, or **syndiotactic**, depending on whether the chiral centers are **random**, **identical**, or **alternating** in configuration as one proceeds down the polymer chain. Ziegler–Natta catalysts (one example is a mixture of trialkylaluminum and titanium tetrachloride) usually produce stereoregular polymers, whereas free-radical catalysts generally give stereorandom polymers.

Step-growth, or **condensation**, **polymers** are usually formed in a reaction between two monomers, each of which is at least difunctional. **Polyesters**, **polyamides**, **polyurethanes**, and **epoxy resins** are typical examples of step-growth polymers. These polymers grow by steps or leaps rather than one monomer unit at a time.

Reaction Summary

Free-Radical Chain-Growth Polymerization

Initiation: $In\cdot + CH_2{=}CH(X) \longrightarrow In{-}CH_2{-}\dot{C}H(X)$

Propagation: $In{-}CH_2{-}\dot{C}H(X) \xrightarrow{CH_2=CHX} In{-}CH_2CH(X){-}CH_2{-}\dot{C}H(X) \xrightarrow{CH_2=CHX} \text{etc.}$

Termination: $\sim HC\cdot(X) + \cdot CH(X)\sim \longrightarrow \sim HC(X){-}CH(X)\sim$ (radical coupling)

$\sim H_2CH\dot{C}(X) + \dot{C}(X)HCH_2\sim \xrightarrow{\text{disproportionation}} \sim CH_2C(X)H + C(X)H{=}CH\sim$

Chain Transfer (Hydrogen Abstraction)

$\sim H\dot{C}(X) + \sim C(H)(X)\sim \longrightarrow \sim H_2C(X) + \sim \dot{C}(X)\sim$

Cationic Chain-Growth Polymerization

$R^+ + H_2C{=}CH(X) \longrightarrow RCH_2{-}\overset{+}{C}H(X) \xrightarrow{H_2C=CH(X)} \text{and so on}$

Anionic Chain-Growth Polymerization

$R^- + H_2C{=}CH(X) \longrightarrow RCH_2{-}\overset{-}{\ddot{C}}H(X) \xrightarrow{H_2C=CH(X)} \text{and so on}$

Step-Growth Polymerization (illustrated for polyesters)

$$\text{HO}-\overset{\overset{\text{O}}{\|}}{\text{C}}\sim\sim\overset{\overset{\text{O}}{\|}}{\text{C}}-\text{OH} + \text{HO}\sim\sim\text{OH} \xrightarrow{-\text{H}_2\text{O}} \underbrace{\text{HO}\sim\sim\text{O}}_{\text{from diol}}-\underbrace{\overset{\overset{\text{O}}{\|}}{\text{C}}\sim\sim\overset{\overset{\text{O}}{\|}}{\text{C}}-\text{OH}}_{\text{from diacid}} \quad \text{(two units)}$$

$$\text{(two units)} + \text{HO}-\overset{\overset{\text{O}}{\|}}{\text{C}}\sim\sim\overset{\overset{\text{O}}{\|}}{\text{C}}-\text{OH} \text{ or } \text{HO}\sim\sim\text{OH} \longrightarrow \text{three units}$$

$$\longrightarrow \text{HO}\sim\sim\text{O}-\overset{\overset{\text{O}}{\|}}{\text{C}}\sim\sim\overset{\overset{\text{O}}{\|}}{\text{C}}-\text{OH} \quad \text{four units}$$

Learning Objectives

1. Know the meaning of: monomer, polymer, macromolecule, polymerization, average molecular weight.
2. Know the meaning of: homopolymer, copolymer, linear, branched, and cross-linked polymer. For copolymers, know the meaning of: alternating, random, block, and graft.
3. Know the meaning of: thermoplastic and thermosetting polymers.
4. Know the meaning of and illustrate the difference between chain-growth (addition) and step-growth (condensation) polymerization.
5. Write the mechanism for an addition polymerization via a radical, cationic, or anionic intermediate. In each case, predict the direction of addition to the monomer if it is an unsymmetrical alkene.
6. Write resonance structures for the reactive intermediate in addition polymerization, to show how it is stabilized by the alkene substituent.
7. Know the meaning of: radical coupling, radical disproportionation, and chain transfer. Illustrate each process with examples.
8. Know the meaning of: Ziegler–Natta polymerization, atactic, isotactic, and syndiotactic polymers. Illustrate each process or type of polymer with examples.
9. Write a mechanism for a step-growth polymerization, as in the formation of a polyester, polyamide, polyurethane, epoxy resin of phenol-formaldehyde polymer.
10. Write the structure of the repeating unit of a chain-growth polymer derived from a given alkene monomer.
11. Write the structure of the repeating unit of a step-growth polymer derived from two given monomers.
12. Write the structure of a polyether derived from polymerization of a given epoxide.

ANSWERS TO PROBLEMS

Problems Within the Chapter

14.1 The odd electron can be delocalized to the *ortho* and *para* carbons of the phenyl group.

= polymer chain

14.2 Sulfonation of polystyrene occurs by an electrophilic aromatic substitution mechanism (see Sec. 4.9c). Since the phenyl groups in polystyrene are substituted with alkyl groups, which are *ortho*,*para* directing (see Sec. 4.11), sulfonation occurs mainly at the *para* position (the *ortho* position is disfavored due to steric hindrance).

14.3

$$\text{phthalic anhydride} + 2\ CH_3(CH_2)_3CH(CH_2CH_3)CH_2OH \xrightarrow[-H_2O]{\text{heat}} C_6H_4[C(=O)OCH_2CH(CH_2CH_3)(CH_2)_3CH_3]_2$$

14.4

a. $-CH_2CH(CH_3)CH_2CH(CH_3)CH_2CH(CH_3)-$

b. $-CH_2CH(OAc)CH_2CH(OAc)CH_2CH(OAc)-$

where $Ac = CH_3C(=O)-$

c. $-CH_2C(E)(CH_3)CH_2C(E)(CH_3)CH_2C(E)(CH_3)-$

where $E = CH_3OC(=O)-$

d. $-CH_2CH(CN)CH_2CH(CN)CH_2CH(CN)-$

14.5 a.

Cl_2 | $FeCl_3$

Cl Cl Cl

b.

HNO_3 | H_2SO_4

NO_2 NO_2 NO_2

14.6 1,4-Addition of *n*-butyllithium to the unsaturated ester initiates polymerization:

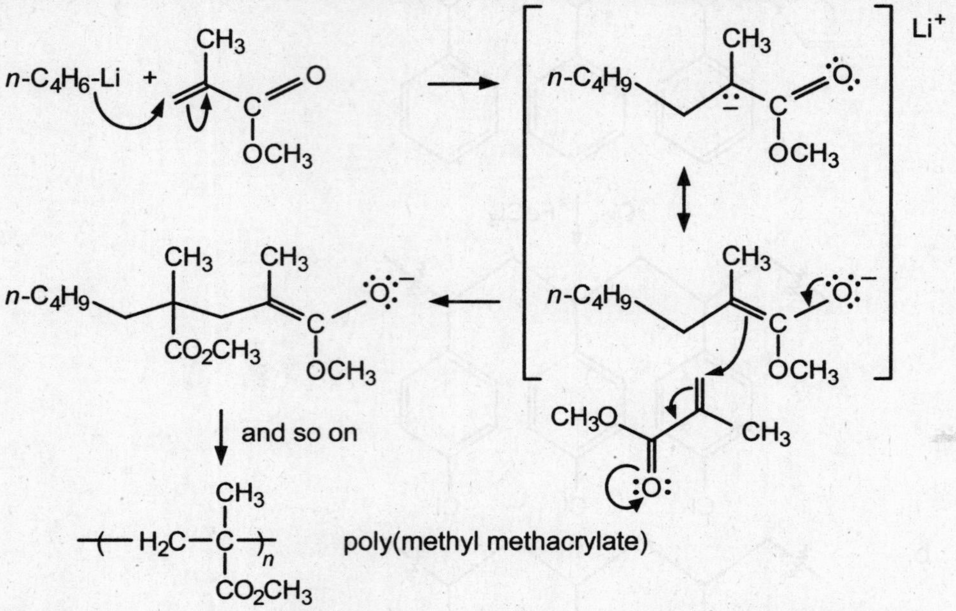

n-C_4H_6-Li + $CH_2=C(CH_3)C(=O)OCH_3$ → [n-C_4H_9–CH_2–$\ddot{C}^-(CH_3)$–C(=O)OCH_3 ↔ n-C_4H_9–CH_2–$C(CH_3)=C(O^-)OCH_3$] Li^+

+ CH_3O–C(=O)–$C(CH_3)=CH_2$ → n-C_4H_9–CH_2–$C(CH_3)(CO_2CH_3)$–CH_2–$C(CH_3)=C(O^-)OCH_3$

↓ and so on

$-(-H_2C-C(CH_3)(CO_2CH_3)-)_n-$ poly(methyl methacrylate)

The intermediate resonance-stabilized ester enolate adds to a molecule of methyl methacrylate to give another ester enolate and so on.

14.7 Hydroxide ion opens the ethylene oxide by a nucleophilic displacement to initiate the polymerization.

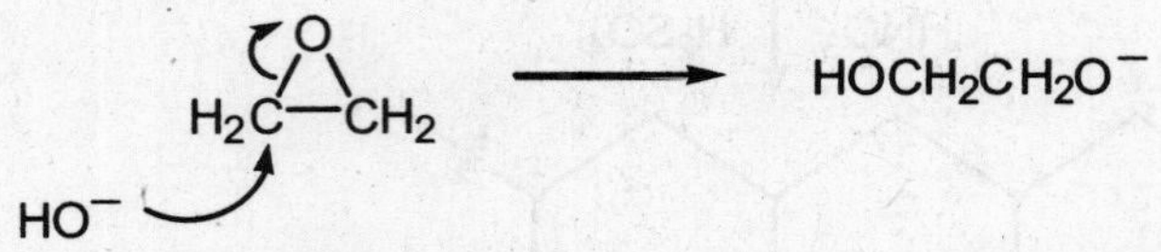

HO^- + $H_2C—CH_2$ (epoxide, O) → $HOCH_2CH_2O^-$

The resulting alkoxide reacts with another molecule of ethylene oxide.

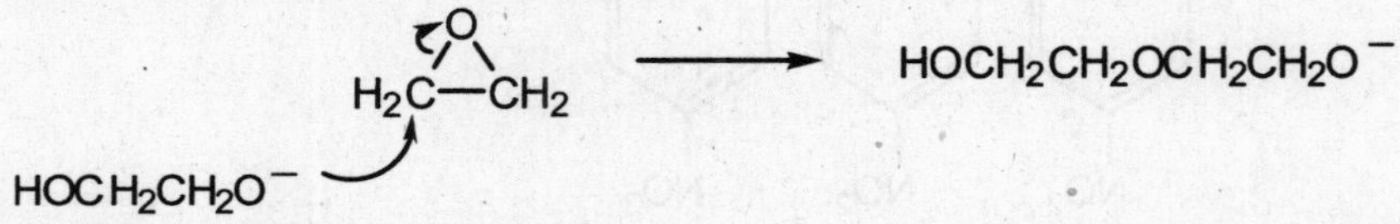

$HOCH_2CH_2O^-$ + $H_2C—CH_2$ (epoxide, O) → $HOCH_2CH_2OCH_2CH_2O^-$

and so on to give carbowax.

$HOCH_2CH_2O—(CH_2CH_2O)_n—$

carbowax

14.8 a.

CH_3 H H CH_3 CH_3 H H CH_3

syndiotactic polypropylene: stereogenic carbons alternate configuration

b.

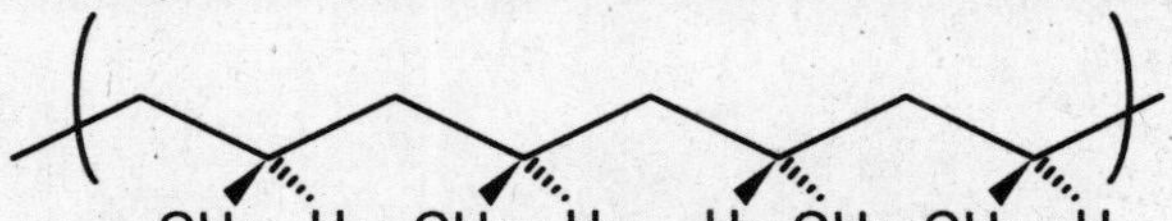

CH_3 H CH_3 H H CH_3 CH_3 H

atactic polypropylene: stereogenic carbons have random configurations

14.9

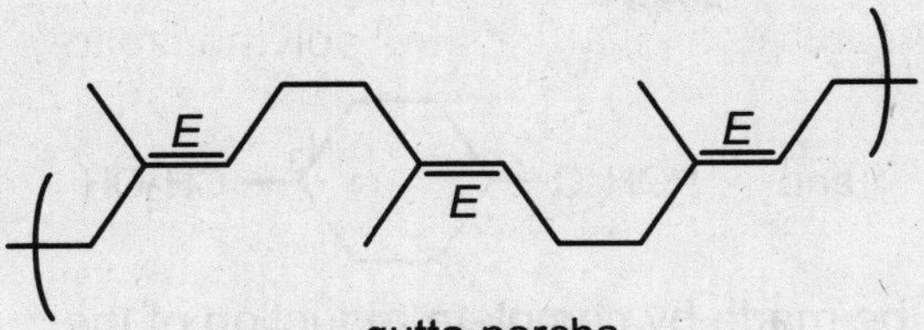

gutta-percha

14.10 a.

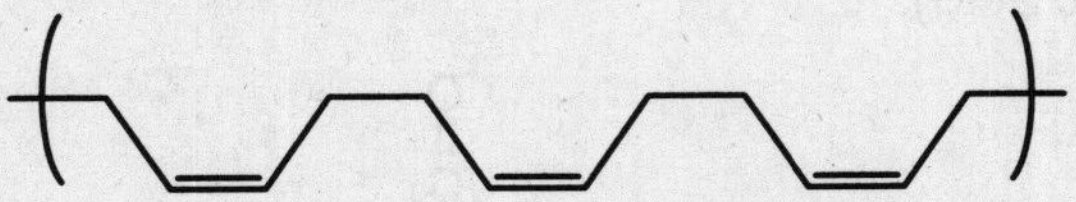

Z-poly(1,3-butadiene)

b.

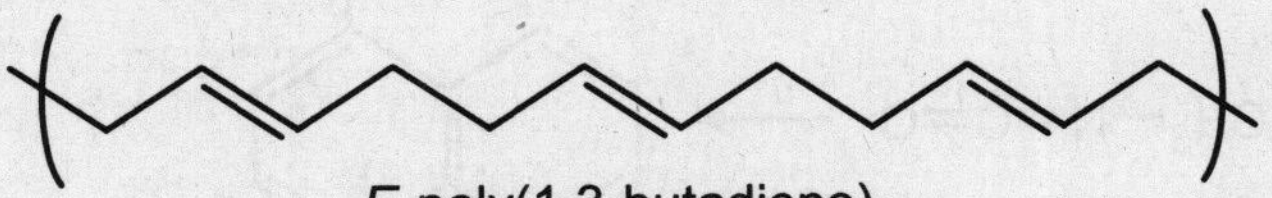

E-poly(1,3-butadiene)

c.

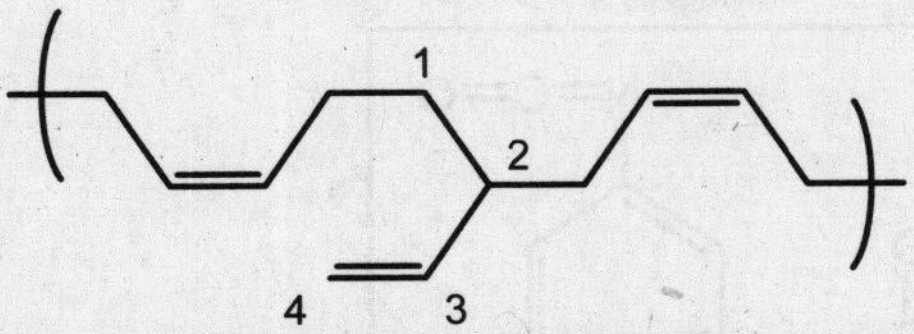

The origin of the middle unit is indicated by the numbers.

14.11

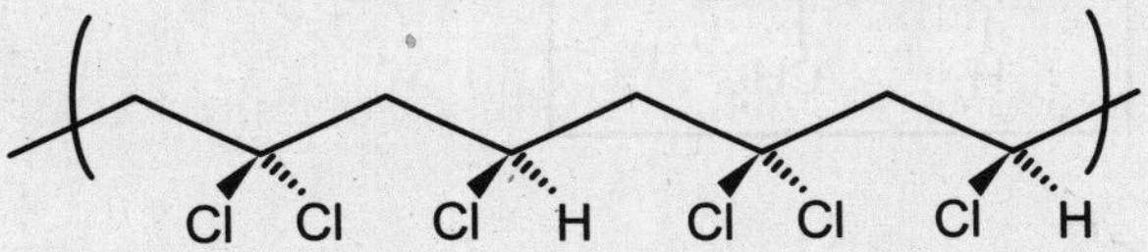

The first and third units (from left to right) come from 1,1-dichloroethene. The second and fourth units come from vinyl chloride.

14.12 Each contains three monomer units, so the next product will contain six monomer units. Its structure is

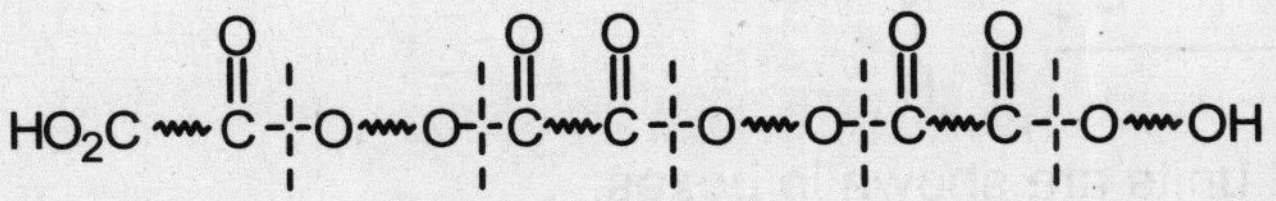

The monomer units are marked off by dashed lines. If the diester-diol reacted only with monomeric diacid, or if the diester-diacid reacted only with monomeric diol, the product would contain only four monomer units. By reacting with one another, the two compounds engage in step growth to six monomer units.

14.13

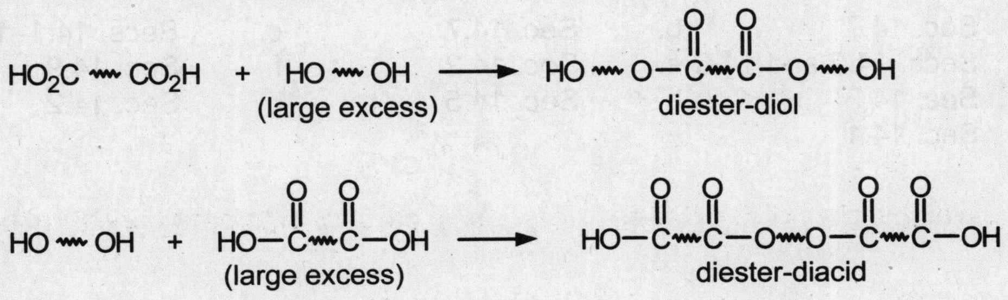

HO_2C ~~ CO_2H + HO ~~ OH (large excess) ⟶ HO ~~ O—C(=O)~~C(=O)—O ~~ OH
diester-diol

HO ~~ OH + HO—C(=O)~~C(=O)—OH (large excess) ⟶ HO—C(=O)~~C(=O)—O~~O—C(=O)~~C(=O)—OH
diester-diacid

14.14 The monomers are:

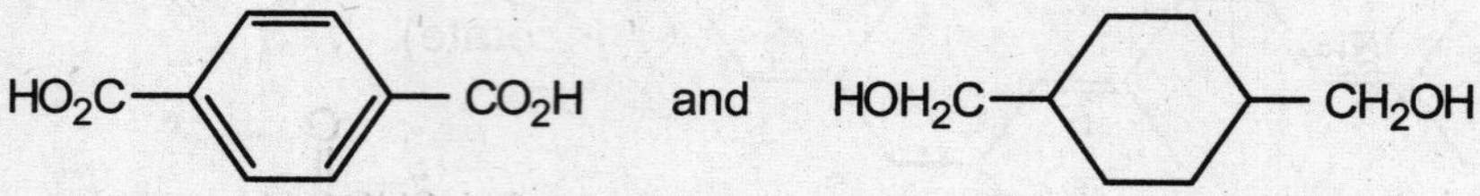

HO_2C–(benzene ring)–CO_2H and HOH_2C–(cyclohexane ring)–CH_2OH

Note that the diol used for Kodel can be made by complete reduction of the dicarboxylic acid (terephthalic acid).

14.15

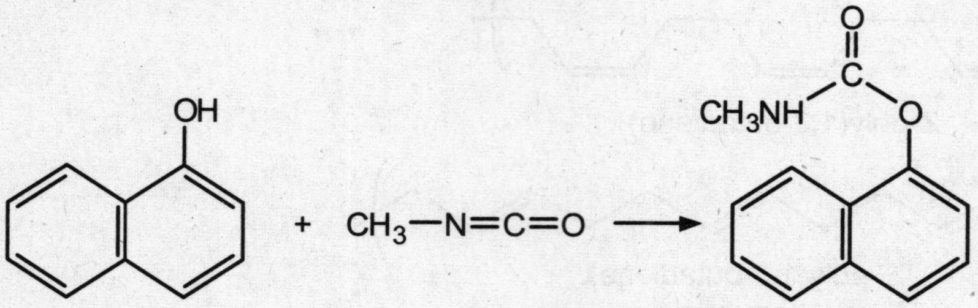

OH (1-naphthol) + CH_3—N=C=O ⟶ CH_3NH–C(=O)–O–(1-naphthyl)

14.16

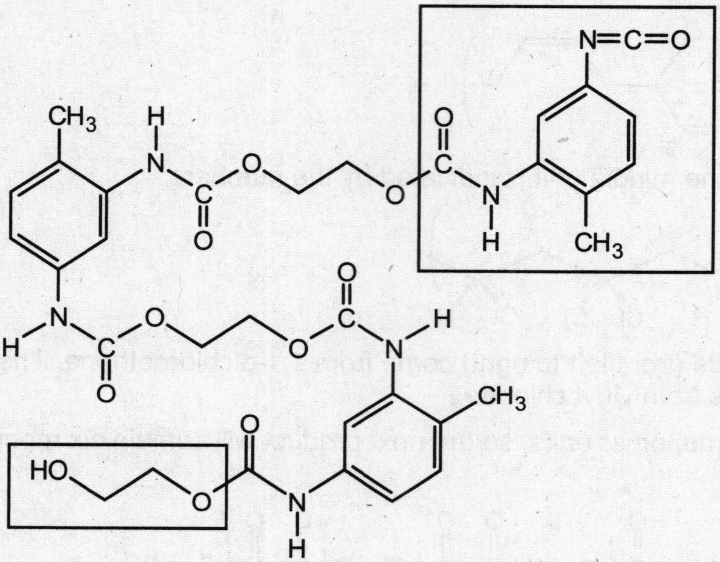

The added units are shown in boxes.

ADDITIONAL PROBLEMS

14.17 For the definitions and examples see the indicated sections in the text:

a.	Sec. 14.7	b.	Sec. 14.7	c.	Secs. 14.1–14.4
d.	Secs. 14.2 and 14.6	e.	Sec. 14.2	f.	Sec. 14.9
g.	Sec. 14.5	h.	Sec. 14.5	i.	Sec. 14.2
j.	Sec. 14.1				

14.18 Follow eqs. 14.5–14.6, with L = $O-\overset{O}{\overset{\|}{C}}CH_3$ (acetate).

$$R\cdot + H_2C{=}CHO\overset{O}{\overset{\|}{C}}CH_3 \longrightarrow RCH_2\dot{C}HO\overset{O}{\overset{\|}{C}}CH_3$$

$$RCH_2\dot{C}HO\overset{O}{\overset{\|}{C}}CH_3 + H_2C{=}CHO\overset{O}{\overset{\|}{C}}CH_3 \longrightarrow RCH_2\underset{O\overset{O}{\overset{\|}{C}}CH_3}{\underset{|}{C}H}CH_2\dot{C}HO\overset{O}{\overset{\|}{C}}CH_3$$

14.19 The vinyl monomer presumably would be $CH_2{=}CHOH$, but this is the *enol* of acetaldehyde, which exists almost completely as $CH_3CH{=}O$, and thus cannot be a vinyl monomer. Polyvinyl acetate can, however, serve as the precursor to polyvinyl alcohol:

$$-CH_2\underset{OAc}{\underset{|}{C}H}\left(CH_2\underset{OAc}{\underset{|}{C}H}\right)_n CH_2\underset{OAc}{\underset{|}{C}H}- \xrightarrow[\text{hydrolysis}]{\text{ester}} -CH_2\underset{OH}{\underset{|}{C}H}\left(CH_2\underset{OH}{\underset{|}{C}H}\right)_n CH_2\underset{OH}{\underset{|}{C}H}-$$

14.20 The growing polymer chain can abstract a hydrogen atom from the methyl group of a propylene monomer:

$$\sim CH_2-\underset{CH_3}{\underset{|}{\dot{C}H}} + CH_3-CH{=}CH_2 \longrightarrow \sim CH_2-\underset{CH_3}{\underset{|}{C}H_2} + \dot{C}H_2-CH{=}CH_2$$

This process terminates the chain growth. It is an energetically favorable process because the resulting radical is allylic and resonance-stabilized:

$$\left[\dot{C}H_2-CH{=}CH_2 \longleftrightarrow H_2C{=}CH-\dot{C}H_2\right]$$

14.21 $CH_3\underset{CH_3}{\underset{|}{C}H}CH_2\underset{CH_3}{\underset{|}{C}H}CH_2\underset{CH_3}{\underset{|}{C}H}CH_2CH{=}CH_2$

The cationic polymerization terminates by loss of a proton to afford the alkene.

14.22 The polymer has the structure:

$$\left(CH_2-\overset{CH_3}{\overset{|}{C}H}-O\right)_n$$

which is formed by S_N2 displacements on the primary carbon of the epoxide,

$$Nu:^- + H_2C\overset{O}{—}CHCH_3 \longrightarrow Nu-CH_2-\overset{CH_3}{\overset{|}{C}H}-\ddot{\underset{..}{O}}:^- \xrightarrow{H_2C\overset{O}{—}CHCH_3}$$

$$Nu-CH_2-\overset{CH_3}{\overset{|}{C}H}-O-CH_2-\overset{CH_3}{\overset{|}{C}H}-\ddot{\underset{..}{O}}:^- \text{ and so on.}$$

14.23 The polymer has the following structure:

$$\text{-}\!\left(CH_2-\underset{CO_2CH_3}{\overset{CN}{C}}\right)_n\!\text{-}$$

The monomer is susceptible to anionic polymerization because the intermediate anion is stabilized by two electron-withdrawing groups.

$$\text{polymer}-\underset{O=C-OMe}{\overset{N\equiv C}{C^-}} \longleftrightarrow \text{polymer}-\underset{O=C-OMe}{\overset{N\equiv C}{C^-}} \longleftrightarrow \text{polymer}-\underset{^-O-C-OMe}{\overset{N\equiv C}{C}}$$

14.24 The structure of a styrene-methyl methacrylate copolymer is:

$$\text{-}\!\left(CH_2\underset{C_6H_5}{CH}CH_2\underset{CO_2CH_3}{\overset{CH_3}{C}}\right)_n\!\text{-}$$

14.25 Poly(acrylamide) and poly(*N*-vinylformamide) are addition polymers. The molecule in parentheses is the monomer used to prepare the respective polymers:

$$\underset{\text{acrylamide}}{CH_2=CH-\overset{O}{\overset{\|}{C}}-NH_2} \longrightarrow \underset{\text{poly(acrylamide)}}{-(CH_2\overset{O=C-NH_2}{CH})_n-}$$

$$\underset{N\text{-vinylformamide}}{CH_2=CH-NH-\overset{O}{\overset{\|}{C}}-H} \longrightarrow \underset{\text{poly(}N\text{-vinylformamide)}}{-(CH_2\overset{HN-\overset{O}{\overset{\|}{C}}-H}{CH})_n-}$$

Note the structural relationship between the polymer and monomer. The CH_2=CH unit in the monomer becomes a —CH_2CH— unit in the polymer.

14.26

Ph H Ph H Ph H Ph H Ph H Ph H

isotactic

= Ph

Ph H H Ph Ph H H Ph Ph H H Ph

syndiotactic

14.27 No. Poly(isobutylene) has the following structure:

$$\left[CH_2-C(CH_3)_2 \right]_n$$

It has no stereogenic carbons and thus does not exist in stereoisomeric forms.

14.28 Polyethylenes obtained by free-radical polymerization have highly branched structures as a consequence of chain-transfer reactions (see eq. 3.42 and the structure below it). Ziegler–Natta polyethylene is mainly linear: $(CH_2CH_2)_n$. It has a higher degree of crystallinity and a higher density than the polyethylene obtained by the free-radical process.

14.29 The structure of natural rubber is

It will be cleaved by ozone at each double bond. The product of ozonolysis is, therefore,

H CH$_3$ ≡ H CH$_3$

(levulinic aldehyde)

14.30 The reaction may be initiated by addition of a free radical from the catalyst to either butadiene or styrene:

$$R\cdot + H_2C{=}CHCH{=}CH_2 \longrightarrow \left[RCH_2\dot{C}H-CH{=}CH_2 \longleftrightarrow RCH_2CH{=}CH\dot{C}H_2 \right]$$

$$R\cdot + H_2C{=}CHC_6H_5 \longrightarrow RCH_2\dot{C}HC_6H_5$$

These radicals may then add to either butadiene or styrene. The allylic radical from butadiene may add in either a 1,2- or a 1,4-manner. Only one of several possible addition sequences is shown:

$$RCH_2CH{=}CH\dot{C}H_2 + H_2C{=}CHCH{=}CH_2 \longrightarrow \left[RCH_2CH{=}CHCH_2CH_2\dot{C}HCH{=}CH_2 \longleftrightarrow RCH_2CH{=}CHCH_2CH_2CH{=}CH\dot{C}H_2 \right]$$

$$RCH_2CH{=}CHCH_2CH_2CH{=}CH\dot{C}H_2 + H_2C{=}CHC_6H_5 \longrightarrow$$

$$RCH_2CH{=}CHCH_2CH_2CH{=}CHCH_2CH_2\dot{C}HC_6H_5 \xrightarrow{H_2C{=}CH{-}CH{=}CH_2}$$

$$RCH_2CH{=}CHCH_2CH_2CH{=}CHCH_2CH_2CH(C_6H_5)CH_2CH{=}CH\dot{C}H_2 \quad \text{and so on}$$

14.31 The structure is analogous to that of natural rubber, except that the methyl group in each isoprene unit is replaced by a chlorine atom.

$$\left(CH_2C(Cl){=}CHCH_2 \right)_n$$

14.32 a.

$$\left(\overset{O}{\overset{\|}{C}}(CH_2)_8\overset{O}{\overset{\|}{C}}NH(CH_2)_6NH \right)_n$$

The other product of the polymerization is hydrochloric acid (HCl).

b.

$$\left(\overset{O}{\overset{\|}{C}}{-}NH{-}C_6H_4{-}CH_2{-}C_6H_4{-}NH{-}\overset{O}{\overset{\|}{C}}{-}OCH_2CH_2O \right)_n$$

c.

$$\left(\overset{O}{\overset{\|}{C}}(CH_2)_4\overset{O}{\overset{\|}{C}}OCH_2CH_2O \right)_n$$

The other product of the polymerization is methanol (CH_3OH).

14.33

$$\left(\overset{O}{\overset{\|}{C}}{-}O{-}C_6H_4{-}C(CH_3)_2{-}C_6H_4{-}O \right)_n$$

The other product of the polymerization is phenol.

14.34 The "unzipping" reaction, which gives formaldehyde, is expressed by the following curved arrows:

$$H{-}O{-}CH_2{-}O{-}CH_2{-}O{-}CH_2\ldots \longrightarrow H^+ + n\ O{=}CH_2$$

This reaction is made possible by dissociation of a proton from a terminal hydroxyl group (an alcohol; see eq. 7.11). If the polymer is treated with acetic anhydride, the terminal hydroxyl groups are esterified, and unzipping is no longer possible:

$$CH_3{-}\overset{O}{\overset{\|}{C}}{-}O{-}CH_2{-}(OCH_2)_n{-}O{-}CH_2{-}O{-}\overset{O}{\overset{\|}{C}}{-}CH_3$$

14.35

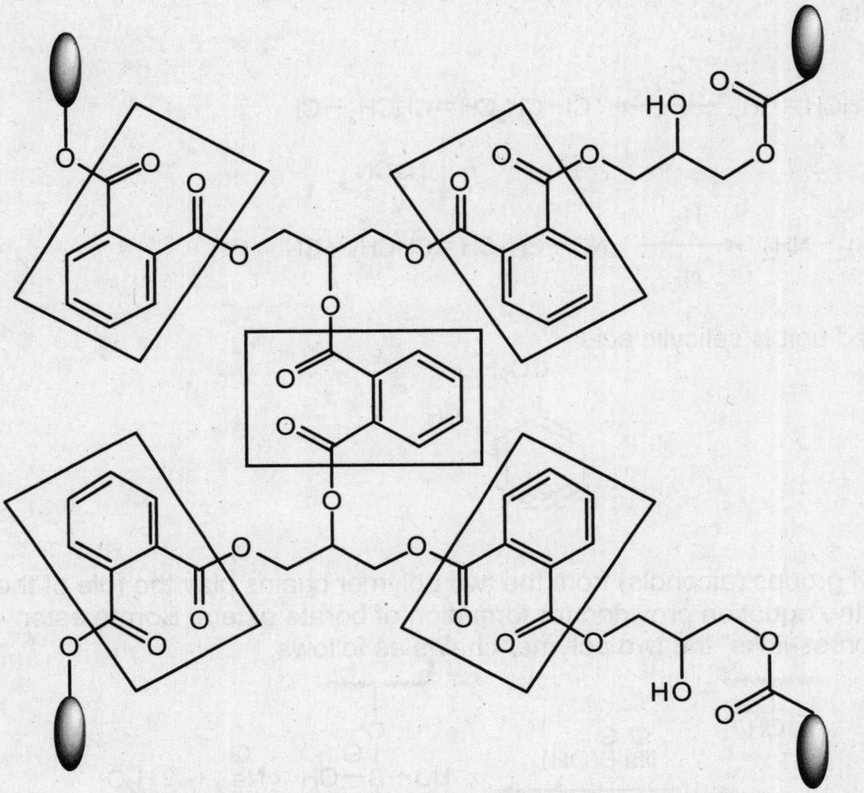

HO
O
O
O
O
O
O
O
O
O
O
O
O
O
O
O
O
HO
O

Portions of the polymer derived from phthalic anhydride are in boxes. The remainder of the polymer partial structure is derived from glycerol.

14.36 With the *para* position "blocked" by a methyl substituent, condensation can occur only *ortho* to the phenolic hydroxyl, leading to a linear (*not* a cross-linked) polymer.

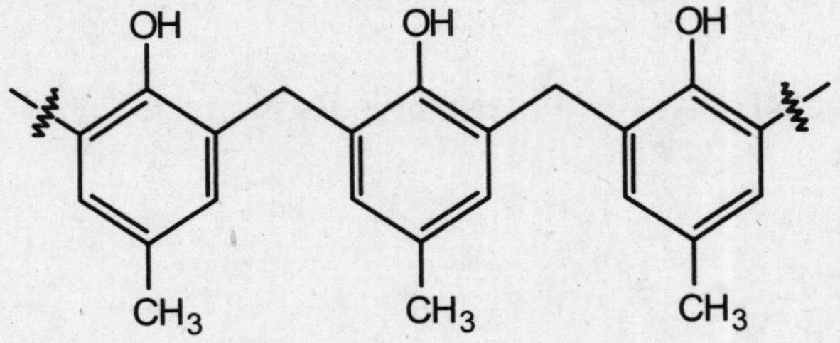

OH OH OH

CH_3 CH_3 CH_3

14.37 The repeating unit of PEN is

PEN is more rigid than PETE because naphthalene groups, which are larger than the phenyl groups present in PETE, serve as "spacers" between ethylene glycol-derived units.

14.38

$$H_2C{=}CHCH{=}CH_2 \xrightarrow{Cl_2} Cl{-}CH_2CH{=}CHCH_2{-}Cl$$

$$\xrightarrow{NaCN} NC{-}CH_2CH{=}CHCH_2{-}CN \xrightarrow[Ni]{H_2} H_2N{-}(CH_2)_6{-}NH_2$$

14.39 The released unit is salicylic acid:

CO_2H OH

14.40 Let hydroxyl groups (alcohols) from the two polymer chains play the role of the two alcohols in the equation provided for formation of borate esters. Borate ester formation "cross-links" the two polymer chains as follows:

OH OH $\xrightarrow{Na^{\oplus}\ B(OH)_4^{\ominus}}$ HO—B$^{\ominus}$—OH (with O to each chain) $Na^{\oplus}$ + 2 H_2O

~~~~ = polymer chain cross-linked polymer
~~~~

14.41 The structures are:

all *trans*

all *cis*

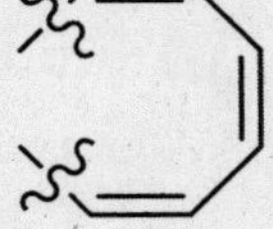

alternating
trans and *cis*

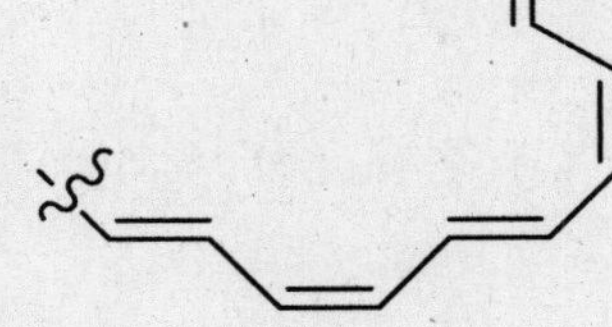

14.42 A repeating structure for Nomex is:

15

Lipids and Detergents

Chapter Summary

Lipids are constituents of plants and animals that are characteristically insoluble in water. **Fats** and **oils** are lipids that are triesters formed by reaction of the triol **glycerol** with long-chain saturated or unsaturated acids called **fatty acids**. The common acids in fats and oils have an even number of carbons (for example, **stearic acid** has 18 carbons), and, if unsaturated (for example, **oleic acid**), they have the *Z* configuration (Table 15.1 lists the fatty acids, with their common names). **Hydrogenation** of oils, which have a high percentage of unsaturated acids, converts them to solid fats in a process called **hardening**.

Saponification of fats and oils by boiling with strong base yields glycerol and the sodium salts of fatty acids. The latter are **soaps**. Soaps contain a long carbon chain that is **lipophilic** and a terminal polar group that is **hydrophilic**. Soap molecules aggregate in water to form **micelles**, which help emulsify droplets of oil or grease.

Two disadvantages of ordinary soaps are their alkalinity and tendency to form insoluble salts with the Ca^{2+} or Mg^{2+} ions present in "hard" water. **Synthetic detergents** or **syndets** overcome these disadvantages. The most widely used syndets at present are **straight-chain alkylbenzenesulfonates**, obtained by successive alkylation and sulfonation of benzene. The polar portion of alkylbenzenesulfonates is anionic, and the straight chains are necessary for biodegradability. Syndets in which the polar group is cationic, neutral, or dipolar are also known.. The manufacture of a detergent is a complex process, and the **surfactant** is usually only one portion of the commercial product. Detergents may also contain builders (to remove calcium and magnesium ions), bleaches, fabric softeners, enzymes (for stain removal), antiredeposition agents (to prevent soil deposition), optical brighteners, antistatic agents, fragrances, and perfumes.

Lipids play a number of important biological roles. **Phospholipids** are triesters of glycerol in which one ester is derived from a phosphatidylamine. They are important structural units in cell membranes. **Prostaglandins** are 20-carbon cyclopentane derivatives of **arachidonic acid** that have profound biological effects, even in minute quantities. **Waxes** are monoesters of long-chain acids and alcohols.

Terpenes are natural products usually obtained from the essential oils of plants. They contain multiples of five-carbon atoms (5, 10, 15, and so on). Each five-carbon arrangement is called an **isoprene unit**, a four-carbon chain with a one-carbon branch at C-2. Terpenes are frequently used in fragrances and perfumes. **Steroids** are lipids that contain a unique four-ring structure and are biosynthetically related to terpenes. Important examples of steroids include **cholesterol**, the **bile acids**, and the **sex hormones**.

Reaction Summary

Saponification of a Triglyceride

$$\begin{array}{l} H_2C{-}O{-}\overset{O}{\overset{\|}{C}}{-}R \\ HC{-}O{-}\overset{O}{\overset{\|}{C}}{-}R \\ H_2C{-}O{-}\overset{O}{\overset{\|}{C}}{-}R \end{array} + 3\,NaOH \longrightarrow \begin{array}{l} H_2C{-}OH \\ HC{-}OH \\ H_2C{-}OH \end{array} + 3\,Na^{+}\ {}^{-}O{-}\overset{O}{\overset{\|}{C}}{-}R$$

Hydrogenation of a Triglyceride (Hardening)

$$\begin{array}{l} H_2C{-}O{-}\overset{O}{\overset{\|}{C}}{-}(CH_2)_nCH{=}CH(CH_2)_mCH_3 \\ HC{-}O{-}\overset{O}{\overset{\|}{C}}{-}(CH_2)_nCH{=}CH(CH_2)_mCH_3 \\ H_2C{-}O{-}\overset{O}{\overset{\|}{C}}{-}(CH_2)_nCH{=}CH(CH_2)_mCH_3 \end{array} \xrightarrow[\text{Ni, heat}]{3\ H_2} \begin{array}{l} H_2C{-}O{-}\overset{O}{\overset{\|}{C}}{-}(CH_2)_nCH_2CH_2(CH_2)_mCH_3 \\ HC{-}O{-}\overset{O}{\overset{\|}{C}}{-}(CH_2)_nCH_2CH_2(CH_2)_mCH_3 \\ H_2C{-}O{-}\overset{O}{\overset{\|}{C}}{-}(CH_2)_nCH_2CH_2(CH_2)_mCH_3 \end{array}$$

Hydrogenolysis of a Triglyceride

$$\begin{array}{l} H_2C{-}O{-}\overset{O}{\overset{\|}{C}}{-}R \\ HC{-}O{-}\overset{O}{\overset{\|}{C}}{-}R \\ H_2C{-}O{-}\overset{O}{\overset{\|}{C}}{-}R \end{array} \xrightarrow[\text{zinc chromite}]{6\ H_2} \begin{array}{l} H_2C{-}OH \\ HC{-}OH \\ H_2C{-}OH \end{array} + HOH_2C{-}R$$

Learning Objectives

1. Know the meaning of: triglyceride, fatty acid, fat, oil, hardening of a vegetable oil, hydrogenolysis, soap, saponification.
2. Know the structures and common names of the acids listed in Table 15.1.
3. Given the name of a glyceride, write its structure.
4. Given the name or structure of a carboxylic acid, write the formula for the corresponding glyceride.

5. Given the structure of a glyceride, write the equation for its saponification.
6. Given the structure of an unsaturated glyceride, write equations (including catalysts) for its hydrogenation and hydrogenolysis.
7. Explain the difference between the structure of a fat and that of a vegetable oil.
8. Describe the structural features essential for a good soap or detergent.
9. Explain, with the aid of a diagram, how a soap emulsifies fats and oils.
10. Explain, with the aid of equations, what happens when an ordinary soap is used in hard water and how synthetic detergents overcome this difficulty.
11. Know the meaning of: lipophilic, hydrophilic, sodium alkyl sulfate, alkylbenzenesulfonate.
12. Explain the difference between anionic, cationic, neutral, and amphoteric detergents.
13. Know the meaning of: lipid, phospholipid, wax, prostaglandin.
14. Know the meaning of: terpene, isoprene unit, monoterpene, sesquiterpene, diterpene triterpene, steroid.
15. Identify the isoprene units in a given terpene.

ANSWERS TO PROBLEMS

Problems Within the Chapter

15.1 Find the condensed formula of linolenic acid in Table 15.1 and use example 15.1 as a guide.

H_3C (skeletal chain with three cis double bonds) $(CH_2)_7CO_2H$

15.2 a.

$$\begin{array}{l} H_2C-O-\overset{O}{\overset{\|}{C}}-(CH_2)_{10}CH_3 \\ \;|\\ HC-O-\overset{O}{\overset{\|}{C}}-(CH_2)_{10}CH_3 \\ \;|\\ H_2C-O-\overset{O}{\overset{\|}{C}}-(CH_2)_{10}CH_3 \end{array}$$

b.

$$\begin{array}{l} H_2C-O-\overset{O}{\overset{\|}{C}}-(CH_2)_{14}CH_3 \\ \;|\\ HC-O-\overset{O}{\overset{\|}{C}}-(CH_2)_7CH=CH(CH_2)_7CH_3 \\ \;|\\ H_2C-O-\overset{O}{\overset{\|}{C}}-(CH_2)_{16}CH_3 \end{array}$$

For mixed triglycerides, the name indicates the order in which the fatty acids are arranged. For example, glyceryl palmitoöleostearate (Problem 15.2b), glyceryl stearopalmitoöleate (Example 15.2) and glyceryl palmitostearoöleate (Sec. 15.1) represent three isomeric mixed triglycerides.

15.3 a. Glycerol and sodium laurate, $CH_3(CH_2)_{10}CO_2^-Na^+$

b. Glycerol and equimolar amounts of the sodium salts of the three carboxylic acids:

$CH_3(CH_2)_{14}CO_2^-Na^+$
sodium palmitate

$CH_3(CH_2)_7CH{=}CH(CH_2)_7CO_2^-Na^+$
sodium oleate

$CH_3(CH_2)_{16}CO_2^-Na^+$
sodium stearate

The triglyceride shown in Example 15.2 would give the same saponification products.

15.4 In general, the ratio of unsaturated to saturated acids is greater in vegetable oils than it is in animal fats.

15.5

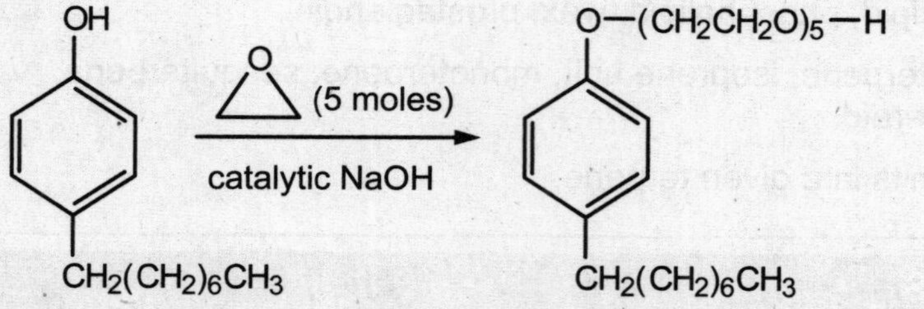

OH-substituted benzene ring with $CH_2(CH_2)_6CH_3$ para; ethylene oxide (5 moles), catalytic NaOH → benzene ring with $O{-}(CH_2CH_2O)_5{-}H$ and $CH_2(CH_2)_6CH_3$ para.

15.6

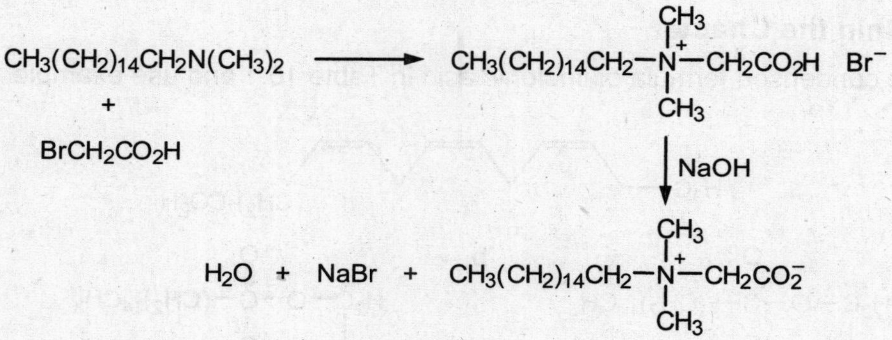

$$CH_3(CH_2)_{14}CH_2N(CH_3)_2 + BrCH_2CO_2H \longrightarrow CH_3(CH_2)_{14}CH_2{-}\overset{+}{N}(CH_3)_2{-}CH_2CO_2H\ \ Br^-$$

$$\xrightarrow{NaOH}\ H_2O + NaBr + CH_3(CH_2)_{14}CH_2{-}\overset{+}{N}(CH_3)_2{-}CH_2CO_2^-$$

15.7

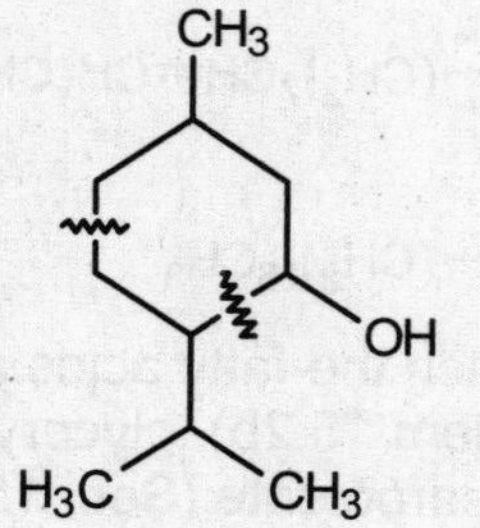

CH3, OH, H3C, CH3

15.8

farnesol

CH_2OH

retinal

CHO

squalene

tail

tail

β-carotene

15.9 There are no stereogenic centers in squalene, while there are seven in lanosterol, as indicated below with asterisks.

lanosterol

ADDITIONAL PROBLEMS

15.10 a. $CH_3(CH_2)_{10}CO_2^-\ {}^+Na$

b. $[CH_3(CH_2)_7CH{=}CH(CH_2)_7CO_2^-]_2\ Mg^{2+}$

c.

$$\begin{array}{l} H_2C-O-\overset{O}{\overset{\|}{C}}-(CH_2)_{12}CH_3 \\ HC-O-\overset{O}{\overset{\|}{C}}-(CH_2)_{12}CH_3 \\ H_2C-O-\overset{O}{\overset{\|}{C}}-(CH_2)_{12}CH_3 \end{array}$$

d.

$$\begin{array}{l} H_2C-O-\overset{O}{\overset{\|}{C}}-(CH_2)_{14}CH_3 \\ | \\ HC-O-\overset{O}{\overset{\|}{C}}-(CH_2)_{10}CH_3 \\ | \\ H_2C-O-\overset{O}{\overset{\|}{C}}-(CH_2)_7CH{=}CH(CH_2)_7CH_3 \end{array}$$

e.

$$CH_3(CH_2)_4CH{=}CHCH_2CH{=}CH(CH_2)_7CH_2-O-\overset{O}{\overset{\|}{C}}-(CH_2)_{16}CH_3$$

f. $CH_3(CH_2)_{18}CO_2CH_3$

15.11 Many answers are possible. Definitions or examples can be found in the indicated sections of the text.

a. Sec. 15.1
b. Secs. 15.1 and 15.2
c. Sec. 15.8
d. Sec. 15.3
e. Sec. 15.5
f. Sec. 15.9
g. Sec. 15.6
h. Sec. 15.9
i. Sec. 15.9

15.12 Saponification:

$$\begin{array}{l} H_2C-O-\overset{O}{\overset{\|}{C}}-(CH_2)_{10}CH_3 \\ | \\ HC-O-\overset{O}{\overset{\|}{C}}-(CH_2)_{10}CH_3 \\ | \\ H_2C-O-\overset{O}{\overset{\|}{C}}-(CH_2)_{10}CH_3 \end{array} \xrightarrow{3\ NaOH} 3\ Na^{+}\ {}^{-}O-\overset{O}{\overset{\|}{C}}-(CH_2)_{10}CH_3 + \begin{array}{l} H_2C-OH \\ | \\ HC-OH \\ | \\ H_2C-OH \end{array}$$

Hydrogenation:

$$\begin{array}{l} H_2C-O-\overset{O}{\overset{\|}{C}}-(CH_2)_{10}CH_3 \\ | \\ HC-O-\overset{O}{\overset{\|}{C}}-(CH_2)_{10}CH_3 \\ | \\ H_2C-O-\overset{O}{\overset{\|}{C}}-(CH_2)_{10}CH_3 \end{array} \xrightarrow[\text{Ni, heat}]{H_2} \text{no reaction}$$

Hydrogenolysis:

$$\begin{array}{l} H_2C-O-\overset{O}{\overset{\|}{C}}-(CH_2)_{10}CH_3 \\ | \\ HC-O-\overset{O}{\overset{\|}{C}}-(CH_2)_{10}CH_3 \\ | \\ H_2C-O-\overset{O}{\overset{\|}{C}}-(CH_2)_{10}CH_3 \end{array} \xrightarrow[\text{zinc, chromite}]{3\ H_2} HOCH_2-(CH_2)_{10}CH_3 + \begin{array}{l} H_2C-OH \\ | \\ HC-OH \\ | \\ H_2C-OH \end{array}$$

15.13

$$H_2C-O-\overset{O}{\overset{\|}{C}}-(CH_2)_7\overset{9}{C}H=\overset{10}{C}HCH_2\overset{12}{C}H(OH)(CH_2)_5CH_3$$
$$HC-O-\overset{O}{\overset{\|}{C}}-(CH_2)_7CH=CHCH_2CH(OH)(CH_2)_5CH_3$$
$$H_2C-O-\overset{O}{\overset{\|}{C}}-(CH_2)_7CH=CHCH_2CH(OH)(CH_2)_5CH_3$$

All of the double bonds have a *cis* geometry.

15.14 a. Compare with eq. 15.5.

$$C_{15}H_{31}-\overset{O}{\overset{\|}{C}}-O^-\ K^+ + HCl \longrightarrow C_{15}H_{31}CO_2H + KCl$$

b. Compare with eq. 15.6.

$$2\ C_{15}H_{31}-\overset{O}{\overset{\|}{C}}-O^-\ Na^+ + Ca^{2+} \longrightarrow (C_{15}H_{31}-\overset{O}{\overset{\|}{C}}-O^-)_2\ Ca^{2+} + 2\ Na^+$$

15.15 Sodium carbonate can be used as a base to convert a long chain carboxylic acid to the corresponding sodium carboxylate (a soap). If 2 moles of acid are used for each mole of sodium carbonate, the other products are water and carbon dioxide, which are easily removed from the soap.

$$2\ R-\overset{O}{\overset{\|}{C}}-OH + Na_2CO_3 \longrightarrow 2\ R-\overset{O}{\overset{\|}{C}}-O^-\ Na^+ + H_2O + CO_2$$

15.16 Note that in the first step Markovnikov's rule is followed, and alkylation occurs *via* a secondary carbocation.

$$C_6H_6 \xrightarrow[H^+]{CH_3(CH_2)_9CH=CH_2} CH_3(CH_2)_9-\overset{H}{\overset{|}{C}}(C_6H_5)-CH_3 \xrightarrow{H_2SO_4,\ SO_3} CH_3(CH_2)_9-\overset{H}{\overset{|}{C}}(C_6H_4SO_3H)-CH_3 \xrightarrow{NaOH} CH_3(CH_2)_9-\overset{H}{\overset{|}{C}}(C_6H_4SO_3^-\ Na^+)-CH_3$$

15.17 The first steps are analogous to eq. 8.20, and the last steps are analogous to eq. 15.8.

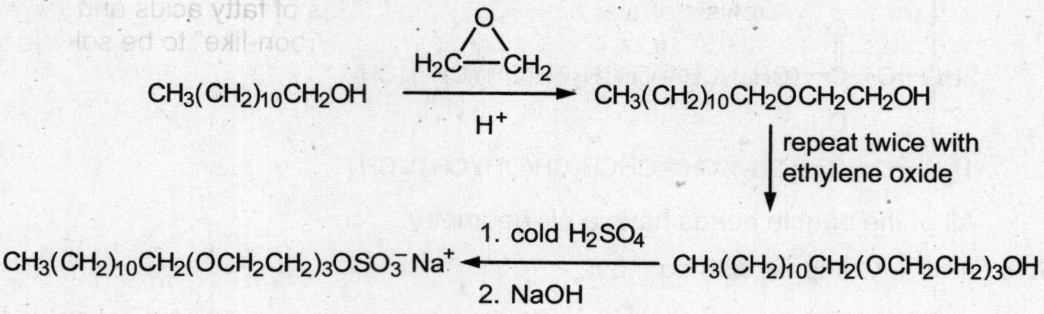

$$CH_3(CH_2)_{10}CH_2OH \xrightarrow[H^+]{H_2C\text{—}CH_2 \text{ (O bridge, ethylene oxide)}} CH_3(CH_2)_{10}CH_2OCH_2CH_2OH$$

repeat twice with ethylene oxide

$$CH_3(CH_2)_{10}CH_2(OCH_2CH_2)_3OH \xrightarrow[2.\ NaOH]{1.\ \text{cold } H_2SO_4} CH_3(CH_2)_{10}CH_2(OCH_2CH_2)_3OSO_3^-\,Na^+$$

15.18 Soaps form insoluble salts with metal cations frequently found in water such as calcium and magnesium. You may have seen these salts in the form of a "tub ring" after taking a bath. The magnesium and calcium salts of a good syndet will not precipitate from water.

15.19 Branched chain syndets are less easily degraded by microorganisms in the environment than unbranched syndets. Therefore, it is for environmental reasons that syndets have mostly unbranched alkyl chains.

15.20 a. The fatty acid from which volicitin is derived is linolenic acid (see Table 15.1). It has been hydroxylated at C-17. All three double bonds have *Z* geometry.

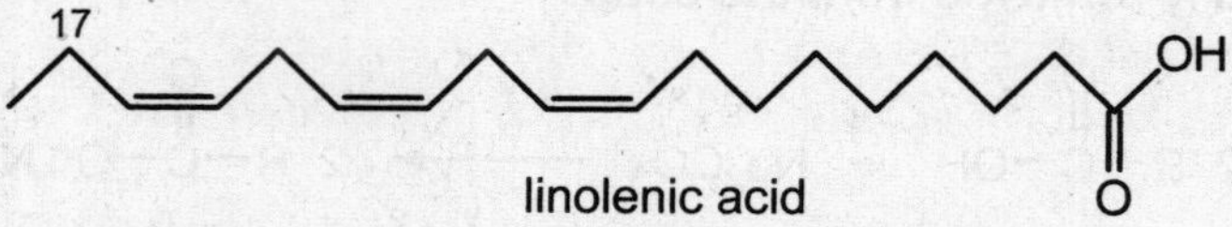

17

OH

O

linolenic acid

b. The stereogenic center has the *S* configuration.

15.21 The IUPAC name for leukotriene B_4 is:

(6*Z*,8*E*,10*E*,14*Z*)-(5*S*,12*R*)-5,12-dihydroxy-eicos-6,8,10,14-tetraenoic acid

15.22 a. There are four stereogenic centers in PGE_2.

b. 8*R*, 11*R*, 12*R*, 15*S*

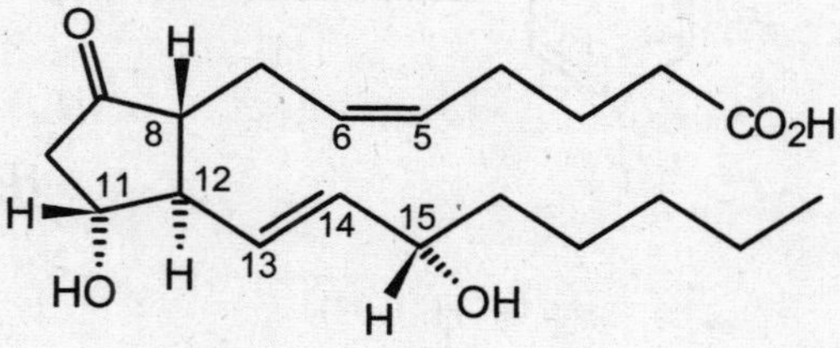

O H 8 6 5 CO_2H 11 12 H 14 15 HO H 13 H OH

c. 5*Z*, 13*E*

d. The side chains are *trans* to one another.

15.23 See Section 15.8.

$$CH_3(CH_2)_{14}\text{—}\overset{O}{\overset{\|}{C}}\text{—}C(CH_2)_{15}CH_3 \xrightarrow{NaOH} CH_3(CH_2)_{14}CO_2^-\,Na^+ \;+\; CH_3(CH_2)_{15}OH$$

15.24 Saponification of fats and oils with strong alkali, such as sodium hydroxide, yields sodium salts of fatty acids and glycerol, both of which are soluble in water. On the other hand, hydrolysis of waxes yields sodium salts of fatty acids and long-chain alcohols. The long-chain alcohols are too "hydrocarbon-like" to be soluble in water and do not dissolve (consult Table 7.1).

15.25 a.

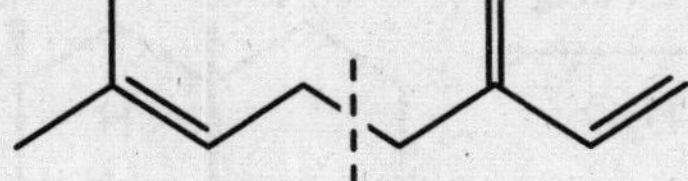

b.

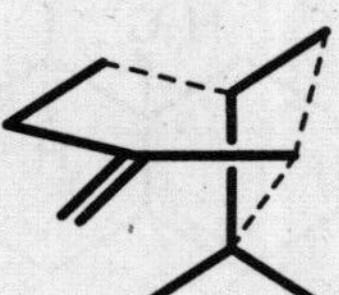

The two isoprene units are outlined in bold and the bonds connecting them are dashed.

15.26 The head-to-tail arrangement of isoprene units breaks down in squalene at the position indicated below:

squalene

This suggests that squalene is biosynthesized by the tail-to-tail connection of two 15-carbon pieces rather than, say, stepwise connection of six isoprene units in a linear fashion. β-Carotene may be biosynthesized in a similar manner by the tail-to-tail connection of two 20-carbon units at the head-to-tail break point:

β-carotene

15.27 The two compounds are enantiomers. They differ only in configuration at the stereogenic center as indicated in the line segment structures shown below.

A B

15.28 a. The ketone will be reduced, but the carbon–carbon double bond will not be reduced. Two stereoisomers are possible.

CH_3 OH H H_3C H H H O — 1. $LiAlH_4$ 2. H_2O → CH_3 OH H H_3C H H H HO

b. The ketone will react with the Grignard reagent. Two stereoisomers are possible.

H O CH_3 H H H H HO H — 1. CH_3MgBr 2. H_2O → OH H CH_3 CH_3 H H H H HO H

c. The alcohol will be converted to an acetate. Only one stereoisomers is formed.

H O CH_3 H H H H HO H — Ac_2O → H O CH_3 H H H H H_3C–C(=O)–O H

d. The secondary alcohol will be oxidized to a ketone.

CH_3 O H_3C H H H H HO H — CrO_3 → CH_3 O H_3C H H H O H

15.29 Cholesterol probably orients such that its hydroxyl group is in the polar (aqueous) portion of the lipid bilayer, while its hydrocarbon portion is in the nonpolar (fatty acid side-chain) region.

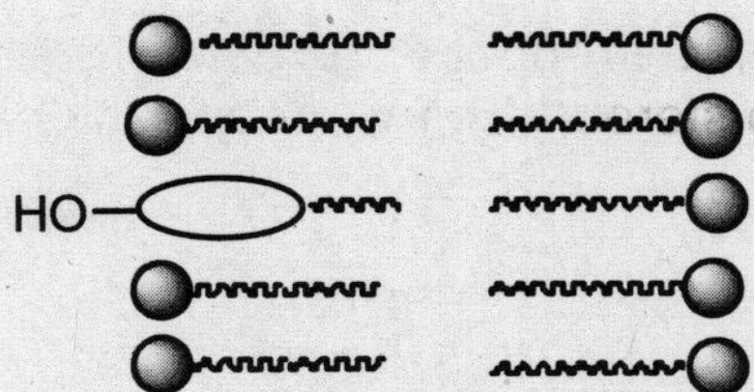

15.30 Reduction of the D-ring ketone to an alcohol must occur. The reduction could occur from the "bottom" to give testosterone (page 454), or from the "top" to give a diastereomer of testosterone.

reduction

androtenedione

testosterone

C_{17} diastereomer of testosterone

Only reduction from the "bottom" gives the muscle-building anabolic steroid testosterone.

16

Carbohydrates

Chapter Summary

Carbohydrates are polyhydroxy aldehydes or polyhydroxy ketones, or substances that give such compounds on hydrolysis. They are classified as monosaccharides, oligosaccharides, and polysaccharides.

Monosaccharides, also called simple sugars, are classified by the number of carbon atoms (triose, tetrose, pentose, and so on) and by the nature of the carbonyl group (**aldose** or **ketose**).

R-(+)-Glyceraldehyde is an aldotriose designated by the following **Fischer projection formula:**

CH=O / H—C—OH / CH_2OH ≡ CH=O / H▸C◂OH / CH_2OH ≡ CH=O / $HOCH_2$—C(H)····OH

This configuration is designated D, whereas the enantiomer with the H and OH positions reversed is L. In larger monosaccharides, the letters *D* and *L* are used to designate the configuration of the stereogenic center with the *highest* number, the stereogenic carbon most remote from the carbonyl group. The D-aldoses through the hexoses are listed in Figure 16.1. **Epimers** are stereoisomers that differ in configuration at only one stereogenic center.

Monosaccharides with four or more carbons usually exist in a cyclic hemiacetal form, in which a hydroxyl group on C-4 or C-5 reacts with the carbonyl group (at C-1 in aldoses) to form a hemiacetal. In this way, C-1 also becomes stereogenic and is called the **anomeric carbon. Anomers** differ in configuration only at this stereogenic center and are designated α or β. The common ring sizes for the cyclic hemiacetals are six-membered (called **pyranoses)** or five-membered (called **furanoses).** The rings contain one oxygen atom and five or four carbon atoms, respectively.

Anomers usually interconvert in solution, resulting in a gradual change in optical rotation from that of the pure anomer to an equilibrium value for the mixture. This rotational change is called **mutarotation.**

Haworth formulas are a useful way of representing the cyclic forms of monosaccharides. The rings are depicted as flat, with hydroxyl groups or other substituents above or below the ring plane.

Monosaccharides can be oxidized at the aldehyde carbon to give carboxylic acids called **aldonic acids.** Oxidation at both ends of the carbon chain gives **aldaric acids.** Reduction of the carbonyl group to an alcohol gives polyols called **alditols.** The –OH groups in sugars, like those in simpler alcohols, can be esterified or etherified.

Monosaccharides react with alcohols (H^+ catalyst) to give **glycosides.** The –OH group at the anomeric carbon is replaced by an –OR group. The product is an acetal. Alcohols and phenols often occur in nature combined with sugars as glycosides, which renders them water-soluble.

Disaccharides consist of two monosaccharides linked by a glycosidic bond between the anomeric carbon of one unit and a hydroxyl group (often on C-4) of the other unit. Examples include **maltose** and **cellobiose** (formed from two glucose units joined by a 1,4-linkage and differing only in configuration at the anomeric carbon, being α and β, respectively), **lactose** (from a galactose and glucose unit linked 1,4 and β), and **sucrose,** or cane sugar (from a fructose and glucose unit, linked at the anomeric carbon of each, or 1,2). Sugars such as fructose, glucose, and sucrose are sweet, but others (for example, lactose and galactose) are not. Some noncarbohydrates, such as **saccharin** and **aspartame,** also taste sweet.

Polysaccharides have many monosaccharide units linked by glycosidic bonds. **Starch** and **glycogen** are polymers of D-glucose, mainly linked 1,4 or 1,6 and α. **Cellulose** consists of D-glucose units linked 1,4 and β.

Monosaccharides with modified structures are often biologically important. Examples include **sugar phosphates, deoxy sugars, amino sugars,** and **ascorbic acid** (vitamin C).

Reaction Summary

Hydrolysis

$$\text{polysaccharide} \xrightarrow{H_3O^+} \text{oligosaccharide} \xrightarrow{H_3O^+} \text{monosaccharide}$$

Acyclic and Cyclic Equilibration

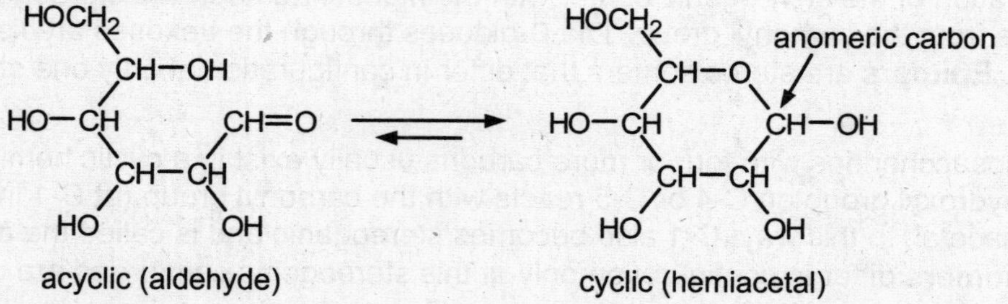

acyclic (aldehyde) cyclic (hemiacetal)

Oxidation

$$CH{=}O\,(CHOH)_n\,CH_2OH \xrightarrow[\text{or } Ag^+ \text{ or } Cu^{2+}]{Br_2,\ H_2O} CO_2H\,(CHOH)_n\,CH_2OH$$

aldose → aldonic acid

$$CH{=}O\,(CHOH)_n\,CH_2OH \xrightarrow{HNO_3} CO_2H\,(CHOH)_n\,CO_2H$$

aldose → aldaric acid

Reduction

$$CH{=}O\,(CHOH)_n\,CH_2OH \xrightarrow[\text{or } NaBH_4]{H_2,\ \text{catalyst}} CH_2OH\,(CHOH)_n\,CH_2OH$$

aldose → alditol

Esterification and Etherification

HOH_2C, HO, O, OH, OH, H, H, H, H, OH $\xrightarrow{Ac_2O}$ $AcOH_2C$, AcO, O, OAc, OAc, H, H, H, H, OAc

$$Ac = CH_3C(=O)-$$

HOH_2C, HO, O, OH, OH, H, H, H, H, OH $\xrightarrow[\text{or } CH_3I,\ Ag_2O]{NaOH,\ (CH_3)_2SO_4}$ CH_3OH_2C, CH_3O, O, OCH_3, OCH_3, H, H, H, H, OCH_3

Formation of Glycosides

HOH_2C, HO, O, OH, OH, H, H, H, H, OH $\xrightarrow[H^+]{ROH}$ HOH_2C, HO, O, OR, OH, H, H, H, H, OH $+ H_2O$

Hydrolysis of Glycosides

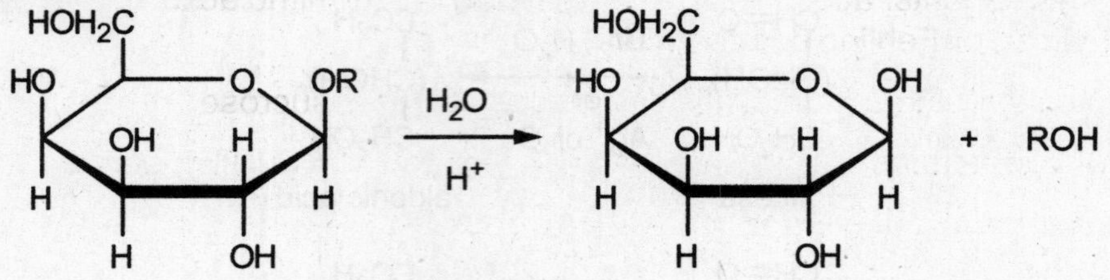

$$\text{glycoside (HOH}_2\text{C, HO, OR, OH, H)} \xrightarrow[H^+]{H_2O} \text{sugar (HOH}_2\text{C, HO, OH, OH, H)} + ROH$$

Learning Objectives

1. Know the meaning of: carbohydrate, monosaccharide, oligosaccharide, polysaccharide, disaccharide, trisaccharide.
2. Know the meaning of: aldose, ketose, triose, tetrose, pentose, hexose, glyceraldehyde, dihydroxyacetone.
3. Know the meaning of: D- or L-sugar, Fischer projection formula, Haworth formula, epimer.
4. Know the meaning of: α and β configurations, anomer, furanose and pyranose forms, mutarotation.
5. Know the meaning of: glycosidic bond, reducing and nonreducing sugar, aldaric acid, aldonic acid, alditol.
6. Learn the formulas for some common monosaccharides, especially the D forms of glucose, mannose, galactose, and fructose.
7. Draw the Fischer projection formula for a simple monosaccharide.
8. Convert the Fischer projection formula for a tetrose to a sawhorse or Newman projection formula, and vice versa.
9. Tell whether two structures are epimers or anomers.
10. Given the acyclic formula for a monosaccharide, draw its cyclic structure in either the pyranose or furanose form and either α or β configuration.
11. Given the rotations of two pure anomers and their equilibrium mixture, calculate the percentage of each anomer present at equilibrium.
12. Given the formula for a monosaccharide, draw the formula for its glycoside with a given alcohol or with a given additional monosaccharide.
13. Draw the cyclic structures (Haworth projection and conformational structure) for α-D- and β-D-glucose and the corresponding methyl glycosides.
14. Write all of the steps in the mechanism for the formation of a glycoside from a given sugar and alcohol.
15. Write all of the steps in the mechanism for the hydrolysis of a given glycoside to the corresponding sugar and alcohol.
16. Write all of the steps in the mechanism for the hydrolysis of a given disaccharide to the component monosaccharides.

17. Given the structure of a sugar, write equations for its reaction with each of the following reagents: acetic anhydride, bromine water, nitric acid, sodium borohydride, and Tollens' or Fehling's reagent.

18. Know the structures of: maltose, cellobiose, lactose, sucrose.

19. Write the structures for the repeating units in starch and cellulose.

20. Know the meaning of: sugar phosphate, deoxy sugar, amino sugar, ascorbic acid (vitamin C).

ANSWERS TO PROBLEMS

Problems Within the Chapter

16.1 The L isomer is the mirror image of the D isomer. The configuration at *every* stereogenic center is reversed.

a.
CH=O
H—OH
HO—H
HO—H
CH_2OH

b.
CH=O
H—OH
H—OH
HO—H
HO—H
CH_2OH

16.2 Follow Example 16.2 as a guide.

CH=O; HO—H; H—OH; CH_2OH ≡ HO, H, CH=O, H, OH, CH_2OH ≡ CH=O, H, OH, H, OH, CH_2OH ≡ CH=O, H, OH, H, OH, CH_2OH

If you view the second formula from the top, you will see that it is just a three-dimensional representation of the Fischer projection. Horizontal groups at each stereogenic center come up toward you, and vertical groups recede away from you. The second formula represents an eclipsed conformation of D-threose. The third and fourth formulas represent sawhorse and Newman projections, respectively, of a staggered conformation of D-threose.

16.3 For each D-aldohexose there will be, as one more stereogenic center is added, two aldoheptoses. Altogether, there are 16 D-aldoheptoses.

16.4 D-Ribose and D-xylose are identical except for the configuration at C-3. So, too, are D-arabinose and D-lyxose (see Figure 16.1).

16.5 D-Galactose is the C-4 epimer of D-glucose (see Figure 16.1). Therefore, the Haworth projection will be identical to that for D-glucose, except that the C-4 hydroxyl group will be up rather than down.

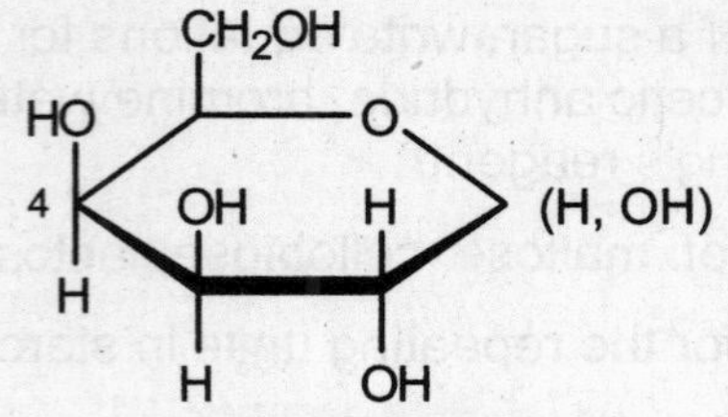

D-galactose

16.6 The anomeric hydroxyl group is up in the β-anomer and down in the α-anomer.

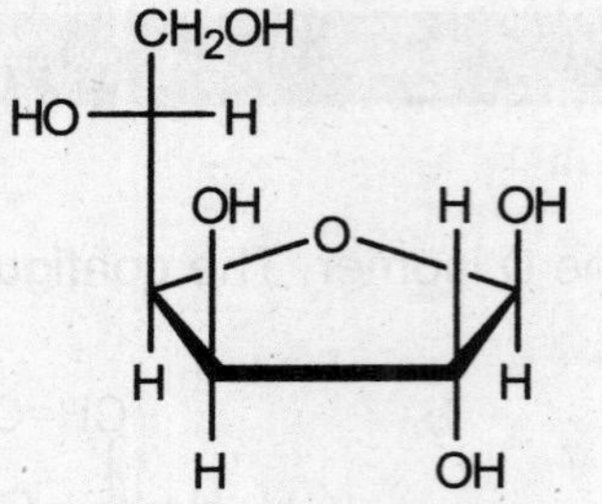

β-D-glucofuranose 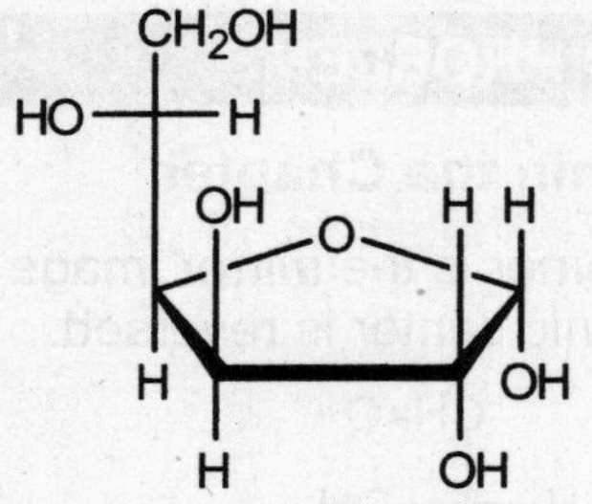 α-D-glucofuranose

16.7 D-Erythrose, a tetrose, has only four carbons. Therefore, it cannot form a pyranose that would require the presence of at least five carbons. Bonding between the C-4 hydroxyl group and the carbonyl carbon, however, does lead to two anomeric furanoses.

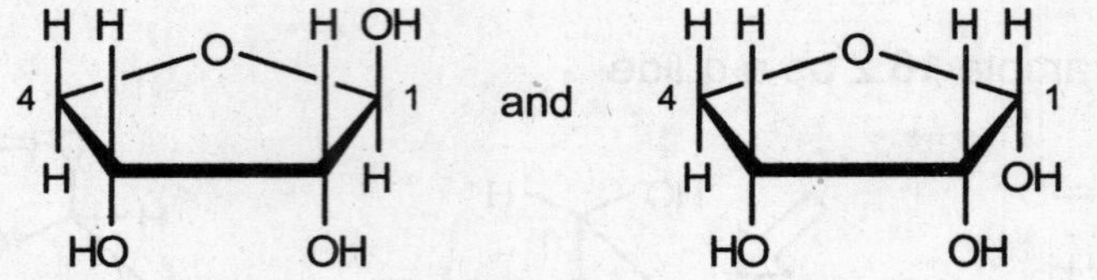

The structure on the right is α-D-erythrofuranose.

16.8 Use eq. 16.6 as a guide. Note that β-D-galactopyranose is the C-4 epimer of β-D-glucopyranose.

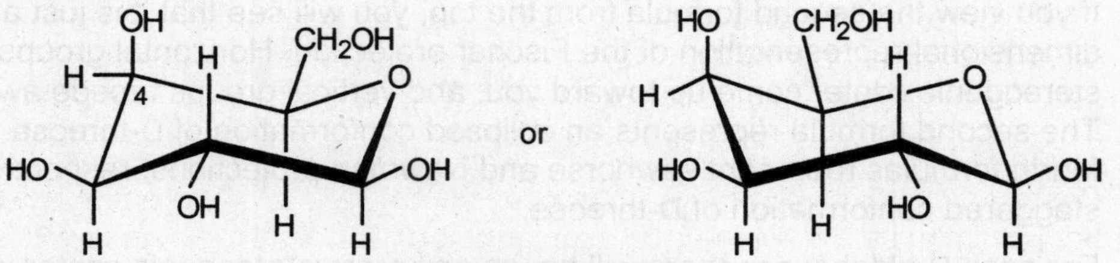

β-D-galactopyranose

16.9

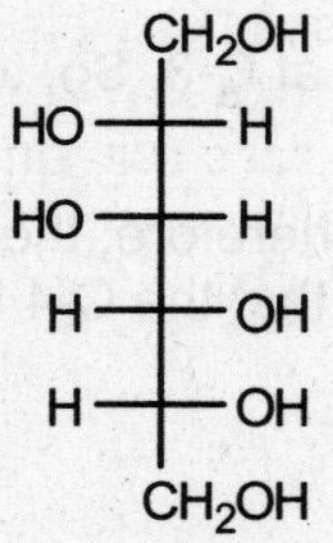

16.10 Use eq. 16.11 as a guide.

CH=O
HO—H
HO—H
H—OH
H—OH
CH_2OH

$+ 2Cu^{+2} + 5HO^- \longrightarrow$

CO_2^-
HO—H
HO—H
H—OH
H—OH
CH_2OH

$+ Cu_2O + 3\ H_2O$

16.11 Use eq. 16.12 as a guide.

CO_2H
HO—H
HO—H
H—OH
H—OH
CO_2H

16.12 Using Haworth formulas and eq. 16.13 as a guide, we have the following:

HOH_2C; HO; O; OH; OH; H; H; H; H; OH

$+ CH_3OH \xrightarrow{H^+}$

HOH_2C; HO; O; OCH_3; OH; H; H; H; H; OH

$+ H_2O$

Although we show only the methyl β-D-galactoside as a product, some of the α epimer will also be formed because the carbocation intermediate can be attacked at either face by the methanol.

16.13 Carbon-1 of the second (right hand) glucose unit in maltose is a hemiacetal carbon. Thus, the α and β forms at this carbon atom can equilibrate *via* the open-chain aldehyde form. Mutarotation is, therefore, possible. Note that carbon-1 of the first (left-hand) glucose unit is an acetal carbon. Its configuration is, therefore, fixed (as α).

16.14 Carbon-1 of the glucose unit in lactose is a hemiacetal carbon and will be in equilibrium with the open-chain aldehyde form. Therefore, lactose will be oxidized by Fehling's solution and will mutarotate.

16.15 Carbon-1 of methyl β-D-glucopyranoside is an acetal carbon. In solution, it is not in equilibrium with an open-chain aldehyde. Therefore, it cannot reduce Ag^+ or Cu^{2+} and does not give positive tests with Tollens', Fehling's, or Benedict's reagent.

ADDITIONAL PROBLEMS

16.16 If you have difficulty with any part of this problem, consult the indicated section in the text, where the term is defined and/or illustrated.

a.	Sec. 16.2	b.	Sec. 16.2
c	Secs. 16.1 and 16.2	d.	Secs. 16.1 and 16.12
e.	Secs. 16.1 and 16.13	f.	Sec. 16.6
g.	Sec. 16.6	h.	Sec. 16.11
i.	Sec. 16.5		

16.17 D-sugars have the same configuration at the carbon atom adjacent to the primary alcohol function as D-(+)-glyceraldehyde (which is *R*).

CH=O / H—C—OH / CH_2OH — D-(+)-glyceraldehyde

CH=O / $(CHOH)_n$ / H—C—OH / CH_2OH — a D-aldose

CH_2OH / C=O / $(CHOH)_n$ / H—C—OH / CH_2OH — a D-ketose

L-Sugars have the opposite configuration at the stereogenic center adjacent to the primary alcohol function.

CH=O / HO—C—H / CH_2OH — L-(−)-glyceraldehyde

CH=O / $(CHOH)_n$ / HO—C—H / CH_2OH — an L-aldose

CH_2OH / C=O / $(CHOH)_n$ / HO—C—H / CH_2OH — an L-ketose

16.18 Although the hydroxyl groups at carbons 2, 3, and 4 of D-talose are on the left (Figure 16.1), the hydroxyl group on *the highest-numbered stereogenic carbon* (C-5) *is on the right*. Only the stereochemistry at that carbon determines whether a sugar belongs to the D- or L-series.

16.19 D-Xylose and D-lyxose differ only in the configuration at C-2. They are epimers at C-2.

16.20 The D-ketoses:

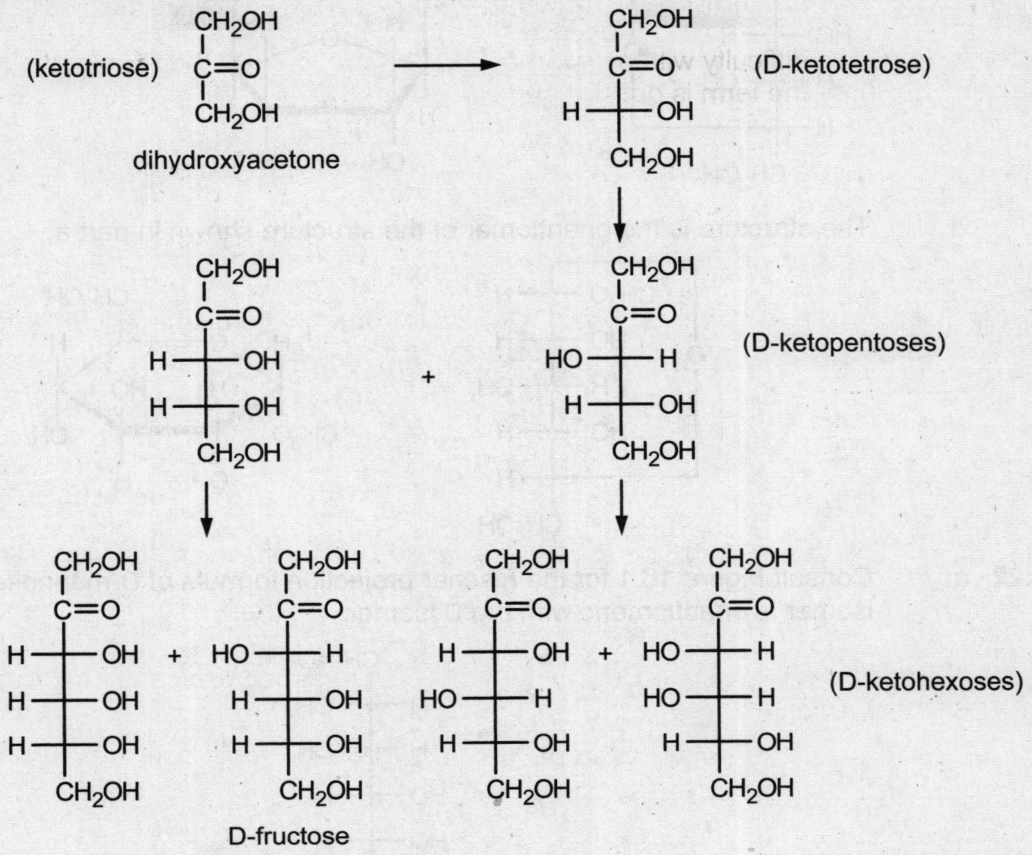
(ketotriose)
dihydroxyacetone
(D-ketotetrose)
(D-ketopentoses)
(D-ketohexoses)
D-fructose

16.21 a.

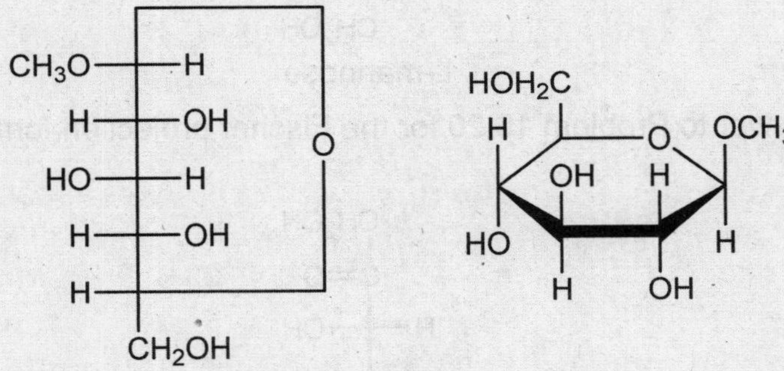

b.

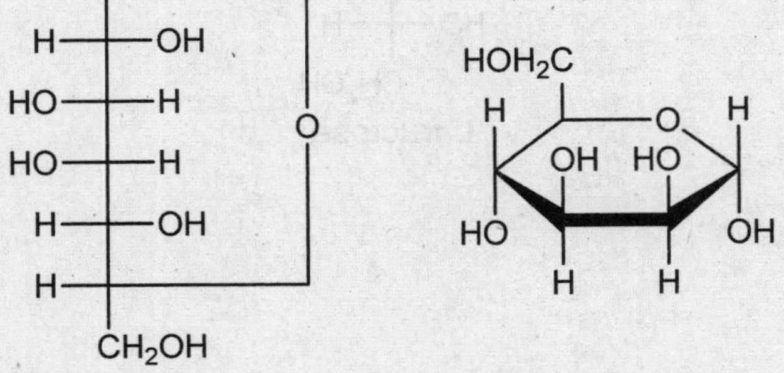

c.

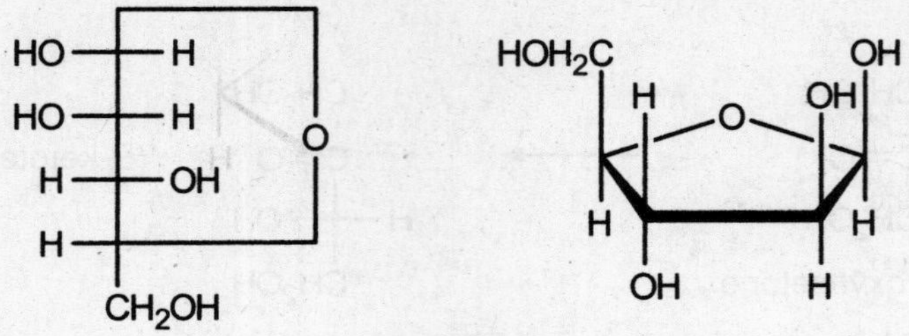

d. The structure is the enantiomer of the structure shown in part a.

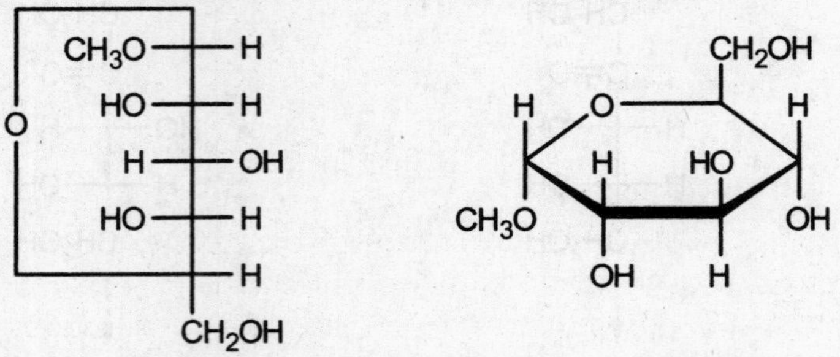

16.22 a. Consult Figure 16.1 for the Fischer projection formula of D-mannose. The L isomer is enantiomeric with the D isomer.

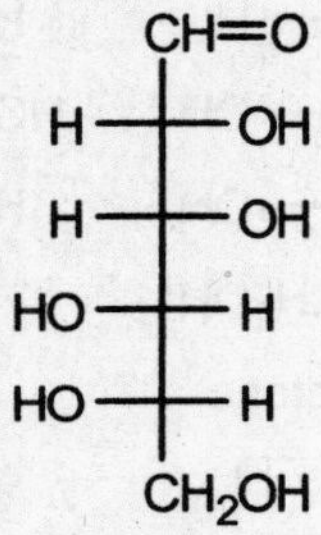

L-mannose

b. See the answer to Problem 16.20 for the Fischer projection formula of D-fructose.

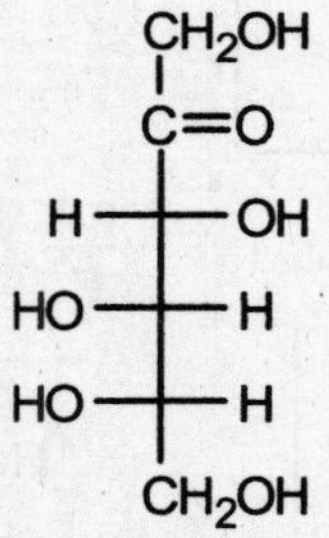

L-fructose

16.23

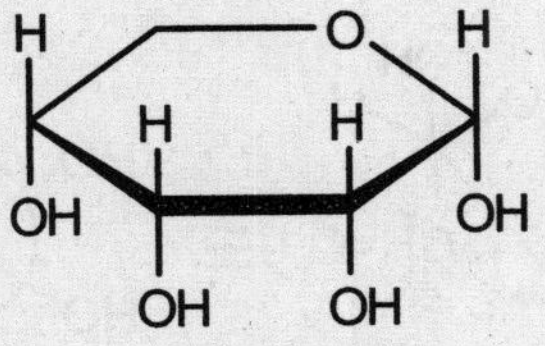

α-pyranose

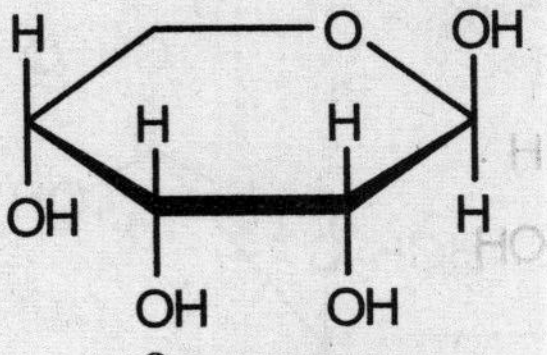

β-pyranose

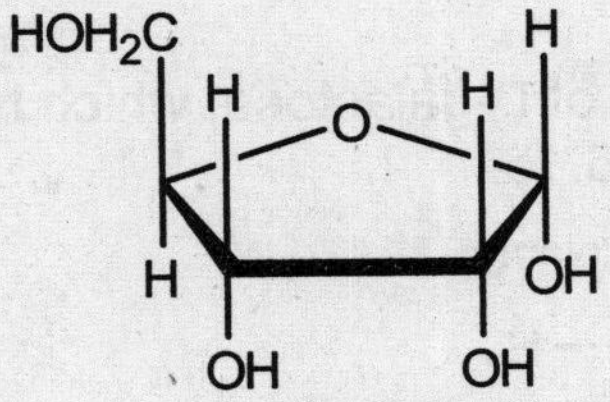

α-furanose

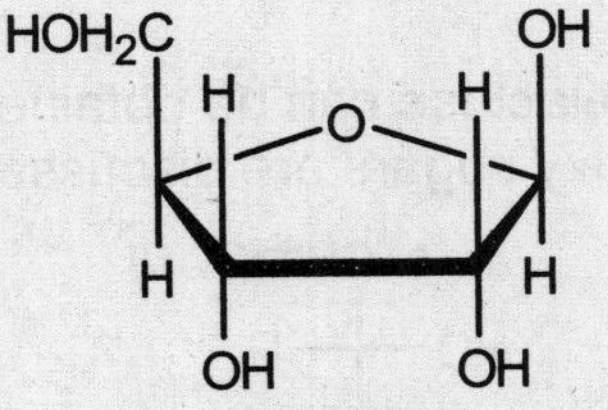

β-furanose

16.24 One way of approaching this problem is to relate the structure to something familiar, for example the structure of glucose. Inspection of the ball-and-stick structure reveals that it represents a β-hexopyranose (6 carbons; 6-membered ring with an OH group bonded to the anomeric carbon). Recall that β-D-glucopyranose has all of the six-membered ring substituents on equatorial sites (see Secs. 16.7 and 16.11). The ball-and-stick structure has all substituents equatorial with the exception of the C-3 hydroxyl group, which is axial. Thus, this structure is the C-3 epimer of β-D-glucopyranose. Inspection of Table 16.1 reveals that D-allose is the C-3 epimer of D-glucose. Thus, the structure represents β-D-allopyranose. The Fischer, Haworth and line-segment structures of this compound are shown below. Notice that the line segment structure most closely resembles the ball-and-stick structure.

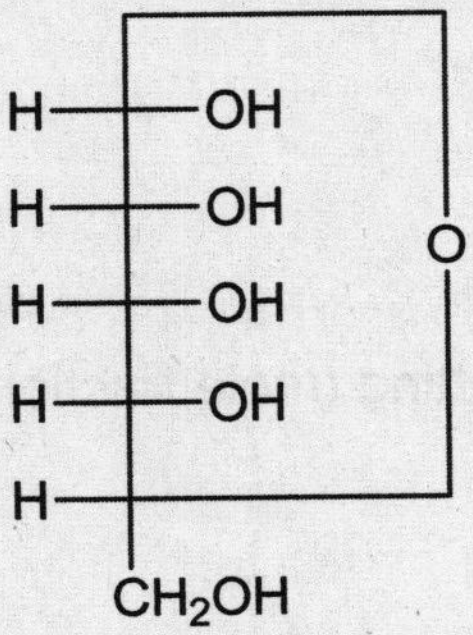

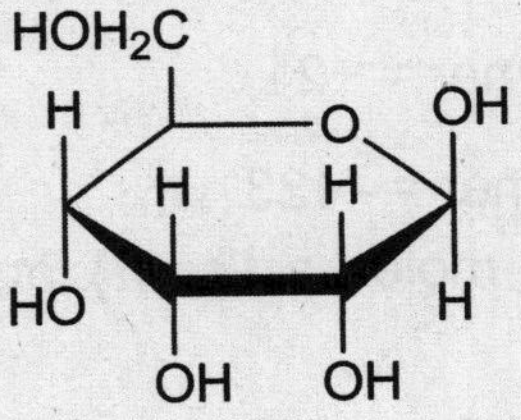

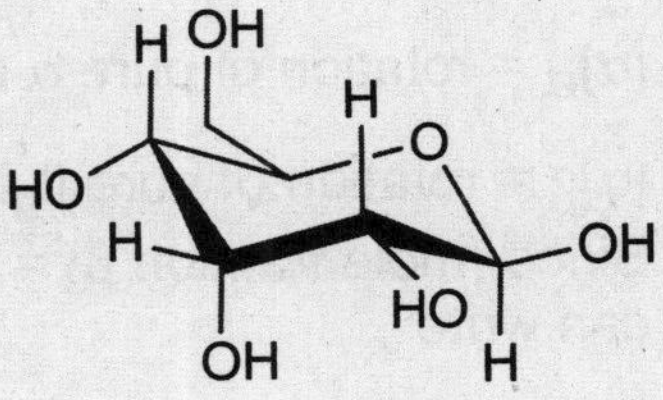

16.25 Consult the answer to Problem 16.7.

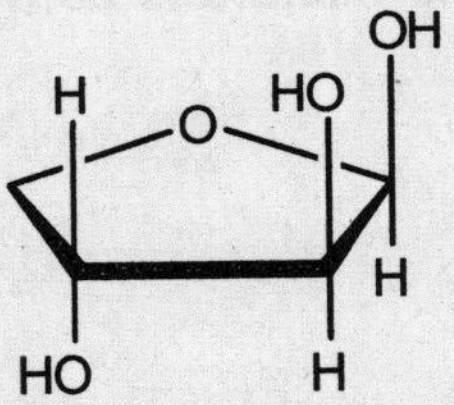

β-D-threofuranose

16.26

L-threose (Fischer projection: CH=O; H—C—OH; HO—C—H; CH_2OH) ≡ ≡ ≡ (Newman projection)

16.27 The formula of L-galactose can be obtained from that of D-galactose, which is shown in Figure 16.1. Deoxy sugars are discussed in Sec. 16.15.

L-galactose: 1 CH=O; HO—H; H—OH; H—OH; HO—H; 6 CH_2OH

L-fucose (6-deoxy-L-galactose): CH=O; HO—H; H—OH; H—OH; HO—H; CH_3

16.28 The compounds are diastereomers. That is, they are stereoisomers but not mirror images. They are, therefore, expected to differ in all properties, including water solubility.

16.29 $[\alpha]_e = [\alpha]_\alpha$ (mole fraction α) + $[\alpha]_\beta$ (mole fraction β)

where $[\alpha]_e$ = rotation at equilibrium = –92

$[\alpha]_\alpha$ = rotation of pure α anomer = +21

$[\alpha]_\beta$ = rotation of pure β anomer = –133

Since (mole fraction β) = (1 – mole fraction α), and letting (mole fraction α) = X, we can write:

$[\alpha]_e = [\alpha]_\alpha X + [\alpha]_\beta (1 - X) = [\alpha]_\beta + X([\alpha]_\alpha - [\alpha]_\beta)$ or

$X = [\alpha]_e - [\alpha]_\beta / [\alpha]_\alpha - [\alpha]_\beta = (-92) - (-133)/(21) - (-133) = 41/154 = 0.266$

Therefore, at equilibrium the mixture is 26.6% α and 73.4% β (also see Example 16.4).

16.30

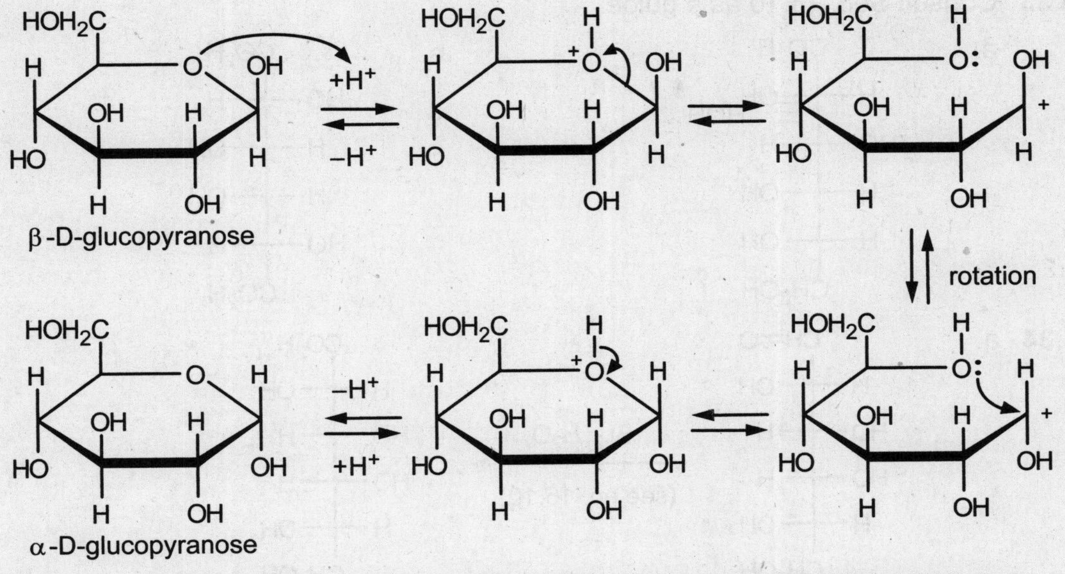

β-D-glucopyranose $\underset{-H^+}{\overset{+H^+}{\rightleftharpoons}}$ [protonated ring oxygen] $\rightleftharpoons$ [ring-opened oxocarbenium ion] $\overset{\text{rotation}}{\rightleftharpoons}$ [rotated ion] $\rightleftharpoons$ [protonated α form] $\underset{+H^+}{\overset{-H^+}{\rightleftharpoons}}$ α-D-glucopyranose

16.31 a. For the structure, see Sec. 16.12c. The α form is shown. The β form is identical with it, except for the configuration at C-1 in the "right-hand" unit, where the OH group is equatorial instead of axial.

b. Follow Example 16.4 or Prob. 16.29 as guides for the calculation.

$$\%\text{ of }\beta = \frac{92.6 - 52}{92.6 - 34} \times 100 = \frac{40.6}{58.6} \times 100 = 69.3\%$$

16.32 The structures are shown in Figure 16.1. Oxidation of D-erythrose gives optically inactive *meso*-tartaric acid:

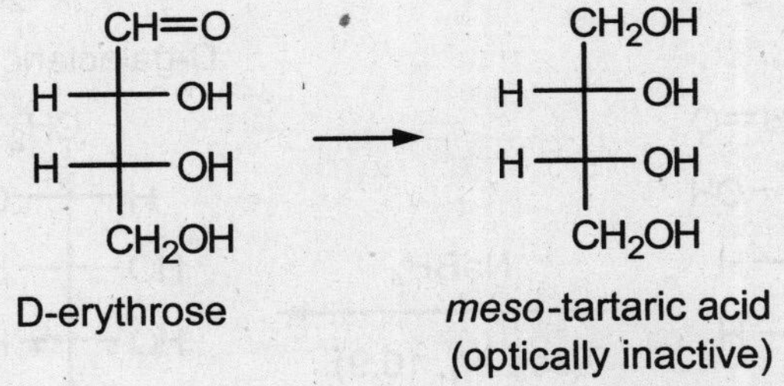

D-erythrose (CH=O; H–C–OH; H–C–OH; CH_2OH) → *meso*-tartaric acid (optically inactive) (CH_2OH; H–C–OH; H–C–OH; CH_2OH)

Analogous oxidation of D-threose gives an optically active tartaric acid:

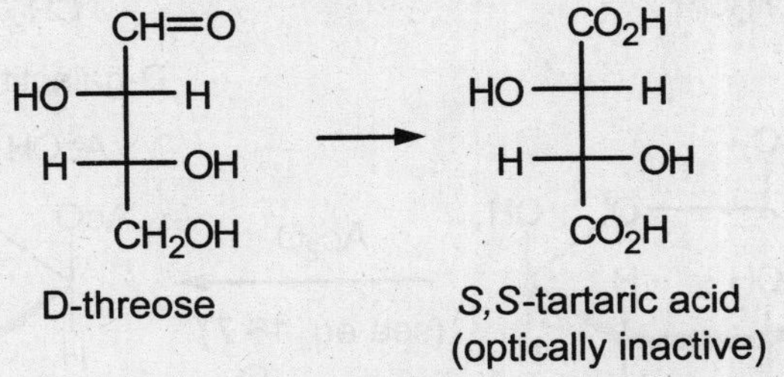

D-threose (CH=O; HO–C–H; H–C–OH; CH_2OH) → *S,S*-tartaric acid (optically inactive) (CO_2H; HO–C–H; H–C–OH; CO_2H)

In this way, we can readily assign structures to the two tetroses.

16.33 Consult Sec. 16.10 as a guide.

a. CO_2H; H–C–OH; HO–C–H; H–C–OH; H–C–OH; CH_2OH

b. CO_2H; HO–C–H; H–C–OH; H–C–OH; HO–C–H; CO_2H

16.34 a. CH=O; H–C–OH; HO–C–H; HO–C–H; H–C–OH; CH_2OH $\xrightarrow[\text{(see eq. 16.10)}]{Br_2, H_2O}$ CO_2H; H–C–OH; HO–C–H; HO–C–H; H–C–OH; CH_2OH

D-galactonic acid

b. CH=O; H–C–OH; HO–C–H; HO–C–H; H–C–OH; CH_2OH $\xrightarrow[\text{(see eq. 16.12)}]{HNO_3}$ CO_2H; H–C–OH; HO–C–H; HO–C–H; H–C–OH; CO_2H

D-galactaric acid

c. CH=O; H–C–OH; HO–C–H; HO–C–H; H–C–OH; CH_2OH $\xrightarrow[\text{(see eq. 16.9)}]{NaBH_4}$ CH_2OH; H–C–OH; HO–C–H; HO–C–H; H–C–OH; CH_2OH

D-galactitol

d. β-D-galactose (HOH_2C, HO, OH, OH, H, H, H, H, OH) $\xrightarrow[\text{(see eq. 16.7)}]{Ac_2O}$ β-D-galactose pentaacetate ($AcOH_2C$, AcO, OAc, OAc, H, H, H, H, OAc)

Ac = $CH_3C(=O)–$

β-D-galactose

β-D-galactose pentaacetate

16.35 Using Fischer projection formulas, we can write the structures of D-glucitol (eq 16.9) and D-mannitol (see the answer to Problem 16.9).

D-glucitol: CH_2OH; C-2: H–OH; C-3: HO–H; C-4: H–OH; C-5: H–OH; CH_2OH

D-mannitol: CH_2OH; C-2: HO–H; C-3: HO–H; C-4: H–OH; C-5: H–OH; CH_2OH

Note that the configurations are identical at C-3, C-4, and C-5. The fact that D-fructose gives both of these polyols on reduction tells us that it must also have the same configuration as they do at C-3, C-4, and C-5. Thus, the *keto* group of D-fructose must be at C-2.

D-fructose (CH_2OH; C=O; HO–H; H–OH; H–OH; CH_2OH) $\xrightarrow{NaBH_4}$ (CH_2OH; H–OH; HO–H; H–OH; H–OH; CH_2OH) + (CH_2OH; HO–H; HO–H; H–OH; H–OH; CH_2OH)

16.36 See eq. 16.12.

(CH=O; H–OH; HO–H; HO–H; H–OH; CH_2OH) $\xrightarrow{HNO_3}$ (CO_2H; H–OH; HO–H; - - - - - -; HO–H; H–OH; CO_2H)

The dicarboxylic acid obtained as the product is a *meso* form, with a plane of symmetry shown by the dashed line. Therefore, this acid is achiral. (Experiments such as this were helpful in assigning configurations to the various monosaccharides.)

16.37

(glucoside of 2-(hydroxymethyl)phenol: HOH_2C, O, OH, H, HO, H, OH, CH_2OH) $\xrightarrow{H_3O^+}$ (glucose: HOH_2C, O, OH, H, HO, H, OH, OH) + (2-(hydroxymethyl)phenol: HO, $HOCH_2$)

16.38 a.

$$\begin{array}{c} CH{=}O \\ H-\!\!\!-OH \\ HO-\!\!\!-H \\ HO-\!\!\!-H \\ H-\!\!\!-OH \\ CH_2OH \end{array} \xrightarrow{NH_2OH} \begin{array}{c} CH{=}NOH \\ H-\!\!\!-OH \\ HO-\!\!\!-H \\ HO-\!\!\!-H \\ H-\!\!\!-OH \\ CH_2OH \end{array} + H_2O$$

b.

$$\begin{array}{c} CH{=}O \\ H-\!\!\!-OH \\ HO-\!\!\!-H \\ HO-\!\!\!-H \\ H-\!\!\!-OH \\ CH_2OH \end{array} \xrightarrow[H^+]{CH_3CH_2OH} \alpha\text{ (ethyl glycoside, } OCH_2CH_3) + \beta\text{-anomer} + H_2O$$

c.

$$\begin{array}{c} CH{=}O \\ H-\!\!\!-OH \\ HO-\!\!\!-H \\ HO-\!\!\!-H \\ H-\!\!\!-OH \\ CH_2OH \end{array} \xrightarrow{Br_2,\ H_2O} \begin{array}{c} CO_2H \\ H-\!\!\!-OH \\ HO-\!\!\!-H \\ HO-\!\!\!-H \\ H-\!\!\!-OH \\ CH_2OH \end{array}$$

d.

$$\begin{array}{c} CH{=}O \\ H-\!\!\!-OH \\ HO-\!\!\!-H \\ HO-\!\!\!-H \\ H-\!\!\!-OH \\ CH_2OH \end{array} \xrightarrow{NaBH_4} \begin{array}{c} CH_2OH \\ H-\!\!\!-OH \\ HO-\!\!\!-H \\ HO-\!\!\!-H \\ H-\!\!\!-OH \\ CH_2OH \end{array}$$

e.

$$\begin{array}{c} CH{=}O \\ H-\!\!\!-OH \\ HO-\!\!\!-H \\ HO-\!\!\!-H \\ H-\!\!\!-OH \\ CH_2OH \end{array} \xrightarrow[NaOH]{HCN} \begin{array}{c} CN \\ H-\!\!\!-OH \\ H-\!\!\!-OH \\ HO-\!\!\!-H \\ HO-\!\!\!-H \\ H-\!\!\!-OH \\ CH_2OH \end{array} + \begin{array}{c} CN \\ HO-\!\!\!-H \\ H-\!\!\!-OH \\ HO-\!\!\!-H \\ HO-\!\!\!-H \\ H-\!\!\!-OH \\ CH_2OH \end{array}$$

f.

CH=O, H—OH, HO—H, HO—H, H—OH, CH_2OH $\xrightarrow[\text{(excess)}]{Ac_2O}$ (α-pentaacetate, pyranose form: AcO, OAc, H, O, H, AcO, H, AcO, OAc, H, α) + β-anomer

g.

CH=O, H—OH, HO—H, HO—H, H—OH, CH_2OH $\xrightarrow{Ag^+, NH_3}$ CO_2H, H—OH, HO—H, HO—H, H—OH, CH_2OH

16.39 a. For the structure of maltose, see Sec. 16.12a.

(structure: HOH_2C, O, OH, HO, OH, $\overset{+}{O}$–H, HOH_2C, O, OH, OH, OH →)

protonated maltose

(HOH_2C, O^+, OH, HO, OH + D-glucose; $:\ddot{O}H_2$) $\xrightarrow{-H^+}$ (HOH_2C, O, OH, HO, OH, OH)

D-glucose

b. For the structure of lactose, see Sec. 16.12c.

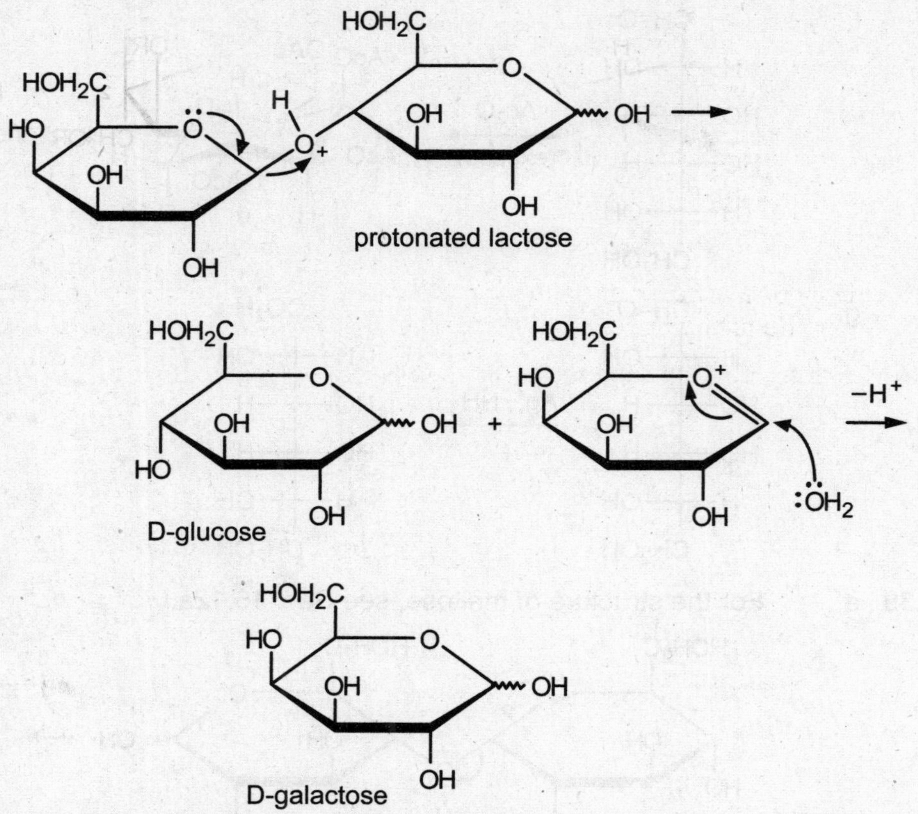

HOH$_2$C HO OH O H O+ HOH$_2$C OH O OH OH OH

protonated lactose

HOH$_2$C OH O OH HO OH + HOH$_2$C HO O+ OH OH :OH$_2$ $-H^+$

D-glucose

HOH$_2$C HO O OH OH OH

D-galactose

c. For the structure of sucrose, see Sec. 16.12d.

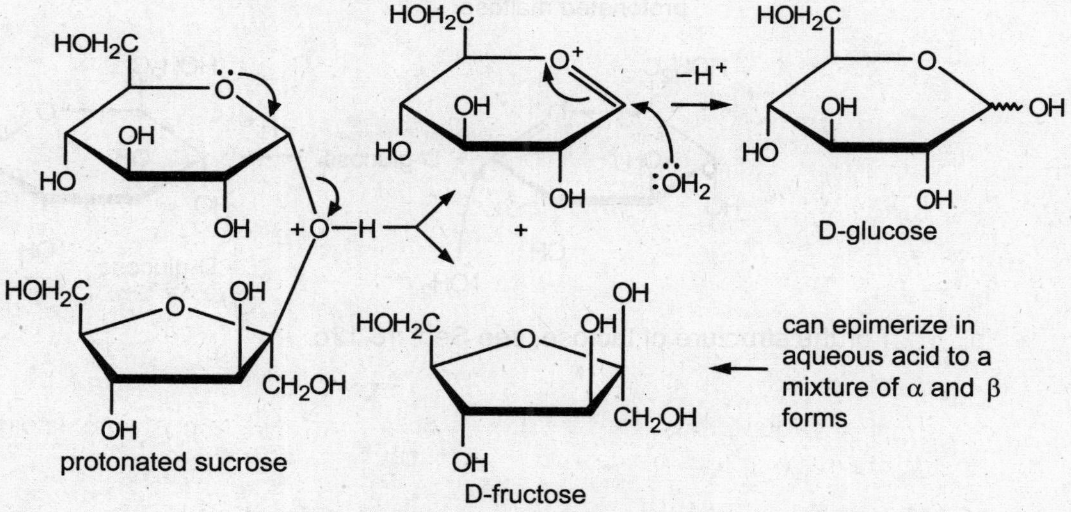

HOH$_2$C O OH HO OH +O—H HOH$_2$C OH O CH$_2$OH OH

protonated sucrose

HOH$_2$C O+ OH HO OH :OH$_2$ $-H^+$ HOH$_2$C O OH HO OH OH

D-glucose

+ HOH$_2$C O OH OH CH$_2$OH OH

D-fructose

can epimerize in aqueous acid to a mixture of α and β forms

The mechanism shown is one of two that are possible. The alternative mechanism would break the other glycoside bond to give glucose and the carbocation from the fructose unit. Both mechanisms undoubtedly occur simultaneously. The products, of course, are the same from both paths—namely, a mixture of the α and β forms of D-glucose and D-fructose.

16.40 Hydrolysis of the glycosidic bond provides the following derivatives of glucose and fructose:

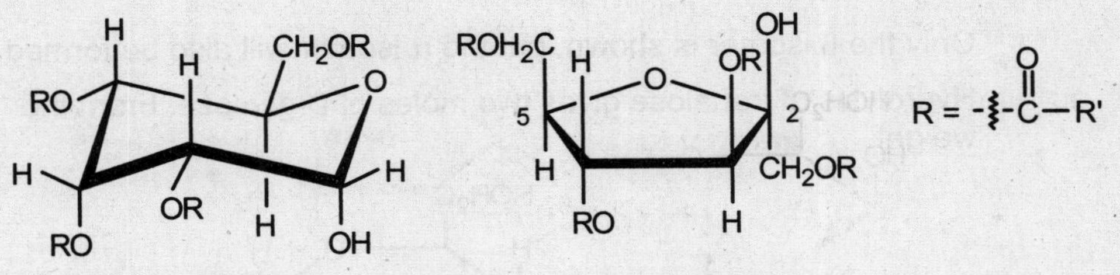

acylated glucose acylated fructose

$R = -\text{C}(=\text{O})-R'$

Only one anomer of each acylated monosaccharide is shown, but the other would also be formed.

16.41 The formula for maltose is given in Sec. 16.12a.

a. maltose $\xrightarrow{Ag^+}$ (see eq. 16.10) + $Ag°\downarrow$

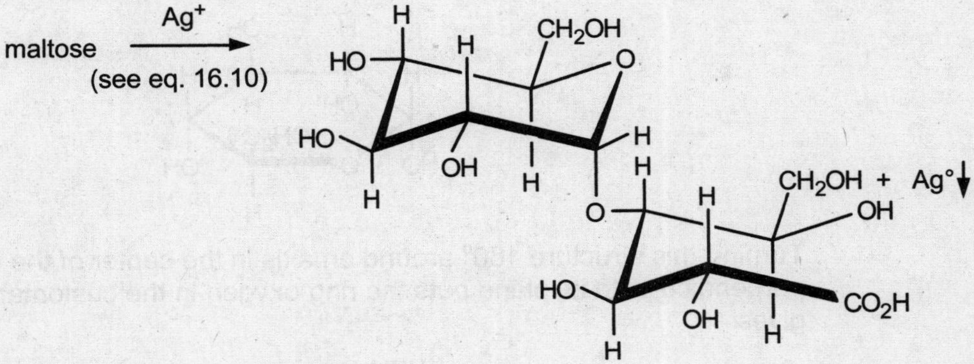

b. The reaction of maltose with bromine water will yield the same product as in part a (see eq. 16.10).

c. maltose $\xrightarrow{Ac_2O}$ (see eq. 16.7)

$Ac = CH_3\overset{O}{\overset{\|}{C}}-$

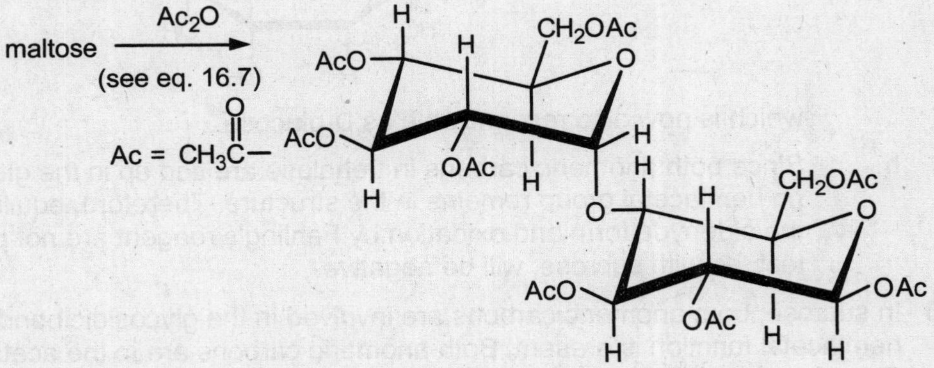

Eight equivalents of acetic anhydride are required to acetylate all eight hydroxyl groups of maltose.

d. maltose $\xrightarrow[H^+]{CH_3CH_2OH}$ (see eq. 16.13) + H_2O

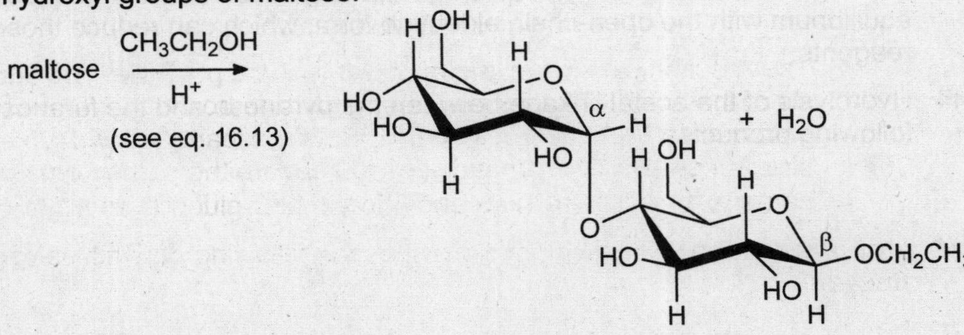

Only the β isomer is shown, but the α isomer will also be formed.

16.42 a. Hydrolysis of trehalose gives two moles of D-glucose. From the "left" portion, we get

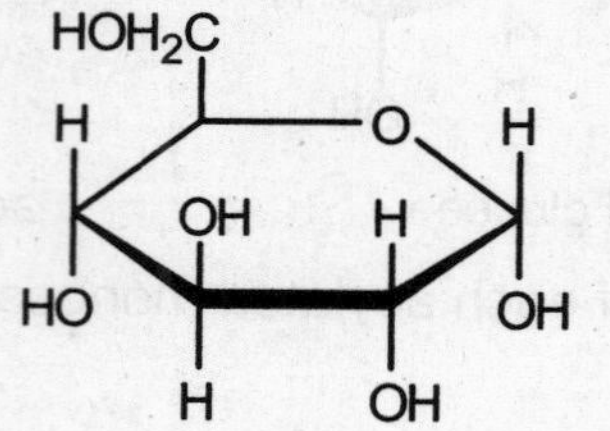

which is easily recognizable as D-glucose. From the "right" portion, we get

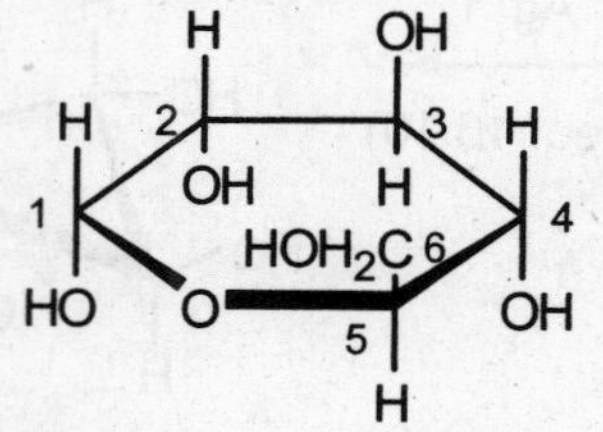

Turning this structure 180° around an axis in the center of the ring and perpendicular to its plane puts the ring oxygen in the customary position and gives

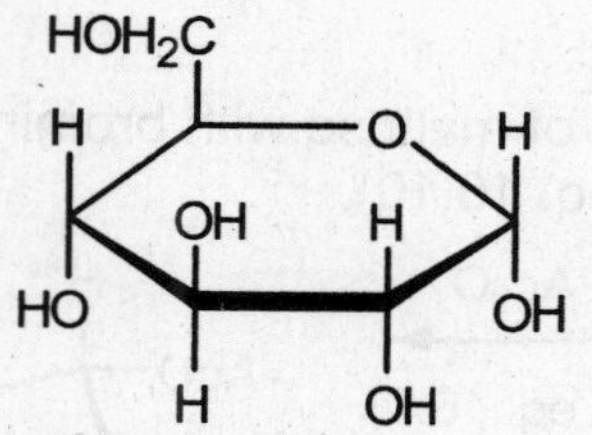

which is now also recognizable as D-glucose.

b. Since both anomeric carbons in trehalose are tied up in the glucosidic bond, no hemiacetal group remains in the structure. Therefore, equilibration with the aldehyde form and oxidation by Fehling's reagent are not possible. The test, as with sucrose, will be negative.

16.43 In sucrose, both anomeric carbons are involved in the glycosidic bond. No hemiacetal function is present. Both anomeric carbons are in the acetal (ketal) form. Therefore, equilibration with an acyclic aldehyde form is not possible, and the sugar cannot reduce Tollens', Fehling's, or Benedict's reagent. In maltose (Sec. 16.12a), on the other hand, carbon-1 of the right-hand glucose unit is a hemiacetal carbon, in equilibrium with the open-chain aldehyde form, which can reduce those same reagents.

16.44 Hydrolysis of the acetal linkage between the pyranose and the furanose gives the following products:

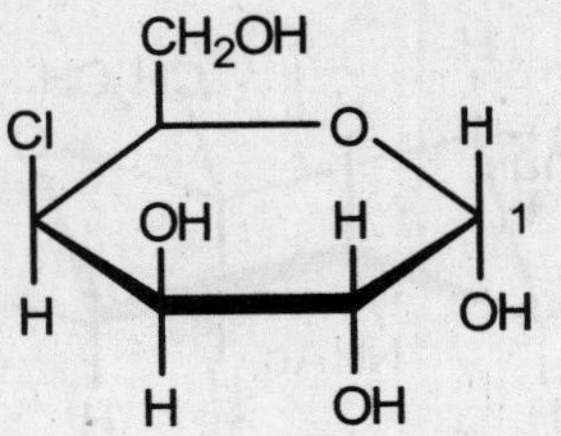

Only the α-anomer is shown. The β-anomer at C-1 will also be formed.

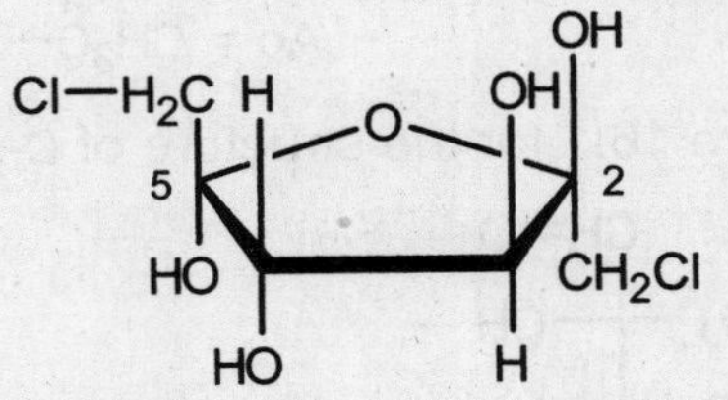

Only one anomer is shown. C-2 and C-5 are both hemiacetals and both anomers will be formed at these stereogenic centers.

16.45 Olestra can be synthesized from sucrose by treating sucrose with the anhydride of the desired acyl unit or the corresponding acyl chloride.

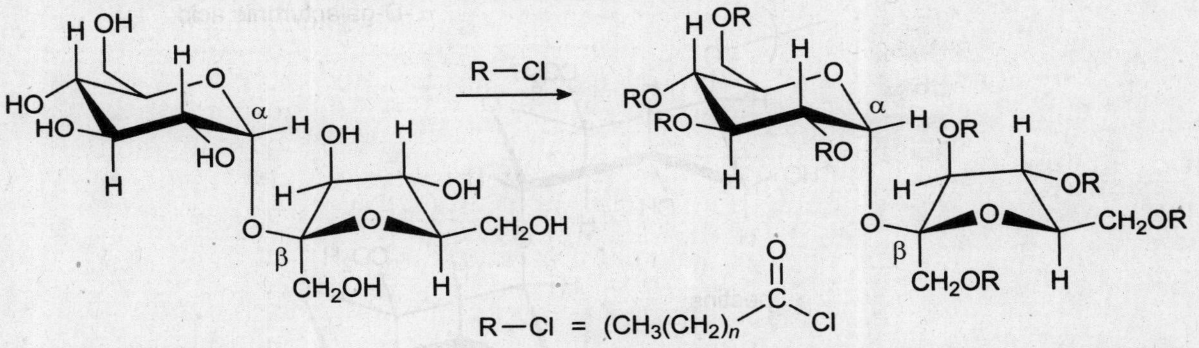

$$R{-}Cl = CH_3(CH_2)_nC(=O)Cl$$

16.46 The disaccharide could exist as either of two anomers at C-1 of the *N*-acetylmuramic acid unit.

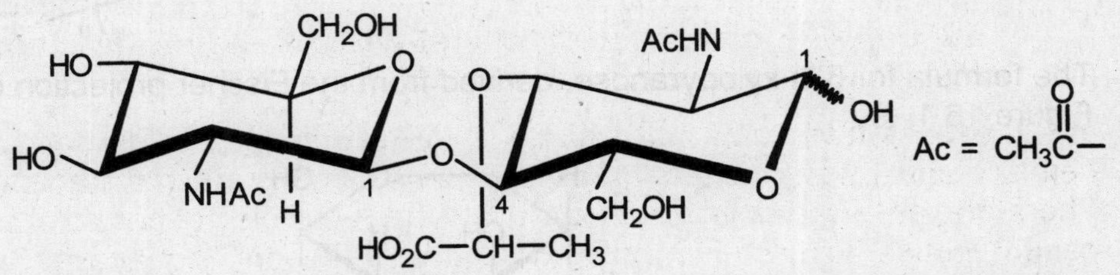

$$Ac = CH_3C(=O){-}$$

16.47 a. Use the formula for cellulose (Figure 16.6) as a guide, and replace the OH at C-2 by $NHCOCH_3$:

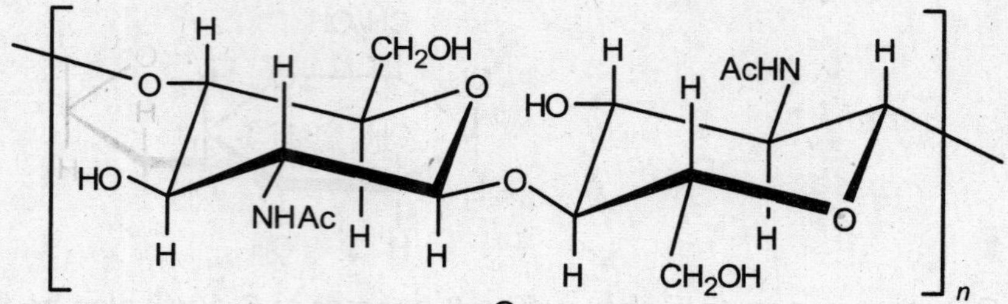

$$Ac = CH_3\overset{O}{\overset{\|}{C}}-$$

b. Consult Figure 16.1 for the structure of D-galactose.

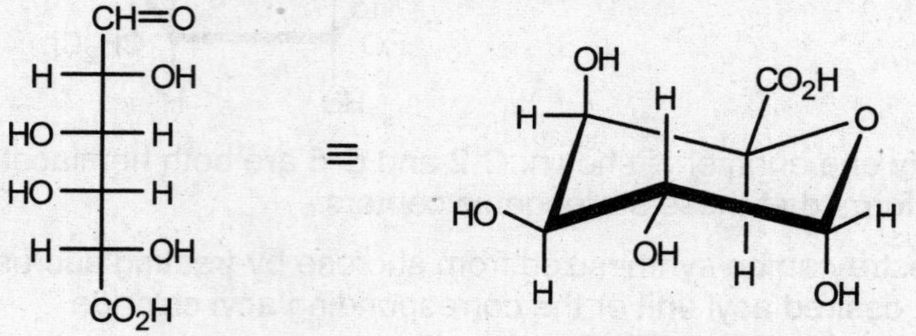

α-D-galacturonic acid

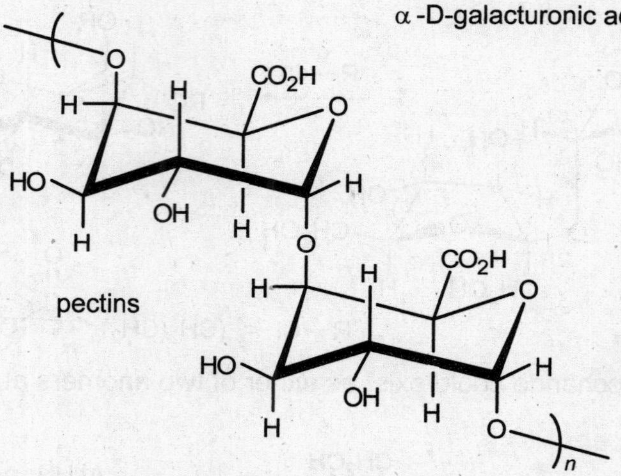

pectins

16.48 The formula for β-D-xylopyranose, derived from the Fischer projection of D-xylose in Figure 16.1, is

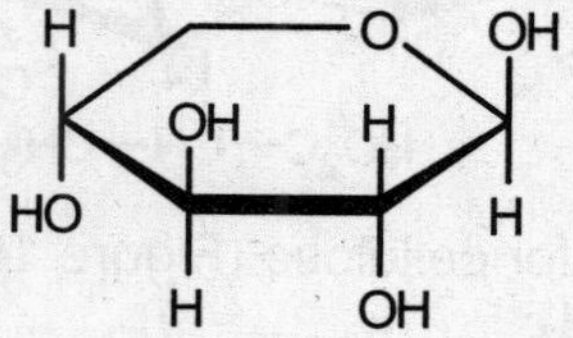

Since the xylans have these units linked 1,4, their structure is

H O 4 OH H H O H OH H OH H 1 ← β O H OH n

16.49 The anomeric carbon of the glucose unit on the "right" is in equilibrium with the corresponding aldehyde. This will be oxidized by the Br_2 to the corresponding acid.

CH_2OH HO O HO OH O CH_2OH O HO OH OH

⇌ [CH_2OH HO O HO OH O CH_2OH OH HO CHO OH]

$\xrightarrow[H_2O]{Br_2}$ CH_2OH HO O HO OH O CH_2OH OH HO CO_2H OH

16.50 Treatment of **A** with Br_2/H_2O followed by acid hydrolysis (H_3O^+) of the glycosidic bond would provide D-galactose (from the sugar on the left) and D-gluconic acid (from the sugar on the right).

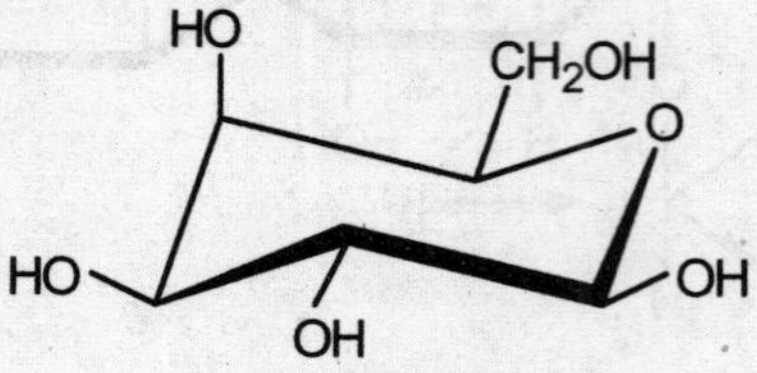

D-galactose

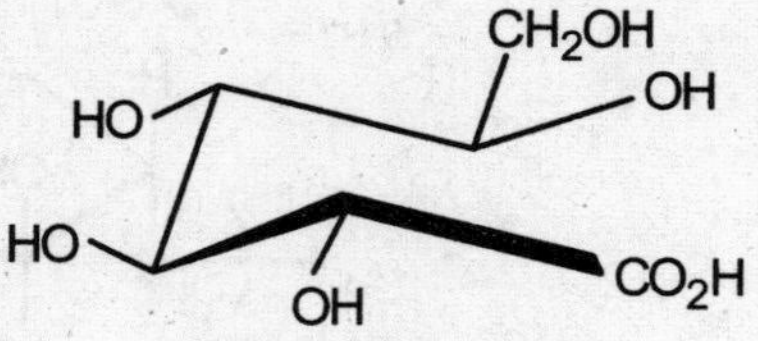

D-gluconic acid

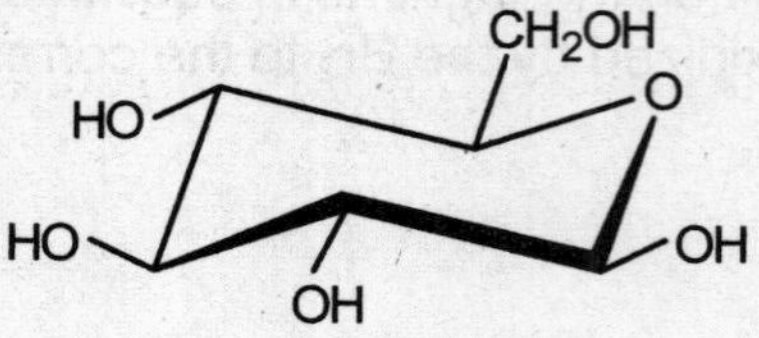

D-glucose

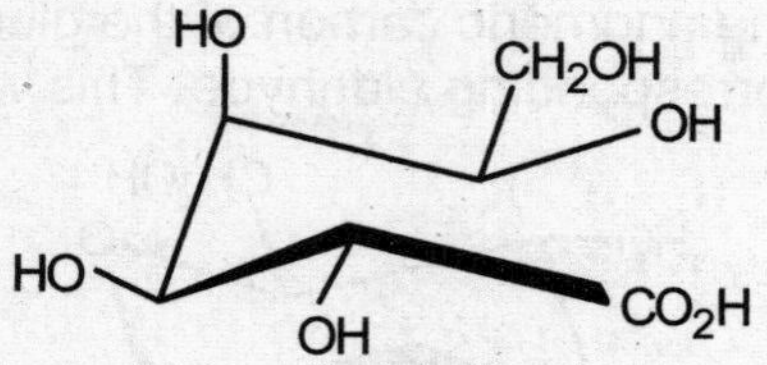

D-galactonic acid

Application of the same two-reaction sequence to disaccharide **B** will provide D-glucose and D-galactonic acid. Each disaccharide provides a different pair of products. This clearly distinguishes one disaccharide from the other.

16.51 See eq. 16.15 for the formula of ascorbic acid.

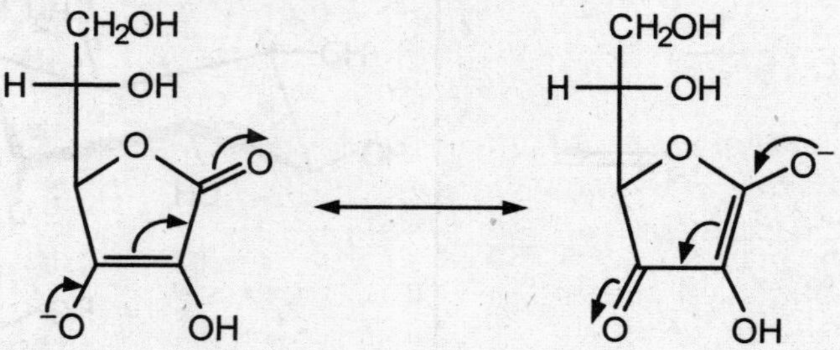

The negative charge can be almost equally spread over the two oxygen atoms.

16.52 Be systematic in deducing all of the structures.

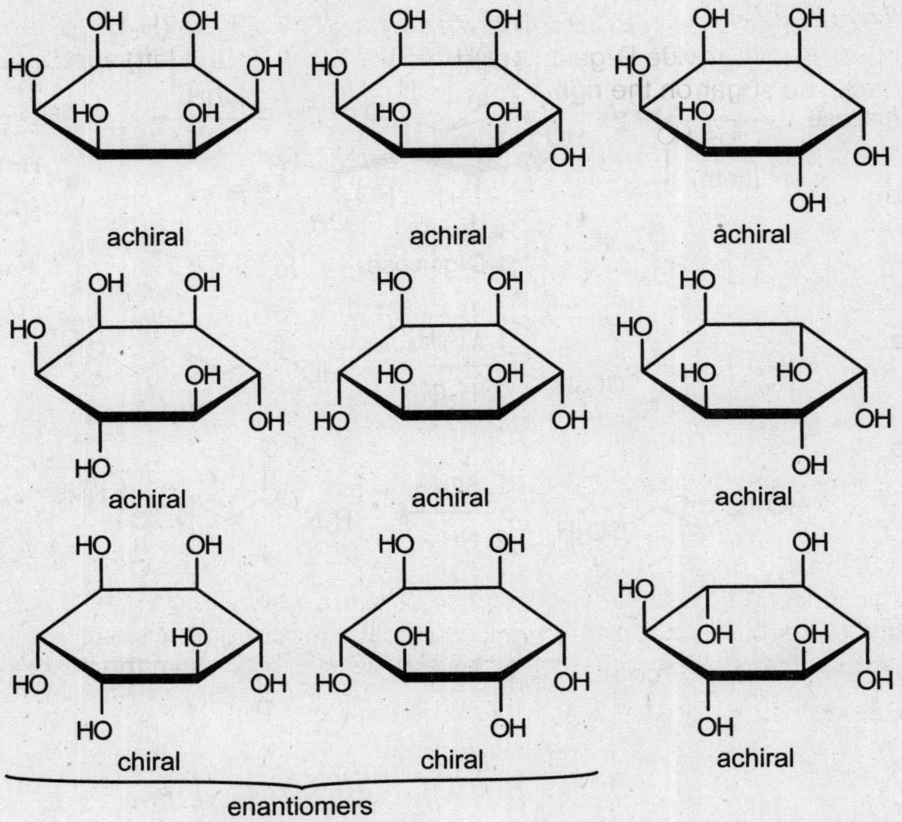

achiral achiral achiral

achiral achiral achiral

chiral chiral achiral

enantiomers

The system to use in writing these formulas is to begin with all hydroxyl group *cis*. Then, we move one hydroxyl to the opposite face, two hydroxyls to the opposite face (1,2; 1,3; 1,4), and finally three hydroxyls to the opposite face (1,2,3; 1,2,4; 1,3,5).

16.53 One representation of aspartame is shown below.

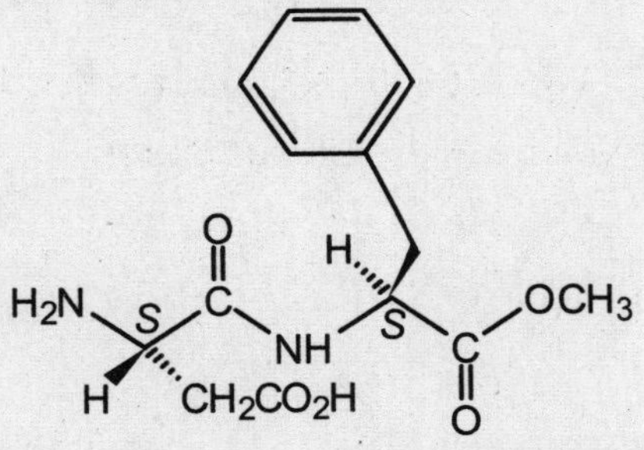

16.54 A reasonable synthesis converts maltose to two independent glucose units by hydrolyzing the glycosidic linkages. Both α and β forms of glucose would be formed, and then reduction of the anomeric mixture by $NaBH_4$ would provide D-glucitol.

maltose $\xrightarrow[\text{H}^+ \text{ (heat)}]{H_2O}$ 2 α-D-glucose $\xrightarrow{NaBH_4}$ 2 D-glucitol

α-D-glucose

D-glucitol

16.55 a. $HO_2C-CH_2CH_2-CO_2H$ $\xrightarrow[\text{2. } H_2O]{\text{1. } LiAlH_4}$ $HO-CH_2CH_2CH_2CH_2-OH$

b. $HO_2C-CH_2CH_2-CO_2H$ $\xrightarrow[\text{2. } NH_3]{\text{1. } SOCl_2}$ $H_2N-C(=O)-CH_2CH_2-C(=O)-NH_2$

c. $HO_2C-CH_2CH_2-CO_2H$ $\xrightarrow[\text{2. } NH_3]{\text{1. } SOCl_2}$ $H_2N-C(=O)-CH_2CH_2-C(=O)-NH_2$ $\xrightarrow[\text{2. } H_2O]{\text{1. } LiAlH_4}$ $H_2N-CH_2CH_2CH_2CH_2-NH_2$

17

Amino Acids, Peptides, and Proteins

Chapter Summary

Proteins are natural polymers composed of α**-amino acids** linked by **amide (peptide) bonds**. Except for **glycine** (aminoacetic acid), protein-derived amino acids are chiral and have the L configuration. Table 17.1 lists the names, one- and three-letter abbreviations, and structures of the 20 common amino acids. Of these, eight (the **essential amino acids**) cannot be synthesized in the bodies of adult humans and must be ingested in food.

Amino acids with one amino and one carboxyl group exist as **dipolar ions**. Amino acids are **amphoteric**: in strong acid, they are protonated and become positively charged (ammonium carboxylic acids); in base, they lose a proton and become negatively charged (amine carboxylates). If placed in an electric field, amino acids migrate toward the cathode (negative electrode) at low pH and toward the anode (positive electrode) at high pH. The intermediate pH at which they do not migrate toward either electrode is the **isoelectric point.** (The isoelectric points for the 20 common amino acids are listed in Table 17.1.) For amino acids with one amino group and one carboxyl group, the isoelectric pH is about 6. Amino acids with two carboxyl groups and one amino group have isoelectric points at a low pH (about 3), whereas amino acids with two amino groups and one carboxyl group have isoelectric points at a high pH (about 9). **Electrophoresis** is a process that makes use of the dependence of charge on pH to separate amino acids and proteins.

Amino acids undergo reactions that correspond to each functional group. In addition to exhibiting both acidic and basic behavior, for example, the carboxyl group can be esterified and the amino group can be acylated.

Amino acids react with **ninhydrin** to give a violet dye. This reaction is useful for detection and quantitative analysis of amino acids.

An amide bond between the carboxyl group of one amino acid and the amino group of another is called a **peptide bond**, which links amino acids to one another in peptides and proteins. The amino acid (often abbreviated **aa**) at one end of the peptide chain will have a free amino group (the **N-terminal amino acid**), and the amino acid at the other end of the chain (the **C-terminal amino acid**) will have a free carboxyl group. By convention, we write these structures from left to right starting with the N-terminal end amino acid.

Another type of covalent bond in proteins is the **disulfide bond,** formed by oxidative coupling of the –SH groups in **cysteine**. **Oxytocin** is an example of a cyclic peptide with a S–S bond.

By the **primary structure** of a peptide or protein, we mean its **amino acid sequence**. Complete hydrolysis gives the amino acid content. The N-terminal amino acid can be identified by the **Sanger method**, using **2,4-dinitrofluorobenzene.** The **Edman degradation** uses **phenyl isothiocyanate** to clip off one amino acid at a time from the N-terminus. Other reagents selectively cleave peptide chains at certain amino acid links. A combination of these methods can "sequence" a protein, and the methods have been automated.

Polypeptides are usually synthesized by the **Merrifield solid-phase technique**. An N-protected amino acid is linked by an ester bond to a benzyl chloride-type unit in a polystyrene. The protecting group is then removed, and the next N-protected amino acid is linked to the polymer-bound one. The cycle is repeated until the peptide is assembled, after which it is detached from the polymer. The most common N-protecting group is the ***t*-butoxycarbonyl (Boc) group:** $(CH_3)_3COC(=O)$–, which is connected to the amino group using di-*t*-butyldicarbonate and can be removed from the amino group using mild acid. The reagent used to attach each N-protected amino acid to the growing peptide chain is **dicyclohexylcarbodiimide**. Detachment of the peptide chain from the polymer is accomplished using HF.

Two features that affect **secondary protein structure** (molecular shape) include the rigid, planar geometry and restricted rotation of the peptide bond, and interchain or intrachain hydrogen bonding of the type C=O···H–N. The α **helix** and the **pleated sheet** are common protein shapes.

Proteins may be **fibrous** or **globular.** The structure and polarity of the particular amino acid R groups and their sequence affect the solubility properties and **tertiary structure** of proteins. **Quaternary structure** refers to the aggregation of similar protein subunits.

Reaction Summary

Dissociation of Amino Acids

$$\mathrm{R{-}CH(NH_3^+){-}CO_2H} \underset{H^+}{\overset{HO^-}{\rightleftharpoons}} \mathrm{R{-}CH(NH_3^+){-}CO_2^-} \underset{H^+}{\overset{HO^-}{\rightleftharpoons}} \mathrm{R{-}CH(NH_2){-}CO_2^-}$$

dipolar ion

Esterification

$$\mathrm{R{-}CH(NH_3^+){-}CO_2^-} + \mathrm{R'OH} + \mathrm{H^+} \longrightarrow \mathrm{R{-}CH(NH_3^+){-}CO_2R'} + \mathrm{H_2O}$$

Acylation

$$\mathrm{R{-}CH(NH_3^+){-}CO_2^-} + \mathrm{R'{-}C(=O){-}Cl} \xrightarrow{2\ HO^-} \mathrm{R'{-}C(=O){-}NH{-}CH(R){-}CO_2^-} + 2\ \mathrm{H_2O} + \mathrm{Cl^-}$$

Ninhydrin Reaction

2 (ninhydrin) + $R-CH(NH_3^+)-CO_2^-$ ⟶ (purple anion) + RCHO + CO_2 + $3H_2O$ + H^+

Sanger's Reagent

(1-fluoro-2,4-dinitrobenzene) + $R-CH(NH_2)-CO_2H$ —base→ $R-CH(NH-C_6H_3(NO_2)_2)-CO_2H$ + F^-

Edman Degradation

$C_6H_5-N{=}C{=}S$ + $H_2N-CHR-C(=O)-NH-$(peptide chain) ⟶

$C_6H_5-NH-C(=S)-NH-CHR-C(=O)-NH-$(peptide chain) —HCl, H_2O→

(phenylthiohydantoin) + H_2N-(peptide chain)

(oval) = peptide chain

Peptide Synthesis

a. N-protection:

$$(CH_3)_3CO-\overset{O}{\overset{\|}{C}}-O-\overset{O}{\overset{\|}{C}}-OC(CH_3)_3 + H_3\overset{+}{N}-\underset{H}{\underset{|}{\overset{R_1}{\overset{|}{C}}}}-CO_2^- \xrightarrow{\text{base}} (CH_3)_3CO-\overset{O}{\overset{\|}{C}}-NH-\underset{H}{\underset{|}{\overset{R_1}{\overset{|}{C}}}}-CO_2^-$$

di-*t*-butyl dicarbonate

BOC or P

b. Polymer attachment:

$$P-NH-\underset{H}{\underset{|}{\overset{R_1}{\overset{|}{C}}}}-CO_2^- + Cl-CH_2-C_6H_4-\text{polymer} \xrightarrow{-Cl^-} P-NH-\underset{H}{\underset{|}{\overset{R_1}{\overset{|}{C}}}}-\overset{O}{\overset{\|}{C}}-O-CH_2-C_6H_4-\text{polymer}$$

= polymer

c. Deprotection (removal of the protecting group):

$$(CH_3)_3CO-\overset{O}{\overset{\|}{C}}-NH-\underset{H}{\underset{|}{\overset{R_1}{\overset{|}{C}}}}-\overset{O}{\overset{\|}{C}}-O-CH_2-C_6H_4-\text{polymer} \xrightarrow[CH_3CO_2H]{HCl} H_2N-\underset{H}{\underset{|}{\overset{R_1}{\overset{|}{C}}}}-\overset{O}{\overset{\|}{C}}-O-CH_2-C_6H_4-\text{polymer} + CO_2 + (CH_3)_2C{=}CH_2$$

d. Amino acid coupling with DDC:

$$\text{(P)}-NH-CH(R_2)-CO_2H + H_2N-CH(R_1)-C(=O)-O-CH_2-C_6H_4-\text{polymer}$$

dicyclohexylcarbodiimide (DCC)

$$\text{(P)}-NH-CH(R_2)-C(=O)-NH-CH(R_1)-C(=O)-O-CH_2-C_6H_4-\text{polymer}$$

+

dicyclohexylurea

e. Detachment from the polymer:

$$\text{(P)}-NH-CH(R_2)-C(=O)-NH-CH(R_1)-C(=O)-O-CH_2-C_6H_4-\text{polymer} \xrightarrow{HF}$$

$$H_2N-CH(R_2)-C(=O)-NH-CH(R_1)-C(=O)-OH + F-CH_2-C_6H_4-\text{polymer}$$

Mechanism Summary

Nucleophilic Aromatic Substitution (Addition-Elimination)

$$C_6H_5X + Nu:^- \longrightarrow [\,\text{Nu, X adduct (carbanion)}\,] \xrightarrow{-X^-} C_6H_5Nu$$

Exemplified by Sanger's reaction (eq. 17.11) and facilitated by electron-withdrawing (carbanion-stabilizing) groups *ortho* or *para* to the leaving group X.

Learning Objectives

1. Know the meaning of: α-amino acid, essential amino acid, dipolar ion, amphoteric, L configuration, isoelectric point (p*I*), electrophoresis, pK_a.
2. Learn the names, structures, and abbreviations of the amino acids listed in Table 17.1.
3. Write the structure of a given amino acid as a function of the pH of the solution.
4. Write an equation for the reaction of an amino acid (dipolar form) with strong acid or strong base.
5. Write the form of a given amino acid that is likely to predominate at high or low pH and at the isoelectric pH.
6. Given the isoelectric points of several amino acids, predict the directions of their migration during electrophoresis at a given pH.
7. Given the isoelectric points of several amino acids, select a pH at which they can be separated by electrophoresis.
8. Write the equation for the reaction of a given amino acid with ninhydrin reagent.
9. Write the equations for the reaction of a given amino acid with (a) a given alcohol and H^+, and (b) acetic anhydride or another activated acyl derivative.
10. Know the meaning of: peptide bond, peptide, dipeptide, tripeptide, and so on; N-terminal and C-terminal amino acid; cysteine unit; disulfide bond.
11. Given the structures of the component amino acids and the name of a di-, tri-, or polypeptide, draw its structure.
12. Identify the N-terminal and C-terminal amino acids for a given peptide.
13. Given the one- or three-letter abbreviated name for a peptide, write its structure.
14. Write the equation for the hydrolysis of a given di- or polypeptide.
15. Know the meaning of: amino acid sequence, Sanger's reagent, Edman degradation, selective peptide cleavage.
16. Write the equation for the reaction of an amino acid with 2,4-dinitrofluorobenzene.
17. Write the equations for the Edman degradation of a given tri- or polypeptide.
18. Given information on selective peptide cleavage and the sequences of fragment peptides, deduce the sequence of the original polypeptide.
19. Know the meaning of: protecting group, solid-phase technique, *t*-butoxycarbonyl (Boc) group, dicyclohexylcarbodiimide (DCC).
20. Write an equation for the protection of a given amino acid using di-*t*-butyl dicarbonate and for its deprotection using acid.
21. Write an equation for the linking of two protected amino acids with dicyclohexylcarbodiimide (DCC).

22. Write an equation for the linking of the N-protected C-terminal amino acid to a polymer for peptide synthesis, and write an equation for detachment of the peptide from the polymer.

23. Describe the geometry of the peptide bond.

24. Know the meaning of: primary, secondary, tertiary, and quaternary protein structure; α helix; pleated sheet; fibrous and globular protein.

ANSWERS TO PROBLEMS

Problems Within the Chapter

17.1 These reactions occur if acid is added to the product of eq. 17.3. That is, they are essentially the reverse of the process shown in eqs. 17.3 and 17.2.

$$H_3C-CH(NH_2)-CO_2^-\ Na^+ + HCl \longrightarrow H_3C-CH(^+NH_3)-CO_2^- + NaCl$$

$$H_3C-CH(^+NH_3)-CO_2^- + HCl \longrightarrow H_3C-CH(^+NH_3)-CO_2H + Cl^-$$

17.2 The $-CO_2H$ group is more acidic than the $-NH_3^+$ group. Thus, on treatment with base, a proton is removed from the $-CO_2H$ group (see eq. 17.2). Then, with a second equivalent of base, a proton is removed from the $-NH_3^+$ group (see eq. 17.3).

17.3 The $-NH_2$ group is more basic than the $-CO_2^-$ group. Thus, on treatment with acid (as in Problem 17.1), a proton adds first to the $-NH_2$ group. Then, with a second equivalent of acid, a proton adds to the $-CO_2^-$ group.

17.4 a. The dipolar ion will not migrate toward either electrode.

$$HOCH_2-CH(\overset{+}{N}H_3)-CO_2^-$$

b. The ion will move toward the cathode (negative electrode).

$$CH_3SCH_2-CH(\overset{+}{N}H_3)-CO_2H$$

c. The ion will migrate toward the anode (positive electrode).

$$HO-C_6H_4-CH_2-CH(NH_2)-CO_2^-$$

17.5 The least acidic group in aspartic acid is the ammonium group, as shown in eq. 17.5. Carboxyl groups are more acidic than ammonium groups.

17.6 The equilibria for arginine are:

$$\underset{\textbf{A}}{H_2N{-}\underset{\underset{NH_2^+}{\|}}{C}{-}NH(CH_2)_3{-}\overset{\overset{+}{N}H_3}{|}{CH}{-}CO_2H} \underset{}{\overset{pK_1 = 2.17}{\rightleftharpoons}} \underset{\textbf{B}}{H_2N{-}\underset{\underset{NH_2^+}{\|}}{C}{-}NH(CH_2)_3{-}\overset{\overset{+}{N}H_3}{|}{CH}{-}CO_2^-}$$

$$\textbf{B} \overset{pK_2 = 9.04}{\rightleftharpoons} \textbf{C}$$

$$\underset{\textbf{D}}{H_2N{-}\underset{\underset{NH}{\|}}{C}{-}NH(CH_2)_3{-}\overset{NH_2}{|}{CH}{-}CO_2^-} \overset{pK_3 = 12.48}{\rightleftharpoons} \underset{\textbf{C}}{H_2N{-}\underset{\underset{NH_2^+}{\|}}{C}{-}NH(CH_2)_3{-}\overset{NH_2}{|}{CH}{-}CO_2^-}$$

The dipolar ion has structure **C** and the isoelectric point (pI) will be the average of the pK_as of **B** and **D**. Specifically,

$$pI = [pK_2 + pK_3] / 2 = [9.04 + 12.48] / 2 = 10.76 \qquad \text{(see Table 17.1)}$$

17.7 a. At pH 7.0, glycine (isoelectric point = 6.0) is present mainly in the form $H_3N^+{-}CH_2{-}CO_2^-$ with some $H_2N{-}CH_2{-}CO_2^-$ and will migrate toward the anode (positive electrode), whereas lysine (isoelectric point = 9.7) is present as:

$$H_3\overset{+}{N}CH_2CH_2CH_2CH_2{-}\underset{\underset{+NH_3}{|}}{\overset{\overset{H}{|}}{C}}{-}CO_2H \quad \text{and} \quad H_3\overset{+}{N}CH_2CH_2CH_2CH_2{-}\underset{\underset{+NH_3}{|}}{\overset{\overset{H}{|}}{C}}{-}CO_2^-$$

and will migrate toward the cathode (negative electrode).

b. At pH 6.0, phenylalanine (isoelectric point = 5.5) will migrate toward the anode. Its structure is

$$C_6H_5{-}CH_2{-}\underset{\underset{NH_2}{|}}{\overset{\overset{H}{|}}{C}}{-}CO_2^-$$

Leucine (isoelectric point = 6.0) will not migrate. Its structure is

$$(CH_3)_2CHCH_2{-}\underset{\underset{+NH_3}{|}}{\overset{\overset{H}{|}}{C}}{-}CO_2^-$$

Proline (isoelectric point = 6.3) will migrate toward the cathode. Its structure is

(structure: pyrrolidine ring with CO_2H and $\overset{+}{N}H_2$)

To generalize, if the solution pH is less than the isoelectric pH, the net charge on the amino acid will be positive and it will migrate toward the cathode. If the solution pH is greater than the isoelectric pH, the net charge on the amino acid will be negative and it will migrate toward the anode.

17.8 a.

$$HO_2CCH_2CH_2-\underset{+NH_3}{\overset{H}{C}}-CO_2^- + 2\ CH_3OH + HCl \longrightarrow$$

$$CH_3O_2CCH_2CH_2-\underset{+NH_3}{\overset{H}{C}}-CO_2CH_3 + 2\ H_2O + Cl^-$$

b.

(proline zwitterion: ring $\overset{H}{C}-CO_2^-$, $\overset{+}{N}H_2$) + $C_6H_5-\overset{O}{\overset{\|}{C}}-Cl$ + 2 HO^- $\longrightarrow$

(N-benzoylproline anion: ring $\overset{H}{C}-CO_2^-$, $N-\overset{O}{\overset{\|}{C}}-C_6H_5$) + 2 H_2O + Cl^-

c.

$$PhCH_2-\underset{+NH_3}{\overset{H}{C}}-CO_2^- + H_3C-\overset{O}{\overset{\|}{C}}-O-\overset{O}{\overset{\|}{C}}-CH_3 \longrightarrow$$

Ph = phenyl group

$$H_3C-\overset{O}{\overset{\|}{C}}-NH\overset{CH_2Ph}{CH}CO_2H + CH_3CO_2H$$

17.9 Rewrite eq. 17.9, with R = CH_3.

17.10 Both Gly–Ala and Ala–Gly have one acidic and one basic group. Using Table 17.2 as a guide, the expected p*I* (the pH at which the dipolar forms shown in Example 17.5 will predominate for these dipeptides) is approximately 6. Gly–Ala is expected to be positively charged at pH 3.

$$H_3\overset{+}{N}CH_2-\overset{O}{\overset{\|}{C}}-NH-\underset{CH_3}{CH}-CO_2H \longleftarrow pK_a \cong 2.3$$

(arrow to $H_3\overset{+}{N}$: $pK_a \cong 9.4$)

and predominantly negatively charged at pH 9.

$$H_2NCH_2-\overset{O}{\overset{\|}{C}}-NH-\underset{CH_3}{\underset{|}{\overset{H}{\overset{|}{C}}}}-CO_2^-$$

17.11 a.

$$H_3\overset{+}{N}-\underset{\underset{\underset{H_3C\quad CH_3}{CH}}{|}}{\underset{CH_2}{\underset{|}{\overset{H}{\overset{|}{C}}}}}-\overset{O}{\overset{\|}{C}}-\overset{H}{N}-\underset{H}{\underset{|}{\overset{CH_3}{\overset{|}{C}}}}-CO_2^-$$

b.

$$H_3\overset{+}{N}-\underset{CH_3}{\underset{|}{\overset{H}{\overset{|}{C}}}}-\overset{O}{\overset{\|}{C}}-\overset{H}{N}-\underset{H}{\underset{|}{\overset{\overset{CH_2CH_3}{|}}{\overset{HC-CH_3}{\overset{|}{C}}}}}-CO_2^-$$

17.12

$$H_2N-CH_2-\overset{O}{\overset{\|}{C}}-NH-\underset{CH_3}{\underset{|}{\overset{H}{\overset{|}{C}}}}-\overset{O}{\overset{\|}{C}}-NH-\underset{CH_2OH}{\underset{|}{\overset{H}{\overset{|}{C}}}}-CO_2H$$

Note that, by convention, in writing peptide structures, each peptide bond is written in the following direction:

$$-\overset{O}{\overset{\|}{C}}-NH-$$

17.13 Leu–Met–Ala (LMA) Ala–Leu–Met (ALM) Ala–Met–Leu (AML)
Met–Leu–Ala (MLA) Met–Ala–Leu (MAL)

There are six isomeric structures altogether, including Leu–Ala–Met (LAM).

17.14 The products will be glycine, alanine, and serine:

$$H_2N-\underset{H}{\underset{|}{\overset{CO_2H}{\overset{|}{C}}}}-H \qquad H_2N-\underset{CH_3}{\underset{|}{\overset{CO_2H}{\overset{|}{C}}}}-H \qquad H_2N-\underset{CH_2OH}{\underset{|}{\overset{CO_2H}{\overset{|}{C}}}}-H$$

Gly (G) Ala (A) Ser (S)

17.15 For alanylglycine, the reactions are as follows:

$$CH_3-\overset{H}{\underset{NH_2}{C}}-\overset{O}{\overset{\|}{C}}-NHCH_2CO_2H + \text{(1-fluoro-2,4-dinitrobenzene: F, }NO_2\text{, }NO_2) \xrightarrow{HO^-}$$

$$CH_3-\overset{H}{\underset{NH-C_6H_3(NO_2)_2}{C}}-\overset{O}{\overset{\|}{C}}-NHCH_2CO_2H \xrightarrow[H^+]{H_2O} CH_3-\overset{H}{\underset{NH-C_6H_3(NO_2)_2}{C}}-\overset{O}{\overset{\|}{C}}-OH + H_2NCH_2CO_2H$$

(glycine)

(DNP-alanine)

For glycylalanine, we have the following:

$$H_2\underset{NH_2}{C}-\overset{O}{\overset{\|}{C}}-NH-\overset{CH_3}{CH}-CO_2H + \text{(1-fluoro-2,4-dinitrobenzene: F, }NO_2\text{, }NO_2) \xrightarrow{HO^-}$$

$$H_2\underset{NH-C_6H_3(NO_2)_2}{C}-\overset{O}{\overset{\|}{C}}-NH-\overset{CH_3}{CH}-CO_2H \xrightarrow[H^+]{H_2O} H_2\underset{NH-C_6H_3(NO_2)_2}{C}-\overset{O}{\overset{\|}{C}}-OH + H_2NCH(CH_3)CO_2H$$

(DNP-glycine) (alanine)

17.16 The first cycle of the Edman degradation of Phe–Ala–Ser will give the phenylthiohydantoin of Phe because it is the N-terminal amino acid.

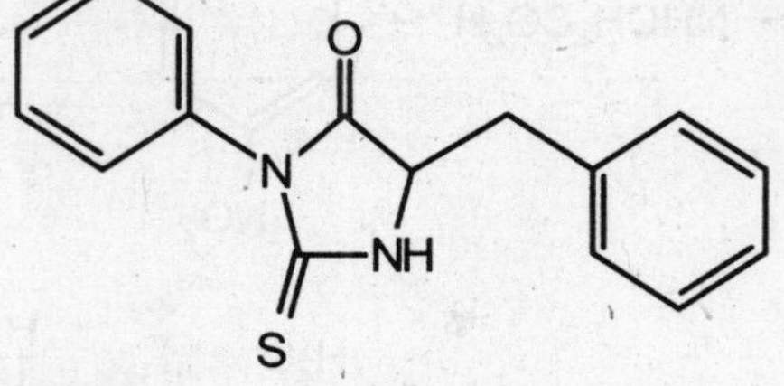

17.17 a. Trypsin cleaves peptides on the *carboxyl* side of lysine and arginine. Bradykinin contains no lysines and two arginines. Cleavage will occur after the first arginine (carboxyl side) but *not* in front of the second arginine (amino side). The cleavage products will be

Arg and Pro–Pro–Gly–Phe–Ser–Pro–Phe–Arg

b. Chymotrypsin will cleave bradykinin on the *carboxyl* sides of Phe to give three products:

Arg–Pro–Pro–Gly–Phe and Ser–Pro–Phe and Arg

17.18 Reduction of the disulfide bonds will chemically separate the A and B chains (see eq. 17.10).

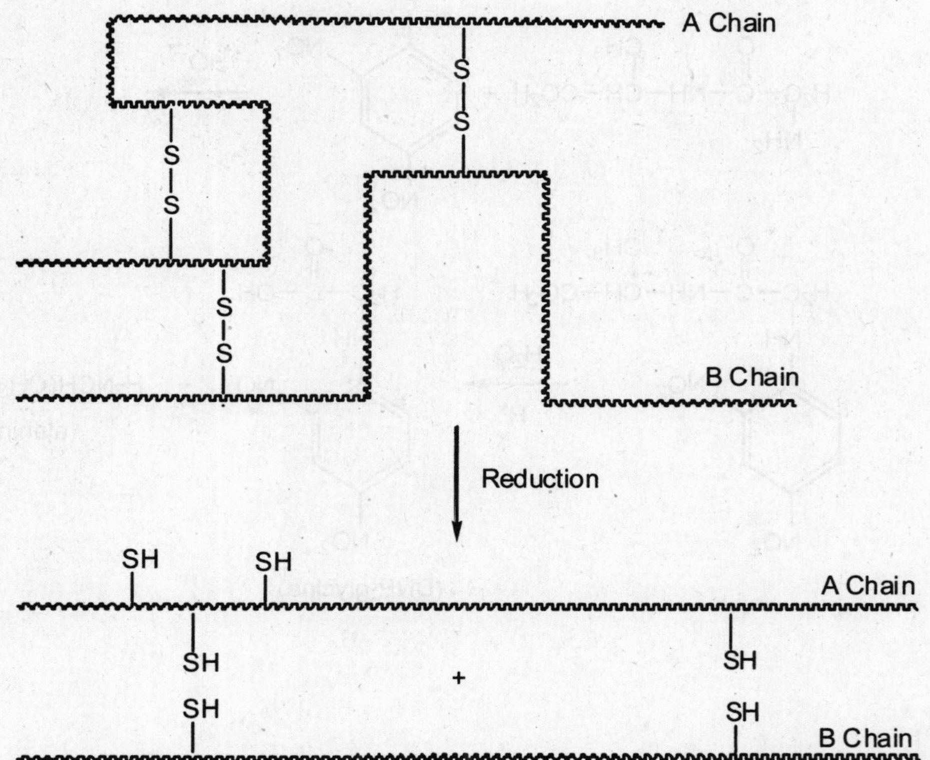

The two chains could then be physically separated using a chromatographic technique—for example, electrophoresis (Figure 17.3).

17.19 First protect the amino group of the C-terminal amino acid (Phe) as described in eq. 17.15:

$$H_3\overset{+}{N}-CH(CH_2C_6H_5)-C(=O)-O^- \xrightarrow[\text{base}]{[(CH_3)_3CO]_2C=O} \text{(P)}-NH-CH(CH_2C_6H_5)-C(=O)-OH$$

$$\text{(P)} = -C(=O)-OC(CH_3)_3$$

Next attach the N-protected phenylalanine to the chloromethylated polystyrene:

$$\text{(polystyrene)}-C_6H_4-CH_2Cl + \text{(P)}-NH-CH(CH_2C_6H_5)-C(=O)-O^- \longrightarrow$$

$$\text{(P)}-NH-CH(CH_2C_6H_5)-C(=O)-O-CH_2-C_6H_4-\text{(polystyrene)}$$

polystyrene backbone

Then deprotect the amino group of the Phe:

$$\text{(P)}-NH-CH(CH_2C_6H_5)-C(=O)-O-CH_2-C_6H_4-\text{(polystyrene)} \xrightarrow[CH_2Cl_2]{CF_3CO_2H}$$

$$H_2N-CH(CH_2C_6H_5)-C(=O)-O-CH_2-C_6H_4-\text{(polystyrene)}$$

Next couple the polymer bound phenylalanine with N-protected alanine (Ala):

$$\text{(P)}-NH-CH(CH_3)-C(=O)-OH + H_2N-CH(CH_2C_6H_5)-C(=O)-O-CH_2-C_6H_4-\text{(polystyrene)}$$

$$\downarrow \text{DCC}$$

$$\text{(P)}-NH-CH(CH_3)-C(=O)-NH-CH(CH_2C_6H_5)-C(=O)-O-CH_2-C_6H_4-\text{(polystyrene)}$$

Then sequentially deblock the Ala, couple with N-blocked glycine (Gly), and deblock the Gly:

$$(\text{polymer})-NH-\underset{CH_3}{CH}-\overset{O}{\overset{\|}{C}}-NH-\underset{CH_2C_6H_5}{CH}-\overset{O}{\overset{\|}{C}}-O-CH_2-C_6H_4-(\text{polymer})$$

1. CF_3CO_2H, CH_2Cl_2
2. (polymer)$-NH-CH_2CO_2H$, (DCC)
3. CF_3CO_2H, CH_2Cl_2

$$H_2N-CH_2-\overset{O}{\overset{\|}{C}}-NH-\underset{CH_3}{CH}-\overset{O}{\overset{\|}{C}}-NH-\underset{CH_2C_6H_5}{CH}-\overset{O}{\overset{\|}{C}}-O-CH_2-C_6H_4-(\text{polymer})$$

Then remove the peptide from the polymer support:

$$H_2N-CH_2-\overset{O}{\overset{\|}{C}}-NH-\underset{CH_3}{CH}-\overset{O}{\overset{\|}{C}}-NH-\underset{CH_2C_6H_5}{CH}-\overset{O}{\overset{\|}{C}}-O-CH_2-C_6H_4-(\text{polymer})$$

HF

$$H_2N-CH_2-\overset{O}{\overset{\|}{C}}-NH-\underset{CH_3}{CH}-\overset{O}{\overset{\|}{C}}-NH-\underset{CH_2C_6H_5}{CH}-\overset{O}{\overset{\|}{C}}-OH$$

(H_2N—Gly—Ala—Phe—CO_2H)

Notice that the peptide is built up from the C-terminal amino acid toward the N-terminal amino acid.

17.20 Hydrocarbon-like side chains are nonpolar:

Gly, Ala, Val, Leu, Ile, Phe, and even Met (a thioether)

Alcohol, thiol, amine, amide, carboxylic acid, guanidine, and imidazole side chains are polar:

Ser, Thr, Cys, Pro, Asp, Glu, Asn, Gln, Lys, Arg, Trp, Tyr and His

Aromatic side chains are relatively flat:

Phe, Tyr, Trp, His

ADDITIONAL PROBLEMS

17.21 Definitions and/or examples are given in the text, where indicated.

a.	Sec. 17.7	b.	Sec. 17.2
c.	Sec. 17.7	d.	Sec. 17.1
e.	Sec. 17.1	f.	Table 17.1, entries 1–5, 9, 10
g.	Table 17.1, entries 6–8, 11–12	h.	Sec. 17.2
i.	Sec. 17.2	j.	Sec. 17.6

17.22 The configuration is *S*.

CO_2H / H_2N—+—H / $CH_2CH(CH_3)_2$ ≡ CO_2H / H_2N▶—◀H / $CH_2CH(CH_3)_2$ or $(CH_3)_2CHCH_2$''''C(CO_2H)(H) / H_2N

L-leucine

priority order: $NH_2 > CO_2H > CH_2CH(CH_3)_2 > H$

17.23 Use Figure 17.1 and Table 17.1 as guides.

a. CO_2H / $H_2N—C—H$ / CH_2 / (benzene ring) / OH

b. CO_2H / $H_2N—C—H$ / $HC—CH_3$ / CH_2CH_3

17.24 a.

$$(CH_3)_2CH—\underset{\overset{+}{NH_3}}{\overset{H}{C}}—CO_2^- + HCl \longrightarrow (CH_3)_2CH—\underset{\overset{+}{NH_3}}{\overset{H}{C}}—CO_2H + Cl^-$$

b.

$$(CH_3)_2CH—\underset{\overset{+}{NH_3}}{\overset{H}{C}}—CO_2^- + NaOH \longrightarrow (CH_3)_2CH—\underset{NH_2}{\overset{H}{C}}—CO_2^-\ Na^+ + H_2O$$

17.25 a.

$$CH_3\underset{HO}{CH}—\underset{\overset{+}{NH_3}}{\overset{H}{C}}—CO_2^-$$

b.

$$HSCH_2—\underset{\overset{+}{NH_3}}{\overset{H}{C}}—CO_2^-$$

c. (pyrrolidine ring) $\overset{H}{C}—CO_2^-$ / $\overset{+}{N}H_2$

d.

$$C_6H_5—CH_2—\underset{\overset{+}{NH_3}}{\overset{H}{C}}—CO_2^-$$

17.26 a. The most acidic proton is on the carboxyl group nearest the ammonium ion. The product of deprotonation is

$$HO_2CCH_2CH_2-\underset{+NH_3}{\overset{H}{\underset{|}{\overset{|}{C}}}}-CO_2^-$$

b. The ammonium ion is more acidic than the alcohol function. The product of deprotonation is

$$HOCH_2-\underset{NH_2}{\overset{H}{\underset{|}{\overset{|}{C}}}}-CO_2^-$$

c. The carboxyl group is a stronger acid than the ammonium group. The product of deprotonation is

$$(CH_3)_2CH-\underset{+NH_3}{\overset{H}{\underset{|}{\overset{|}{C}}}}-CO_2^-$$

d. The proton on the positive nitrogen is the only appreciably acidic proton. The product of deprotonation is

$$HN=\underset{NH_2}{\underset{|}{C}}-NH(CH_2)_3-\underset{NH_2}{\overset{H}{\underset{|}{\overset{|}{C}}}}-CO_2^-$$

17.27 a. The most basic group is CO_2^-. The product of protonation is

$$H_3C-\underset{HO}{\underset{|}{CH}}-\underset{+NH_3}{\overset{H}{\underset{|}{\overset{|}{C}}}}-CO_2H$$

b. The carboxylate ion remote from the ammonium group is most basic. The product of protonation is

$$HO_2CCH_2-\underset{+NH_3}{\overset{H}{\underset{|}{\overset{|}{C}}}}-CO_2^-$$

17.28 Because of its positive charge, the ammonium group is an electron-attracting substituent. As such, it enhances the acidity of the nearby carboxyl group (in protonated alanine).

17.29

$$HO_2CCH_2CH_2-\underset{}{CH}(\overset{+}{N}H_3)-CO_2H \underset{}{\overset{pK_1 = 2.19}{\rightleftharpoons}} HO_2CCH_2CH_2-CH(\overset{+}{N}H_3)-CO_2^-$$

$$\overset{pK_2 = 4.25}{\rightleftharpoons}$$

$$^-O_2CCH_2CH_2-CH(NH_2)-CO_2^- \overset{pK_3 = 9.67}{\rightleftharpoons} {}^-O_2CCH_2CH_2-CH(\overset{+}{N}H_3)-CO_2^-$$

17.30

$$H_2N-C(=NH)-NH(CH_2)_3-CH(NH_2)-CO_2^- \overset{pK_3 = 12.48}{\rightleftharpoons} H_2N-C(=\overset{+}{N}H_2)-NH(CH_2)_3-CH(NH_2)-CO_2^-$$

$$\overset{pK_2 = 9.04}{\rightleftharpoons}$$

$$H_2N-C(=\overset{+}{N}H_2)-NH(CH_2)_3-CH(\overset{+}{N}H_3)-CO_2H \overset{pK_1 = 2.17}{\rightleftharpoons} H_2N-C(=\overset{+}{N}H_2)-NH(CH_2)_3-CH(\overset{+}{N}H_3)-CO_2^-$$

The order of basicities (and therefore the sequence of protonation of groups) is guanidine > $-NH_2$ > $-CO_2^-$.

17.31 In strong acid (pH 1), histidine has two resonance contributors:

(Structures: imidazolium ring with $\overset{+}{N}$–H and N–H, side chain $-CH_2-C(H)(+NH_3)-CO_2H$ $\longleftrightarrow$ resonance form with positive charge on the other ring nitrogen, $-CH_2-C(H)(+NH_3)-CO_2H$)

17.32 When an amino acid is in aqueous solution at a pH greater than its isoelectric point (pI), it will be negatively charged and will migrate toward the anode (positive electrode) of an electrophoresis apparatus. If the pH is less than the pI, the amino acid will migrate toward the cathode (negative electrode). The p*I* of Gln, His, and Glu are 5.7, 7.6, and 3.2, respectively (consult Table 17.1). Therefore, at pH 6, His will migrate toward the cathode, while Gln and Glu will migrate toward the anode. Glu will migrate faster than Gln because it will be more negatively charged at pH 6.

17.33 a. Use eq. 17.7 with R = $CH_2CH(CH_3)_2$ and R' = CH_3CH_2–.

$$(CH_3)_2CHCH_2-\underset{NH_2}{\overset{H}{C}}-CO_2H + CH_3CH_2OH \xrightarrow{H^+} (CH_3)_2CHCH_2-\underset{NH_2}{\overset{H}{C}}-CO_2CH_2CH_3 + H_2O$$

b. Use eq. 17.8 with R = $CH_2CH(CH_3)_2$ and R' = C_6H_5.

$$(CH_3)_2CHCH_2-\underset{NH_2}{\overset{H}{C}}-CO_2H + C_6H_5COCl \xrightarrow{\text{base}} (CH_3)_2CHCH_2-\underset{NH-C(=O)C_6H_5}{\overset{H}{C}}-CO_2H + H_2O + Cl^-$$

c. Use eq. 17.15 with R = $CH_2CH(CH_3)_2$ and substitute $(CH_3)_3CO$ with CH_3.

$$(CH_3)_2CHCH_2-\underset{NH_2}{\overset{H}{C}}-CO_2H + (CH_3CO)_2O \longrightarrow (CH_3)_2CHCH_2-\underset{NH-C(=O)CH_3}{\overset{H}{C}}-CO_2H + CH_3CO_2H$$

17.34 a. The amino group and the hydroxyl group react with acetic anhydride.

$$HOCH_2-\underset{NH_2}{\overset{H}{C}}-CO_2H + 2\,(CH_3CO)_2O \longrightarrow$$

$$\underset{\text{ester}}{H_3C-\overset{O}{\overset{\|}{C}}-O}-CH_2-\underset{\underset{\text{amide}}{NH-\overset{}{C}(=O)-CH_3}}{\overset{H}{C}}-CO_2H + 2\,CH_3CO_2H$$

b. Both the hydroxyl and amino groups react with benzoyl chloride.

$$H_3C-\underset{HO}{CH}-\underset{NH_2}{\overset{H}{C}}-CO_2H + 2\,C_6H_5-C(=O)-Cl \longrightarrow H_3C-\underset{O-C(=O)-C_6H_5}{CH}-\underset{NH-C(=O)-C_6H_5}{\overset{H}{C}}-CO_2H + 2HCl$$

c. Both carboxyl groups are esterified.

$$HO_2CCH_2CH_2-\underset{NH_2}{\overset{H}{C}}-CO_2H + 2\ CH_3OH \xrightarrow{H^+} CH_3O_2CCH_2CH_2-\underset{NH_2}{\overset{H}{C}}-CO_2CH_3 + +2\ H_2O$$

d. Follow eq. 17.15

$$(CH_3)_2CH-\underset{NH_2}{\overset{H}{C}}-CO_2H \xrightarrow{[(CH_3)_3CO]_2C=O} (CH_3)_2CH-\underset{NHCOC(CH_3)_3}{\overset{H}{C}}-CO_2H + (CH_3)_3COH$$

(in the product, the carbonyl C of NHCOC(CH₃)₃ bears =O)

17.35 Follow eq. 17.9, with R = $HOCH_2$–.

17.36 Review Sec. 17.7 if necessary, and use Table 17.1 for the structures of the amino acids. Write the structures in neutral form, recognizing that dipolar ion structures are possible and that the exact form and degree of ionization depend on the pH of the solution.

a. $H_2N-CH(CH_2OH)-C(=O)-NH-CH(CH_3)-CO_2H$

b. $H_2N-CH(CH(CH_3)_2)-C(=O)-NH-CH(CH_2C_6H_5)-CO_2H$

c. $H_2N-CH(CH_2\text{-indol-3-yl})-C(=O)-NH-CH(CH(CH_3)CH_2CH_3)-CO_2H$

d. $H_2N-CH_2-C(=O)-NH-CH(CH_3)-C(=O)-NH-CH(CH_2OH)-CO_2H$

17.37 a.

$$H_2N-CH(CH_2CH(CH_3)_2)-C(=O)-NH-CH(CH(OH)CH_3)-C(=O)-OH \xrightarrow{H^+} H_2N-CH(CH_2CH(CH_3)_2)-C(=O)-OH + H_2N-CH(CH(OH)CH_3)-C(=O)-OH$$

b.

$$H_2N-CH(CH_2OH)-C(=O)-NH-CH(CH_2CH_2SCH_3)-C(=O)-OH \xrightarrow{H^+} H_2N-CH(CH_2OH)-C(=O)-OH + H_2N-CH(CH_2CH_2SCH_3)-C(=O)-OH$$

c.

$$H_2N-CH(CH_2SH)-C(=O)-NH-CH(CH_2C_6H_4OH)-C(=O)-NH-CH(CH_2CH_2SCH_3)-CO_2H \xrightarrow{H^+} H_2N-CH(CH_2SH)-COOH + H_2N-CH(CH_2C_6H_4OH)-COOH + H_2N-CH(CH_2CH_2SCH_3)-CO_2H$$

17.38

$$H_2N-CH(CH_2CO_2H)-C(=O)-NH-CH(CH_2C_6H_5)-C(=O)-OCH_3 \xrightarrow[\text{heat}]{H_2O,\ H^+} H_2N-CH(CH_2CO_2H)-C(=O)-OH + H_2N-CH(CH_2C_6H_5)-C(=O)-OH + CH_3OH$$

Asp Phe

17.39 The behavior of a simple dipeptide like this, with no acidic or basic functions in the R group of each unit, is very much like that of a simple amino acid.

$$H_3\overset{+}{N}-CH(CH(CH_3)_2)-CO-NH-CH_2-CO-OH \xrightarrow{HO^-} H_3\overset{+}{N}-CH(CH(CH_3)_2)-CO-NH-CH_2-CO-O^-$$

most acidic (pH 1) about pH 6 (dipolar ion)

$$\xrightarrow{HO^-} H_2N-CH(CH(CH_3)_2)-CO-NH-CH_2-CO-O^-$$

most basic (pH 10)

17.40 With four different amino acids there are 4!, or 4 x 3 x 2 x 1 = 24, possible structures. They are as follows:

CASL	ACSL	SCAL	LCAS
CALS	ACLS	SCLA	LCSA
CSAL	ASCL	SACL	LACS
CSLA	ASLC	SALC	LASC
CLSA	ALCS	SLCA	LSCA
CLAS	ALSC	SLAC	LSAC

17.41 Hydrogen peroxide oxidizes thiols to disulfides (Sec. 7.17). See also eq. 17.10.

$$2\; H_2N-CH(CH_3)-CO-NH-CH(CH_2SH)-CO-OH \xrightarrow{H_2O_2} [H_2N-CH(CH_3)-CO-NH-CH(CO-OH)-CH_2-S-]_2$$

17.42 a. Use eq. 17.11, with R_1 = H and ... = OH:

$$\text{2,4-}(NO_2)_2C_6H_3F + H_2NCH_2CO_2H \xrightarrow[\text{base}]{\text{mild}} \text{2,4-}(NO_2)_2C_6H_3-NHCH_2CO_2H$$

b. Lysine contains two primary amino groups. Each group can react with 2,4-dinitrofluorobenzene:

$$H_2NCH_2CH_2CH_2CH_2-\underset{NH_2}{\overset{H}{C}}-\overset{O}{\overset{\|}{C}}-OH + 2\ \text{(2,4-dinitrofluorobenzene)} \xrightarrow[\text{base}]{\text{mild}}$$

$$O_2N-C_6H_3(NO_2)-NHCH_2CH_2CH_2CH_2-\underset{NH-C_6H_3(NO_2)_2}{\overset{H}{C}}-\overset{O}{\overset{\|}{C}}-OH$$

17.43 a. There are 11 amino acid units in cyclosporin A.

b. There are three leucine units:

H_2N–CH(CO_2H)–$CH_2CH(CH_3)_2$ leucine

c. The two unusual amino acids follow one another in the cyclic peptide sequence. They are:

H_2N–CH(CO_2H)–CH(OH)–CH(CH_3)–CH_2–CH=CH–CH_3 and H_2N–CH(CO_2H)–CH_2CH_3

d. There are two alanines, two glycines and two valines.

H_2N–CH(R)–CO_2H R = CH_3 (alanine)
R = H (glycine)
R = $CH(CH_3)_2$ (valine)

17.44 Use the one-letter abbreviations in Table 17.1. The one-letter representation of substance P is RPKPQQFFGLM–NH_2. The –NH_2 indicates that the C-terminal amino acid is present as the primary amide.

17.45 The first result tells us that the N-terminal amino acid is methionine. The structure at this point is

Met (2 Met, Ser, Gly)

From dipeptide D, we learn that one sequence is

Ser–Met

From dipeptide C, we learn that one sequence is

Met–Met

From tripeptide B, we learn that one sequence is

Met (Met, Ser)

And, in view of the result from dipeptide D, tripeptide B must be

Met–Ser–Met

Since two methionines must be adjacent (dipeptide C), the possibilities are

Met–Met–Ser–Met–Gly and Met–Ser–Met–Met–Gly

Tripeptide A allows a decision since two methionines and one glycine must be joined together. Therefore, the correct structure is

Met–Ser–Met–Met–Gly

17.46 Follow the scheme in Figure 17.5, with $R_1 = CH_3$, $R_2 = H$, and $R_3 = (CH_3)_2CH$, and with an –OH attached to the carbonyl group at the right. The final products are the phenylthiohydantoin of alanine and the dipeptide glycylvaline.

17.47 Use the one-letter abbreviations in Table 17.1:

FVNQHLCGSHLVDALYLVCGDRGFFYTPKA

The space-saving value of the one-letter abbreviations is obvious!

17.48 When insulin is subjected to the Edman degradation, the N-terminal amino acids of the A and B chains (glycine and phenylalanine, respectively) will both be converted to their phenylthiohydantoins. Their structures are:

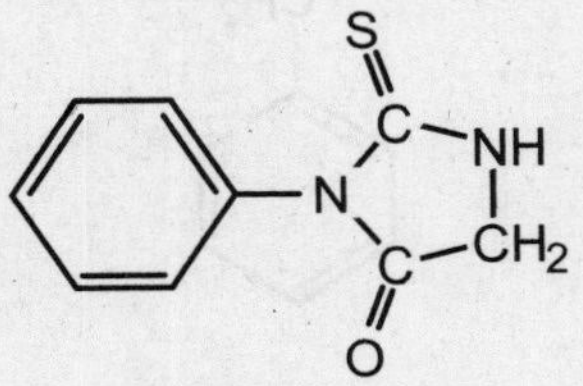

(from Gly)

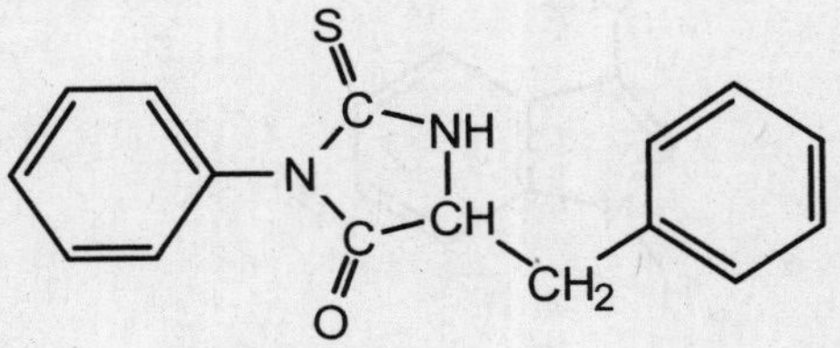

(from Phe)

17.49 The hydrolysis products must overlap as shown below:

Ala–Gly
Gly–Val
Gly–Val–Tyr
Val–Tyr–Cys
Tyr–Cys–Phe
Cys–Phe–Leu
Phe–Leu–Try

The peptide must be Ala–Gly–Val–Tyr–Cys–Phe–Leu–Try, an octapeptide. The

N-terminal amino acid is alanine, the C-terminal one is tryptophan, and the name is alanylglycylvalyltyrosylcystylphenylalanylleucyltryptophan.

17.50 The linear sequence of vasopressin is:

(N terminus) Cys–Tyr–Phe–Gln–Asn–Cys–Pro–Arg–Gly (C terminus)

Each of the cysteine amino acid residues has the $-CH_2SH$ moiety as the R group.

a. In the presence of a mild oxidizing agent (H_2O_2), a disulfide linkage will form between both cysteine residues.

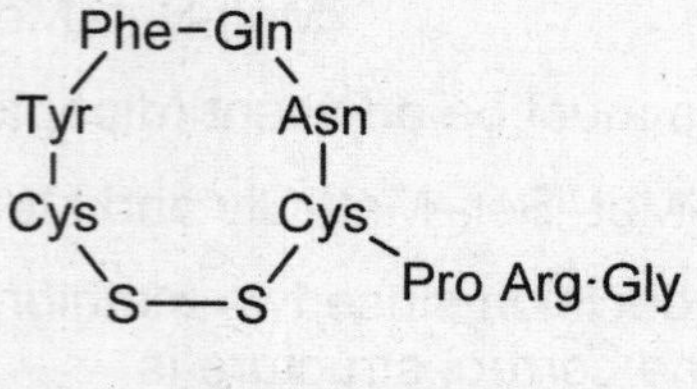

```
      Phe−Gln
     /       \
   Tyr       Asn
    |         |
   Cys       Cys
     \       /   \
      S —— S      Pro Arg·Gly
```

b. The linear form of this short peptide would be less compact than the cyclic form. Also, relative to the open chain form, the ring formed by the disulfide version would present the various R groups on the other amino acids in unique ways to other proteins for regulating biological processes.

17.51 The point of this problem is to illustrate that it is deceptively easy to use the three-letter abbreviations in designating peptide structures. Although Try–Gly–Gly–Phe–Met appears quite simple, the full structure is not. With a molecular formula of $C_{29}H_{36}O_6N_6S$, this pentapeptide has a molecular weight of 596 and is quite a complex polyfunctional molecule.

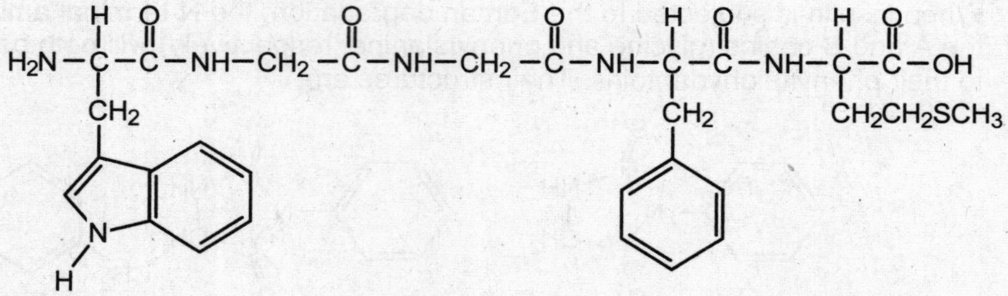

$H_2N-CH(CH_2\text{-indolyl})-C(=O)-NH-CH_2-C(=O)-NH-CH_2-C(=O)-NH-CH(CH_2C_6H_5)-C(=O)-NH-CH(CH_2CH_2SCH_3)-C(=O)-OH$

17.52 The complete hydrolysis tells us what eight amino acids are in angiotensin II. Reaction with Sanger's reagent indicates that Asp is the N-terminal amino acid. Partial digestion with carboxypeptidase indicates that Phe is the C-terminal amino acid. We can now write the following partial structure:

Asp–(Arg, His, Ile, Pro, Tyr, Val)–Phe

Trypsin cleaves peptides at the carboxyl side of Arg and chymotrypsin cleaves peptides at the carboxyl side of Tyr and Phe. Since we already know that Phe is the C-terminal amino acid, the chymotrypsin treatment tells us that Tyr is the fourth amino acid from the N-terminus, and we can write the following partial structure:

Asp–(Arg, Val)–Tyr–Ile–His–Pro–Phe

The trypsin treatment tells us that Arg is the second amino acid from the N-terminus. Therefore, angiotensin II is:

Asp–Arg–Val–Tyr–Ile–His–Pro–Phe

17.53 Trypsin cleaves peptides at the carboxyl side of Lys and Arg. The only fragment that does not end in Lys is the dipeptide. Therefore, it must come at the C-terminus of endorphin, and the C-terminal amino acid is Gln.

Cyanogen bromide cleaves peptides at the carboxyl side of Met. There is only one Met in the structure (see the fifth trypsin fragment), and the cyanogen bromide cleavage gives a hexapeptide consisting of the first six amino acids of this fragment. The N-terminal amino acid must be Tyr. We now know the first ten and the last two amino acids of β-endorphin.

Chymotrypsin cleaves peptides at the carboxyl side of Phe, Tyr, and Trp. Working backward in the 15-unit fragment, we see that it starts at the ninth amino acid in the last fragment from the trypsin digestion. We see that the first amino acid in this fragment (Ser) must be connected to the Lys at the end of the fifth fragment from the trypsin cleavage. Since this fragment begins with the N-terminal amino acid, we now know the sequence of the first twenty and the last two amino acids. The partial structure of β-endorphin must be

Tyr–Gly–Gly–Phe–Leu–Met–Thr–Ser–Glu–Lys–Ser–Gln–Thr–Pro–Leu–Val

Thr–Leu–Phe–Lys–(Lys)(Asn–Ala–His–Lys)(Asn–Ala–Ile–Val–Lys)–Gly–Gln

We cannot deduce, from the data given, the sequence of the amino acids in parentheses. To put them in order, we must seek peptides from partial hydrolysis that contain overlapping sequences of these amino acids.

17.54 Use Table 17.3 as a guide.

a. With trypsin, we expect four fragments:

HSEGTFTSDYSK
YLDSR
R
AQDFVQWLMNT

b. With chymotrypsin, we expect six fragments:

HSEGTF
TSDY
SKY
LDSRRAQDF
VQW
LMNT

17.55 The nucleophile is the carboxylate anion; the leaving group is chloride ion from the benzyl chloride. Benzyl halides react rapidly in S_N2 displacements.

$$\text{(protecting group)}-NH-CH(R_1)-C(=O)-O^- + ClCH_2-C_6H_4-\text{(polystyrene backbone)} \xrightarrow{S_N2} \text{(protecting group)}-NH-CH(R_1)-C(=O)-O-CH_2-C_6H_4-\text{(polystyrene)}$$

protecting group

polystyrene backbone

17.56 The protonated peptide is the leaving group. The nucleophile is fluoride ion.

$$H_2N\text{—[peptide chain]—}C(=O)\text{—}O\text{—}CH_2\text{—}C_6H_4\text{—[polystyrene backbone]} \xrightarrow{H^+}$$

peptide chain

polystyrene backbone

$$H_2N\text{—[peptide]—}C(=\overset{+}{O}H)\text{—}O\text{—}CH_2\text{—}C_6H_4\text{—[polymer]} \;\; F^- \longrightarrow$$

$$H_2N\text{—[peptide]—}C(OH)\text{=}O \; + \; F\text{—}CH_2\text{—}C_6H_4\text{—[polymer]}$$

17.57 First, protect the C-terminal amino acid (proline):

$$\text{Pro (H—N—CH—C(=O)—OH)} \xrightarrow[\text{base (see eq. 17.15)}]{[(CH_3)_3CO]_2C=O} \text{(protecting group)—N—CH—C(=O)—OH}$$

(protecting group) $= (CH_3)_3CO\text{—}C(=O)\text{—}$

Then attach it to the polymer:

$$\text{(protecting group)—Pro—C(=O)—OH} + ClCH_2\text{—}C_6H_4\text{—[polymer]} \xrightarrow[\text{(see Fig. 17.7)}]{\text{base }(-Cl^-)}$$

polystyrene backbone

$$\text{(protecting group)—N—CH—C(=O)—O—}CH_2\text{—}C_6H_4\text{—[polymer]}$$

Next, deprotect the attached proline:

$$\text{(protecting group)—N—CH—C(=O)—O—}CH_2\text{—}C_6H_4\text{—[polymer]} \xrightarrow[\text{(see eq. 17.16)}]{CF_3CO_2H,\, CH_2Cl_2}$$

$$\text{H—N—CH—C(=O)—O—}CH_2\text{—}C_6H_4\text{—[polymer]}$$

Couple protected leucine to the deprotected, polymer-bound proline:

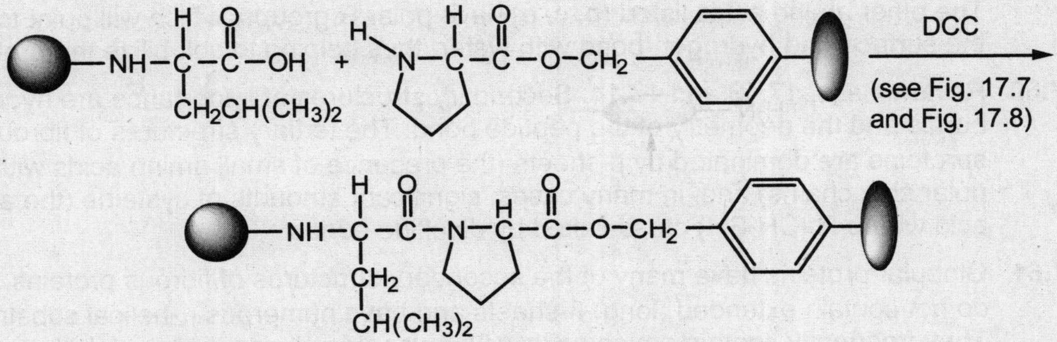

DCC

(see Fig. 17.7 and Fig. 17.8)

Then deprotect again.

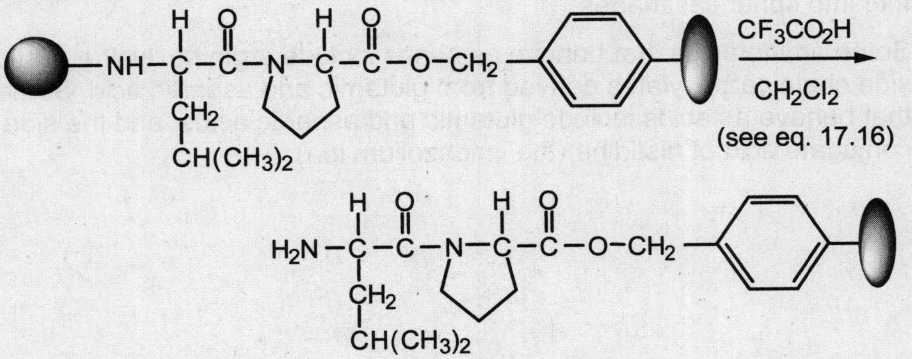

CF_3CO_2H

CH_2Cl_2

(see eq. 17.16)

At this stage, the dipeptide is constructed, but it must be detached from the polymer:

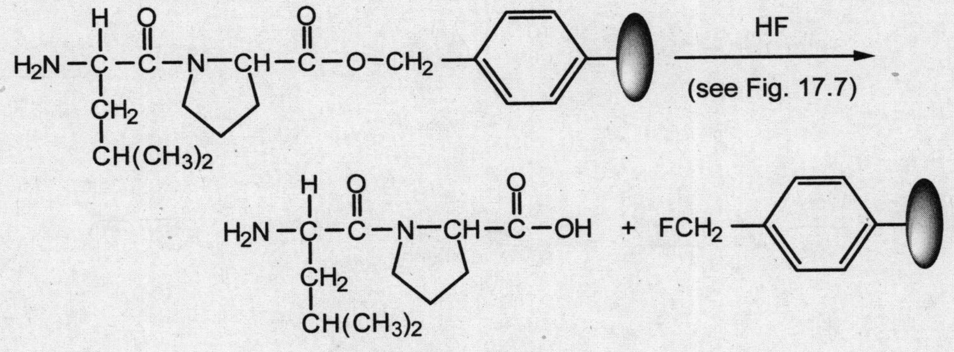

HF

(see Fig. 17.7)

(Leu–Pro)

This problem, which involves only the construction of a simple dipeptide, should give you some sense of the number of operations required in solid-phase peptide syntheses.

17.58

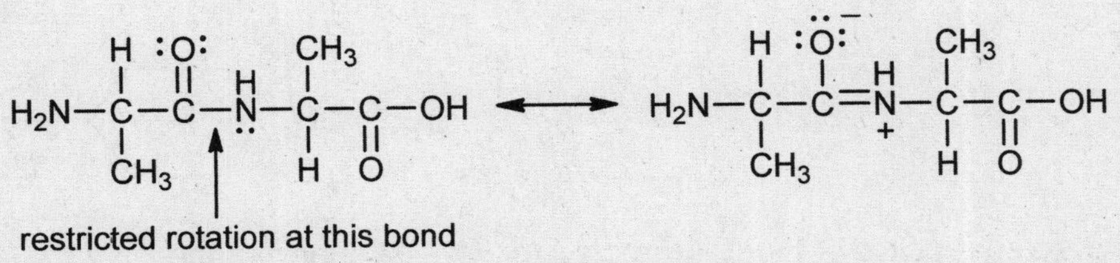

restricted rotation at this bond

17.59 Amino acids with nonpolar R groups will have those groups pointing toward the center of a globular protein. These are leucine (a), valine (d), and phenylalanine (f).

The other amino acids listed (b, c, e) have polar R groups, which will point toward the surface and hydrogen-bond with water, thus helping to solubilize the protein.

17.60 Review Secs. 17.13 and 17.14. Secondary structures of importance are hydrogen bonds and the geometry of the peptide bond. The tertiary structures of fibrous sproteins are dominated by β-sheets (the presence of small amino acids with non-polar side chains) and, in many cases, significant amounts of cysteine (the amino acid where $R{=}CH_2SH$) cross-linked by disulfide bonds.

17.61 Globular proteins have many of the secondary structures of fibrous proteins, but they do not contain extended (long) β-sheets and have numerous α-helical substructures. They frequently contain amino acids with polar side chains that contribute to their water solubility. In globular proteins, the primary sequence of amino acids tend to fold into spherical shapes.

17.62 Some amino acids that behave as bases include arginine, lysine, histidine and the side chain carboxylates derived from glutamic and aspartic acid. Some amino acids that behave as acids include glutamic and aspartic acids, and the side chains conjugate acid of histidine (the imidazolium ion).

18

Nucleotides and Nucleic Acids

Chapter Summary

Nucleic acids, the carriers of genetic information, are macromolecules that are composed of and can be hydrolyzed to **nucleotide units**. Hydrolysis of a nucleotide gives one equivalent each of a **nucleoside** and phosphoric acid. Further hydrolysis of a nucleoside gives one equivalent each of a sugar and a heterocyclic base.

The DNA sugar is **2-deoxy-D-ribose**. The four heterocyclic bases in DNA are **cytosine**, **thymine**, **adenine**, and **guanine**. The first two bases are **pyrimidines**, and the latter two are **purines**. In nucleosides, the bases are attached to the anomeric carbon (C-1) of the sugar as β-**N-glycosides**. In nucleotides, the hydroxyl group (–OH) at C-3 or C-5 of the sugar is present as a phosphate ester.

The primary structure of DNA consists of nucleotides linked by a phosphodiester bond between the 5'–OH of one unit and the 3'–OH of the next unit. To fully describe a DNA molecule, the **base sequence** must be known. Methods for sequencing have been developed, and, at present, over 150 bases can be sequenced per day. The counterpart of sequencing, the synthesis of oligonucleotides having known base sequences, is also highly developed.

The secondary structure of DNA is a **double helix.** Two helical polynucleotide chains coil around a common axis. In **B-DNA**, the predominant form, each helix is right-handed, and the two strands run in opposite directions with respect to their 3' and 5' ends. The bases are located inside the double helix, in planes perpendicular to the helix axis. They are paired (A–T and G–C) by hydrogen bonds, which hold the two chains together. The sugar-phosphate backbones form the exterior surface of the double helix. Genetic information is passed on when the double helix uncoils and each strand acts as a template for binding and linking nucleotides to form the next generation. Other forms of DNA include **A-DNA** (a right-handed double helix in which base pairs are tilted to the helical axis) and **Z-DNA** (a left-handed double helix).

RNA differs from DNA in three ways: the sugar is D-ribose, the pyrimidine **uracil** replaces thymine (the other three bases are the same), and the molecules are mainly single-stranded. The three principal types of RNA are **messenger RNA** (involved in transcribing the genetic code), **transfer RNA** (which carries a specific amino acid to the site of protein synthesis), and **ribosomal RNA**.

The **genetic code** involves sequences of three bases called **codons**, each of which translates to a specific amino acid. The code is degenerate (that is, there is more than one codon per amino acid), and some codons are "stop" signals that terminate synthesis. **Protein biosynthesis** is the process by which the message carried in the base sequence is transformed into an amino acid sequence in a peptide or protein.

In addition to their role in genetics, nucleotides play other important roles in biochemistry. Key enzymes and coenzymes such as **nicotinamide adenine dinucleotide** (NAD), **flavin adenine dinucleotide** (FAD), and **vitamin B_{12}** also include nucleotides as part of their structures. Also, the major component of **viruses** is DNA.

Reaction Summary

Hydrolysis of Nucleic Acids

DNA segment $\xrightarrow[\text{enzyme}]{H_2O}$ nucleotide $\xrightarrow[HO^-]{H_2O}$ nucleoside $\xrightarrow{H_2O,\ H^+}$ 2-deoxy-D-ribose + HN-Base (heterocyclic base)

Learning Objectives

1. Know the meaning of: nucleic acid, nucleotide, nucleoside.
2. Know the structures of: cytosine, thymine, adenine, guanine, uracil, 2-deoxy-D-ribose, D-ribose.
3. Know the meaning of: DNA, RNA, N-glycoside, pyrimidine base, purine base.
4. Given the name, draw the structure of a specific nucleoside.
5. Write an equation for the hydrolysis of a specific nucleoside by aqueous acid. Write the steps in the reaction mechanism.
6. Given the name, draw the structure of a specific nucleotide.
7. Write an equation for the hydrolysis of a specific nucleotide by aqueous base.
8. Draw the structure of an N-glycoside and an O-glycoside.
9. Given the name or abbreviation for a DNA or RNA nucleotide or nucleoside, draw its structure.
10. Draw the primary structure of a segment of an RNA or DNA chain.
11. Explain why only pyrimidine-purine base pairing is permissible in the double helix structure.

12. Describe the main features of the secondary structure of DNA.
13. Explain, with the aid of structures, the role of hydrogen bonding in nucleic acid structures.
14. Describe the main features of DNA replication.
15. Given the base sequence in one strand of a DNA molecule, write the base sequence in the other strand, or in the derived *m*RNA. Conversely, given a base sequence for *m*RNA, write the base sequence in one strand of the corresponding DNA.
16. Given a synthetic polyribonucleotide and the peptide sequence in the resulting polypeptide, deduce the codons for the amino acids.
17. Describe the main features of protein biosynthesis.
18. Explain the different functions of messenger, ribosomal, and transfer RNA.
19. Know the meaning of: codon, anticodon, genetic code, transcription, polymerase chain reaction.
20. Draw the structure of: AMP, ADP, ATP, and cAMP.
21. Describe the main features of DNA profiling.

ANSWERS TO PROBLEMS

Problems Within the Chapter

18.1

2'-deoxythymidine

2'-deoxyguanosine

18.2 a. The products are thymidine and 2'-deoxyribose. Both anomers of 2'-deoxyribose are produced although only the α-anomer is shown.

H_3O^+

2'-deoxyribose

thymine

b. The products are guanine and 2'-deoxyribose.

HO, H, O, H_3O^+, 2'-deoxyribose, guanine

18.3 a. b.

18.4 Consult eqs. 18.3 and 18.2.

H_2O, HO^-, $+ HPO_4^{-2}$, dTMP, thymidine, H_3O^+, thymine, 2-deoxyribose

18.5 –TCGGTACA– (written from the 3' end to the 5' end).

18.6 a. Note that C-2' has a hydroxyl group. AMP is an RNA mononucleotide.

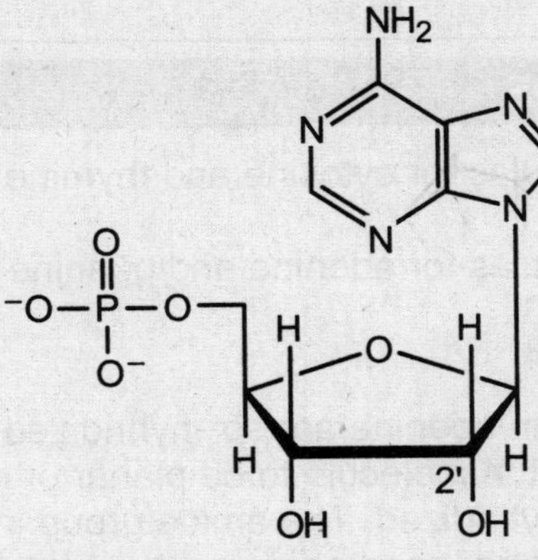

b. Note that there is a hydroxyl group at C-2' of each unit and that the 5' and 3' ends do not have phosphate groups attached.

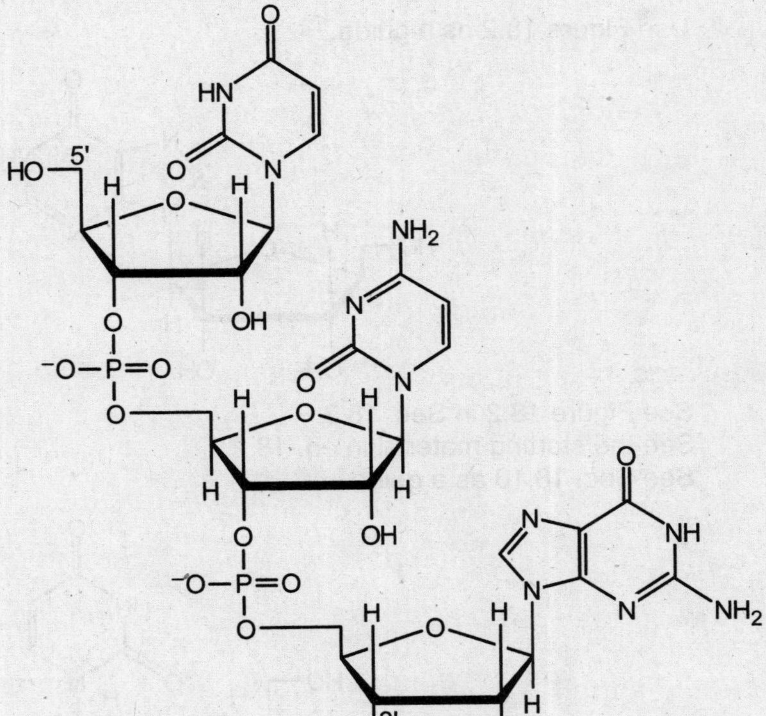

18.7 A polynucleotide made from UA will have the codons UAU and AUA:

<u>UAU</u> <u>AUA</u> <u>UAU</u> ...

UAU is the codon for Tyr, and AUA is a codon for Ile. Note that this codon differs from the codon in Example 18.3 only in the last letter. Such differences are common. The first two letters of codons are frequently more important than the third letter.

18.8 Consult Table 18.2. The codon UUU translates to the amino acid phenylalanine. If a UUU sequence were to be mutated to UCU (the codon for serine), the protein produced would have a Ser residue in place of the Phe. Since UCU and UCC both code for Ser, a UCU $\rightarrow$ UCC mutation would not lead to a change in the protein sequence. Thus, the advantage of a redundant code is that not all mutations cause disadvantageous changes in protein structure.

ADDITIONAL PROBLEMS

18.9 a. See the formulas for cytosine and thymine (Figure 18.1) and uracil (Sec. 18.10).
b. See the formulas for adenine and guanine (Figure 18.1).
c. See Sec. 18.3
d. See Sec. 18.4

18.10 All of the ring atoms in adenine are sp^2-hybridized and planar. Since both rings are aromatic, we expect the molecule to be planar or nearly so. In guanine, the ring atoms are also sp^2-hybridized. The amide group in the six-membered ring (N-1, C-6) is also planar. Therefore, once again we expect the molecule to be planar, or very nearly so. In the pyrimidine bases cytosine and thymine, the rings will also be essentially planar, for the same reasons.

18.11 a. Use Figure 18.2 as a guide.

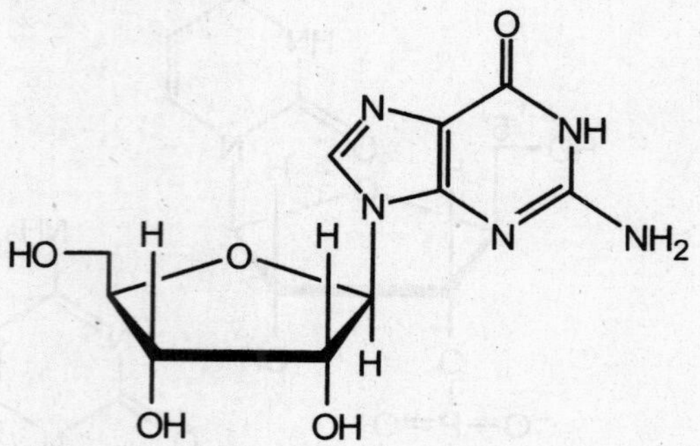

O NH N NH2 N N HO H O H H OH OH

b. See Figure 18.2 in Sec. 18.3.
c. See the starting material in eq. 18.2.
d. See Sec. 18.10 as a guide.

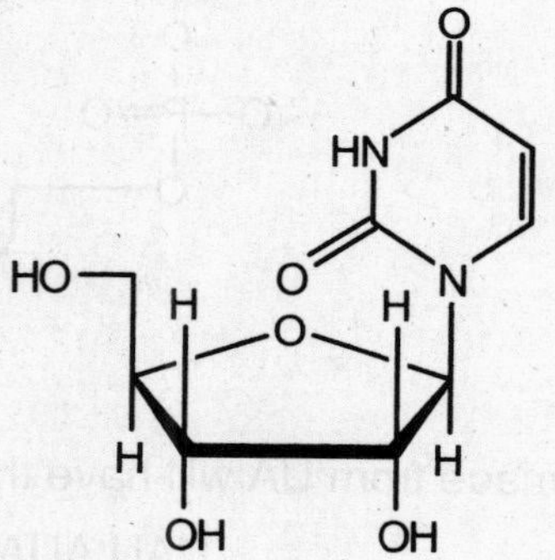

O HN O N HO H O H H H OH OH

18.12 The products are adenine, ribose, and inorganic phosphate.

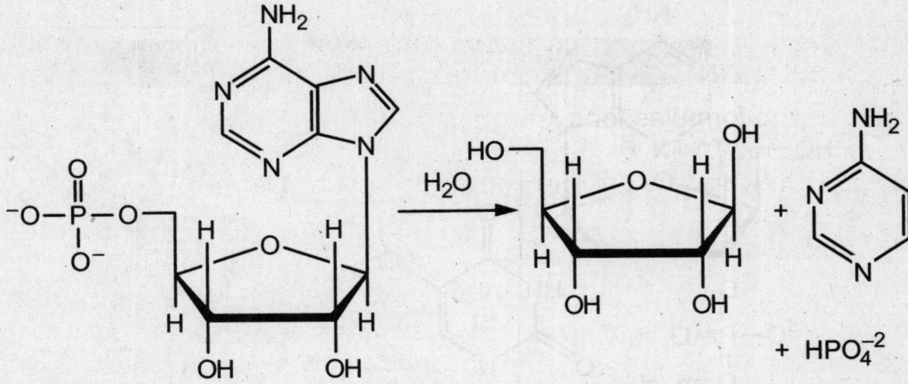

H_2O

+ HPO_4^{-2}

18.13 a. The sugar part of the molecule lacks a hydroxyl group at C-2' (2-deoxy-D-ribose).

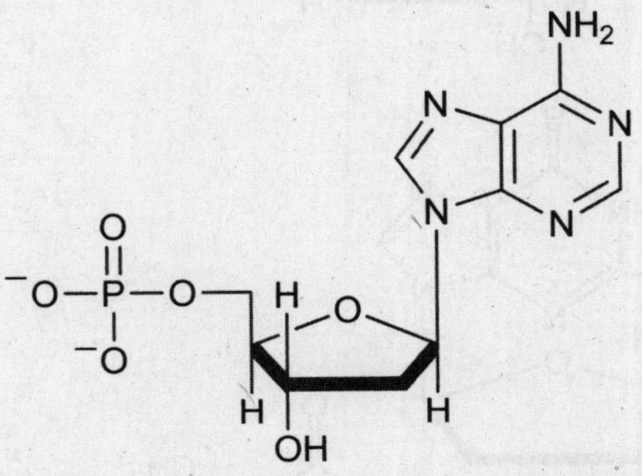

b. The sugar part of the molecule has a hydroxyl group at C-2' (D-ribose).

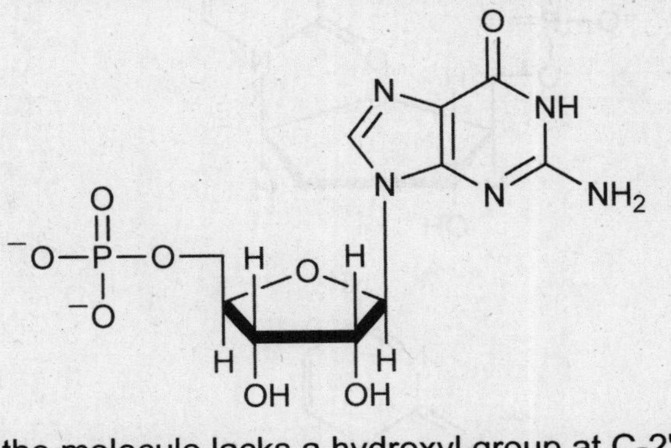

c. The sugar part of the molecule lacks a hydroxyl group at C-2' (2-deoxy-D-ribose).

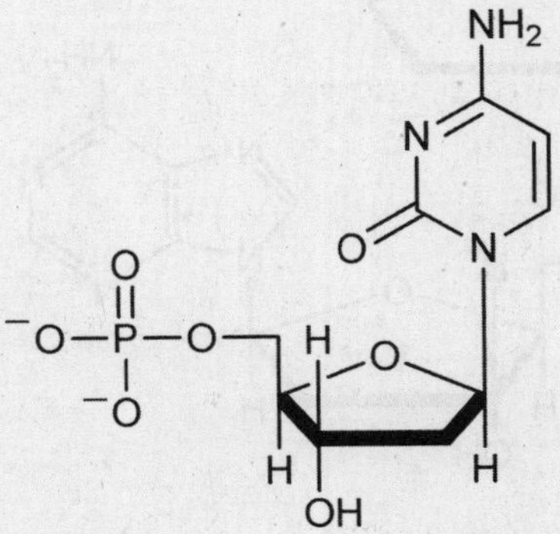

18.14 a.

NH$_2$ N N N N HO H O O CH$_3$ HN O $^-$O–P=O O N O H O H H OH

b.

O HN N H$_2$N N N HO H O O CH$_3$ O HN $^-$O–P=O O N O H O H H OH

c.

NH$_2$ N O N HO H O NH$_2$ O N N $^-$O–P=O O N N H O H H OH

18.15 a. b.

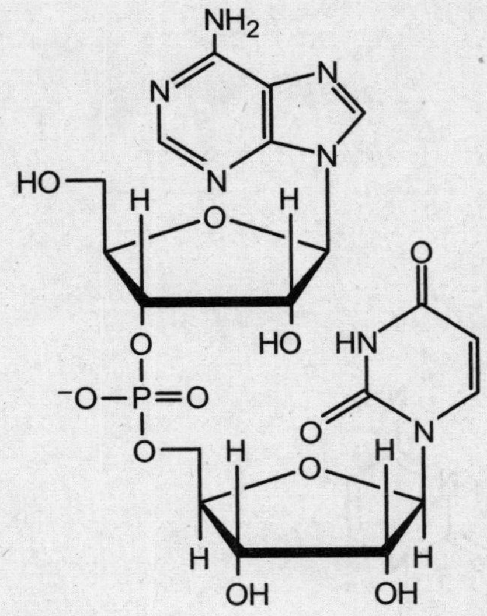

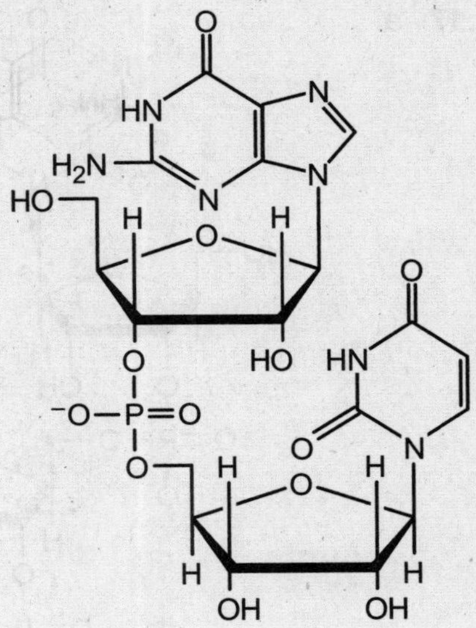

c.

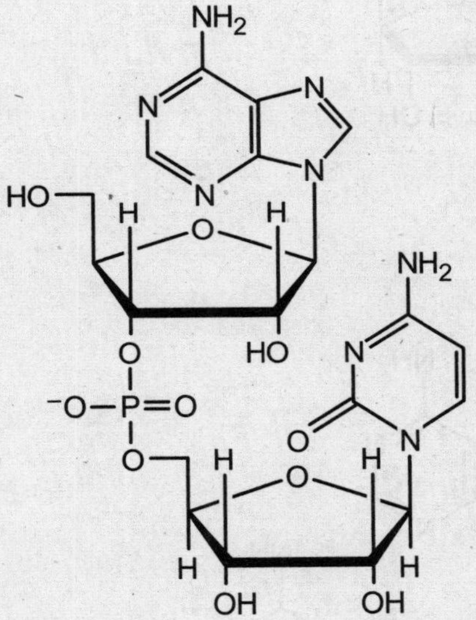

18.16 a. The products are one equivalent of 2'-deoxyadenosine-3'-monophosphate, one equivalent of 2'-deoxyguanosine-3'-monophosphate, one equivalent of 2'-deoxycytidine and one equivalent of 2'-deoxythymidine.

b. The products are one equivalent of adenine, one equivalent of guanidine, one equivalent of cytosine, one equivalent of thymine, three equivalents of inorganic phosphate, and four equivalents of 2-deoxy-D-ribose.

18.17 a.

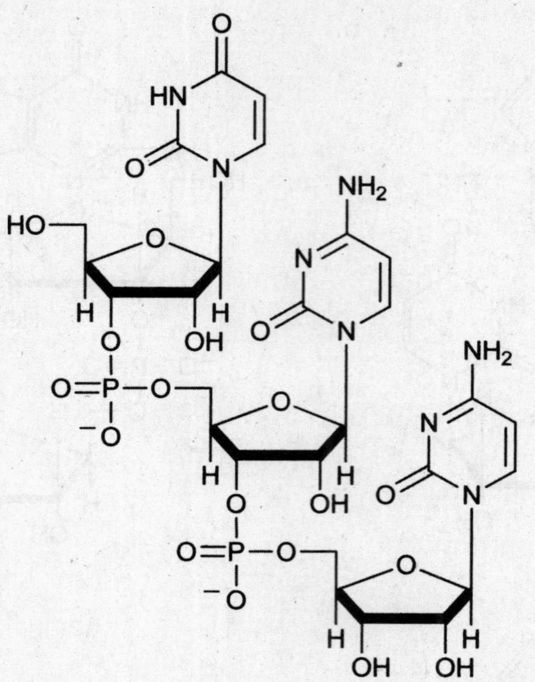

b.

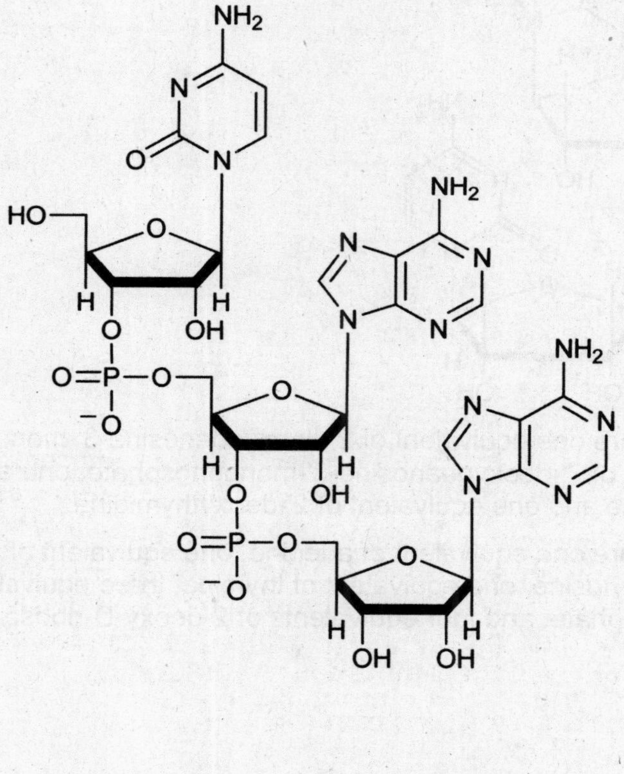

c.

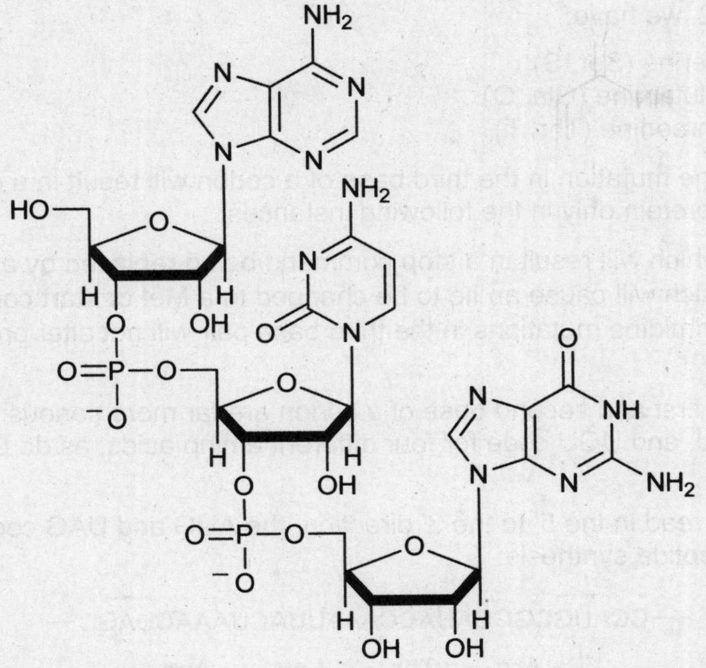

18.18 The structure is identical to that of the T-A base pair shown in Sec. 18.8 of the text, except that the methyl group in the thymine unit is replaced by a hydrogen.

18.19 3' T–A–C–C–A–G–A–T–C 5'

18.20 3' U–A–C–C–A–G–A–U–C 5'

The only difference between this sequence and the answer to Problem 18.19 is that each T is replaced by U.

18.21 Given *m*RNA sequence: 5' A–G–C–U–C–G–U–A–C 3'

DNA strand from which *m*RNA was transcribed:

```
3'  T—C—G—A—G—C—A—T—G  5'
    |  |  |  |  |  |  |  |  |     } DNA double helix
5'  A—G—C—T—C—G—T—A—C  3'
```

Note that the DNA strand that was *not* transcribed (the last segment shown above) is identical with the given *m*RNA segment, except that each U is replaced by T.

18.22 For each T, there must be an A. For each G, there must be a C. As a consequence of this base pairing, the mole percentages of T in any sample of DNA will be equal to the mole percentages of A, and the same will be true for G and C. It is *not* necessary, however, that there be any special relationship between the percentages of the two pyrimidines (T and C) or of the two purines (A and G).

18.23 The code reads from the 5' to the 3' end of *m*RNA.

5' C–A–C 3'	*m*RNA
3' G–T–G 5'	DNA-transcribed chain
5' C–A–C 3'	DNA complement

18.24 From Table 18.2, we have:

UCC	serine (Ser, S)
CAA	glutamine (Gln, Q)
ACG	threonine (Thr, T)

18.25 A purine → purine mutation in the third base of a codon will result in a change of biosynthesized protein only in the following instances:

UGA → UGG, which will result in a stop command being replaced by a Trp, and AUA →AUG, which will cause an Ile to be changed to a Met or start command. Pyrimidine →pyrimidine mutations in the third base pair will not alter protein biosynthesis.

18.26 Mutations in the first and second base of a codon are far more serious. For example, UUU, UCU, UAU, and UGU code for four different amino acids, as do UUU, CUU, AUU, and GUU.

18.27 As the *m*RNA is read in the 5' to the 3' direction, the AUG and UAG codons would start and stop peptide synthesis:

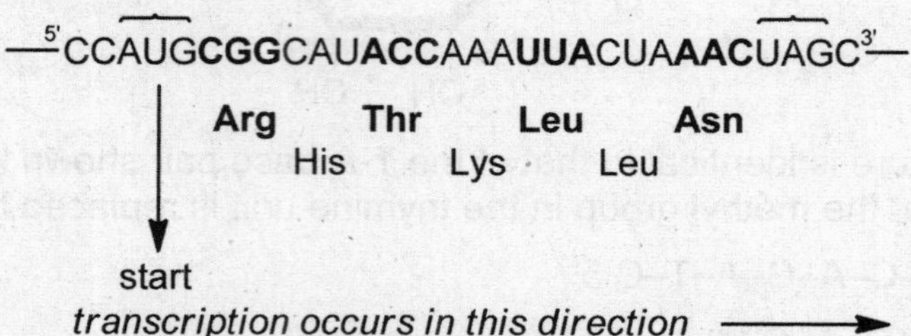

Thus, the *m*RNA strand codes for synthesis of the heptapeptide:
Arg–His–Thr–Lys–Leu–Leu–Asn

18.28 For this modified sequence with the first U being deleted, there would not be an AUG start codon. Thus, the strand would not produce any peptide sequence.

18.29

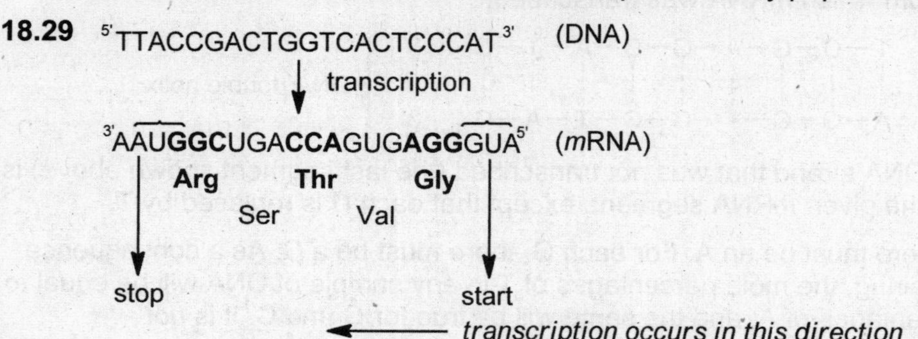

Transcription occurs in the 5' to the 3' direction for the *m*RNA. The AUG codon would start peptide synthesis and Gly would be the C-terminal amino acid. Peptide synthesis would continue until a stop codon (in this case, UAA) appears. The peptide fragment from the DNA strand would be:

–Gly–Val–Thr–Ser–Arg

18.30 The CODIS and United Kingdom databases are growing rapidly, but as of 2007, CODIS had over 5 million profiles and the UK database had over 4 million.

18.31 Hydrolysis occurs at all glycosidic, ester, and amide linkages. The products are as follows:

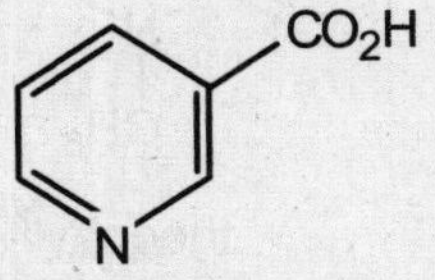

nicotinic acid ammonia ribose (2 mol)

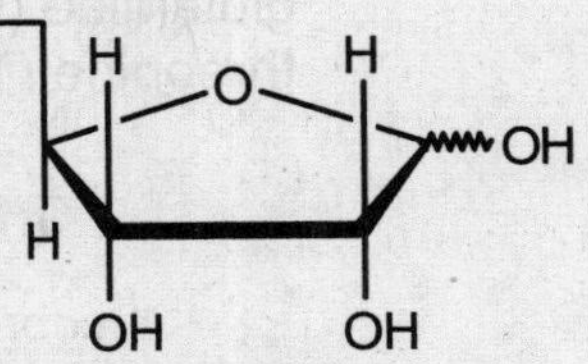

CO_2H NH_3

NH_2

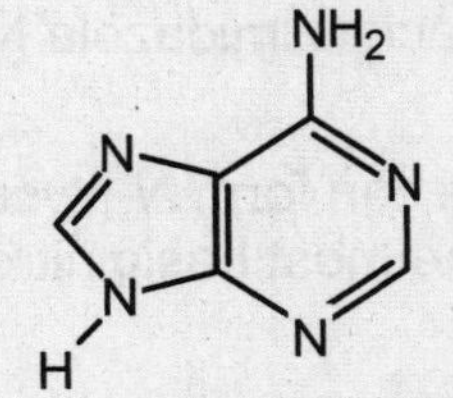

adenine H_3PO_4 phosphoric acid (2 mol)

18.32 The formula for uridine *mono*phosphate is shown in Sec. 18.10 of the text, and the formula for α-D-glucose is shown in eq. 16.3. The structure of UDP-glucose is

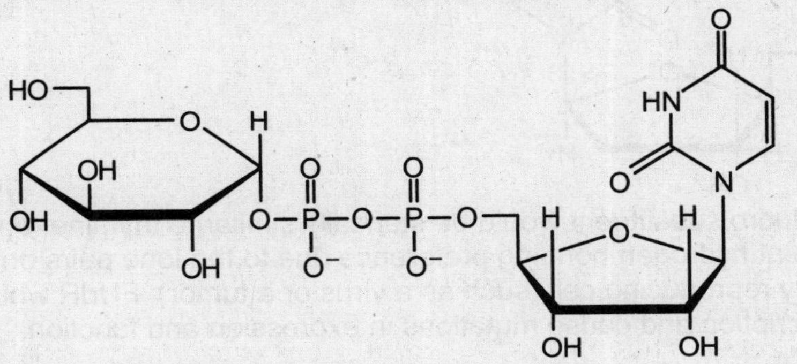

18.33 There are no N–H bonds in caffeine. Therefore (unlike adenine and guanine), caffeine cannot form N-glycosides. Basic sites are indicated by the red in ESP maps. Thus the non-methylated nitrogen in the 5-membered ring is most basic. Suppose you did not have a color-coded ESP map. How could you determine which nitrogen is most basic? First, remember that a base is an electron pair donor. Next, examine the nitrogens. Two of the nitrogens are amides (see Sec. 11.7). One nitrogen has its lone pair delocalized as part of an aromatic ring (see Sec. 13.6). Only one nitrogen has an electron pair that is not involved in bonding. Thus, this will be the most basic nitrogen.

amide — lone pair part of aromatic ring

amide — non-bonded lone pair in sp^2-hybridized orbital

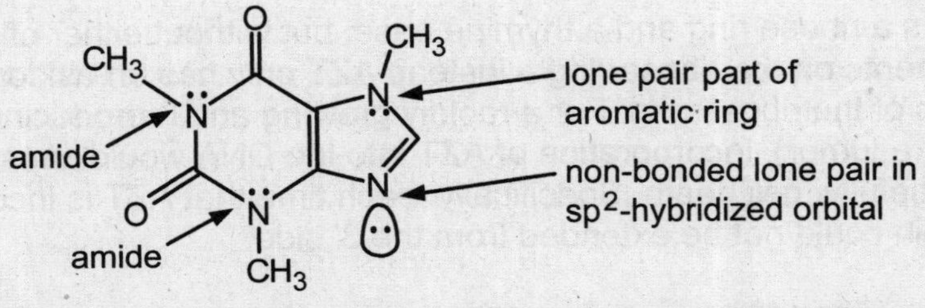

18.34 a. For theobromine and theophylline, the most basic nitrogen is the one with the non-bonded lone pair available in an sp^2-hybridized orbital.

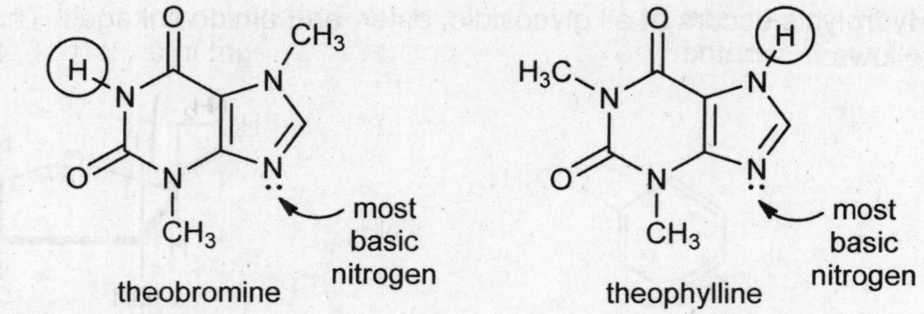

b. Theobromine has one acidic hydrogen for the amide unit (shown above, circled), and theophylline has a much less acidic imidazole N–H bond (also circled above).

c. Both theobromine and theophylline can form *N*-glycosides with sugars at the same N position noted above as the most basic nitrogen.

18.35

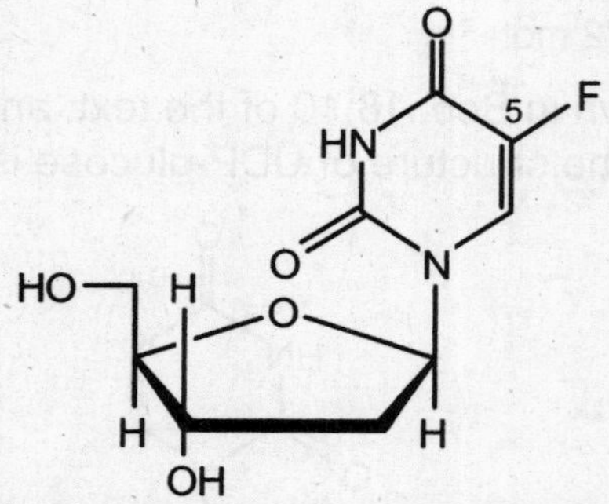

The fluoro substituent would be sterically similar to thymine, but would have a different hydrogen bonding preference due to the lone pairs on fluorine. So, for a rapidly reproducing cell (such as a virus or a tumor), FUdR would disrupt transcription and cause mutations in expression and function.

18.36

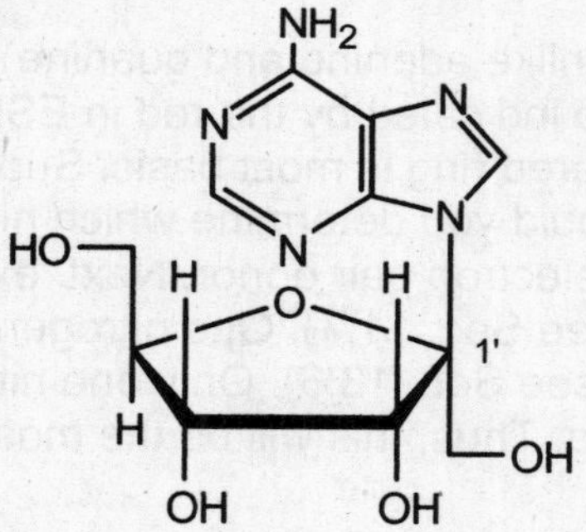

18.37 AZT has a ribose ring and a thymine base, but without either of the oxygen substituents on the ribose ring – instead AZT only has an azido ($-N_3$) unit on the 3' position of the ribose ring. For a rapidly growing and reproducing cell (such as a virus or a tumor), incorporation of AZT into the DNA would cause chain termination and hopefully, cell death. Specifically, each time that AZT is incorporated into DNA, the chain could not be extended from the 3' side.

18.38 See the structure shown below. The base is adenine and its linkage to ribose is β. The ribose is sulfated on C-5 and the 6-deoxyhexose (see Sec 16.15) is an L-sugar (focus on C-5, the highest numbered stereogenic center of the hexose, and determine whether the oxygen is on the left or right in a Fischer projection).

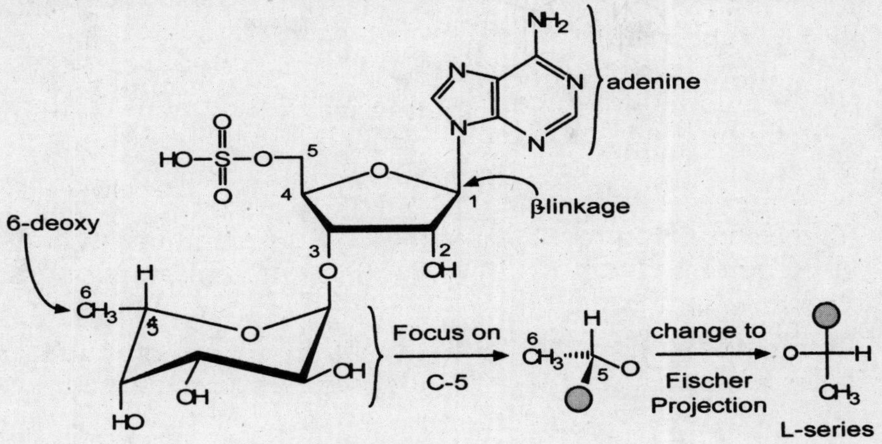

NH2
adenine
HO–S–O
5
4
1
β linkage
6-deoxy
3
2
OH
H
6
CH3
5
O
OH
OH
HO
Focus on
C-5
H
6
CH3
5
O
change to
Fischer
Projection
O–H
CH3
L-series

Summary of Synthetic Methods

In the text, methods for preparing the particular classes of compound are presented in two ways. In certain chapters, methods for synthesizing the class of compound with which the chapter deals are enumerated and discussed. More frequently, however, syntheses are presented less formally as a consequence of various reactions. In this summary of the main synthetic methods for each important class of compound, a general equation for each reaction is given. Also, section numbers, given in parentheses, refer you to the place or places in the book where the reaction is described. References to "A Word About" sections are abbreviated AWA.

1. Alkanes and Cycloalkanes

a. Alkenes + H_2 (3.1; 3.14; 4.7)

$$\mathrm{>C{=}C<} + H_2 \xrightarrow[\text{catalyst}]{\text{Ni or Pt}} -\underset{H}{\overset{|}{C}}-\underset{H}{\overset{|}{C}}-$$

b. Alkynes + H_2 (3.1)

$$-C{\equiv}C- + 2\ H_2 \xrightarrow[\text{catalyst}]{\text{Ni or Pt}} -\underset{H}{\overset{H}{C}}-\underset{H}{\overset{H}{C}}-$$

c. Cyclohexanes from aromatic compounds + H_2 (4.7)

$$\text{benzene} + 3\ H_2 \xrightarrow[\text{catalyst}]{\text{Ni or Pt}} \text{cyclohexane}$$

d. Grignard reagent + H_2O (or D_2O) (8.4)

$$R-MgX + H-OH \longrightarrow R-H + Mg(OH)X$$

2. Alkenes and Cycloalkenes

a. Alkynes + H_2 (3.20)

$$-C{\equiv}C- + H_2 \xrightarrow[\text{catalyst}]{\text{Lindlar's}} \mathrm{\overset{H}{C}{=}\overset{H}{C}} \quad (\textit{cis}\text{-alkene})$$

b. Dehydration of alcohols (7.8)

$$-\underset{HO}{\overset{|}{C}}-\underset{H}{\overset{|}{C}}- \xrightarrow{H^+} \mathrm{>C{=}C<} + H_2O$$

c. Elimination reaction; alkyl halide + strong base (6.7; 6.8)

$$-\underset{X}{\overset{|}{C}}-\underset{H}{\overset{|}{C}}- \xrightarrow[(B:)]{\text{base}} \mathrm{>C{=}C<} + BH^+ + X^-$$

d. *cis–trans* Isomerism (3.5; AWA on "The Chemistry of Vision," Ch. 3)

$$\underset{B\quad\;\; B}{\overset{A\quad\;\; A}{C{=}C}} \xrightleftharpoons{\text{heat or light}} \underset{B\quad\;\; A}{\overset{A\quad\;\; B}{C{=}C}}$$

e. Cracking of alkanes (AWAs on "Ethylene" and "Petroleum," Ch. 3)

$$CH_3CH_3 \xrightarrow{700\text{–}900°C} H_2C{=}CH_2 + H_2$$

$$C_nH_{2n+2} \xrightarrow{\text{catalyst}} C_mH_{2m} + C_pH_{2p+2} \quad (m + p = n)$$

3. <u>Alkynes</u>

a. From acetylides and alkyl halides (3.21; 6.2)

$$R{-}C{\equiv}C{-}H \xrightarrow[NH_3]{NaNH_2} R{-}C{\equiv}C^- \;Na^+ \xrightarrow[\text{(best for R' = primary)}]{R'X} R{-}C{\equiv}C{-}R'$$

4. <u>Aromatic Compounds</u>

a. Alkylbenzenes from Friedel–Crafts reactions (4.8; 4.9d; 15.5)

$$R{-}X + Ar{-}X \xrightarrow{AlCl_3} R{-}Ar + HX$$

$$\rangle C{=}C\langle + Ar{-}H \xrightarrow{H^+} Ar{-}\overset{|}{\underset{|}{C}}{-}\overset{|}{\underset{|}{C}}{-}H$$

b. Acylbenzenes from Friedel–Crafts acylations (4.8; 4.9d)

$$Ar{-}H + R{-}\overset{O}{\overset{\|}{C}}{-}Cl \xrightarrow{AlCl_3} R{-}\overset{O}{\overset{\|}{C}}{-}Ar + HCl$$

c. Aromatic nitro compounds, by nitration (4.8; 4.9b; 4.11; 4.12; 7.14; 13.2)

$$Ar{-}H + HONO_2 \xrightarrow{H^+} Ar{-}NO_2 + H_2O$$

d. Aromatic sulfonic acids, by sulfonation (4.8; 4.9c; 15.5)

$$Ar{-}H + HOSO_3H \xrightarrow{H^+} Ar{-}SO_3H + H_2O$$

e. Aromatic halogen compounds, by halogenation (4.1; 4.8; 4.9a; 4.11; 4.12; 7.14; 13.2)

$$Ar{-}H + X_2 \xrightarrow{FeX_3} Ar{-}X + HX \quad (X = Cl, Br)$$

f. Aromatic halogen compounds from diazonium salts (11.12)

$$Ar-N_2^+ + HX \xrightarrow{Cu_2X_2} Ar-X + N_2 \quad (X = Cl, Br)$$

$$Ar-N_2^+ + KI \longrightarrow Ar-I + N_2 + K^+$$

g. Alkylbenzenes from alkanes (AWA on "Petroleum," Ch. 3)

h. Pyridines (13.2; 13.3)

i. Quinolines (13.3)

j. Pyrroles (13.5)

k. Furans (13.4; 13.5)

l. Thiophenes (13.5)

5. Alcohols

a. Hydration of alkenes (3.7b; 3.9)

$$>C=C< + H-OH \xrightarrow{H^+} HO-\overset{|}{\underset{|}{C}}-\overset{|}{\underset{|}{C}}-H$$

(follow Markovnikov's rule)

b. Hydroboration–oxidation of alkenes (3.13)

$$RCH=CH_2 \xrightarrow[2.\ H_2O_2, HO^-, H_2O]{1.\ BH_3} RCH_2CH_2OH$$

c. Alkyl halides + aqueous base (6.1; 6.2; 6.5; 6.6)

$$R-X + HO^- \longrightarrow R-OH + X^- \quad \text{(best for R = primary)}$$

d. Grignard reagent + carbonyl compound (9.9)

$$>C=O + RMgX \longrightarrow R-\overset{|}{\underset{|}{C}}-OMgX \xrightarrow[H^+]{H_2O} R-\overset{|}{\underset{|}{C}}-OH$$

For primary alcohols, use formaldehyde. For secondary alcohols, use other aldehydes. For tertiary alcohols, use a ketone.

e. Reduction of aldehydes or ketones (9.12; 9.19; 16.9)

$$>C=O \xrightarrow[\text{or } NaBH_4,\ LiAlH_4]{H_2\text{, catalyst}} H-\overset{|}{\underset{|}{C}}-OH$$

f. Grignard reagent + ethylene oxide (8.8)

$$RMgX + H_2C\overset{O}{-}CH_2 \longrightarrow R-CH_2CH_2OMgX \xrightarrow[H^+]{H_2O} RCH_2CH_2OH$$

Useful only for primary alcohols.

g. Grignard reagent (excess) + ester (10.15)

$$\underset{R'O}{\overset{R}{>}}C{=}O + 2\ R''MgX \longrightarrow R''{-}\underset{R''}{\overset{R}{\overset{|}{\underset{|}{C}}}}{-}OMgX \xrightarrow[H^+]{H_2O} R''{-}\underset{R''}{\overset{R}{\overset{|}{\underset{|}{C}}}}{-}OH$$

Useful for tertiary alcohols with at least two identical R groups.

h. Saponification of esters (10.13)

$$R{-}\overset{O}{\overset{\|}{C}}{-}OR' + Na^+\ {}^-OH \longrightarrow R{-}\overset{O}{\overset{\|}{C}}{-}O^-\ Na^+ + R'OH$$

i. Reduction of esters (10.16; 15.5)

$$R{-}\overset{O}{\overset{\|}{C}}{-}OR' \xrightarrow[\text{or } LiAlH_4]{H_2\text{, catalyst}} RCH_2OH + R'OH$$

j. Methanol from carbon monoxide and hydrogen (AWA on "Industrial Alcohols," Ch. 7)

$$CO + 2\ H_2 \xrightarrow{\text{catalyst}} CH_3OH$$

6. Phenols

a. From diazonium salts and base (11.12)

$$Ar{-}N_2^+\ X^- + NaOH \xrightarrow[\text{warm}]{H_2O} Ar{-}OH + N_2 + Na^+\ X^-$$

b. From phenoxides and acid (7.6)

$$Ar{-}O^-\ Na^+ + H^+X^- \longrightarrow Ar{-}OH + Na^+X^-$$

7. Glycols

a. Ring opening of epoxides (8.8)

$$-\overset{|}{C}\underset{O}{-}\overset{|}{C}- + H{-}OH \xrightarrow{H^+} -\underset{HO}{\overset{|}{\underset{|}{C}}}-\underset{OH}{\overset{|}{\underset{|}{C}}}-$$

b. Oxidation of alkenes (3.17)

$$>C{=}C< \xrightarrow{KMnO_4} -\underset{HO}{\overset{|}{\underset{|}{C}}}-\underset{OH}{\overset{|}{\underset{|}{C}}}-$$

c. Hydrolysis of a fat or oil to give glycerol (15.1; 15.3)

$$\begin{array}{l} H_2C-O-\overset{O}{\overset{\|}{C}}-R \\ HC-O-\overset{O}{\overset{\|}{C}}-R \\ H_2C-O-\underset{O}{\underset{\|}{C}}-R \end{array} \xrightarrow[\text{heat}]{3\ NaOH} \begin{array}{l} H_2C-O-H \\ HC-O-H \\ H_2C-O-H \end{array} + 3\ Na^+\ {}^-O-\overset{O}{\overset{\|}{C}}-R$$

8. Ethers and Epoxides

a. From alkoxides and alkyl halides; Williamson synthesis (8.5; 16.8)

$$\left.\begin{array}{l} RO^-\ Na^+ + R'X \longrightarrow ROR' + Na^+\ X^- \\ ArO^-\ Na^+ + R'X \longrightarrow ArOR' + Na^+\ X^- \end{array}\right\}\ \text{Best for R' = primary or secondary}$$

b. Dehydration of alcohols (8.5)

$$2\ ROH \xrightarrow{H^+} ROR + H_2O$$

Most useful for symmetric ethers.

c. Ethylene oxide from ethylene and air (8.7)

$$H_2C{=}CH_2 + O_2 \xrightarrow[\text{catalyst}]{Ag} \text{H}_2C\overset{O}{—}CH_2 \text{ (ethylene oxide)}$$

d. Alkenes and peracids (8.7)

$$\gt C{=}C\lt + R-\overset{O}{\overset{\|}{C}}-O-OH \longrightarrow -\overset{|}{C}\overset{O}{-}\overset{|}{C}- + R-\overset{O}{\overset{\|}{C}}-OH$$

R is usually CH_3–, C_6H_5–, or *m*-ClC_6H_4–

e. Ring opening of epoxides with alcohols (8.8)

$$-\overset{|}{C}\overset{O}{-}\overset{|}{C}- + R-OH \xrightarrow{H^+} -\underset{HO}{\overset{|}{C}}-\underset{OR}{\overset{|}{C}}-$$

9. Alkyl Halides

a. Halogenation of alkanes (2.12b)

$$-\overset{|}{\underset{|}{C}}-H + X_2 \xrightarrow{\text{heat or light}} -\overset{|}{\underset{|}{C}}-X + H-X$$

b. Alkenes (or dienes) + hydrogen halides (3.7c; 3.15; 5.11)

$$\text{>C=C<} \xrightarrow{H-X} -\underset{H}{\overset{|}{C}}-\underset{X}{\overset{|}{C}}-$$

c. Vinyl halides from alkynes + hydrogen halides (3.20)

$$-C{\equiv}C- \;+\; H-X \longrightarrow \text{>C=C<} \;\; (\text{H and X on the two carbons})$$

d. Alcohols + hydrogen halides (7.9)

$$R-OH \;+\; H-X \longrightarrow R-X \;+\; H_2O$$

(Catalysts such as ZnX_2 are required when R is primary.)

e. Alcohols + thionyl chloride or phosphorus halides (7.10)

$$R-OH \;+\; SOCl_2 \longrightarrow R-Cl \;+\; SO_2 \;+\; HCl$$

$$3\ R-OH \;+\; PX_3 \longrightarrow 3\ R-X \;+\; H_3PO_3 \quad (X = Cl, Br)$$

f. Cleavage of ethers with hydrogen halides (8.6)

$$R-O-R' \;+\; 2\ H-X \longrightarrow R-X \;+\; R'-X \;+\; H_2O$$

g. Alkyl iodides from alkyl chlorides (6.2)

$$R-Cl \;+\; NaI \xrightarrow{\text{acetone}} R-I \;+\; NaCl$$

10. Polyhalogen Compounds

a. Halogenation of alkanes (2.12; 6.9)

b. Addition of halogen to alkenes, dienes, and alkynes (3.7a; 3.20)

11. Aldehydes and Ketones

a. Oxidation of alcohols (7.12; 9.3)

$$-\underset{H}{\overset{OH}{C}}- \xrightarrow[H^+]{CrO_3} -\overset{O}{\overset{\|}{C}}-$$

Primary alcohols give aldehydes. The best oxidant is pyridinium chlorochromate (PCC). Secondary alcohols give ketones; the reagent is usually CrO_3, H^+.

b. Hydration of alkynes (3.20; 9.3)

$$R-C{\equiv}C-H \;+\; H_2O \xrightarrow[H_2SO_4]{HgSO_4} R-\overset{O}{\overset{\|}{C}}-CH_3$$

c. Ozonolysis of alkenes (3.17b)

$$\text{>C=C<} + O_3 \longrightarrow \text{>C=O} + \text{O=C<}$$

d. Hydrolysis of acetals (9.7)

$$-\underset{OR}{\overset{OR}{\underset{|}{\overset{|}{C}}}}- \xrightarrow[H^+]{H_2O} -\overset{O}{\overset{\|}{C}}- + 2\ ROH$$

e. Acylation of benzenes (4.8; 4.9d; 9.3)

$$C_6H_6 \xrightarrow[AlCl_3]{R-\overset{O}{\overset{\|}{C}}-Cl} C_6H_5-\overset{O}{\overset{\|}{C}}-R$$

f. Acetaldehyde from ethylene (9.2)

$$2\ H_2C{=}CH_2 + O_2 \xrightarrow[100\text{–}130°C]{Pd\text{-}Cu} 2\ CH_3CH{=}O$$

g. Formaldehyde from methanol (9.2)

$$CH_3OH \xrightarrow[600\text{–}700°C]{Ag} H_2C{=}O + H_2$$

h. Deuterated aldehydes or ketones (9.16)

$$-\overset{H}{\overset{|}{\underset{|}{C}}}-\overset{O}{\overset{\|}{C}}- \xrightarrow[CH_3OD]{CH_3O^-\ Na^+} -\overset{D}{\overset{|}{\underset{|}{C}}}-\overset{O}{\overset{\|}{C}}-$$

Only α-hydrogens exchange. Acid catalysis is also used, especially with aldehydes.

i. β-Hydroxy carbonyl compounds and α,β-unsaturated carbonyl compounds by way of the aldol condensation (9.17; 9.18; 9.19)

$$2\ -\overset{|}{\underset{H}{\underset{|}{C}}}-\overset{O}{\overset{\|}{C}}- \xrightarrow{HO^-} -\overset{|}{\underset{H}{\underset{|}{C}}}-\overset{OH}{\overset{|}{C}}-\overset{|}{\underset{H}{\underset{|}{C}}}-\overset{O}{\overset{\|}{C}}- \xrightarrow[\text{or heat } HO^-]{H^+} -\overset{|}{\underset{H}{\underset{|}{C}}}-C=\overset{|}{C}-\overset{O}{\overset{\|}{C}}-$$

12. Carboxylic Acids

a. Hydrolysis of nitriles (cyanides) (10.7 d)

$$R-C{\equiv}N \xrightarrow[H^+ \text{ or } HO^-]{H_2O} R-\overset{O}{\overset{\|}{C}}-OH \quad (+NH_3 \text{ or } NH_4^+)$$

b. Grignard reagents + carbon dioxide (10.7c)

$$R{-}MgX + O{=}C{=}O \longrightarrow R{-}\overset{O}{\overset{\|}{C}}{-}OMgX \xrightarrow[H^+]{H_2O} R{-}\overset{O}{\overset{\|}{C}}{-}OH$$

c. Oxidation of aromatic side chains (10.7b)

$$Ar{-}CH_3 \xrightarrow{KMnO_4} Ar{-}\overset{O}{\overset{\|}{C}}{-}OH$$

d. Oxidation of aldehydes (9.13; 10.7a; 16.10)

$$R{-}CH{=}O \xrightarrow{Ag^+ \text{ or other oxidant}} R{-}\overset{O}{\overset{\|}{C}}{-}OH$$

e. Oxidation of primary alcohols (10.7a; 16.10)

$$RCH_2OH \xrightarrow[H^+]{K_2Cr_2O_7} RCO_2H$$

f. Saponification of esters (10.13)

$$R{-}\overset{O}{\overset{\|}{C}}{-}OR' + NaOH \longrightarrow R'OH + R{-}\overset{O}{\overset{\|}{C}}{-}O^-\,Na^+ \xrightarrow{H^+} R{-}\overset{O}{\overset{\|}{C}}{-}OH$$

g. Hydrolysis of acid derivatives (10.18; 10.19; 10.20)

$$R{-}\overset{O}{\overset{\|}{C}}{-}Cl + H_2O \longrightarrow R{-}\overset{O}{\overset{\|}{C}}{-}OH + HCl$$

$$R{-}\overset{O}{\overset{\|}{C}}{-}O{-}\overset{O}{\overset{\|}{C}}{-}R + H_2O \longrightarrow 2\ R{-}\overset{O}{\overset{\|}{C}}{-}OH$$

$$R{-}\overset{O}{\overset{\|}{C}}{-}NH_2 + H_2O \xrightarrow[\text{or } HO^-]{H^+} R{-}\overset{O}{\overset{\|}{C}}{-}OH + NH_3$$

h. Dicarboxylic acids from cyclic ketones (9.13)

$$\text{cyclohexanone} + HNO_3 \xrightarrow{V_2O_5} HOOC(CH_2)_4COOH$$

13. <u>Esters</u>

a. From an alcohol and an acid (10.10; 10.11;14.8; 17.5)

$$R{-}\overset{O}{\overset{\|}{C}}{-}OH + R'OH \xrightarrow{H^+} R{-}\overset{O}{\overset{\|}{C}}{-}OR' + H_2O$$

b. From an alcohol and an acid derivative (10.18; 10.19; 10.21; 16.8;16.13b)

$$R-\overset{O}{\overset{\|}{C}}-Cl + R'OH \xrightarrow{H^+} R-\overset{O}{\overset{\|}{C}}-OR' + HCl$$

$$R-\overset{O}{\overset{\|}{C}}-O-\overset{O}{\overset{\|}{C}}-R + R'OH \xrightarrow{H^+} R-\overset{O}{\overset{\|}{C}}-OR' + RCO_2H$$

c. Salt + alkyl halide (6.2; 17.12)

$$R-\overset{O}{\overset{\|}{C}}-O^- Na^+ + R'X \longrightarrow R-\overset{O}{\overset{\|}{C}}-OR' + Na^+ X^-$$

d. Lactones from hydroxy acids (10.12)

$$\text{HO}(CH_2)_3\text{COOH} \xrightarrow{\text{heat or } H^+} \gamma\text{-butyrolactone} + H_2O$$

e. β-Keto esters by way of the Claisen condensation (10.22)

$$2\ RCH_2-\overset{O}{\overset{\|}{C}}-OR' \xrightarrow{\text{base}} RCH_2-\overset{O}{\overset{\|}{C}}-\underset{R}{\underset{|}{CH}}-\overset{O}{\overset{\|}{C}}-OR' + R'OH$$

14. Amides

a. Acyl halides + ammonia (or primary or secondary amines) (10.18; 11.10; 17.5)

$$R-\overset{O}{\overset{\|}{C}}-Cl + H-\underset{R''}{\underset{|}{N}}-R' \longrightarrow R-\overset{O}{\overset{\|}{C}}-\underset{R''}{\underset{|}{N}}-R' + HCl$$

(R' and R" = H, alkyl, or aryl)

b. Acid anhydrides and ammonia (or primary or secondary amines) (10.19; 11.10)

$$R-\overset{O}{\overset{\|}{C}}-O-\overset{O}{\overset{\|}{C}}-R + H-\underset{R''}{\underset{|}{N}}-R' \longrightarrow R-\overset{O}{\overset{\|}{C}}-\underset{R''}{\underset{|}{N}}-R' + RCO_2H$$

c. Esters and ammonia (or primary or secondary amines) (10.14)

$$R-\overset{O}{\overset{\|}{C}}-OR' + H-\underset{R''}{\underset{|}{N}}-R''' \longrightarrow R-\overset{O}{\overset{\|}{C}}-\underset{R''}{\underset{|}{N}}-R''' + R'OH$$

d. From acids and ammonia (or primary or secondary amines) (10.20)

$$R-\overset{O}{\overset{\|}{C}}-OH + H-\underset{R'}{\underset{|}{N}}-R' \xrightarrow[\text{(R' = H or alkyl)}]{\text{heat}} R-\overset{O}{\overset{\|}{C}}-\underset{R'}{\underset{|}{N}}-R' + H_2O$$

e. Polyamides from diamines and dicarboxylic acids or their derivatives (14.1; 14.9; AWA on “Aramids,” Ch. 14)

$$H_2N\sim\sim NH_2 + HOOC\sim\sim COOH \longrightarrow \left(\sim \overset{O}{\overset{\|}{C}} - \underset{H}{\underset{|}{N}} \sim\sim \underset{H}{\underset{|}{N}} - \overset{O}{\overset{\|}{C}} \sim \right)_n$$

f. Peptides (17.12)

15. Other Carboxylic Acid Derivatives

a. Salts from acids and bases (10.8)

$$R-\overset{O}{\overset{\|}{C}}-OH + NaOH \longrightarrow R-\overset{O}{\overset{\|}{C}}-O^- Na^+ + H_2O$$

b. Salts (soaps) by saponification of esters (10.13; 15.3)

$$R-\overset{O}{\overset{\|}{C}}-OR' + NaOH \xrightarrow{\text{heat}} R-\overset{O}{\overset{\|}{C}}-O^- Na^+ + R'OH$$

c. Acyl halides from acids (10.18)

$$R-\overset{O}{\overset{\|}{C}}-OH \xrightarrow{SOCl_2 \text{ or } PCl_3} R-\overset{O}{\overset{\|}{C}}-Cl$$

d. Anhydrides from diacids (10.19)

$$\text{phthalic acid } (C_6H_4(COOH)_2) \xrightarrow{\text{heat or } Ac_2O} \text{phthalic anhydride}$$

16. Nitriles (Cyanides)

a. Nitriles from alkyl halides and inorganic cyanides (6.2; 10.7d)

$$R-X + Na^+ CN^- \longrightarrow R-CN + Na^+ X^-$$

(R = primary or secondary)

b. Aryl nitriles from diazonium ions (11.12)

$$Ar-N_2^+ + KCN \xrightarrow{Cu_2(CN)_2} Ar-CN + N_2 + Cu_2X_2$$

17. Amines and Related Compounds

a. Alkylation of ammonia or amines (6.2; 11.4)

$$2\ -\underset{|}{\overset{H}{\overset{|}{N}}} + R-X \longrightarrow -\underset{|}{\overset{R}{\overset{|}{N}}} + -\underset{|}{\overset{H}{\overset{|}{N^+}}}-H\ X^-$$

b. Reduction of nitriles (11.5)

$$R-C\equiv N \xrightarrow{LiAlH_4 \text{ or } H_2\text{, Ni catalyst}} RCH_2NH_2$$

c. Reduction of nitro compounds (11.5)

$$Ar-NO_2 \xrightarrow{SnCl_2,\ HCl\ \text{or}\ H_2,\ \text{catalyst}} Ar-NH_2\ (+\ 2H_2O)$$

d. Reduction of amides (10.20; 11.5)

$$R-\overset{O}{\overset{\|}{C}}-\underset{R''}{\underset{|}{N}}-R' \xrightarrow{LiAlH_4} R-\underset{H}{\underset{|}{\overset{H}{\overset{|}{C}}}}-\underset{R''}{\underset{|}{N}}-R'$$

e. Hydrolysis of amides (10.20)

$$R-\overset{O}{\overset{\|}{C}}-\underset{R''}{\underset{|}{N}}-R' + H_2O \xrightarrow{HO^-} R-\overset{O}{\overset{\|}{C}}-O^- + H-\underset{R''}{\underset{|}{N}}-R'$$

(R' and R" = alkyl or aryl)

f. Amine salts from amines (11.8; 13.1)

$$-\underset{|}{\overset{|}{N}} + HX \longrightarrow -\underset{|}{\overset{|}{N^+}}-H\ X^-$$

g. Quaternary ammonium salts (6.2; 11.11)

$$R'-\underset{R''}{\underset{|}{\overset{R}{\overset{|}{N}}}} + R'''-X \longrightarrow R'-\underset{R''}{\underset{|}{\overset{R}{\overset{|}{N^+}}}}-R'''\quad X^-$$

(R''' = primary or secondary)

18. <u>Miscellaneous Nitrogen Compounds</u>

a. Oximes, hydrazones, and imines from carbonyl compounds and ammonia derivatives (9.11)

$$\rangle C=O + H_2N-R \longrightarrow \rangle C=N-R + H_2O$$

R = alkyl, aryl, OH, NH_2

b. Cyanohydrins from carbonyl compounds and hydrogen cyanide (9.10)

$$\rangle C=O + HCN \longrightarrow \rangle C\langle^{CN}_{OH}$$

c. Diazonium compounds from primary aromatic amines and nitrous acid (11.12)

$$Ar-NH_2 + HONO + HX \longrightarrow Ar-N_2^+\ X^- + 2\ H_2O$$

d. Azo compounds, *via* diazonium coupling reactions (11.13)

$$Ar{-}N_2^+\ X^- + C_6H_4{-}OH\ (\text{or } NR_2) \longrightarrow Ar{-}N{=}N{-}C_6H_4{-}OH\ (\text{or } NR_2)$$

e. Alkyl nitrates (7.13; 16.13b)

$$ROH + HONO_2 \longrightarrow RONO_2 + H_2O$$

f. Ureas (14.9)

g. Carbamates (urethanes) (14.9)

h. Piperidine (13.2)

19. Organic Sulfur Compounds

a. Thiols from alkyl halides and sodium hydrosulfide (6.2; 7.17)

$$R{-}X + Na^+\ {}^-SH \longrightarrow R{-}SH + Na^+\ X^-$$

(best when R is primary)

b. Thioethers from alkyl halides and sodium mercaptides (6.2)

$$R{-}X + Na^+\ {}^-SR' \longrightarrow R{-}S{-}R' + Na^+\ X^-$$

(best when R is primary)

c. Disulfides from thiols (7.17; AWA on "Hair," Ch. 7; 17.8)

$$2\ R{-}SH \xrightarrow{H_2O_2} R{-}S{-}S{-}R$$

d. Sulfonium salts (6.2)

$$R'{-}\ddot{S}{-}R + R''{-}X \longrightarrow R''{-}\overset{+}{S}(R')(R)\ X^-$$

e. Alkyl hydrogen sulfates from alcohols or from alkenes (3.7c; 3.9; 15.5)

$$R{-}OH + HOSO_3H \xrightarrow{\text{cold}} ROSO_3H + H_2O$$

$$>C{=}C< + HOSO_3H \xrightarrow{\text{cold}} {-}\underset{H}{C}{-}\underset{OSO_3H}{C}{-}$$

f. Sulfonic acids by sulfonation (4.8; 4.9c; 15.5)

20. Miscellaneous Classes of Compounds

a. Grignard reagents (8.4)

$$R{-}X + Mg \xrightarrow{\text{ether}} RMgX$$

b. Organolithium compounds (8.4)

$$R{-}X + 2\ Li \longrightarrow RLi + LiX$$

c. Acetylides (3.21)

$$R{-}C{\equiv}C{-}H + NaNH_2 \xrightarrow{NH_3} R{-}C{\equiv}C^-\ Na^+ + NH_3$$

d. Alkoxides and phenoxides (7.6)

$$2\ ROH + 2\ Na \longrightarrow 2\ RO^-\ Na^+ + H_2$$

$$Ar{-}OH + Na^+\ {}^-OH \longrightarrow Ar{-}O^-\ Na^+ + H_2O$$

e. Hemiacetals and acetals (9.7; 16.4; 16.5; 16.11)

$$\rangle C{=}O + ROH \underset{}{\overset{H^+}{\rightleftharpoons}} \rangle C(OH)(OR) \underset{H^+}{\overset{ROH}{\rightleftharpoons}} \rangle C(OR)(OR) + H_2O$$

f. Quinones (7.15, AWA on "Quinones and the Bombardier Beetle," Ch. 7)

$$HO{-}C_6H_4{-}OH \xrightarrow[H^+]{Na_2Cr_2O_7} O{=}C_6H_4{=}O$$

g. Vinyl polymers (3.16; 6.9; 14.1; 14.2; 14.3; 14.4)

$$\rangle C{=}C\langle_X \xrightarrow{\text{catalyst}} \left(\overset{|}{\underset{|}{C}} - \overset{|}{\underset{X}{C}} \right)_n$$

Summary of Reaction Mechanisms

Although a substantial number of reactions are described in the text, they belong to a relatively modest number of mechanistic types. The preparation of alkyl halides from alcohols and HX, the cleavage of ethers, and the preparation of amines from alkyl halides and ammonia (and many other reactions) all, for example, occur by a nucleophilic substitution mechanism. The following is a brief review of the main mechanistic pathways discussed in the text.

1. <u>Substitution Reactions</u>

   a. Free-radical chain reaction (2.13)

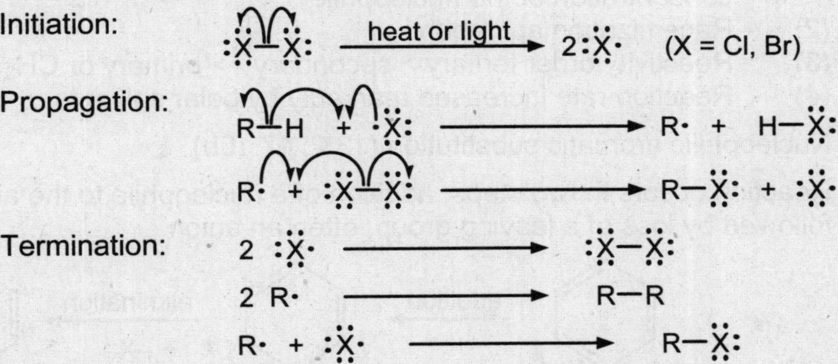

   The sum of the propagation steps gives the overall reaction.

   b. Electrophilic aromatic substitution (4.9)

   Reaction occurs in two steps: addition of an electrophile to the aromatic π electron system followed by loss of a ring proton.

   E^+ —addition→ benzenonium ion (E, H, +) —elimination→ (E-substituted benzene) + H^+

   Substituents already on the aromatic ring affect the reaction rate and the orientation of subsequent substitutions (4.10; 4.11).

   c. Nucleophilic aliphatic substitution (6.3; 6.4; 6.5; 6.6)
   S_N2 (substitution, nucleophilic, bimolecular):

   Nu: (nucleophile) + C—L (substrate) → $\overset{+}{Nu}$—C + $^-$:L (leaving group)

(1) Rate depends on the concentration of both reactants, that is, the nucleophile and the substrate.
(2) Inversion of configuration at carbon
(3) Reactivity order CH_3 > primary > secondary >> tertiary
(4) Rate only mildly dependent on solvent polarity

S_N1 (substitution, nucleophilic, unimolecular):

slow ⇌ (carbocation) + ⁻:L —fast, Nu:→ C–Nu + Nu–C

(1) Rate depends on the concentration of the substrate but *not* on the concentration of the nucleophile.
(2) Racemization at carbon
(3) Reactivity order tertiary > secondary >> primary or CH_3
(4) Reaction rate increased markedly by polar solvents

d. Nucleophilic aromatic substitution (13.2; 17.10b)

Reaction occurs in two steps: addition of a nucleophile to the aromatic ring followed by loss of a leaving group, often an anion.

Nu: + Ar–X —addition→ (Nu, X intermediate) —elimination, $-X^-$→ Ar–Nu

The reaction is facilitated by electron-withdrawing substituents *ortho* or *para* to X since they stabilize the intermediate carbanion.

2. Addition Reactions

a. Electrophilic additions to C=C and C≡C

Addition of acids (3.9; 3.10)

C=C —H^+→ –C(H)–C⁺– (carbocation) —$Nu:^-$→ –C(H)–C(Nu)–

$Nu:^-$ = R–Ö–H, H–Ö–H, $^-OSO_3H$, and so on

Addition proceeds via the most stable carbocation intermediate (Markovnikov's rule).

Addition to alkynes (3.20)

$$-C\equiv C- \xrightarrow{E^+} \underset{E}{>}C=\overset{+}{C}< \xrightarrow{:Nu^-} \underset{E}{>}C=C<^{Nu}$$

1,4-Addition to conjugated dienes (3.15a)

$$C=C-C=C \xrightarrow{E^+} \left[\underset{|}{\underset{E}{C}}-\overset{+}{C}-C=C \longleftrightarrow \underset{|}{\underset{E}{C}}-C=C-\overset{+}{C} \right] \text{ allylic carbocation}$$

$$\xrightarrow{Nu:^-} \underset{E}{C}-C=C-\underset{Nu}{C}$$

(+ normal 1,2-addition)

b. Free-radical addition (3.16; 14.2)

Initiation:

$$\text{Initiator} \xrightarrow{\text{heat or light}} 2\ R\cdot$$

Propagation:

$$R\cdot + H_2C=\underset{X}{CH} \longrightarrow R-\overset{H}{\underset{H}{C}}-\overset{H}{\underset{X}{C}}\cdot \xrightarrow{H_2C=\underset{X}{CH}} \text{and so on}$$

Termination:

$$\sim\overset{H}{\underset{X}{C}}\cdot + \cdot\overset{H}{\underset{X}{C}}\sim \xrightarrow[\text{(radical coupling)}]{} \sim\overset{H}{\underset{X}{C}}-\overset{H}{\underset{X}{C}}\sim$$

$$\sim\overset{H}{\underset{H}{C}}-\overset{H}{\underset{X}{C}}\cdot + \cdot\overset{H}{\underset{X}{C}}-\overset{H}{\underset{H}{C}}\sim \xrightarrow[\text{(disproportionation)}]{} \sim\overset{H}{\underset{H}{C}}-\overset{H}{\underset{X}{C}}-H + \overset{H}{\underset{X}{C}}=\overset{H}{C}\sim$$

Another reaction type associated with such reactions is chain-transfer (or hydrogen abstraction) (14.2).

$$\sim H\dot{C}- + \sim\overset{H}{\underset{X}{C}}\sim \longrightarrow \sim H_2\underset{X}{C}- + \sim\underset{X}{\dot{C}}\sim$$

Cationic and anionic chain-growth polymerizations occur by chain reactions similar to those for free-radical polymerizations but involving charged intermediates (14.3; 14.4).

c. Hydroboration *via* a cyclic transition state (3.13)

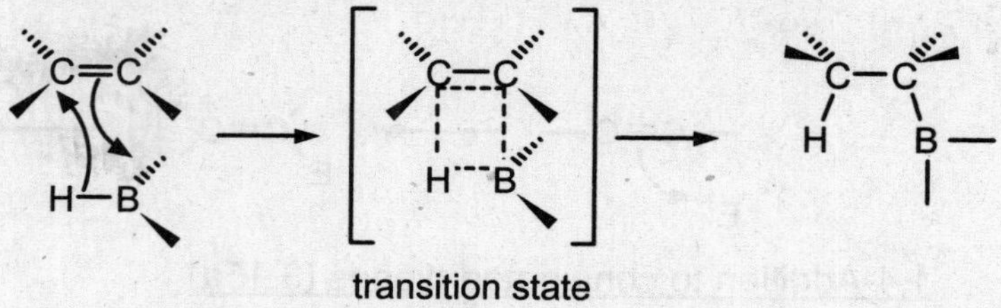

transition state

The boron adds to the least crowded carbon; both the boron and hydrogen add to the same "face" of the double bond. Subsequent oxidation with H_2O_2 and HO^- places a hydroxyl group in the exact position previously occupied by boron:

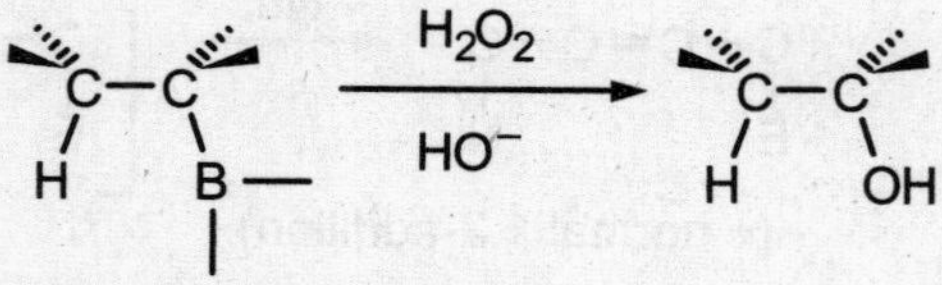

H_2O_2, HO^-

d. Nucleophilic addition to C=O (9.6; 10.11)

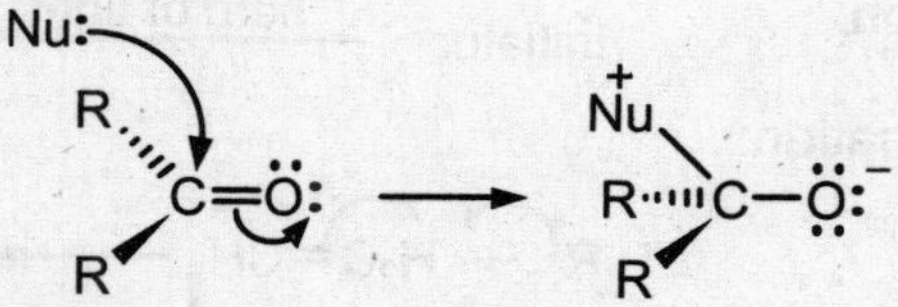

The reaction is often followed by protonation of the oxygen. The C=O group may be present in an aldehyde, ketone, ester, anhydride, acyl halide, amide, and so forth. Acid catalysts may be necessary with weak nucleophiles. The acid protonates the oxygen, making the carbon more positive and thus, more electrophilic.

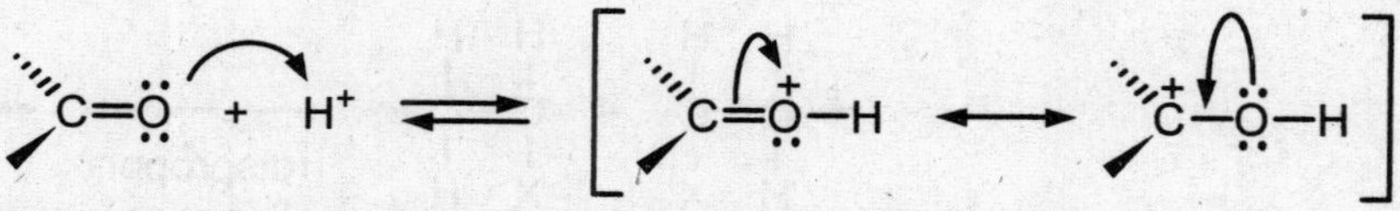

Nucleophilic *substitution* at a carbonyl group (10.11) takes place by an addition-elimination mechanism (10.16):

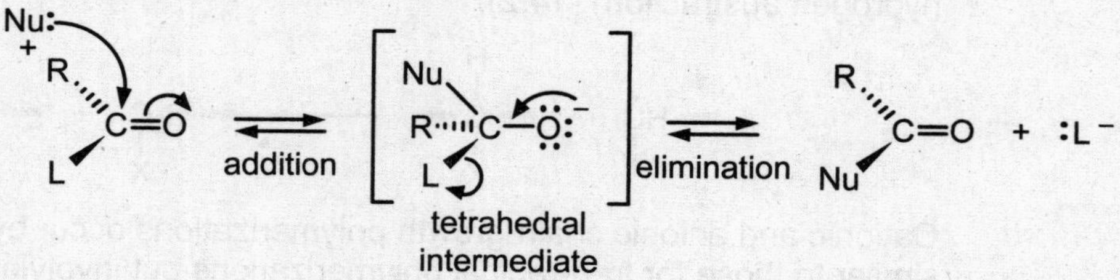

addition

elimination

tetrahedral intermediate

3. Elimination Reactions

a. The E2 mechanism (6.7)

$$B: + H-C-C-L \longrightarrow BH^+ + C=C + ^-:L$$

A planar, *anti* arrangement of the eliminated groups (H and L) is preferred. The reaction rate depends on the concentration of both the base (B) and the substrate. This mechanism can be important regardless of whether L is attached to a primary, secondary, or tertiary carbon.

b. The E1 mechanism (6.7)

$$H-C-C-L \rightleftharpoons H-C-C^+ \; ^-:L \xrightarrow[-H^+]{\text{fast}} C=C$$

The first step is the same as the first step of the S_N1 mechanism. The reaction rate depends only on the substrate concentration. This mechanism is most important when L is attached to a tertiary carbon.

Review Problems On Synthesis

The following problems are designed to give you practice in multi-step synthesis. They should be helpful in preparing for quizzes or examinations. To add to the challenge and make this preparation more realistic, no answers are given. If you do have difficulty with some of the problems, consult your textbook first, then your instructor.

The best technique for solving a multi-step synthesis problem is to work backward from the final goal. Keep in mind the structure of the available starting material, however, so that eventually you can link the starting material with the product. The other constraint on synthesis problems is that you must use combinations of known reactions to achieve your ultimate goal. Although research chemists do try to discover and develop new reactions, you cannot afford that luxury until you have already mastered known reactions. All of the following problems can be solved using reactions in your text.

1. Show how each of the following can be prepared from propene:

a. propane ($CH_3CH_2CH_3$)
b. 2-bromopropane
c. 1,2-dichloropropane
d. 2-propanol [$CH_3CH(OH)CH_3$]
e. 1-propanol ($CH_3CH_2CH_2OH$)

2. Each of the following conversions requires two reactions, in the proper sequence. Write equations for each conversion.

a. *n*-propyl bromide to propene to 1,2-dibromopropane
b. isopropyl alcohol to propene to 2-iodopropane
c. 1-bromobutane to 1-butene to 2-chlorobutane
d. 2-butanol to butenes to butane
e. bromocyclopentane to cyclopentene to 1,2-dibromocyclopentane
f. *t*-butyl chloride to isobutylene to isobutyl alcohol

3. Starting with acetylene, write equations for the preparation of:

a. ethane
b. ethyl iodide
c. 1-butyne
d. 1,1-diiodoethane
e. 2,2-dibromobutane
f. 3-hexyne
g. 1,1,2,2-tetrabromoethane
h. *cis*-3-hexene

4. Write equations for each of the following conversions:

a. 2-butene to 1,3-butadiene (two steps)
b. 2-propanol to 1-propanol (three steps)
c. 1-bromopropane to 1-aminobutane (two steps)
d. 1,3-butadiene to 1,4-dibromobutane (two steps)

In solving problems of this type, carefully examine the structures of the starting material and the final product. Seek out similarities and differences. Note the types of bonds that must be made or broken in order to go from one structure to the other. Sometimes it is profitable to work backward from the product and forward from the starting material simultaneously, with the goal of arriving at a common intermediate.

5. Using benzene or toluene as the only organic starting material, devise a synthesis for each of the following:

a. *m*-chlorobenzenesulfonic acid
b. 2,4,6-tribromotoluene
c. 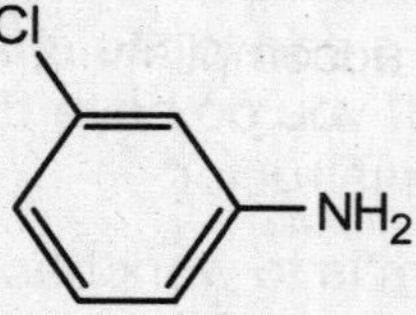

d. Cl — NH2
e. 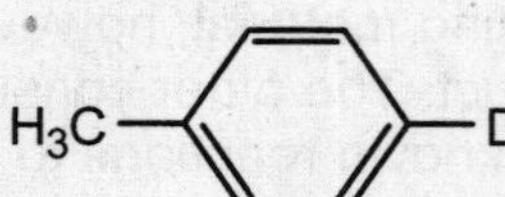

f. CN

6. Write equations for the preparation of:

a. 1-phenylethanol from styrene
b. 2-phenylethanol from styrene
c. 1-butanol from 1-bromobutane
d. sodium 2-butoxide from 1-butene
e. 1-butanethiol from 1-butanol
f. 2,4,6-tribromobenzoic acid from toluene
g. ethyl cyclohexyl ether from ethanol and phenol
h. di-*n*-butyl ether from 1-butanol

7. Write equations that show how 2-propanol can be converted to each of the following:

a. isopropyl chloride
b. 1,2-dibromopropane
c. 2-methoxypropane
d. isopropylbenzene

8. 2-Bromobutane can be obtained in one step from each of the following precursors: butane, 2-butanol, 1-butene, and 2-butene. Write an equation for each method. Describe the advantages or disadvantages of each.

9. Starting with an unsaturated hydrocarbon, show how each of the following can be prepared:

a. 1,2-dibromobutane
b. 1,1-dichloroethane
c. 1,2,3,4-tetrabromobutane
d. cyclohexyl iodide
e. 1,4-dibromo-2-butene
f. 1,1,2,2-tetrachloropropane
g. 1-bromo-1-phenylethane
h. 1,2,5,6-tetrabromocyclooctane

10. Give equations for the preparation of the following carbonyl compounds:

a. 2-pentanone from an alcohol
b. pentanal from an alcohol
c. cyclohexanone from phenol (two steps)
d. acetone from propyne

11. Complete each of the following equations, giving the structures of the main organic products:

a. benzoic acid + ethylene glycol + H^+
b. $C_6H_5CH_2MgBr$ + CO_2, followed by H_3O^+
c. *n*-propylamine + acetic anhydride
d. *p*-hydroxybenzoic acid + acetic anhydride
e. *n*-propylbenzene + $K_2Cr_2O_7$ + H^+
f. phthalic anhydride + methanol + H^+

g. pentanedioic acid + thionyl chloride
h. cyclopropanecarboxylic acid + NH_4OH, then heat
i. methyl 3-butenoate + $LiAlH_4$
j. ethyl propanoate + NaOH, heat

12. Show how each of the following conversions can be accomplished:

a. butanoyl chloride to methyl butanoate
b. propanoic anhydride to propaneamide
c. butanoic acid to 1-butanol
d. 1-pentanol to pentanoic acid
e. propanoyl bromide to *N*-ethyl propaneamide
f. oxalic acid to diethyl oxalate
g. urea to ammonia and CO_2
h. benzoyl chloride to *N*-methylbenzamide

13. Show how each of the following compounds can be prepared from the appropriate acid:

a. propionyl bromide
b. ethyl pentanoate
c. *n*-butanamide
d. phthalic anhydride
e. calcium oxalate
f. phenylacetamide
g. isopropyl benzoate
h. *m*-nitrobenzoyl chloride

14. Write equations to describe how each of the following conversions might be accomplished:

a. *n*-butyl chloride to *n*-butyltrimethylammonium chloride
b. *o*-toluidine to *o*-toluic acid
c. *o*-toluidine to *o*-bromobenzoic acid
d. 1-butene to 2-methyl-1-aminobutane

15. Starting with benzene, toluene, or any alcohol with four carbon atoms of fewer, and any essential inorganic reagents, outline steps for the synthesis of the following compounds:

a. *n*-butylamine
b. *p*-toluidine
c. 1-aminopentane
d. *N*-ethylaniline
e. *m*-aminobenzoic acid
f. tri-*n*-butylamine
g. 1,4-diaminobutane
h. ethyl cyclohexanecarboxylate

16. Plan a synthesis of each of the following compounds, using starting materials with five or fewer carbon atoms:

a. 1-hexanol
b. 2-hexanol
c. 3-hexanol
d. 1-heptanol
e. ethylcyclopentane
f. 1-ethynylcyclopentene
g. *cis*-3-hexene
h. 2-ethyl-2-hexenal

17. Show how you could accomplish the following conversions:

a.  O=(bicyclic ketone)(H)(OH) to H_3C (HO) (bicyclic alkene)

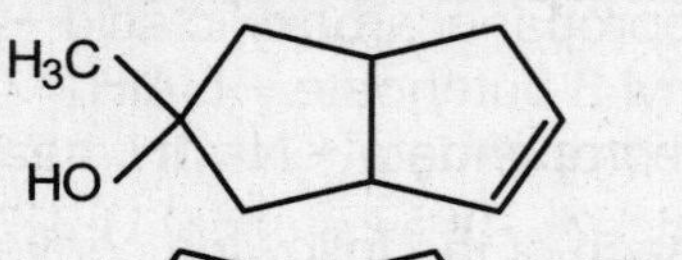

b. 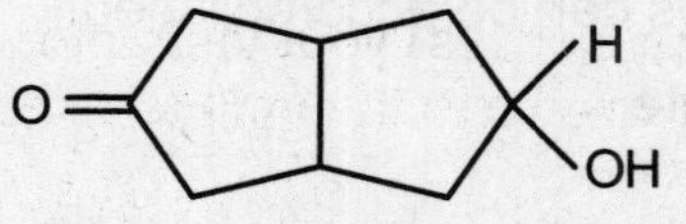 O=(bicyclic ketone)(H)(OH) to O=(bicyclic ketone)(CH_3)(OH)

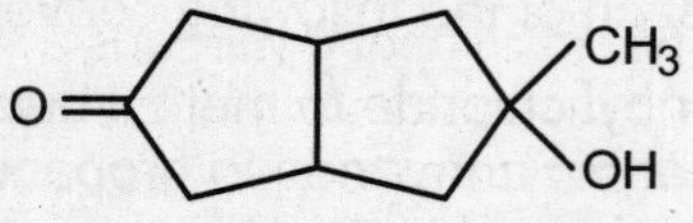

18. Show how each of the following compounds could be made starting with benzene:

a.

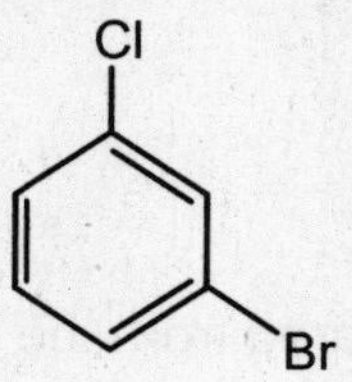

b. 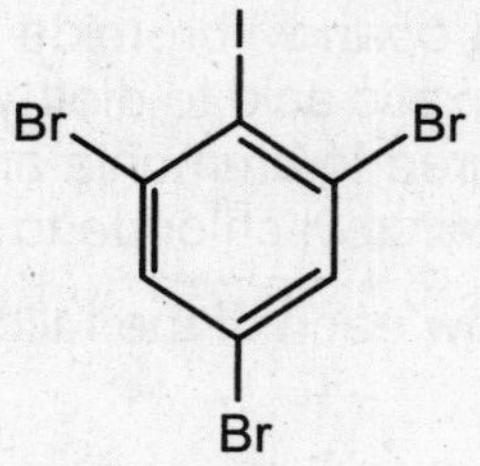 (I, Br, Br, Br)

c. 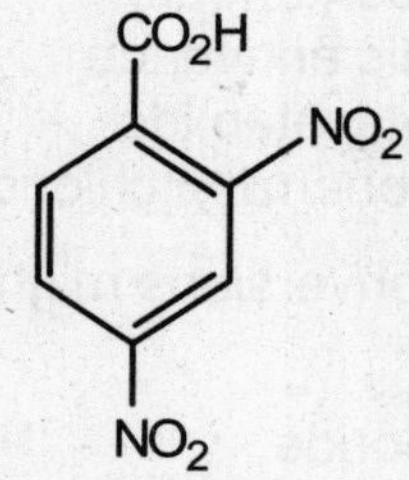 (CO_2H, NO_2, NO_2)

d. 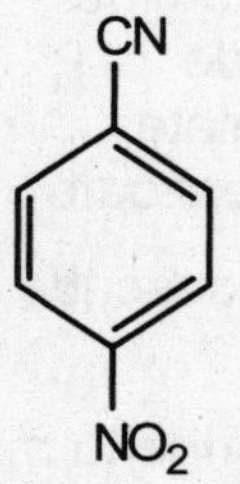 (CN, NO_2)

19. Starting with 4-methylpyridine, write equations for the preparation of:

a. 2-amino-4-methylpyridine b. 3-amino-4-methylpyridine
c. 4-methylpiperidine d. pyridine-4-carboxylic acid

20. Plan a synthesis of the following polymers from appropriate monomers:

a. $\left(CH_2CH(CN)\right)_n$ b. $\left(CH_2CH(OH)\right)_n$ c. $\left(CH_2CH(CH_3)O\right)_n$

d. $\left(NHCH_2CH_2CH_2CH_2CH_2C(=O)\right)_n$

e. $\left(OCH_2CH_2OC(=O)CH_2CH_2CH_2CH_2C(=O)\right)_n$

f. 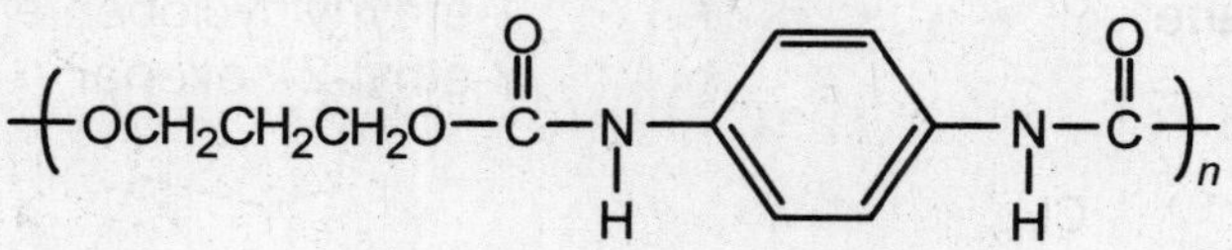 $\left(OCH_2CH_2CH_2O{-}C(=O){-}N(H){-}C_6H_4{-}N(H){-}C(=O)\right)_n$

Sample Multiple Choice Questions

These questions are presented here for your practice. Although they do not cover all of the subject matter in the text, these questions provide a fair review of the material and may give you some idea of what to expect if (because of large classes) your instructor uses multiple-choice exams. Do not just guess at the correct answer. Take the time to write out the structures or equations before you make a choice.

1. Which of the following compounds may exist as *cis–trans* isomers?

a. 1-butene b. 2-butene c. 2-pentene

1. a 2. b 3. a and b 4. b and c
5. a and c

2. What is the molecular formula of the following compound?

O

1. $C_7H_{15}O$ 2. $C_8H_{13}O$ 3. $C_8H_{14}O$ 4. $C_8H_{15}O$
5. $C_9H_{14}O$

3. How many σ (sigma) bonds are there in CH_2=CH–CH=CH_2?
1. 3 2. 9 3. 10 4. 11 5. 12

4. The preferred conformation of butane is:

1. CH_3, H_3C, H, H, H, H

2. CH_3, H, H, H, CH_3, H

3. CH_3, H, CH_3, H, H, H

4. H, H_3C, H, H, CH_3, H

5. none of these

5. The preferred conformation of *cis*-1,3-dimethylcyclohexane is:

1.

2. CH_3, H, CH_3, H

3. (chair structure: H, CH_3, H_3C, H)
4. (chair structure: CH_3, CH_3, H, H)
5. none of these

6. The major reaction product of

(1,2-diethylcyclopentene, CH_2CH_3 / CH_2CH_3) $\xrightarrow[Pt]{H_2}$

is:

1. (cyclopentane with H / CH_2CH_3 and CH_2CH_3 / H)
2. (cyclopentane with H_2 / CH_2CH_3 and CH_2CH_3 / H_2)
3. (cyclopentane with H / CH_2CH_3 and H / CH_2CH_3)
4. (cyclopentane with CH_2CH_3 / H and CH_2CH_3 / H)
5. (cyclopentane with CH_2CH_3, Pt, CH_2CH_3)

7. The structure that corresponds to a reaction intermediate in the reaction

$$H_3C{-}CH_3 \;+\; Br_2 \xrightarrow{\text{heat}}$$

is:

1. $CH_3CH_2^+$
2. $CH_3CH_2\cdot$
3. $CH_3CH_2^-$
4. $CH_2CH_3^+$
5. $\cdot CH_2CH_2\cdot$

8. The best way to prepare $BrCH_2CH_2Br$ is to start with

1. $HC{\equiv}CH + Br_2$ (excess)
2. $CH_2{=}CH_2 + HBr$
3. $HC{\equiv}CH + HBr$ (excess)
4. $ClCH_2CH_2Cl + Br_2$
5. $CH_2{=}CH_2 + Br_2$

9. The IUPAC name (*E*)-2-methyl-3-hexene corresponds to:

1. $CH_3CH_2(H)C{=}C(H)CH_2CH_3$
2. $CH_3CH_2(H)C{=}C(H)CH(CH_3)_2$
3. $CH_3CH_2(H)C{=}C(H)CH(CH_3)_2$
4. $(H_3C)_2C{=}C(H)CH_2CH_3$

5. $(H_3C)_2C{=}C(H)CH_2CH_3$

10. The major product of the reaction

$$H_3C(H)C{=}C(H)CH_3 \xrightarrow[H_2SO_4]{H_2O} \text{is:}$$

1. $H_3C-CH(OH)-CH(OH)-CH_3$
2. $H_3C-CH(OH)-CH(OSO_3H)-CH_3$
3. $H_3C-CH_2-CH(OSO_3H)-CH_3$
4. $H_3C-CH_2-CH_2-CH_3$
5. $H_3C-CH_2-CH(OH)-CH_3$

11. The most likely structure for an unknown that yields only $CH_3CH_2CH{=}O$ on ozonolysis is:

1. 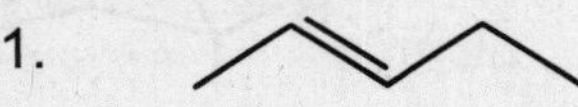
2. 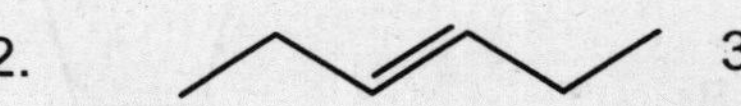
3. 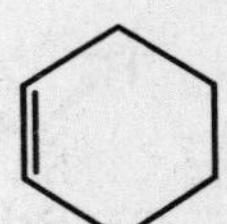
4. 
5. 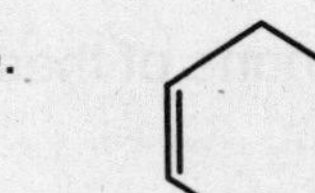

12. Considering the hybridization of carbon orbitals, which of the following structures is *least* likely to exist?

1. 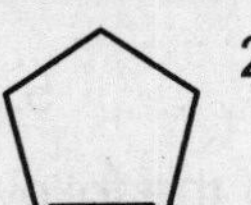
2. 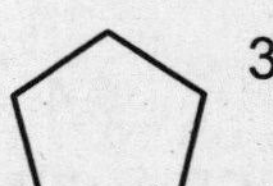
3. 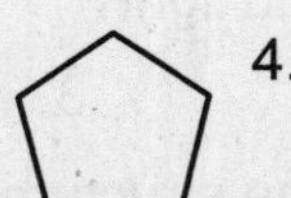
4. 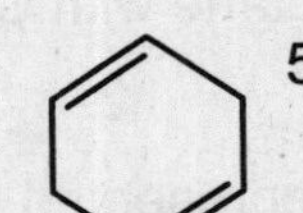
5. 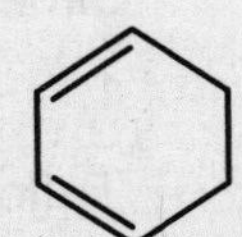

13. What is a correct name for the following structure?

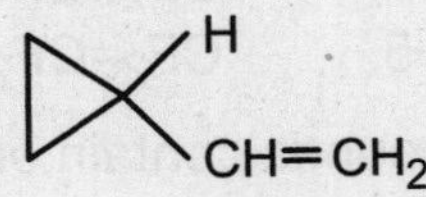

1. cyclopentene
2. cyclopropylpropene
3. vinylcyclopropane
4. vinylcyclobutane
5. 1-cyclopropyl-2-ene

14. H^+ often reacts as a(an):

1. electrophile
2. nucleophile
3. radical
4. electron
5. carbocation

15. The major reaction product of

$$(H_3C)(CH_3CH_2)C{=}C(CH_3)(H) + HCl \longrightarrow$$

is:

1.

2. 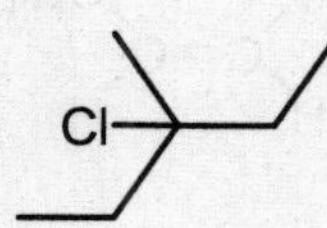

3. 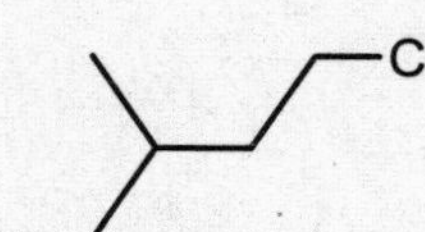

4. 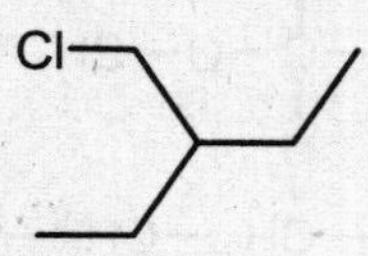

5. 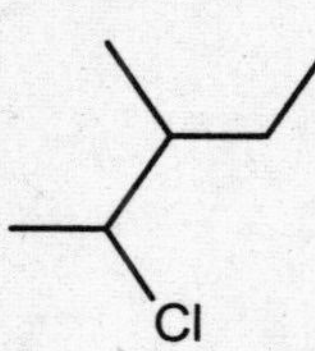

16. Which of the following classes of compounds is unreactive toward sulfuric acid?

a. alkanes b. alcohols c. alkenes d. alkynes

1. only a 2. only b 3. a and b 4. only d
5. a, b, and d

17. Which of the following structures represents a *trans*-dibromo compound?

1. 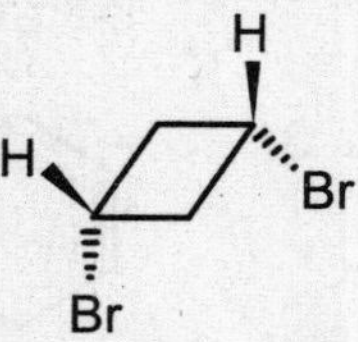

2. 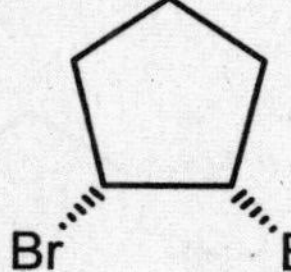

3. 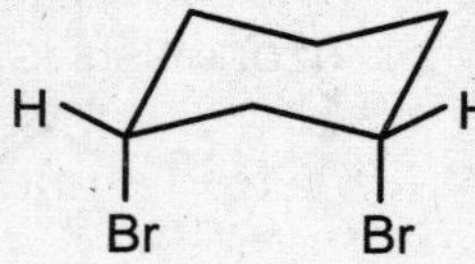

4. 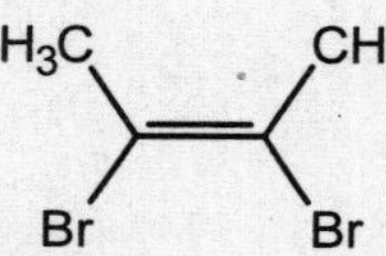

5. none of these

18. How many *mono* chlorination products are possible in the reaction of 2,2-dimethylbutane with Cl_2 and light?

1. 2 2. 3 3. 4 4. 5 5. 6

19. Teflon is represented by the structure $-(CF_2CF_2)_n-$. Which of the following monomers is used to make Teflon?

1. $CF_2{-}CF{=}F$ 2. $CF_2{=}CF{-}CF{=}CF_2$ 3. $CF_2{=}CFCF_3$
4. $CHF_2{-}CHF_2$ 5. $CF_2{=}CF_2$

20. Which of the following compounds contain one or more polar covalent bonds?

a. CH_3CH_2OH b. $CH_3CH_2CH_3$
c. $ClHC{=}CHCl$ d. Cl_2

1. a and b 2. a and c 3. b and c
4. only d 5. a and d

21. What is the relationship between

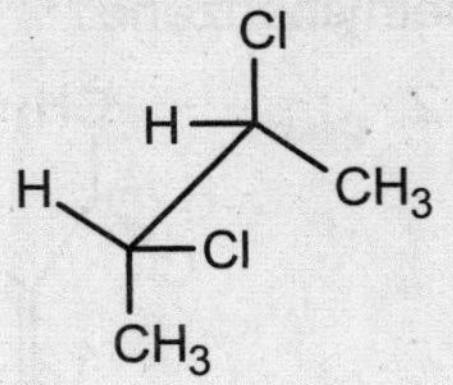

Cl, H, H, CH$_3$, Cl, CH$_3$ and H, H$_3$C, H$_3$C, Cl, H, Cl ?

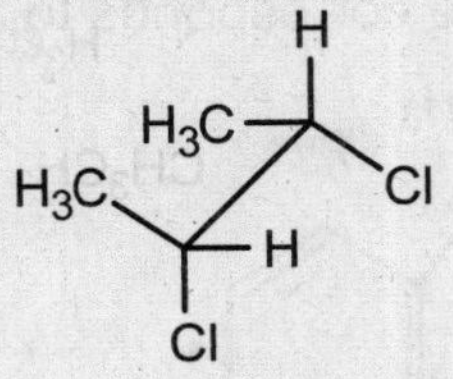

1. structural isomers
2. geometric isomers
3. conformational isomers
4. identical
5. none of these

22. In which of the following structures does nitrogen have a formal charge of +1?

a. 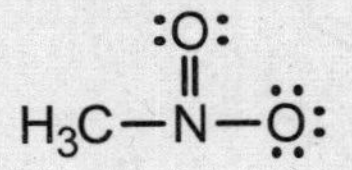 $H_3C-N(=O)-O$ (with lone pairs: :O:, :Ö:)   b. 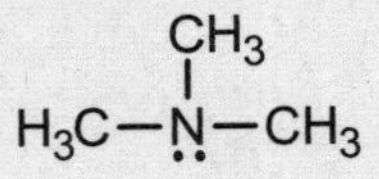 $H_3C-\ddot{N}(CH_3)-CH_3$ c. NH_4^+

1. a and c
2. a and b
3. b and c
4. a, b, and c
5. only a

23. Which of the following statements is(are) true of $CH_3CH_2CH_3$?

a. all bond angles are about 109.5°   b. each carbon is sp^3-hybridized
c. the compound is combustible

1. only a
2. only b
3. a and b
4. a and c
5. a, b, and c

24. Which reagent can accomplish this conversion?

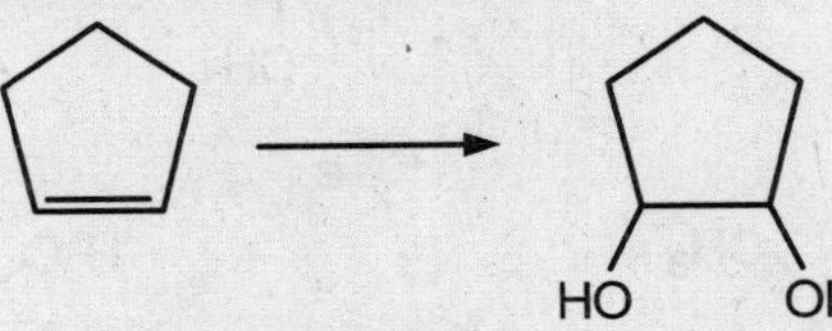

1. $KMnO_4$
2. H_2O, H^+
3. O_3, then Zn, H^+
4. OH^-
5. H_2O, heat

25. How many elimination products are possible for the following reaction?

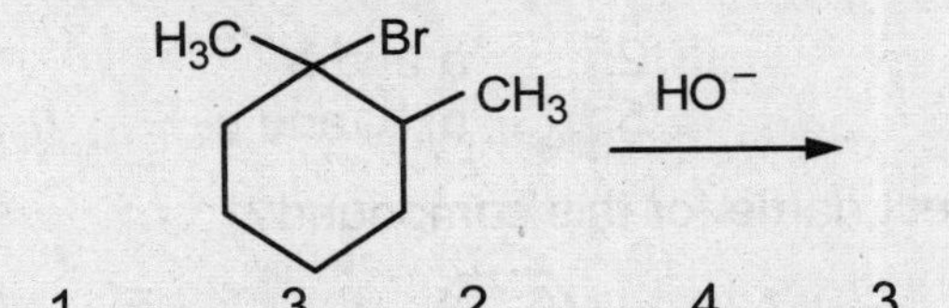

1. 0   2. 1   3. 2   4. 3   5. 4

26. Arrange the following compounds in order of increasing acidity:

a. 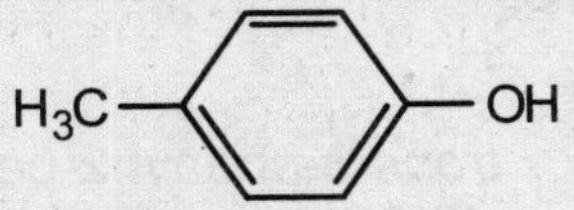 $H_3C-C_6H_4-OH$ b. 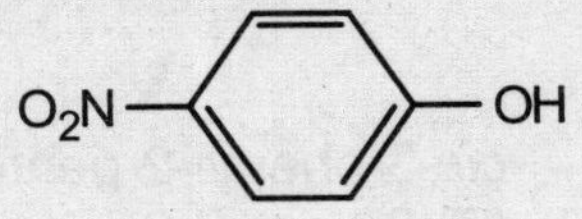 $O_2N-C_6H_4-OH$

c.

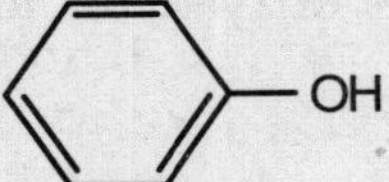

1. a < b < c
2. c < b < a
3. b < a < c
4. a < c < b
5. b < c < a

27. Which structure corresponds to *R*-1-chloroethylbenzene?

1. 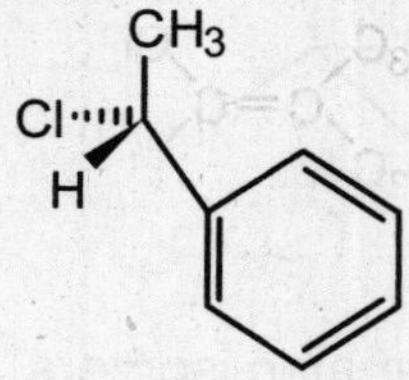

2. 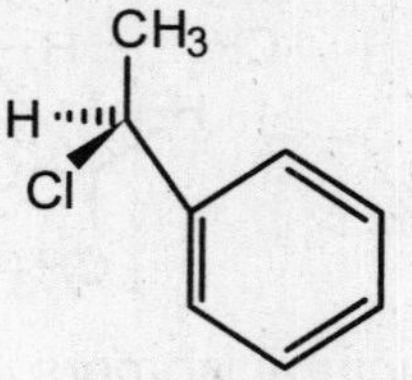

3. 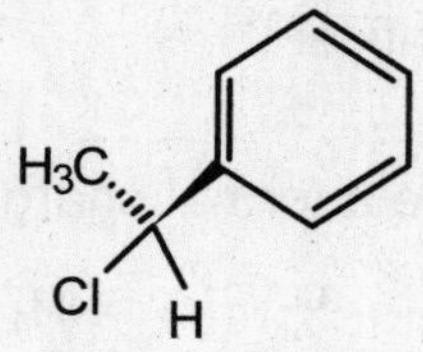

4. 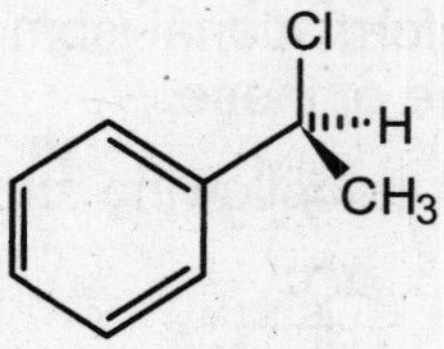

5. 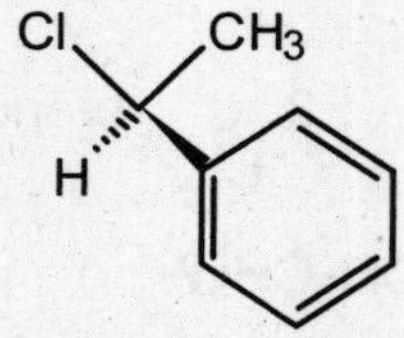

28. Which of the following structures represents a *meso* compound?

a. 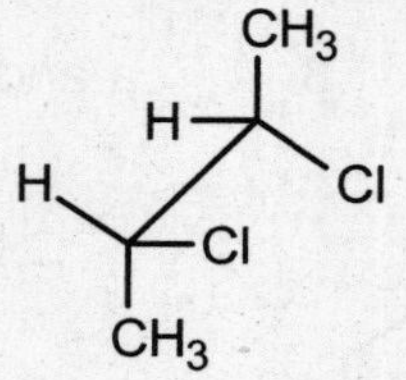

b. 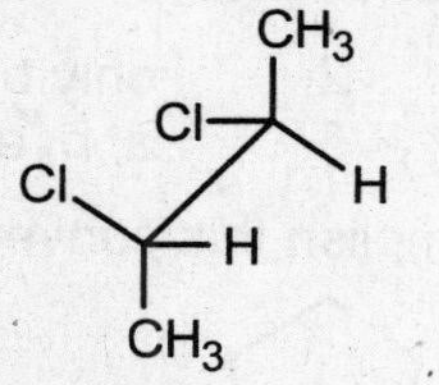

c. 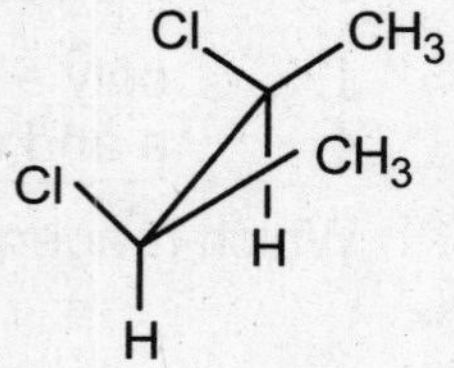

d. 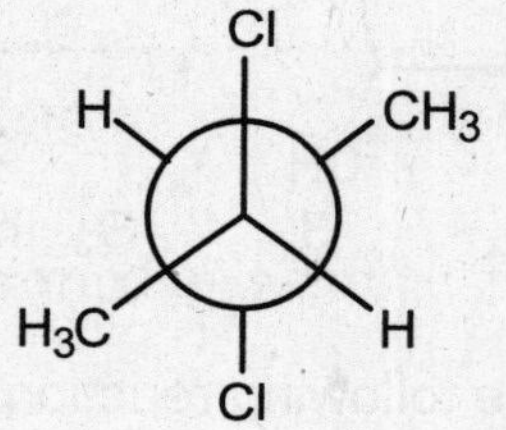

e. 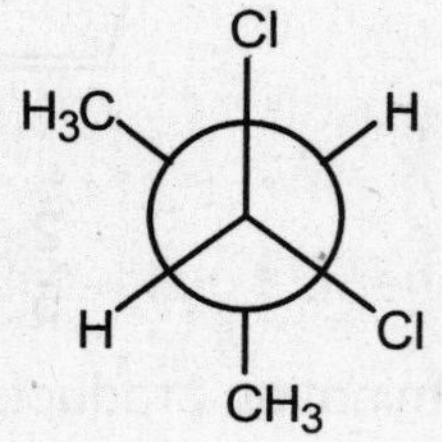

1.	only e	2.	a and b	3.	c and d
4.	d and e	5.	a, b, and c		

29. What is the correct name for this compound?

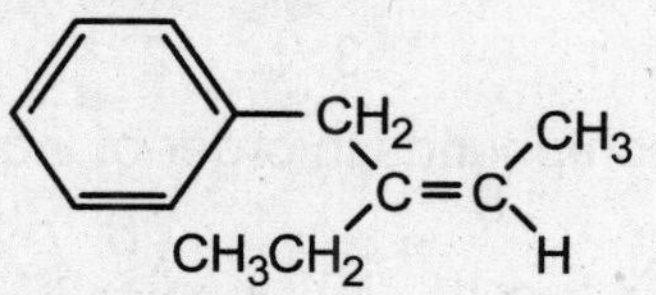

1.	*cis*-3-phenyl-2-pentene	2.	*trans*-3-phenyl-2-pentene
3.	(*Z*)-3-benzyl-2-pentene	4.	(*E*)-3-benzyl-2-pentene
5.	(*Z*)-3-benzyl-3-pentene		

30. Which of the following pairs of compounds can be separated by physical methods (for example, crystallization or distillation)?

a.

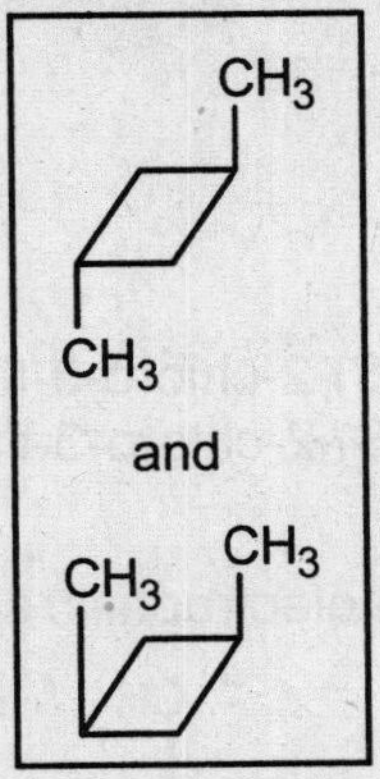
CH3
CH3
and
CH3
CH3

b.

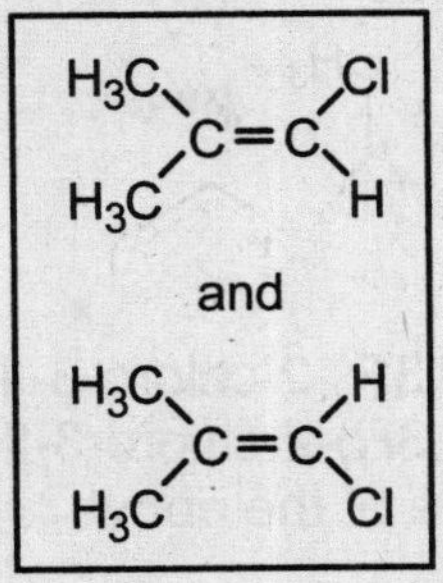
H3C
Cl
C=C
H3C
H
and
H3C
H
C=C
H3C
Cl

c.

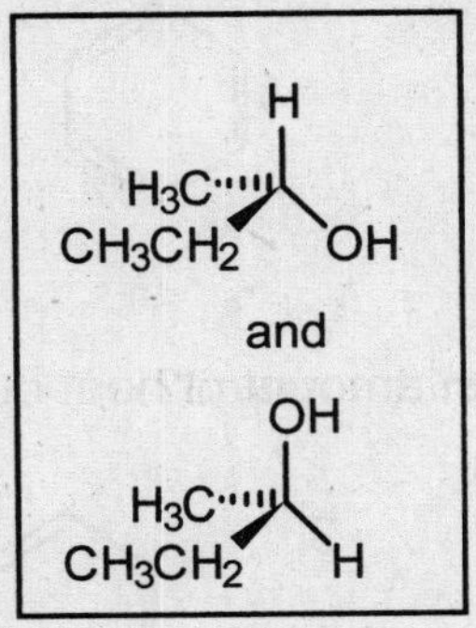
H
H3C
CH3CH2
OH
and
OH
H3C
CH3CH2
H

d.

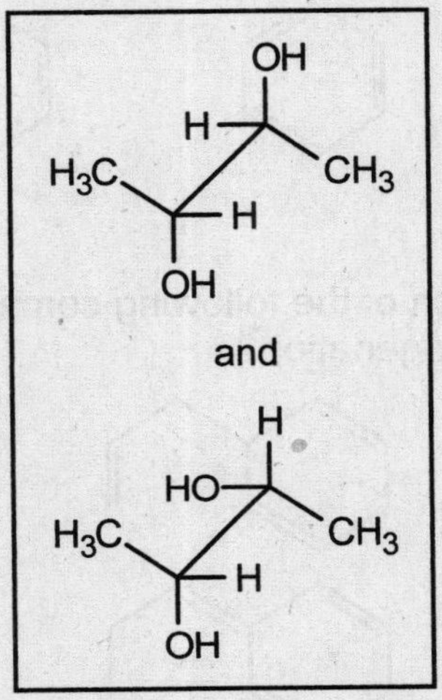
OH
H
H3C
CH3
H
OH
and
H
HO
H3C
CH3
H
OH

1. all of them
2. a and d
3. b and c
4. only d
5. only c

31. To which Fischer projection formula does the following drawing correspond?

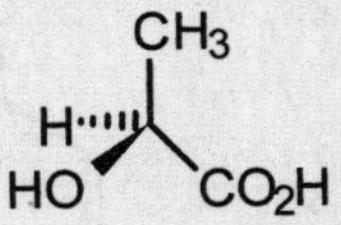
CH3
H
HO
CO2H

1.

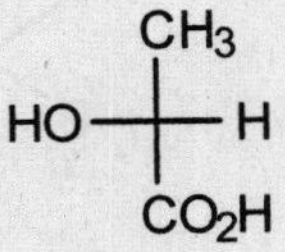
CH3
HO
H
CO2H

2.

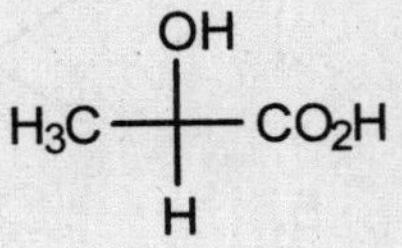
OH
H3C
CO2H
H

3.

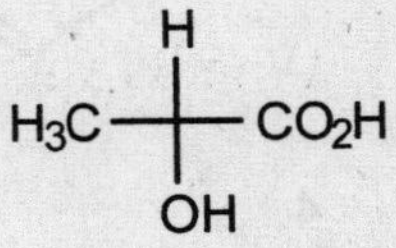
H
H3C
CO2H
OH

4.

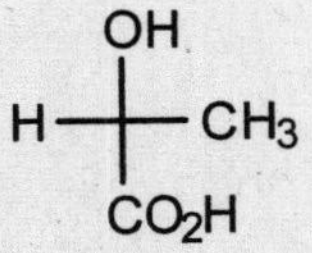
OH
H
CH3
CO2H

5.

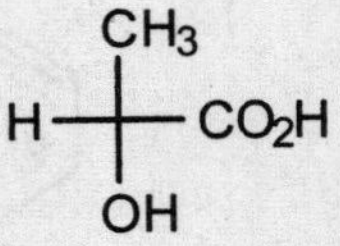
CH3
H
CO2H
OH

32. What is the correct name for the following compound?

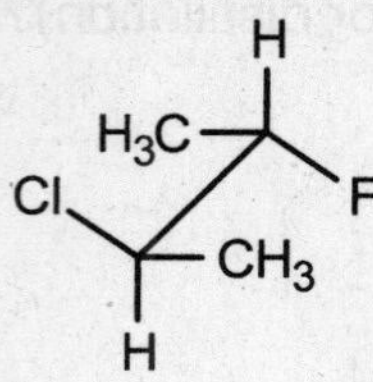

1. (2*R*,3*R*)-2-chloro-3-fluorobutane
2. (2*R*,3*S*)-2-chloro-3-fluorobutane
3. (2*S*,3*R*)-2-chloro-3-fluorobutane
4. (2*S*,3*S*)-2-chloro-3-fluorobutane
5. none of the above

33. Which of the following compounds reacts least rapidly in electrophilic substitutions?

1. 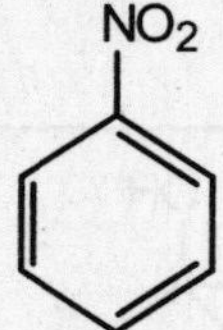

2. 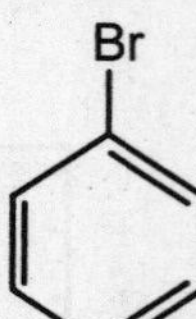

3. 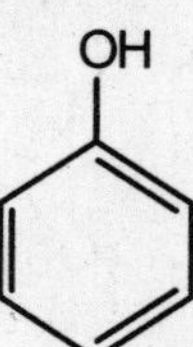

4. 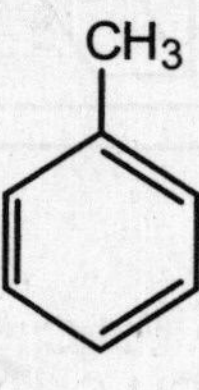

5. 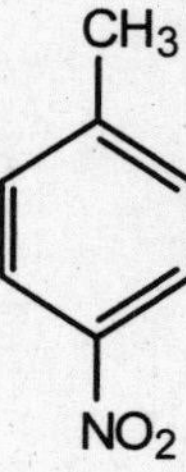

34. Which of the following compounds gives off the *least* amount of heat during hydrogenation?

1. 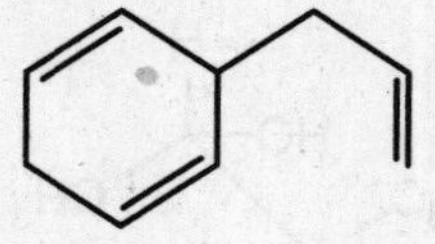
2. 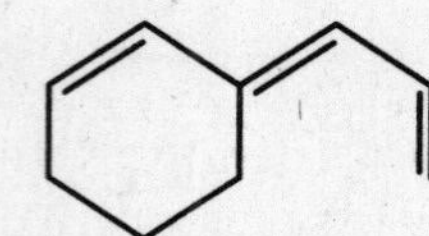
3. 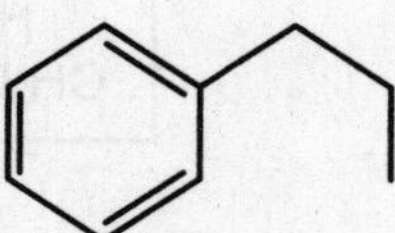
4. 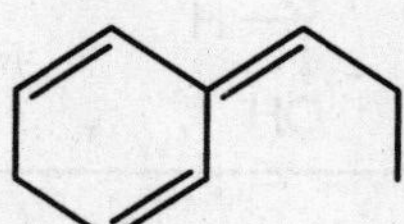
5. 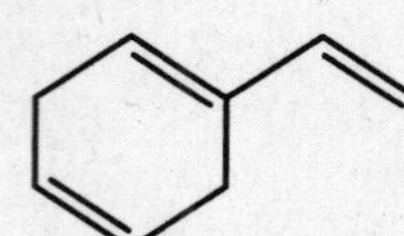

35. The resonance contributor that is most important in the intermediate for *para* electrophilic substitution in toluene is:

1. 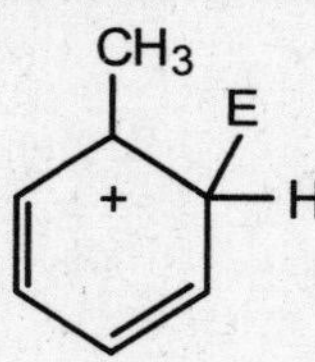

2. 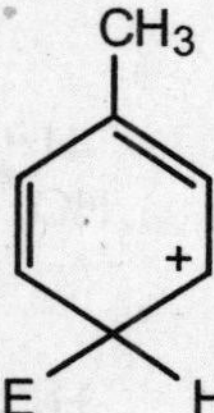

3. 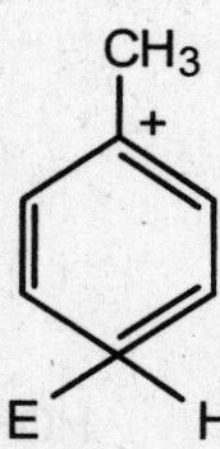

4. 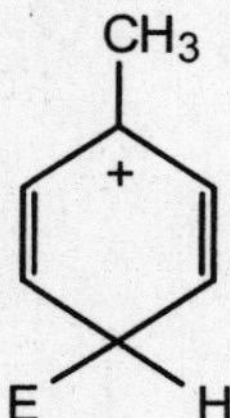

5. 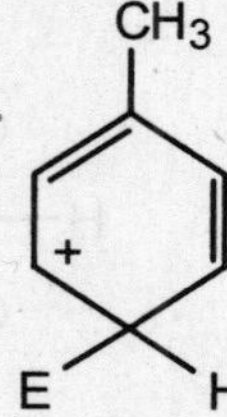

36. How are the following structures related?

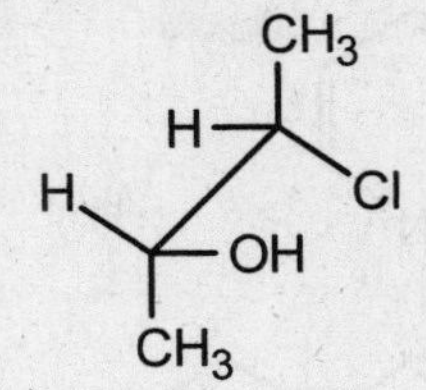

and

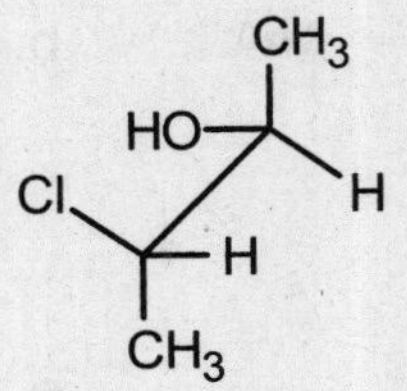

1. enantiomers
2. diastereomers
3. *meso* compounds
4. identical
5. achiral compounds

37. Which reagent would be most useful for carrying out the following transformation?

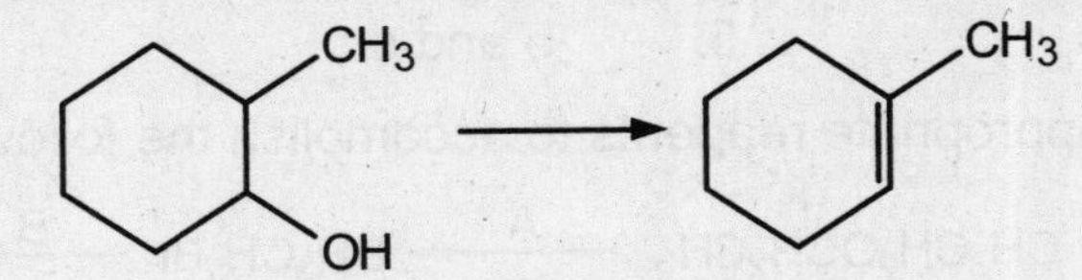

1. $LiAlH_4$
2. conc. H_2SO_4
3. H_2/Pd
4. CrO_3
5. NaOH

38. By which of the indicated mechanisms would the following reaction proceed?

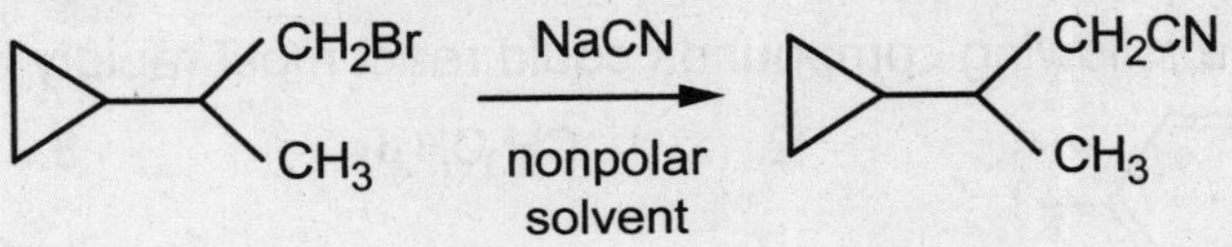

NaCN, nonpolar solvent

1. S_N1
2. S_N2
3. E1
4. E2
5. free-radical

39. What is the most likely product of the following reaction?

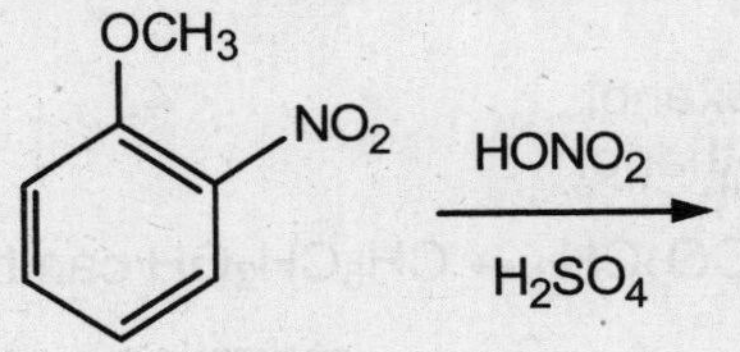

$HONO_2$, H_2SO_4

1. 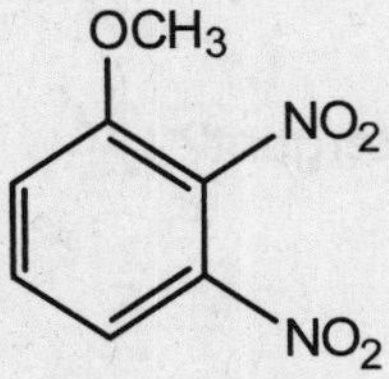
2. 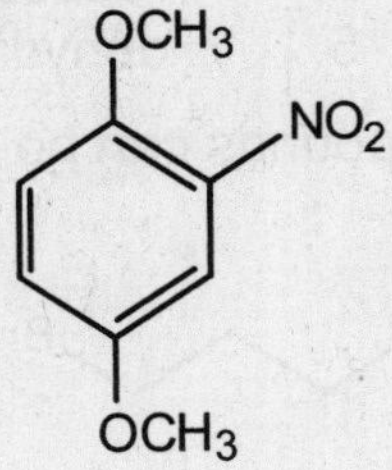
3. 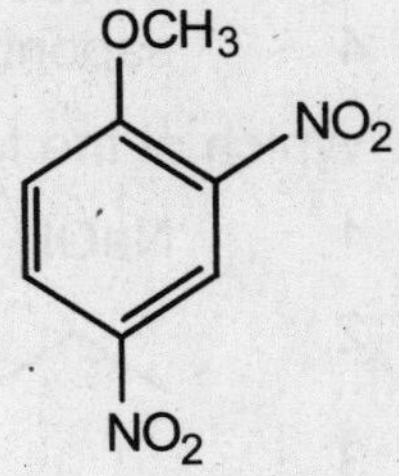
4. 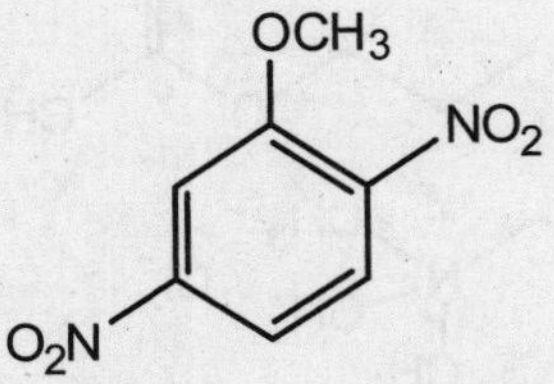
5.

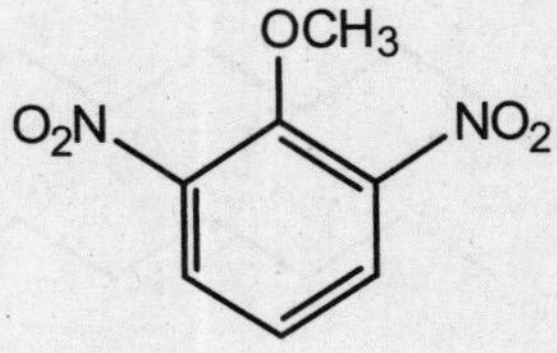

40. Which of the following compound(s) is(are) not aromatic?

a. 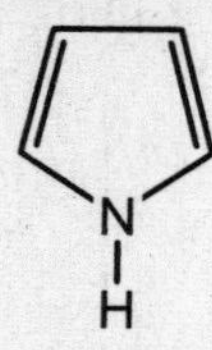

b. 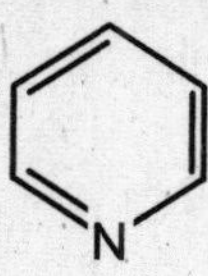

c. 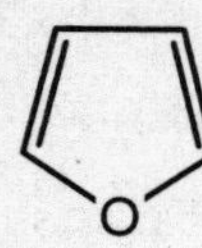

d. 

e. 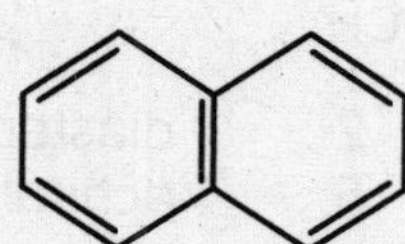

1. only d
2. only e
3. d and e
4. a and c
5. b and d

41. What are the appropriate reagents to accomplish the following transformations?

$$CH_3CH_2OCH_2CH_3 \xrightarrow{A} CH_3CH_2Br \xrightarrow{B} CH_3CH_3$$

A = ____________ B = ____________

1. Br_2
2. HBr
3. Na
4. conc. H_2SO_4
5. Mg, then H_2O
6. CrO_3

42. Which of the following compounds could react most rapidly in an S_N2 reaction?

1. 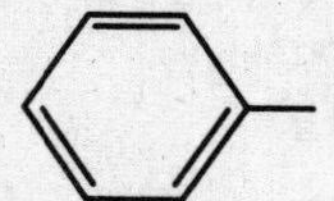

2. CH_3CH_2I
3. $H_2C{=}CH{-}I$
4. $(CH_3)_2CH-I$
5. $(CH_3)_3C-I$

43. Which of the following compounds is a secondary alcohol?

1. 3-hexanol
2. 1-hexanol
3. 2-methyl-2-hexanol
4. 1-ethyl-1-cyclohexanol
5. cyclohexylmethanol

44. The conversion $CH_3CO_2CH_3 \rightarrow CH_3CH_2OH$ can be called:

1. oxidation
2. reduction
3. hydrolysis
4. saponification
5. dehydration

45. Which of the following compounds can be used as a detergent?

1. NaOH
2. Br
3. 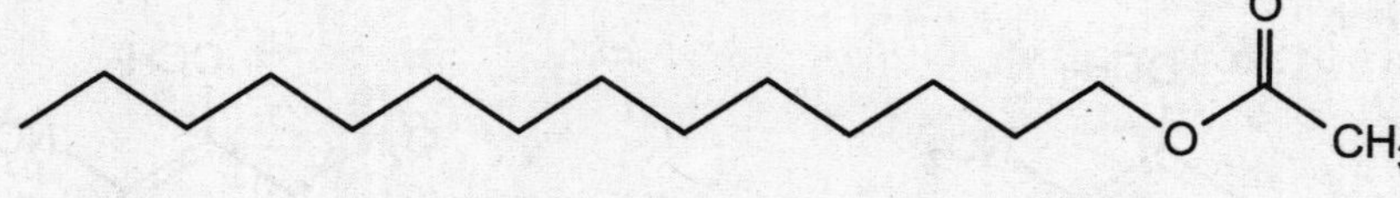

4. 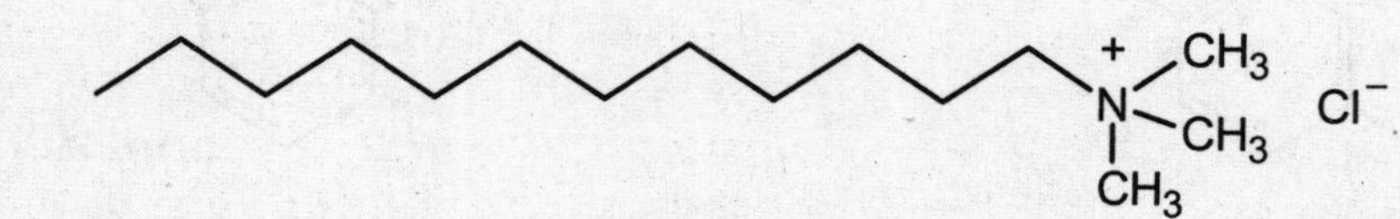

5. CO_2H

46. Which of the following compounds is a tautomer of phenol?

1. 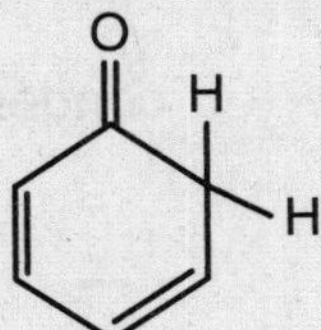

2. 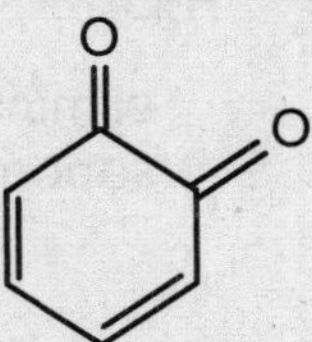

3. 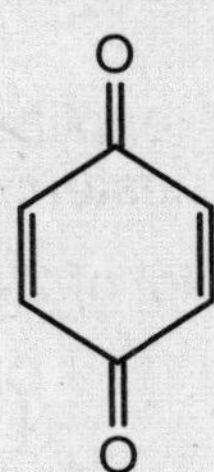

4. 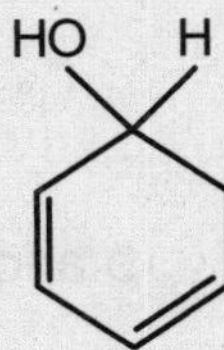

5. OH, CO_2H

47. Which compound is the major product of the following reaction?

$$CH_3CH_2CH{=}O \ + \ CH_3MgBr \longrightarrow \xrightarrow[H^+]{H_2O}$$

1. $CH_3CH_2-C(OH)(CH_3)-CH_3$
2. $CH_3CH_2CH_2OH$
3. $CH_3CH_2-C(H)(CH_3)-CH_3$
4. $CH_3CH_2-C(=O)-OCH_3$
5. $CH_3CH_2-C(OH)(H)-CH_3$

48. The following compound is a (an):

1.	diketone	2.	keto ether	3.	lactone
4.	anhydride	5.	ketal		

49. Nylon has the following structure. It can be called a:

$$\left[-\overset{O}{\overset{\|}{C}}-(CH_2)_4-\overset{O}{\overset{\|}{C}}-NH(CH_2)_6NH- \right]_n$$

1.	polyamide	2.	polyamine	3.	polyester
4.	vinyl polymer	5.	polyurethane		

50. The following reaction can be called:

$$CH_3CH_2CH_2-\overset{O}{\overset{\|}{C}}-OCH_2CH_3 \xrightarrow[H_2O]{NaOH} CH_3CH_2CH_2-\overset{O}{\overset{\|}{C}}-O^-Na^+ + CH_3CH_2OH$$

1. enolization
2. elimination
3. condensation
4. esterification
5. saponification

51. The enol of 2-butanone is:

a. $CH_3\overset{OH}{\overset{|}{C}}HCH=CH_2$

b. $CH_3\overset{OH}{\overset{|}{C}}=CHCH_3$

c. $H_2C=\overset{OH}{\overset{|}{C}}CH_2CH_3$

d. $H_2C=\overset{OH}{\overset{|}{C}}-CH=CH_2$

1. only a
2. a and b
3. b and c
4. a and d
5. a, b, and c

52. What is the major product of the following reaction?

$$C_6H_5-CH=O + (CH_3)_2C=O \xrightarrow[heat]{NaOH}$$

1. $C_6H_5-CH=CH-\overset{O}{\overset{\|}{C}}-CH_3$
2. $C_6H_5-\overset{OH}{\overset{|}{C}}H-\underset{O}{\underset{\|}{C}}-CH_3$
3. $(CH_3)_2\overset{OH}{\overset{|}{C}}CH_2\overset{O}{\overset{\|}{C}}CH_3$
4. $C_6H_5-\overset{O}{\overset{\|}{C}}-CH_2-\overset{OH}{\overset{|}{C}}H-CH_3$
5. $C_6H_5-\overset{O}{\overset{\|}{C}}-\overset{CH_3}{\overset{|}{\underset{OH}{\underset{|}{C}}}}-CH_3$

53. What is the major product of the following reaction?

$$\text{cyclohexanone} + HOCH_2CH_2OH \xrightarrow{H^+}$$

1. cyclohexane with OH and CH_2CH_2OH on the same carbon
2. a polyester
3. cyclohexane with OH and OCH_2CH_2OH on the same carbon
4. cyclohexane with H and OH on the same carbon
5. cyclohexane spiro-fused at one carbon to a ring $O-CH_2-CH_2-O$

54. The structure of malonic acid is:

1. $HO_2C-C_6H_4-CO_2H$ (para-substituted benzene ring)
2. $HO-\overset{O}{\overset{\|}{C}}-\overset{O}{\overset{\|}{C}}-OH$
3. $HO-\overset{O}{\overset{\|}{C}}-CH_2-\overset{O}{\overset{\|}{C}}-OH$
4. $HO-\overset{O}{\overset{\|}{C}}-(CH_2)_4-\overset{O}{\overset{\|}{C}}-OH$
5. $H_3C-\overset{O}{\overset{\|}{C}}-\overset{O}{\overset{\|}{C}}-OH$

55. Which hybrid structure represents the most stable enolate anion formed in the following reaction?

$$CH_3CH_2-\overset{O}{\overset{\|}{C}}-CH_2-\overset{O}{\overset{\|}{C}}-CH_3 \xrightarrow{\text{base}}$$

1. $CH_3CH=\overset{O^-}{\overset{|}{C}}CH_2\overset{O}{\overset{\|}{C}}CH_3 \longleftrightarrow CH_3\overset{-}{C}H\overset{O}{\overset{\|}{C}}CH_2\overset{O}{\overset{\|}{C}}CH_3$
2. $CH_3CH_2\overset{O^-}{\overset{|}{C}}=CH\overset{O}{\overset{\|}{C}}CH_3 \longleftrightarrow CH_3CH_2\overset{O}{\overset{\|}{C}}=\overset{-}{C}H\overset{O}{\overset{\|}{C}}CH_3$
3. $CH_3CH_2\overset{O}{\overset{\|}{C}}CH=\overset{O^-}{\overset{|}{C}}CH_3 \longleftrightarrow CH_3CH_2\overset{O}{\overset{\|}{C}}\overset{-}{C}H\overset{O}{\overset{\|}{C}}CH_3$
4. $CH_3CH_2\overset{O}{\overset{\|}{C}}CH_2\overset{O^-}{\overset{|}{C}}=CH_2 \longleftrightarrow CH_3CH_2\overset{O}{\overset{\|}{C}}CH_2\overset{O}{\overset{\|}{C}}\overset{-}{C}H_2$
5. $CH_3CH_2\overset{O^-}{\overset{|}{C}}=CH\overset{O}{\overset{\|}{C}}CH_3 \longleftrightarrow CH_3CH_2\overset{O}{\overset{\|}{C}}\overset{-}{C}H\overset{O}{\overset{\|}{C}}CH_3 \longleftrightarrow CH_3CH_2\overset{O}{\overset{\|}{C}}CH=\overset{O^-}{\overset{|}{C}}CH_3$

56. Arrange the following compounds in order of decreasing basicity:

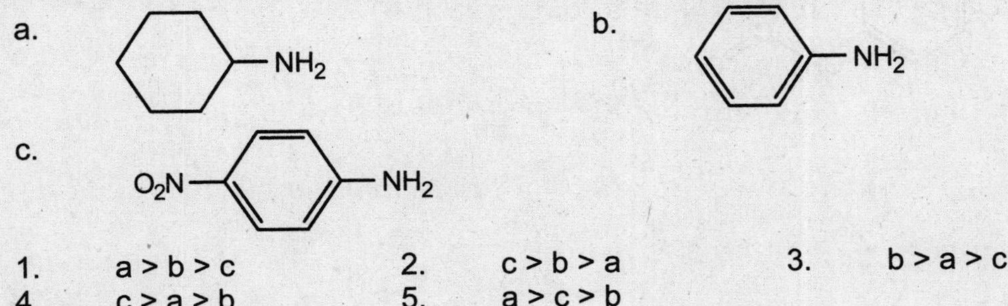

1. a > b > c
2. c > b > a
3. b > a > c
4. c > a > b
5. a > c > b

57. What is the major product of the following reaction?

$C_6H_5NH_2 \xrightarrow[H^+,\ 0°C]{HONO} \xrightarrow[Cu_2(CN)_2]{KCN}$

1. 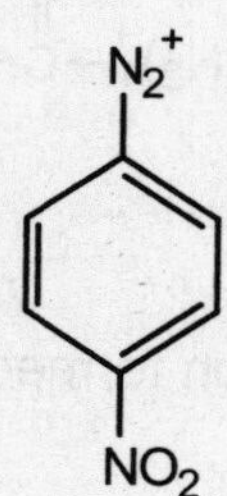

2. C_6H_5CN

3. 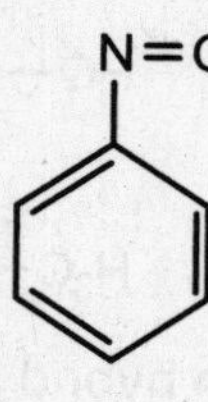

4. 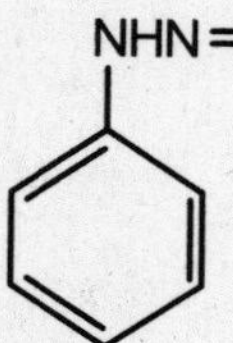

5. $C_6H_5CO_2H$

58. What is the major product of the following reaction?

$C_6H_5CH_2OH + CH_3CO_2H \xrightarrow{H^+}$

1. $H_3C-C(=O)-O-CH_2-C_6H_5$

2. $C_6H_5-CH_2-C(=O)-OCH_3$

3. $H_3C-C(=O)-C_6H_4-CO_2H$

4. $CH_3-C(=O)-CH_2-C_6H_5$

5. $C_6H_4(CH_2-O-C(=O)-CH_2)$ (cyclic ester fused to benzene ring)

59. What is the major product of the following reaction?

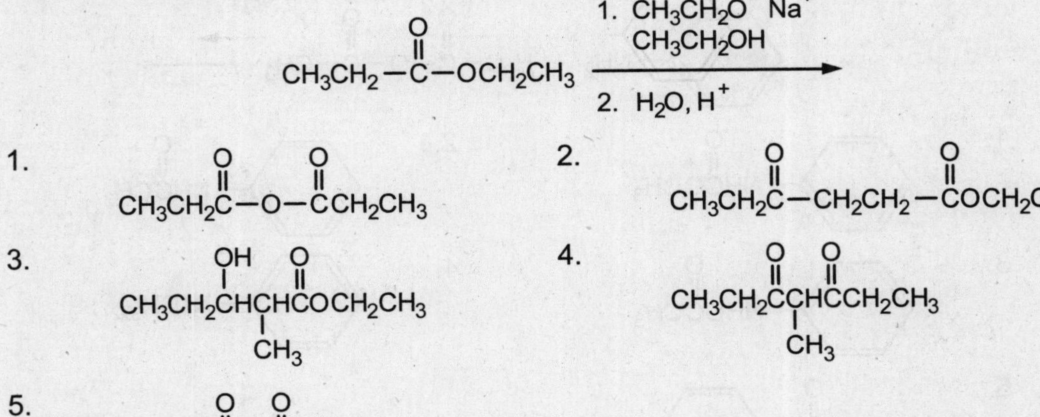

60. Which of the following compounds is a secondary amine?

1. $H_3C-CH(NH_2)-CH_3$
2. $CH_3CH_2NHCH_3$
3. $H_3C-C(=O)-N(NH_2)-CH_3$
4. $H_2NCH_2CH_2NH_2$
5. $(CH_3)_3N$

61. The resonance contributor that is most important in the intermediate anion formed upon reaction of 4-chloropyridine with sodium methoxide is:

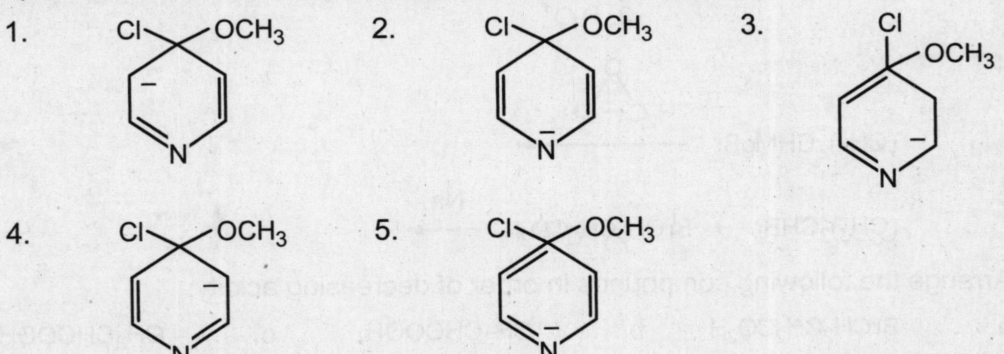

62. What is an IUPAC name for $BrCH_2CH_2CH_2CH{=}O$?

1. γ-bromobutyraldehyde
2. 1-bromo-3-carbonylpropane
3. 1-bromo-4-butanal
4. 4-bromo-1-butanone
5. 4-bromobutanal

63. What is the product of the following reaction?

$$C_6H_5-NH_2 + H_3C-\overset{O}{\overset{\|}{C}}-O-\overset{O}{\overset{\|}{C}}-CH_3 \longrightarrow$$

1. $C_6H_5-NH\overset{O}{\overset{\|}{C}}OCH_3$
2. $C_6H_5-NH\overset{O}{\overset{\|}{C}}CH_3$
3. $C_6H_5-NHO\overset{O}{\overset{\|}{C}}CH_3$
4. $C_6H_5-N(CH_3)_2$
5. $H_3C-\overset{O}{\overset{\|}{C}}-C_6H_4-NH_2$

64. Which reaction is most appropriate for the preparation of $(CH_3)_2CHCH_2CH_2CO_2H$?

1. $(CH_3)_2CHCH_2Br \xrightarrow[\text{2. } CO_2 \;\; \text{3. } H_3O^+]{\text{1. Mg}}$
2. $(CH_3)_2CHCH_2CH{=}O \xrightarrow{Ag(NH_3)_2^+}$
3. $(CH_3)_2CHCH_2CH_2Br \xrightarrow[\text{2. } H_3O^+]{\text{1. KCN}}$
4. $(CH_3)_2CHMgBr \xrightarrow{H_2C-CH_2 \text{ (epoxide, O bridging)}}$
5. $(CH_3)_2CHBr + BrCH_2CH_2CO_2H \xrightarrow{Na}$

65. Arrange the following compounds in order of decreasing acidity:

a. $BrCH_2CH_2CO_2H$ b. $CH_3CHCOOH$ (with Br on the CH) c. $CH_3CHCOOH$ (with F on the CH)

1. a > b > c
2. c > b > a
3. c > a > b
4. b > a > c
5. b > c > a

66. Which structure corresponds to *cis*-3-pentenoic acid?

1.
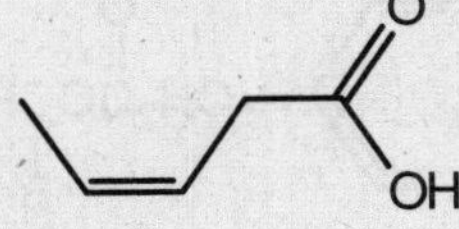

2.

3.
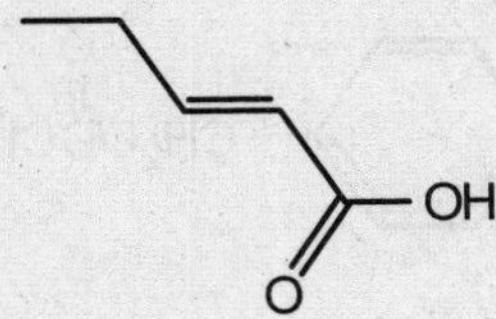

4.
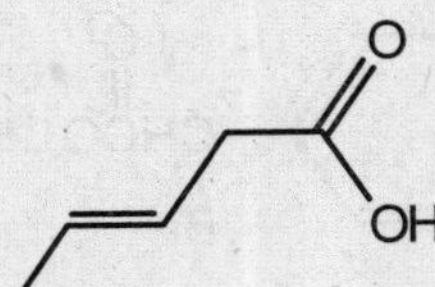

5.
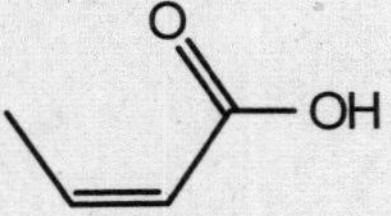

67. What is(are) the product(s) of the following reaction?

$$CH_3CH_2CH(OCH_2CH_3)_2 \xrightarrow[H^+]{H_2O}$$

a. CH_3CH_2OH

b.
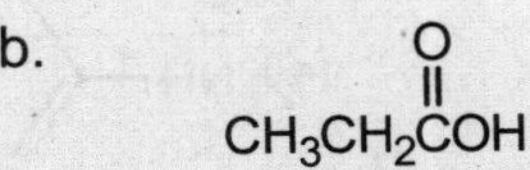

c.
OH
$CH_3CH_2COCH_2CH_3$
OH

d. $CH_3CH_2CH{=}O$

1. a and b
2. a and d
3. a and c
4. only c
5. b and d

68. Cyclic ethers with a three-membered ring

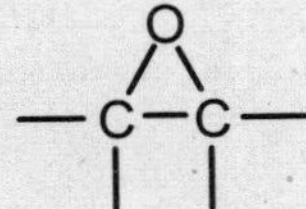

are called:

1. epoxy resins
2. lactones
3. oxiranes
4. lactams
5. alkoxides

69. The appearance of a silver mirror in a Tollens' test indicates the presence of:

1. a monosaccharide
2. an aldehyde
3. starch
4. cellulose
5. glucose

70. The name benzyl α-bromopropionate refers to:

1. C_6H_5–$OC(=O)CH(Br)CH_3$
2. C_6H_5–$CH_2C(=O)OCH(Br)CH_3$
3. C_6H_5–$CH(Br)C(=O)OCH_2CH_2CH_3$
4. C_6H_5–$CH_2OC(=O)CH_2CH_2Br$
5. C_6H_5–$CH_2OC(=O)CH(Br)CH_3$

71. Compound X gives the following test results:

$$X + Ag(NH_3)_2^+ \longrightarrow \text{no silver mirror}$$

$$X + H_2NNH\text{–}C_6H_3(NO_2)_2 \text{ (2,4-dinitrophenylhydrazine)} \longrightarrow \text{orange solid}$$

$$X \xrightarrow[CH_3OD]{CH_3O^-\,Na^+} \text{exchanges } \textit{four} \text{ hydrogens for deuterium}$$

Which of the following compounds is a possible structure for X?

1. $CH_3C(=O)OC(CH_3)_3$
2. $CH_3CH_2CH_2C(=O)CH_2CH_3$
3. $CH_3C(=O)CH_2$–C_6H_5
4. $O{=}CHCH_2CH_2CH{=}O$
5. C_6H_5–$C(=O)$–C_6H_{11} (phenyl cyclohexyl ketone)

72. Which statements regarding the following structure are true?

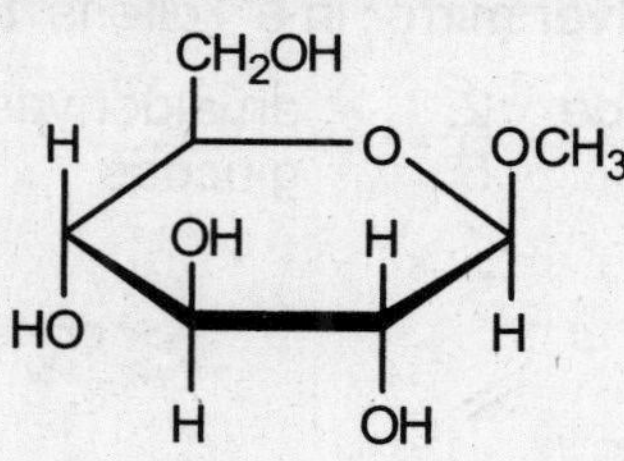

a. It is a hemiacetal.
b. It is an acetal.
c. It can be hydrolyzed by H_3O^+.
d. It gives a positive Tollens' test.

1. b and d
2. a and d
3. only b
4. b and c
5. b, c, and d

73. The monosaccharide obtained from the hydrolysis of starch is:

1. maltose 2. D-ribose 3. 2-deoxy-D-ribose
4. D-galactose 5. D-glucose

74. The products of the complete hydrolysis of DNA are:

1. D-ribose, phosphoric acid, and four heterocyclic bases
2. nucleosides 3. nucleotides
4. glucose and fructose
5. 2-deoxy-D-ribose, phosphoric acid, and four heterocyclic bases

75. Uracil is a:

1. constituent of urine 2. pyrimidine
3. purine 4. heterocyclic base present in DNA
5. potent carcinogen

76. The primary structure of a protein refers to:

1. the amino acid sequence in the polypeptide chain
2. the presence or absence of an α helix
3. the orientation of the amino acid side chains in space
4. interchain cross-links with disulfide bonds
5. whether the protein is fibrous or globular

77. Which of the following compounds is a fat?

1. $CH_3(CH_2)_{16}\overset{O}{\overset{\|}{C}}O(CH_2)_{15}CH_3$

2. 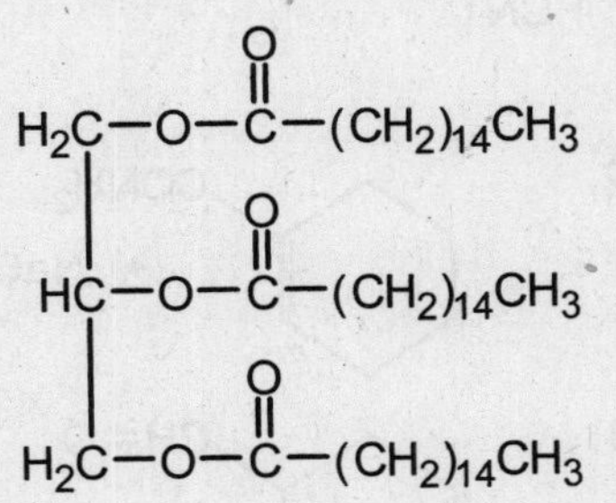 $H_2C-O-\overset{O}{\overset{\|}{C}}-(CH_2)_{14}CH_3$
$HC-O-\overset{O}{\overset{\|}{C}}-(CH_2)_{14}CH_3$
$H_2C-O-\overset{O}{\overset{\|}{C}}-(CH_2)_{14}CH_3$

3. $H_2C-\overset{O}{\overset{\|}{C}}-(CH_2)_{14}CH_3$
$HC-\overset{O}{\overset{\|}{C}}-(CH_2)_{14}CH_3$
$H_2C-\overset{O}{\overset{\|}{C}}-(CH_2)_{14}CH_3$

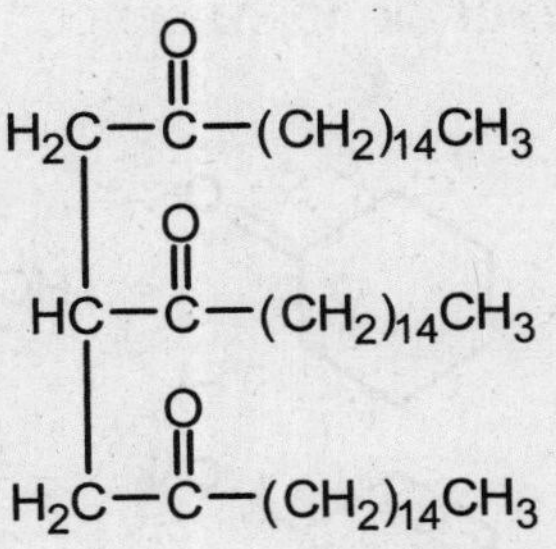

4. 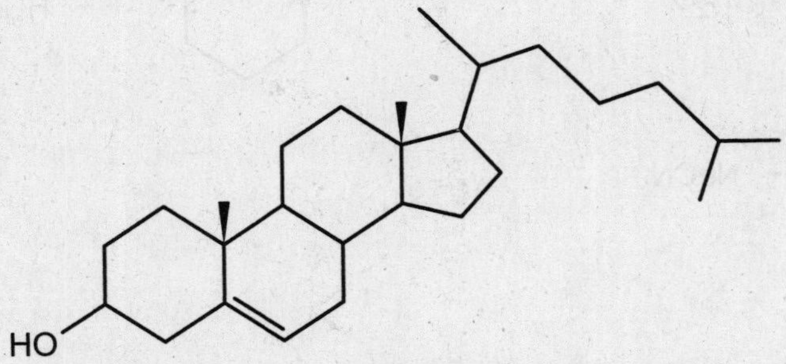 HO

5.

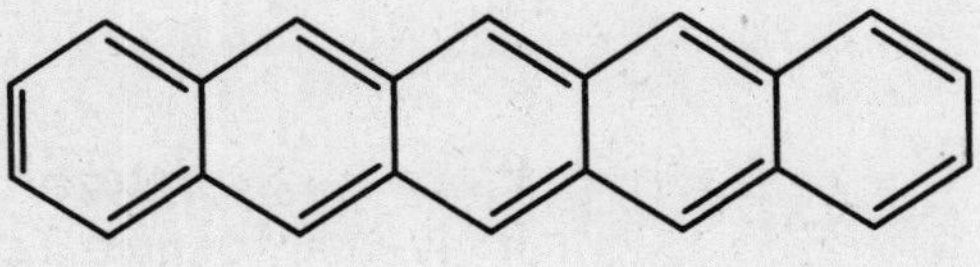

78. Which of the following structures is a zwitterion?

1. $^{-}OC(=O)CH_2CH_2C(=O)O^{-}$
2. Ca^{2+}
3. $H_3\overset{+}{N}CH_2C(=O)O^{-}$
4. $CH_3(CH_2)_{16}C(=O)O^{-}\ Na^{+}$
5. 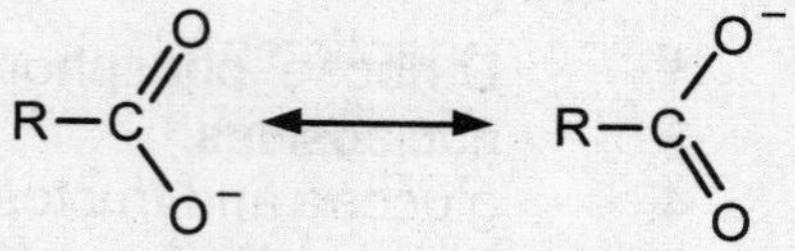 $R{-}C(=O)O^{-} \longleftrightarrow R{-}C(O^{-})=O$

79. Which of the following compounds would show only a single peak in a proton NMR spectrum?

a. $(CH_3)_4Si$ b. CCl_4 c. CH_2Cl_2
d. $CH_3CH_2OCH_2CH_3$

1. only b
2. a and c
3. only d
4. b and c
5. a, b, and c

80. Which of the following spectroscopic techniques is used to determine molecular weight?

1. proton NMR
2. ^{13}C NMR
3. infrared spectroscopy
4. ultraviolet-visible spectroscopy
5. mass spectrometry

81. The following compound can best be prepared from:

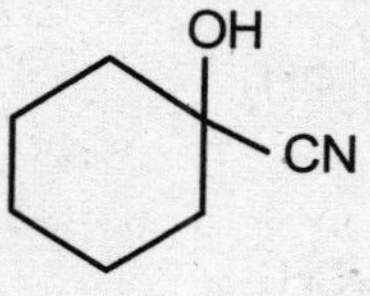

(1-hydroxycyclohexane-1-carbonitrile: cyclohexane ring bearing OH and CN on the same carbon)

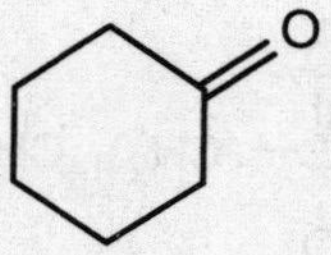

1. cyclohexanone (ring C=O) + HCN
2. cyclohexene-1-carboxamide (ring C=C bearing $CONH_2$) 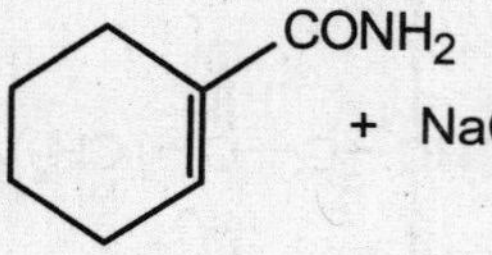 + NaCN
3. 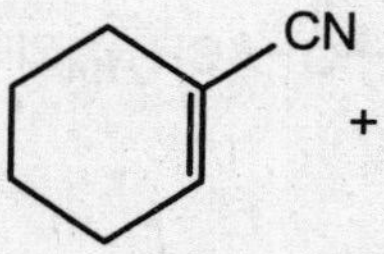 cyclohexene-1-carbonitrile (ring C=C bearing CN) + H_2O
4. 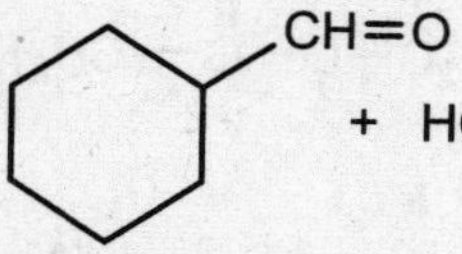 cyclohexyl with CH=O + HCN
5. 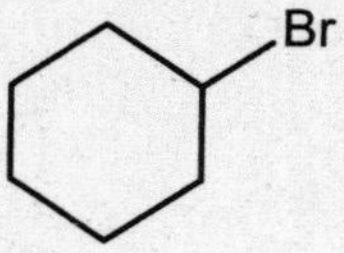 cyclohexyl with Br + NaCN

82. Which of the following terms can be used to describe this pair of compounds?

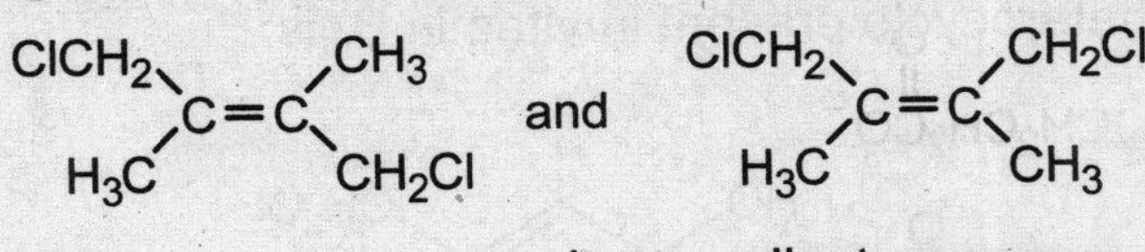

a. enantiomers
b. diastereomers
c. constitutional isomers
d. stereoisomers
e. chiral
f. achiral

1. a, b, and d
2. a, c, and e
3. b, d, and e
4. a, d, and e
5. b, d, and f

83. Which of the following best describes $C_6H_9Br_3$?

1. no rings, no double bonds
2. one ring, no double bonds
3. two double bonds
4. one triple bond
5. one ring and one double bond
6. two rings

84. The polymer segment shown below

H Cl Cl H H Cl Cl H H Cl

is:

1. syndiotactic
2. amorphous
3. atactic
4. block
5. glassy
6. isotactic

85. Which of the following is a step-growth polymer?

1. polystyrene
2. nylon
3. Teflon
4. polyvinyl
5. synthetic rubber

86. How many stereogenic centers are present in cortisone?

O CH2OH CH3 OH O CH3 O

cortisone

1. 0
2. 3
3. 4
4. 6
5. 8

87. The following Fischer projections

CHO / H — OH / CH3 and H / H3C — CHO / OH

represent:

1. the same compound
2. enantiomers
3. diastereomers
4. *meso* forms
5. rotamers

88. The parent heterocycle present in vitamin B_6 is:

vitamin B_6 (pyridine ring bearing R, HO, CH_2OH, H_3C and ring N)

1. pyridine
2. pyrrole
3. pyrimidine
4. purine
5. imidazole

89. Oxidation of ethanol can give:

1. $CH_2{=}O$
2. $HOCH_2CH_2OH$ and $CH_3CH(OH)_2$
3. $CO_2 + H_2$
4. $CH_3CHO + CH_3CO_2H$
5. $CH_3OCH_3 + CH_3CO_2H$

90. The structure of purine is:

(purine ring, atoms numbered 1–9; N-1, N-3, N-7, N-9 with H on N-9)

The least basic nitrogen present is:

1. N-1
2. N-3
3. N-7
4. N-9
5. They all are approximately equally basic.

91. The formulas for the amino acids glycine and alanine are as follows:

glycine: $\overset{+}{N}H_3CH_2CO_2^-$

alanine: $CH_3CH(\overset{+}{N}H_3)CO_2^-$

The correct formula for alanylglycine is:

1. $H_3\overset{+}{N}CH(CH_3)C(=O)\overset{+}{N}H_2CH_2CO_2^-$
2. $H_3\overset{+}{N}CH_2C(=O)NHCH(CH_3)CO_2^-$
3. $H_3\overset{+}{N}CH(CH_3)C(=O)CH(\overset{+}{N}H_3)CO_2^-$
4. $H_3\overset{+}{N}CH(CH_3)C(=O)NHCH_2CO_2^-$
5. $H_3\overset{+}{N}CH(CH_3)C(=O)ONHCH_2CO_2^-$

92. Which of the following can act as a nucleophile?

a. H_2O b. ^{-}OH c. H^+ d. ^{-}CN e. Na^+

1. a, b, and d
2. a and b
3. c and e
4. only d
5. b and d

93. Penicillin is a β-lactam. Its formula is:

1. R–C(=O)–NH– attached to a five-membered lactam ring fused to a thiazolidine ring (S, CH_3, CH_3, N, CO_2H)
2. R–C(=O)–NH– attached to a four-membered β-lactam ring (C=O on carbon adjacent to N) fused to a thiazolidine ring (S, CH_3, CH_3, N, CO_2H)
3. HO_2C– on a four-membered ring fused to a thiazolidine ring (S, CH_3, CH_3, N, CO_2H)
4. Four-membered ring bearing C=O and R–C(=O)–NH– fused to a thiazolidine ring (S, CH_3, CH_3, N)
5. R–C(=O)–NH– on a four-membered ring (no carbonyl) fused to a thiazolidine ring (S, CH_3, CH_3, N, CO_2H)

94. The antihistamine diphenhydramine (Benadryl) has the formula:

$$(C_6H_5)_2CHOCH_2CH_2N(CH_3)_2$$

On heating with aqueous acid, it gives:

1. $(C_6H_5)_2CHOH + HOCH_2CH_2N(CH_3)_2$
2. $(C_6H_5)_2CHOH + CH_3CH_2N(CH_3)_2$
3. $(C_6H_5)_2CH_2 + HOCH_2CH_2N(CH_3)_2$
4. $(C_6H_5)_2CHOCH_2CH_2NH_2 + H_3C{-}CH_3$
5. $(C_6H_5)_2CHOCH_2CH_2\overset{+}{N}(CH_3)_3$

95. How many peaks appear in the ^1H NMR spectrum of the following compound?

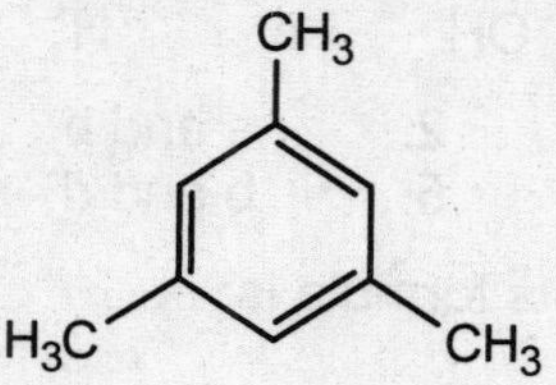

(Structure: benzene ring with CH_3, H_3C, CH_3 substituents at 1,3,5 positions)

1. 1 2. 2 3. 3 4. 6 5. 9

96. How many peaks appear in the proton-decoupled ^{13}C NMR spectrum of the compound in Question 95?

1. 1 2. 2 3. 3 4. 6 5. 9

97. The ^1H NMR spectrum of one isomer of $C_3H_3Cl_5$ consists of a triplet at δ 4.5 and a doublet at δ 6.0, with relative areas 1:2. The structure of the isomer is:

1. $CH_2ClCHClCCl_3$ 2. $CH_2ClCCl_2CHCl_2$ 3. $CHCl_2CH_2CCl_3$
4. $CHCl_2CHClCHCl_2$ 5. $CH_3CCl_2CCl_3$

98. Cocaine is an alkaloid. Its correct structure is:

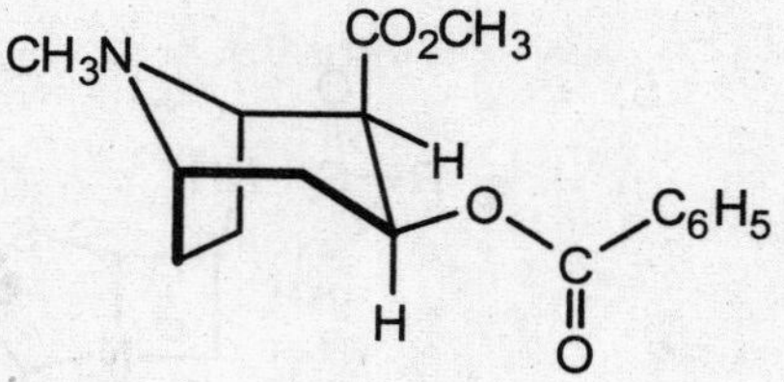

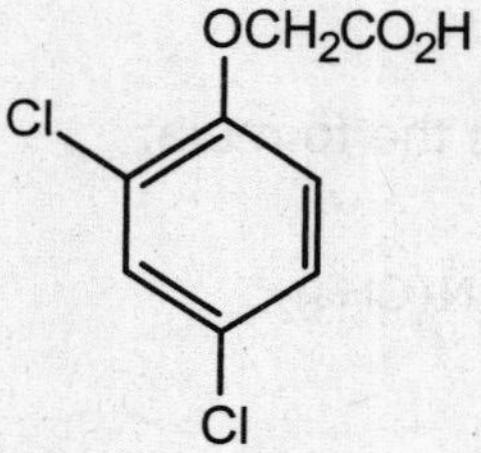

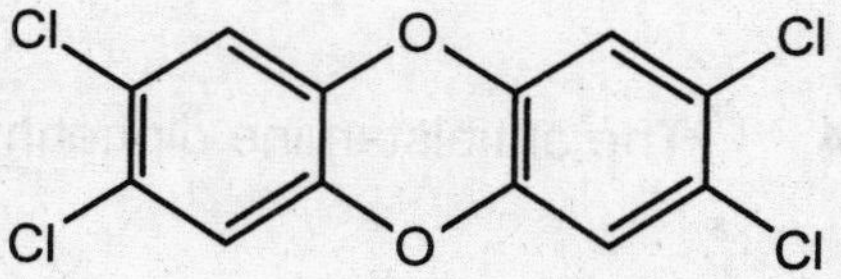

1. $H_2C-CH-CH_2$ with HO, OH, OH

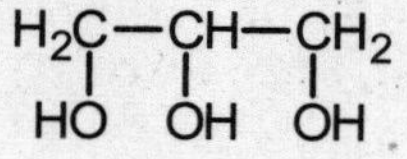

2. (Bicyclic structure with CH_3N, CO_2CH_3, H, O–C(=O)–C_6H_5, H)

3. (Benzene ring with OCH_2CO_2H, Cl, Cl)

4. (Dibenzodioxin with Cl, Cl, Cl, Cl and O, O)

5. (Steroid structure with CH_3, O, HO)

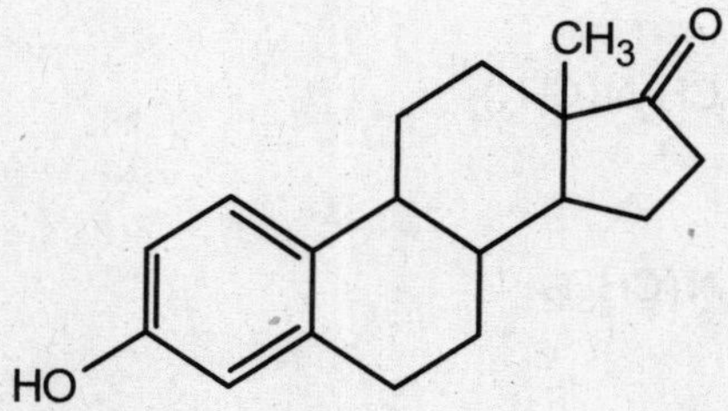

99. Which of the following statements is *not* true?

1. Natural rubber is a hydrocarbon.
2. Natural rubber is made of isoprene units.
3. Natural rubber is a polymer of 1,3-butadiene.
4. Natural rubber has *cis* double bonds.
5. Natural rubber can be vulcanized.

100. What is(are) the product(s) of the following reaction?

$CH_3N{=}C{=}O$ + (benzene ring)–OH ⟶

1. $CH_3NH-\overset{O}{\overset{\|}{C}}-O-$(benzene ring)
2. (benzene ring)–OCH_3 + $HN{=}C{=}O$
3. CO_2 + (benzene ring)–$NHCH_3$
4. a polyurethane
5. $CH_3N{=}\overset{OH}{\overset{|}{C}}-O-$(benzene ring)

101. Which structure is consistent with the following ^{1}H NMR spectrum?

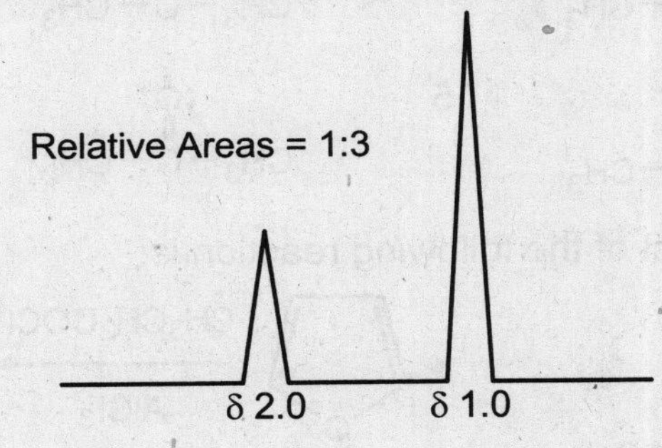

1. $(CH_3)_3CO-\overset{O}{\overset{\|}{C}}-CH_3$
2. $(CH_3)_3CO-\overset{O}{\overset{\|}{C}}-OCH_3$
3. $(CH_3)_3C-\overset{O}{\overset{\|}{C}}-OCH_3$
4. $H_3C-\overset{O}{\overset{\|}{C}}-OCH_3$
5. $H_3C-\overset{O}{\overset{\|}{C}}-CH_3$

102. The infrared spectrum of a compound shows absorptions at 1700 cm^{-1} and 3600 cm^{-1}. A possible structure is:

1. 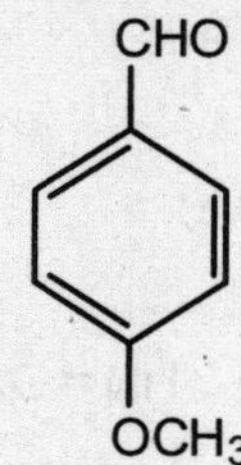

2. CHO

3. CHO
NH_2

4. CHO
OH

5. CO_2H

103. The parent ion in the mass spectrum of 2-propanone has the following structure:

1. $CH_3-\overset{\overset{+}{:\ddot{O}:}}{\overset{\|}{C}}-CH_3$

2. $CH_3-\overset{\overset{\cdot\,-}{:\ddot{O}:}}{\overset{\|}{C}}-CH_3$

3. $CH_3-\overset{\overset{-}{:\ddot{O}:}}{\overset{\|}{C}}\underset{+}{}-CH_3$

4. $CH_3-\overset{:\dot{O}^{+}}{\overset{\|}{C}}-CH_3$

5. $CH_3-\overset{:\ddot{O}:}{\overset{\|}{C}}-\overset{+}{C}H_2$

104. The major product of the following reaction is:

furan $\xrightarrow[AlCl_3]{CH_3CH_2COCl}$

1. furan-2-yl $-O-\overset{O}{\overset{\|}{C}}-CH_2CH_3$

2. furan-3-yl $-\overset{O}{\overset{\|}{C}}-CH_2CH_3$

3. furan-2-yl $-\overset{O}{\overset{\|}{C}}-CH_2CH_3$

4. furan-2-yl $-\overset{O}{\overset{\|}{C}}-OCH_2CH_3$

5. furan $\overset{+}{O}$ $-\overset{-}{A}lCl_3$

105. Which reaction sequence can be used to accomplish the following transformation?

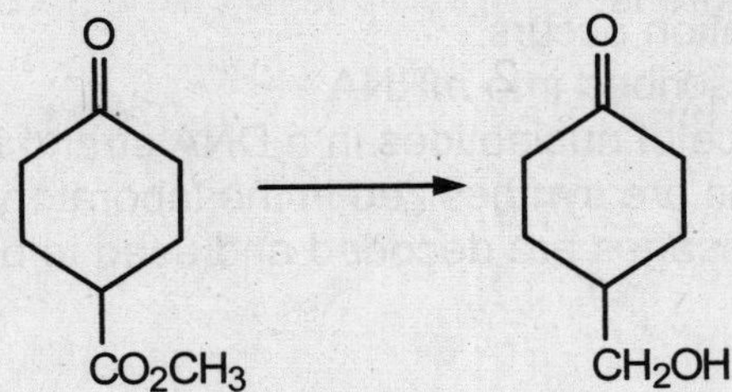

1. (a) $LiAlH_4$ (b) CH_3OH, H^+ (c) H_2O, H^+
2. (a) CH_3OH, H^+ (b) H_2O, H^+ (c) $LiAlH_4$
3. (a) CH_3OH, H^+ (b) $LiAlH_4$ (c) H_2O, H^+
4. (a) CH_3OH, H^+ (b) PCC (c) H_2O, H^+
5. (a) $NaBH_4$ (b) $LiAlH_4$ (c) PCC

106. Arrange the following compounds in order of increasing electrophilic aromatic substitution rates.

a.  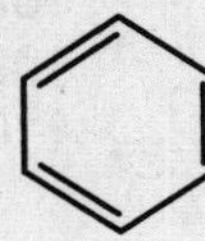 b. 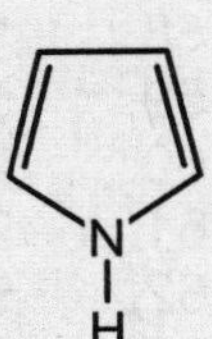

c. 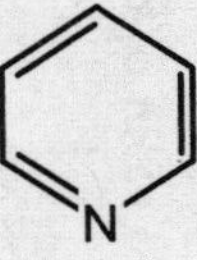

1. a < b < c
2. c < b < a
3. b < c < a
4. c < a< b
5. b < a < c

107. The following compound is a:

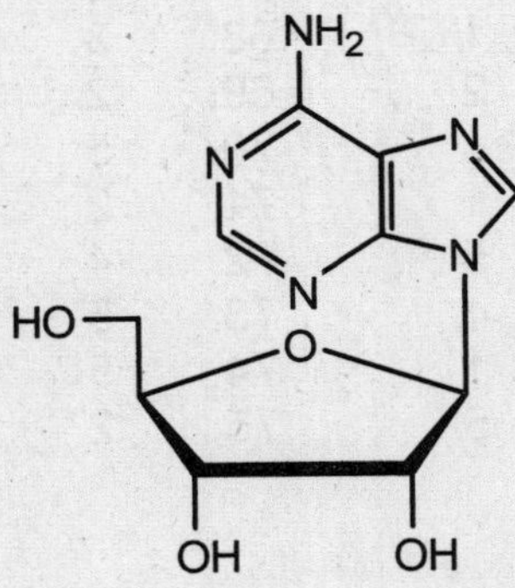

1. nucleotide
2. glycoside
3. nucleoside
4. codon
5. pyrimidine

108. The DNA strand that is complementary to the following sequence of bases is:
5'–TTACCG–3'

1. 3'–AATGGC–5'
2. 3'–UUTGGC–5'
3. 5'–AATGGC–3'
4. 3'–CCGTTA–5'
5. 5'–CCGTTA–3'

109. The *m*RNA strand produced upon transcription of the following DNA strand is:
5'–TTACCG–3'

1. 5'–AAUGGC–3'
2. 3'–AAUGGC–5'
3. 5'–AATGGC–3'
4. 3'–AATGGC–5'
5. 3'–GCCATT–5'

110. Translation is the process by which:

1. DNA replication occurs
2. DNA is transcribed into *m*RNA
3. the sequence of nucleotides in a DNA strand is determined
4. nucleic acids are synthesized in the laboratory
5. genetic messages are decoded and used to build proteins

Key to the Sample Multiple-Choice Questions

1.	4	26.	4	51.	3	76.	1	101.	1
2.	3	27.	1	52.	1	77.	2	102.	4
3.	2	28.	3	53.	5	78.	3	103.	4
4.	4	29.	3	54.	3	79.	2	104.	3
5.	1	30.	2	55.	5	80.	5	105.	3
6.	1	31.	3	56.	1	81.	1	106.	4
7.	2	32.	2	57.	2	82.	5	107.	3
8.	5	33.	1	58.	1	83.	2	108.	1
9.	3	34.	3	59.	5	84.	1	109.	2
10.	5	35.	4	60.	2	85.	2	110.	5
11.	2	36.	1	61.	2	86.	4		
12.	2	37.	2	62.	5	87.	2		
13.	3	38.	2	63.	2	88.	1		
14.	1	39.	3	64.	3	89.	4		
15.	2	40.	1	65.	2	90.	4		
16.	1	41.	2, 5	66.	1	91.	4		
17.	5	42.	2	67.	2	92.	1		
18.	2	43.	1	68.	3	93.	2		
19.	5	44.	2	69.	2	94.	1		
20.	2	45.	4	70.	5	95.	2		
21.	4	46.	1	71.	2	96.	3		
22.	1	47.	5	72.	4	97.	4		
23.	5	48.	4	73.	5	98.	2		
24.	1	49.	1	74.	5	99.	3		
25.	4	50.	5	75.	2	100.	1